Green Energy and Technology

More information about this series at http://www.springer.com/series/8059

Andrej Kitanovski · Jaka Tušek
Urban Tomc · Uroš Plaznik
Marko Ožbolt · Alojz Poredoš

Magnetocaloric Energy Conversion

From Theory to Applications

Springer

Andrej Kitanovski
Urban Tomc
Uroš Plaznik
Marko Ožbolt
Alojz Poredoš
University of Ljubljana
Ljubljana
Slovenia

Jaka Tušek
Department of Energy Conversion and Storage
Technical University of Denmark
Roskilde
Denmark

ISSN 1865-3529 ISSN 1865-3537 (electronic)
ISBN 978-3-319-08740-5 ISBN 978-3-319-08741-2 (eBook)
DOI 10.1007/978-3-319-08741-2

Library of Congress Control Number: 2014954592

Springer Cham Heidelberg New York Dordrecht London

Printed on acid-free paper

Springer is part of Springer Science+Business Media (www.springer.com)

"I believe we are witnessing history in the making."

Karl A. Gschneidner Jr., At the unveiling of the prototype, proof-of-principle, near room temperature magnetic refrigerator at the Astronautics Technology Center in Madison, Wisconsin, on February 20, 1997.

To those that have been and are working in the field: "The limitations of magnetic refrigeration are only in the minds of the individual engineers and scientists."

Karl A. Gschneidner Jr.

"Practical magnetocaloric energy conversion on a large scale has been a grand challenge to the ingenuity of engineers and scientists off and on since the time of Edison and Tesla. Recent years has seen the concept brought ever closer to practicality. Until now there has been no comprehensive book on the subject."

R.E. Rosensweig

Preface

This book provides the basis for engineering a new, alternative form of technology, i.e., magnetocaloric energy conversion. It has been written for postgraduate students, engineers and researchers, especially those working in the areas of refrigeration and heat pumping, power generation and heat transfer. The aim of the book is to provide the reader with a theoretical basis supported by practical examples, facilitate the understanding, design and construction, and to introduce new solutions to engineering problems in future magnetocaloric energy-conversion devices.

The global energy demand for refrigeration and air conditioning is rapidly increasing. Nearly 20 % of all energy is wasted in existing cooling or air-conditioning devices. Here, vapour-compression air-conditioning and refrigeration technologies have dominated the market for more than 100 years.

As the capacity per unit mass of vapour-compression technology has improved over time their dominant position has become even stronger by lowering the manufacturing costs and improving efficiency. However, the trends of the converging S-curve of development show that vapour compression has become a mature technology, having only a small potential for some significant improvements in efficiency.

There is a continuous quest to find new, environmentally friendly refrigerants for vapour compression. Many of the alternative substances are related to a lower energy efficiency, high pressures, flammability or even explosive hazards.

Based on the adopted Kyoto Protocol, the chlorofluorocarbons CFCs, which had an important impact on the ozone layer (with high ozone depletion potential (ODP)), were phased out in the 1990s. Another group of refrigerants, the hydrochlorofluorocarbons (HCFSs) will be phased out by 2020. Despite this, the majority of remaining refrigerants possess a global warming potential (GWP). Moreover, most of the electricity production is based on fossil fuels. In order to reduce the amount of their use, one way is certainly to increase the use of renewable energy sources. On the other hand, one should take care of reducing energy consumption throughout the whole energy chain—from production to consumption. The last is also associated with the efficiency of energy conversion.

Therefore, any improvement in the efficiency of energy-conversion technologies will drastically influence the global energy demand as well as reduce the harmful environmental impacts. Vapour compression represents an important part of the world's energy consumption. Since it represents a mature technology, it is therefore obvious to think about alternative technologies. One such example is magnetocaloric energy conversion, i.e., magnetic refrigeration, magnetic heat pumping, and magnetic power generation, respectively.

Magnetocaloric energy conversion is a technology that is based on the exploitation of the magnetocaloric effect (MCE). The MCE is a physical phenomenon that occurs in magnetic materials under the influence of a varying magnetic field, i.e., magnetization and demagnetisation. Instead of a gas-refrigerant as the working substance, magnetocaloric devices employ a solid magnetic material.

Magnetocaloric materials have a GWP and an ODP equal to zero. In analogy with the polytropic compression and isenthalpic expansion of the gas-refrigerant, the processes of magnetization and demagnetization of some magnetocaloric materials are nearly isentropic. This is why the magnetocaloric energy conversion, as an alternative, offers improvements in efficiency. Note that the earliest prototypes have already demonstrated an exergy efficiency higher than that of existing vapour-compression technologies. Moreover, silent operation without vibration makes this technology attractive for a large number of applications. It is therefore not a coincidence that different technology foresights have characterized magnetocaloric energy conversion as one of the most promising alternatives for future air conditioning and refrigeration.

In 1843, James Prescott Joule observed that heat is evolved in iron samples under an applied magnetic field. In 1860, William Thomson (Lord Kelvin) knew that ferromagnetic materials lose their magnetic properties when heated above a certain temperature (known today as the Curie temperature). Twenty years later in 1881 Emil Warburg published his work explaining magnetic hysteresis. A year later, in 1882, James Alfred Ewing discovered the same phenomenon and was the first to name it "hysteresis". Later in 1917–1918, the physicists Pierre Weiss and Auguste Piccard announced the discovery of the "novel magnetocaloric phenomenon".

In parallel with the discovery of the MCE, scientists were creating ideas to apply the MCE in energy-conversion devices. Towards the end of the 19th century Jožef Stefan had the first ideas that ferromagnetic materials could be applied in power generation. Some years later, but still in the 19th century, Nikola Tesla and Thomas Alva Edison patented ideas about thermomagnetic motors. In 1926 Peter Joseph William Debye and in 1927 William Francis Giauque independently discussed the application of the MCE for cryogenic temperatures under 1 K. The experimental proof came in 1933 from W.F. Giauque and D.P. MacDougall. The first magnetic refrigerator ever built for room-temperature applications was constructed and tested by G.V. Brown in 1976. In 1982, John A. Barclay and William A. Steyert introduced the idea of an active magnetic regenerator (AMR), which is today widely applied.

An important milestone in the development of magnetocaloric energy conversion is the discovery of the so-called giant MCE, which was announced in 1997 by Karl A. Gschneidner, Jr. and Vitalij K. Pecharsky. Since then, there has been an exponential increase in patents, articles, conference contributions as well as individual chapters in a number of different books.

Among the books that concern *Magnetocaloric Energy Conversion*, it is important to mention that on *Ferrohydrodynamics* by Ronald E. Rosensweig, which was published in 1985. Although this book could be taken as the bible for ferrohydrodynamics, a large part of it was dedicated to magnetocaloric energy conversion, not only by providing the basic theory, but also by encouraging researchers to enter the engineering of this particular domain.

Another important work was published by Alexander M. Tishin and Yuri I. Spichkin in 2003. The book *The Magnetocaloric Effect and its Applications* was actually the first that was strictly focused on magnetocalorics, by giving a comprehensive description of magnetocaloric materials and their properties, as well as by pointing out different potential applications.

A very large number of different publications on magnetocaloric materials and systems have been provided in the proceedings of the International Conferences on Magnetic Refrigeration at Room Temperature—THERMAG. Since starting in 2005 in Montreux, Switzerland, these conferences have been organized in 2007 in Portorož, Slovenia, in 2009 in Des Moines, Iowa, USA, in 2010 in Baotou, Inner Mongolia, China, in 2012 in Grenoble, France, and in 2014 in Victoria, British Columbia, Canada.

For the continuous organization of the first five THERMAG events under the umbrella of the *IIR—International Institute of Refrigeration*, we have to thank Peter William Egolf, who in 2004 established the *IIR Magnetic Cooling Working Party* and led it until 2012. Without his efforts and also the efforts of all the contributors to such conferences, the magnetocaloric community would be missing an important part, which led to closer collaborations between engineers, physicists and material scientists as well as leading to better transfer of knowledge.

With about 60 prototypes being developed all over the world, and the involvement of different industries, magnetocaloric energy conversion is today on its way towards the first market applications. For these, strong interdisciplinary engineering knowledge is required. The authors of this book have recognised that there exists no such complete work that would provide the required basis for engineering.

The aim of this book is therefore to provide the most important information for students, engineers and researchers to help them understand the physics behind magnetocaloric energy conversion, and to provide sufficient basis, which will serve for the design and construction of future magnetocaloric energy-conversion devices. Moreover, by adding the latest results in research and completing them with new ideas and concepts, this book should be helpful for researchers and their industrial partners in finding and developing new magnetocaloric market applications.

The first seven chapters provide the fundamentals and practical examples on *Magnetocaloric Energy Conversion* and its various sub-domains. Chapters 8 and 9 provide potential solutions to engineering problems supported by new concept proposals, and economic aspects of magnetocaloric energy conversion. The last chapter is dedicated to other, analogous, alternative technologies of solid-state energy conversion, which might also have a potential for future market applications.

In Chap. 1 on the *Thermodynamics of Magnetocaloric Energy Conversion*, the basic magnetocaloric thermodynamic potentials are presented and described. The state of the art gives an overview of the existing theoretical and experimental approaches to magnetocaloric thermodynamic cycles. Design issues related to different thermodynamic refrigeration cycles are shown as well, described through basic thermodynamic equations, as well as definitions of the coefficient of performance (COP), the exergy efficiency and the theoretical maximum cooling capacity.

Chapter 2 on *Magnetocaloric Materials for Freezing, Cooling and Heat-Pump Applications* introduces different magnetocaloric materials in terms of engineering and their applications. Among those, the most promising magnetocaloric materials are highlighted and presented with their basic thermodynamic properties.

Magnetic Field Source represents one of the most important and also most expensive parts of the magnetocaloric device. Therefore, the optimal design of a magnetic field source is crucial for obtaining a cost-effective and energy-efficient device. In this chapter we briefly describe some of the most important issues that relate to magnetic field sources. Despite the fact that different magnetic field sources with respect to their application in magnetocaloric energy conversion can be applied, the main emphasis of Chap. 3 is on permanent magnets and their assemblies. However, brief information is also provided about electric resistive magnets and superconducting magnets, including their cryogenic parts. A review supported by drawings of different permanent-magnet assemblies is given in the chapter, which includes basic information about their characteristics.

Chapter 4 deals with *Active Magnetic Regeneration*. Note, most of the existing prototype devices and studies are based on applying magnetocaloric materials with the principle of active magnetic regeneration (AMR). In the first part of the chapter, the operation of an AMR within different thermodynamic cycles is explained. Comprehensive information about numerical modelling of an AMR is given, supported by a review of this particular research, and by the necessary mathematical models required for modelling. The impact of the operating conditions (mass-flow rate and operating frequency) and geometrical characteristics on AMR are presented. In addition, results on numerical and experimental investigations of the different AMR thermodynamic cycles are shown, and guidelines for future research on AMR thermodynamic cycles are given. The impact of the heat-transfer fluid on the characteristics of the AMR is discussed. Brief information about some processing and manufacturing techniques is provided. The end of the chapter provides brief information on the practical limitations for the application of AMR cycles.

A special case of magnetocaloric energy conversion relates to *Magnetocaloric Fluids*. These can be further divided into ferrofluids (nanofluids) and magneto-rheological particle suspensions. In Chap. 5, this book provides important information on rheologic constitutive models of magnetic fluids, and introduces the basic governing equations, which relate to the thermo-magneto-fluid dynamics of ferrofluids. The review on existing studies and applications of magnetocaloric fluids is given. The chapter provides a discussion and guidelines for the potential design of devices that concern magnetocaloric energy conversion with magnetocaloric fluids. Since other magnetic fluids can also be successfully applied in magnetocaloric energy conversion, their various potential uses are outlined in the chapter.

The existing practical limitations of AMR principles relate to the power density of magnetocaloric devices. This is associated with the number of thermodynamic cycles performed in a unit of time. Therefore, in the past few years, some research groups have started a serious investigation of the potential in other mechanisms, which led them to research on *Special Heat Transfer Mechanisms—Active and Passive Thermal Diodes*. The thermal diode (heat "semiconductor", thermal switch, heat valve, thermal rectifier) is a physical phenomenon, mechanism or device in which it is possible to manipulate and control the direction of the heat flux and sometimes also the intensity of the heat flux. Chapter 6 presents different kinds of mechanisms and devices that can be applied as thermal diodes. These can be further divided into two main areas: solid-state thermal diodes and microfluidic thermal diodes, respectively. The characteristics of such thermal diodes are compared and presented with respect to different potential applications. A review of existing research with regard to magnetic refrigeration is given and the potential configurations of thermal diodes in magnetic refrigeration as guidelines for future applications are shown.

Chapter 7 provides an *Overview of Existing Magnetocaloric Prototype Devices*. Generally, two different types of such devices have been exploited—linear and rotary devices. Based on information from the literature, as well as due to the help of the magnetocaloric community, the majority of these prototypes are presented in this chapter. The information is supported by drawings, photographs as well as the main operating characteristics. To this we also add contact details of the institutes and people, which may be used for collaboration, potential investments in the technology or industrial development. At the end of the chapter, the summary of all prototypes is given in tables.

As a continuation of Chap. 7, the Chap. 8 on *Design Issues and Future Perspectives for Magnetocaloric Energy Conversion* gives comprehensive information on the different possible designs for particular types of magnetic refrigerators and heat pumps. Whereas some of the configurations were already applied in research, Chap. 8 also provides information about new solutions, which are based on the author's research experiences and which might be applied in future studies and related devices. In addition, a note on power generation is added, with a brief review of the existing work in this particular domain. By pointing out the most successful design approaches, this chapter also serves as a future guideline on magnetic refrigeration and heat pumping.

The commercialization of the presented technology requires analyses of the *Economic Aspects of the Magnetocaloric Energy Conversion*. In Chap. 9, we address the cost issues that relate to magnetic field sources and magnetocaloric materials. With supporting information on global market prices, we have also added a simple calculation of the costs for the magnetocaloric refrigerator. This is followed by a review of the different economic studies covering magnetic refrigeration, including some of the ecological aspects that are related to the carbon footprint and lifecycle analyses (LCAs).

The last chapter in this book is dedicated to other *Alternative Solid-State Energy Conversion* technologies. Because of the analogy with magnetocaloric energy conversion, these "ferroic", solid-state technologies can be applied by engineers having a knowledge of magnetocalorics. The chapter is divided into three sections, which regard the electrocaloric (pyroelectric), barocaloric and elastocaloric energy conversion technologies. In each section, the physical principle behind the technology is presented with an overview of the existing materials and their physical properties. Furthermore, different possibilities for how to design an energy-conversion device using these materials are reviewed (especially for electrocalorics). Since the technology based on these three effects is in an early stage of development, only a few prototypes of energy-conversion devices, as the state of the art, have been presented.

With the presented knowledge and engineering solutions we hope that this book will well serve the reader in exploiting alternative possibilities of energy conversion by learning and designing future magnetocaloric devices for refrigeration, air-conditioning, heating and power generation.

Ljubljana, September 2014

Andrej Kitanovski
Jaka Tušek
Urban Tomc
Uroš Plaznik
Marko Ožbolt
Alojz Poredoš

Acknowledgments

Many thanks go to members of the international magnetocaloric community, who generously provided photographs and important information about their developments and prototypes, giving this book the latest pictorial updates and other details about magnetocaloric energy-conversion technology.

In particular, we would like to thank Steven L. Russek from Astronautics Corporation of America (USA); Farhad Shir from The George Washington University (USA); Michael Benedict and David Beers from GE Appliances (USA); Oliver Gutfleisch from Technical University of Darmstadt (GER); Andrew Rowe from University of Victoria (CAN); Tsuyoshi Kawanami from Kobe University (JAP); Akiko Takahashi Saito from Toshiba Corporation (JAP); Naoki Hirano from Chubu Electric Power Co., Inc. (JAP); Shigeki Hirano from Hokkaido Research Institute (JAP); Yoshiki Miyazaki from Railway Technical Research Institute (JAP); Sanchoru Pe from Sanden Co., Inc. (JAP); Jiaohong Huang from Baotou Research Institute of Rare Earths (PRC); Yongbai Tang from Sichuan University (PRC); Afef Kedous-Lebouc from Grenoble Electrical Engineering Laboratory (FRA); Christian Müller from Cooltech Applications (FRA); Nini Pryds and Christian Bahl from Technical University of Denmark, Risø (DEN); Osmann Sari from the University of Applied Sciences of Western Switzerland (CH); Fouad Rahali from Clean Cooling Systems (CH); Jader R. Barbosa Jr. from Federal University of Santa Catarina (BRA); Xavier Bohigas from Polytechnical University of Catalonia (SPA); Elies Molins from Independent University of Barcelona (SPA); Javier Tejada from University of Barcelona (SPA); Daniel Lewandowski from Wroclaw University of Technology (POL); Qiming Zhang from Pennsylvania State University (USA).

We would further like to thank Peter W. Egolf for supporting this work as well as for all his efforts in establishing the international magnetocaloric community under the umbrella of the IIR—International Institute of Refrigeration.

Thanks also to Paul McGuiness, an expert in the field of magnetic materials, for all his corrections, remarks and suggestions.

We would like to thank Vitalij K. Pecharsky and Karl A. Gschneidner Jr., whose discovery of the Gd–Ge–Si giant-magnetocaloric-effect material in 1997 was the stimulus for so many of the recent advances in magnetocalorics, for supporting this work and for providing the book with important quotes that relate to the development of magnetocaloric technologies.

Finally, we would like to thank Ronald E. Rosensweig, the father of ferrohydrodynamics, for his helpful remarks and encouraging support.

Contents

Chapter 1
The Thermodynamics of Magnetocaloric Energy Conversion

Magnetocaloric energy conversion is a technology based on the exploitation of the magnetocaloric effect (MCE). The MCE is a physical phenomenon that occurs in magnetic materials under the influence of a varying magnetic field. Is it usually expressed as the adiabatic temperature change or isothermal total entropy change of a material. In a ferromagnetic material the entropy can be, for instance, related to the magnetic part and the part related to the temperature of the system (e.g. the lattice entropy). In the absence of a magnetic field, the magnetic moments in the material are disordered. If a magnetic field is applied to the material, the magnetic moments will be forced to align in a higher order. As a consequence, the magnetic entropy will decrease. In isentropic (adiabatic) conditions, the total entropy will remain constant. Therefore, the decreased magnetic entropy will manifest itself in an increased lattice entropy. The atoms in the material will start to vibrate more intensively, and as the consequence, the temperature of the magnetic material will increase. The opposite occurs when the magnetic field is removed: the magnetic entropy is increased and the temperature decreases. On this basis, it is possible to create energy conversion cycles by applying different thermodynamic processes.

In this chapter, the basic magnetocaloric thermodynamic potentials are presented and described. The state of the art gives an overview of the existing theoretical and experimental approaches to magnetocaloric thermodynamic cycles. Different magnetic thermodynamic cycles are described. Besides thermodynamic cycles with conventional simple cycles, an important emphasis is placed on thermodynamic cycles that apply active magnetic regeneration (AMR). Since most of the existing devices apply the AMR principle, a whole chapter (Chap. 4) is dedicated to this topic.

1.1 Introduction

The earliest thermodynamic studies of the magnetocaloric effect near or above room temperature began in the 1950s and 1960s. In addition to a number of cryogenic applications, this work was initially focused on the development of heat engines for the generation of useful power. Researchers were investigating different magnetic

A. Kitanovski et al., *Magnetocaloric Energy Conversion*,
Green Energy and Technology, DOI 10.1007/978-3-319-08741-2_1

power generation thermodynamic cycles and their specific processes. Their work was based on that of Tesla [1] and Edison [2], who had patented ideas on "pyromagnetic generators". At this time, electric coils were used as the sources of the magnetic field. However, there is no evidence that such devices were ever built.

At the end of 1950s, one of the first thermodynamic analyses of magnetocaloric power generation was presented by Brillouin and Iskenderian [3]. This was soon followed by other reports [4–7]. Whereas most of the early investigations considered magnetocaloric materials in their solid form, in the 1960s there was a lot of interest in the idea of producing magnetic power generators by using magnetocaloric suspensions as the working fluids. Most of this pioneering work was performed by Resler and Rosensweig [8, 9]. Subsequent work in the 1980s considered magnetocaloric power generators based on solid working materials [10–12]. There is no evidence that any real prototype device for power generation has been developed.

With the discovery of the giant magnetocaloric effect in 1997 [13], which was followed by a number of prototypes for magnetic refrigerators, magnetocaloric power generation has again become an interesting topic. Furthermore, a knowledge about many new materials and the possibility of layering these materials (in order to have a potentially larger temperature span) also led to new activities in magnetic power generation [14–22]. Despite this, most of the published work in the field of thermodynamics in recent years has been dedicated to magnetic refrigeration. This is also the reason why the thermodynamic cycles presented in this chapter relate to refrigeration. A discussion of aspects relating to power generation is therefore only given in the Chap. 8.

The earlier investigations that considered particular thermodynamic cycles were performed by Resler and Rosensweig in the 1960s [8, 9], who mostly did work on magnetocaloric fluids. Brown in 1976 [23], then Steyert in 1978 [24], analysed the thermodynamics of solid magnetocaloric refrigerants, with the latter focused on a magnetic Stirling cycle. Kirol and Mills in 1984 and 1985 [25] performed analyses of the thermodynamics of magnetocaloric power generators with solid magnetocaloric materials. Rosensweig in 1985 [26] published a book on ferrohydrodynamics, with a comprehensive description of the thermodynamics and the fluid dynamics of magnetocaloric fluids. Barclay [27] in 1990 investigated magnetocaloric heat pumps.

No systematic approach has been employed by the research community to evaluate the various magnetocaloric thermodynamic cycles. The first record of this can be found in Chen et al. 1992 [28]. They performed a theoretical evaluation based on four different magnetocaloric thermodynamic cycles: Carnot, Brayton, Stirling, and the so-called "ideal regenerative" cycle. In the last of these, the authors proposed a thermodynamic cycle that is similar to the Stirling cycle. However, there is no evidence of any experimental results relating to such a cycle.

Since 1990 there have been a large number of publications related to the basics of magnetocaloric thermodynamics [29–38].

1.2 Heat, Work and the Basic Thermodynamic Relations

The thermodynamics in this chapter relates to the magnetocaloric material as the observed system. Because of this, we deal with the thermodynamics described with the internal magnetic field in the magnetocaloric material. This should not be misunderstood as the external magnetic field, which is related to the magnetic field source (e.g. the magnetic field in the "empty" air gap of a permanent magnet).

One way to understand magnetic energy is to consider it as a form of potential energy. Imagine a rock on a mountain, having a potential energy (a magnetocaloric material in a magnetic field, produced by a permanent magnet or induced by an electrical coil). In order to put the rock on the mountain, work has to be performed on it (work is performed on the magnetocaloric material when it is magnetized; therefore, the magnetocaloric material receives magnetic work from the permanent magnet or electric field source). When the rock is rolled downhill, its potential energy decreases and the rock does work, e.g. through kinetic energy (the magnetocaloric material does work during the demagnetization process). Actually, during the demagnetization the magnetocaloric material will have to be pulled out of the magnetic field. Looking at the magnetocaloric material, which in our case is the observed system, the magnetocaloric material performs the work.

In the following text, we will assume conditions of constant pressure p and volume V for a solid magnetocaloric material. For a simpler presentation we have written all the equations in their one-dimensional form and apply notation for the exact differential for specific work (dw instead of δw) and specific heat (dq instead of δq).

Figure 1.1 shows an example of a thermodynamically closed system, which is analogous to a piston compressing a gas in a cylinder. In such a system, there is no transfer of mass over the system boundaries (the magnetocaloric material represents the system boundaries). In a thermodynamically opened system, however, there is a mass flow of the magnetocaloric material in and out of the system boundaries. For instance, as shown in Fig. 1.2, a fixed system boundary is shown around the magnetocaloric material. According to Fig. 1.2, the magnetocaloric material rotates (flows) through such a boundary. The first law of thermodynamics for a closed thermodynamic system states that:

$$\mathrm{d}u = \delta q - \delta w \tag{1.1}$$

The internal energy of the magnetocaloric material will increase if heat is added to a system or if work is performed on the magnetocaloric material. According to Fig. 1.1, the work is performed on the magnetocaloric material by moving the material into a magnetic field, produced by a permanent magnet or by the induction of a magnetic field using an electrical coil. Because of this, the magnetocaloric material is magnetized (Fig. 1.1b), and its internal magnetic field increases.

The specific work required to magnetize the magnetocaloric material in a thermodynamically closed system is equal to (see also [26, 37, 39–46]):

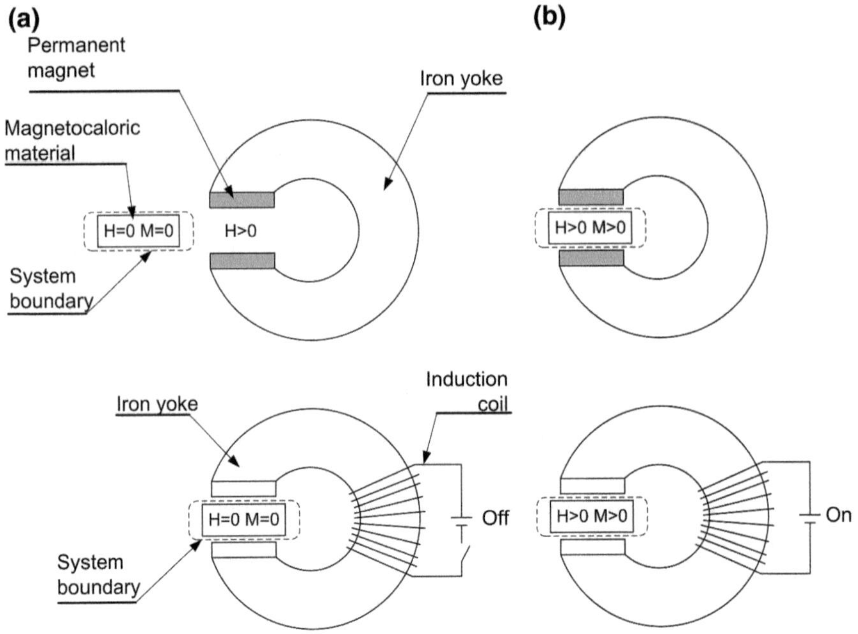

Fig. 1.1 Thermodynamically closed system, **a** Example of a magnetocaloric material in the absence of an external magnetic field (demagnetized state), **b** Example of a magnetocaloric material being magnetized

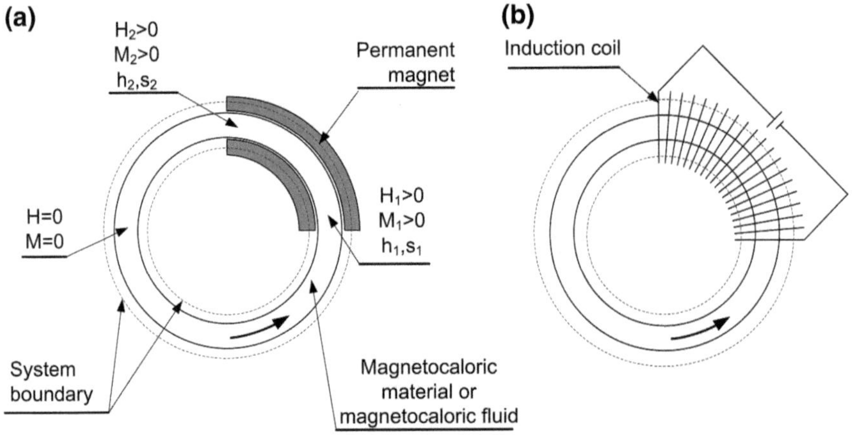

Fig. 1.2 A thermodynamically open system for "flow" of the magnetocaloric material, **a** Example of a magnetocaloric material in the absence of an external magnetic field (demagnetized state) for permanent magnet assembly or electric coil, **b** Example of a magnetocaloric material being magnetized by permanent magnet assembly or electric coil

$$\mathrm{d}w = -\mu_0 H \, \mathrm{d}M \tag{1.2}$$

Now, the first law of thermodynamics can be rewritten as:

$$\mathrm{d}u = \mathrm{d}q + \mu_0 H \, \mathrm{d}M \tag{1.3}$$

And for a reversible process, where

$$\mathrm{d}q = T\mathrm{d}s \tag{1.4}$$

Equation (1.3) takes the following form:

$$\mathrm{d}u = T\,\mathrm{d}s + \mu_0 H \, \mathrm{d}M \tag{1.5}$$

The derivative of the specific total entropy (for isobaric and isochoric conditions) can be defined in our particular case as:

$$\mathrm{d}s(T,H) = \left(\frac{\partial s}{\partial T}\right)_H \mathrm{d}T + \left(\frac{\partial s}{\partial H}\right)_T \mathrm{d}H \tag{1.6}$$

Applying Eqs. (1.4) and (1.6), we can define the derivative of the specific heat and the specific heat capacities as:

$$\mathrm{d}q = c_H(T,H)\,\mathrm{d}T + c_T(T,H)\,\mathrm{d}H \tag{1.7}$$

with:

$$c_H = \left(\frac{\partial q}{\partial T}\right)_H = T\left(\frac{\partial s}{\partial T}\right)_H \tag{1.8}$$

$$c_T = \left(\frac{\partial q}{\partial H}\right)_T = T\left(\frac{\partial s}{\partial H}\right)_T \tag{1.9}$$

The following Maxwell relation can be applied as well:

$$\left(\frac{\partial s}{\partial H}\right)_T = \mu_0 \left(\frac{\partial M}{\partial T}\right)_H \tag{1.10}$$

In an open system (Fig. 1.2) related to the control volume, the magnetocaloric material flows or moves out from the system boundaries. In this particular case, we consider the magnetocaloric fluid or the magnetocaloric solid regenerator to move in such a way (Fig. 1.2).

By applying the Legendre transformation to the first law of thermodynamics (Eq. 1.1), we can derive the definition of enthalpy as:

$$h = u - \mu_0 H M \tag{1.11}$$

Applying Eq. (1.5), the derivative of the specific enthalpy is:

$$\mathrm{d}h(s, H) = \left(\frac{\partial h}{\partial s}\right)_H \mathrm{d}s + \left(\frac{\partial h}{\partial H}\right)_s \mathrm{d}H = T\,\mathrm{d}s - \mu_0 M \mathrm{d}H \tag{1.12}$$

In most publications relating to the characterization of magnetocaloric materials, the **isothermal entropy change** is presented as a function of the temperature and the internal magnetic field. It is one of the most widely published properties related to the magnetocaloric effect. The entropy change in an isothermal process can be defined using Eqs. (1.6–1.10) and written as follows:

$$\mathrm{d}s(T, H) = \left(\frac{\partial s}{\partial H}\right)_T \mathrm{d}H = \mu_0 \left(\frac{\partial M}{\partial T}\right)_H \mathrm{d}H \tag{1.13}$$

For a certain increase (or decrease) in the magnetic field between the two states of different magnetic fields under isothermal conditions, the **isothermal entropy change** is defined as follows:

$$\Delta s = s_2 - s_1 = \int_{H_1}^{H_2} \left(\frac{\partial s}{\partial H}\right)_T \mathrm{d}H \tag{1.14}$$

$$\Delta s = \int_{H_1}^{H_2} \mu_0 \left(\frac{\partial M}{\partial T}\right)_H \mathrm{d}H \tag{1.15}$$

$$\Delta s = \int_{H_1}^{H_2} \frac{c_T}{T} \mathrm{d}H \tag{1.16}$$

Another important parameter that is often used for the characterization of magnetocaloric materials is the **adiabatic temperature change**. It denotes the increase or decrease in the temperature due to the increase or decrease of the magnetic field in the absence of a heat flow (adiabatic-isentropic magnetization or demagnetization). In the adiabatic–isentropic process, the total specific entropy does not alter ($\mathrm{d}s = 0$). From Eqs. (1.6) and (1.9), it follows that:

$$\left(\frac{\partial s}{\partial T}\right)_H \mathrm{d}T = -\left(\frac{\partial s}{\partial H}\right)_T \mathrm{d}H = -\mu_0\left(\frac{\partial M}{\partial T}\right)_H \mathrm{d}H \tag{1.17}$$

By using Eq. (1.17) and the definition of the specific heat at a constant magnetic field from Eq. (1.7), it follows:

$$\mathrm{d}T = -\frac{\mu_0}{c_H} \cdot T \cdot \left(\frac{\partial M}{\partial T}\right)_H \mathrm{d}H \tag{1.18}$$

From Eq. (1.14), we can define the adiabatic temperature change. However, since the measurement of c_H has a much longer characteristic time than the measurement of the magnetization M [46], this kind of calculation is rather problematic. Therefore, the adiabatic temperature change can simply be derived or by using the following equation and knowledge on two temperatures at constant total entropy.

$$\Delta T = \int_{T_1}^{T_2} \mathrm{d}T = T_2(s, H_2) - T_1(s, H_1) \tag{1.19}$$

The derivative of the enthalpy in the adiabatic–isentropic case equals:

$$\mathrm{d}h = -\mu_0 M \, \mathrm{d}H \tag{1.20}$$

We will further denote the process of heat transfer during a constant magnetic field as the **isofield process of cooling or heating**. Such a process is analogous to isobaric heating and cooling and similar to condensation and evaporation, respectively. However, the temperature of the solid magnetocaloric material is not held constant in such a process.

The derivative of the heat and enthalpy in this case equals (Eqs. 1.8 and 1.12):

$$\mathrm{d}q = \mathrm{d}h = c_H \, \mathrm{d}T = T\left(\frac{\partial s}{\partial T}\right)_H \mathrm{d}T \tag{1.21}$$

Figure 1.3 shows the magnetic field–enthalpy–temperature-specific entropy diagram for a reference magnetocaloric material, i.e. gadolinium. The diagram was constructed using the mean-field approximation [47–49]. This led to information about the specific total entropy as a function of the temperature and the internal magnetic field. With the use of Eq. (1.12), the enthalpy values can be defined. It should be pointed out that the specific enthalpy and entropy presented in Fig. 1.3 are given per unit of mass and not per unit of volume. Note, also, that in analogy with conventional vapour compression this diagram is analogous to a *p–h* (log *p–h* diagram), which is usually applied for conventional refrigerants.

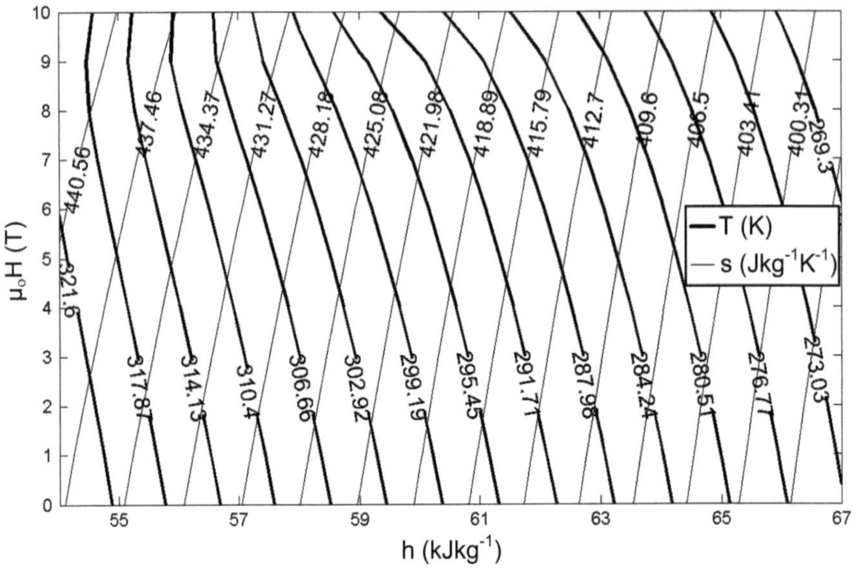

Fig. 1.3 *H–h–T–s* diagram for the magnetocaloric material gadolinium for magnetic fields between 0 and 10 T

1.3 Magnetocaloric Thermodynamic Cycles

We will start here with some basic equations that relate to cyclic thermodynamics. We will also represent basic magnetic thermodynamic cycles without regeneration.

In any thermodynamic cycle, the cyclic integral of the state function equals zero. This holds true for specific enthalpy, the internal energy, as well as for the magnetic "flow" work denoted by product (*M·H*).

Since the internal energy represents the state function, then Eq. (1.1) for a thermodynamic cycle can be rewritten in the following form:

$$\oint \mathrm{d}w = \oint \mathrm{d}q = \oint T\,\mathrm{d}s \tag{1.22}$$

Since the cyclic (net) work in the closed system is equal to:

$$\oint \mathrm{d}w = -\mu_0 \oint H\,\mathrm{d}M \tag{1.23}$$

And the cyclic integral of the state function, the product of the magnetization and the magnetic field (*M H*), equals zero:

$$\oint \mathrm{d}(M\,H) = \oint M\,\mathrm{d}H + \oint H\,\mathrm{d}M = 0 \tag{1.24}$$

then the work of the closed thermodynamic system equals the work of an open thermodynamic system.

$$\oint \mathrm{d}w = -\mu_0 \oint H\,\mathrm{d}M = \mu_0 \oint M\,\mathrm{d}H = \oint \mathrm{d}w_t \tag{1.25}$$

By using the character w_t we keep in mind the technical work.

1.3.1 The Coefficient of Performance (COP) and Exergy Efficiency

The ratio between the cooling capacity q_R of the refrigerator and the work w is defined as the coefficient of performance (COP). This can also be expressed in terms of power:

$$\mathrm{COP} = \frac{q_R}{|w|} \qquad \mathrm{COP} = \frac{\dot{Q}_R}{P} \tag{1.26}$$

The COP of the ideal Carnot cycle operating at the refrigeration temperature T_R and the heat rejection at ambient temperature can be defined as follows:

$$\mathrm{COP}_c = \frac{T_R}{|T_R - T_{\mathrm{amb}}|} \tag{1.27}$$

Sometimes, the thermodynamic cycles are theoretically "Carnotized" in order to compare such cycles with the COP of the Carnot cycle. Equation (1.27) can be assumed for the "Carnotization" of the Ericsson, Stirling or Rankine cycles. However in other cycles, where the cooling temperature of the refrigerant varies (as is the case in, e.g. the Brayton cycle), the "Carnotization" requires an average value of T_R. In this particular case, we define the average refrigeration temperature as [41]:

$$\overline{T_R} = \frac{q_R}{\Delta s_R} = \frac{q_R}{\int_a^b \frac{\mathrm{d}q}{T_R}} \tag{1.28}$$

and the COP of the theoretically Carnotized cycle as:

$$\overline{\mathrm{COP}_c} = \frac{\overline{T_R}}{|\overline{T_R} - T_{\mathrm{amb}}|} \tag{1.29}$$

The exergy concept is well known and often applied by engineers in different domains. The term *exergy* was suggested by Rant in 1956 [50]. According to Rant [50], energy can be divided into two parts, from which the part that is fully transformable (available) into other kinds of energy is named the “exergy”, and the part that is not transformable (e.g. the internal energy of the ambient) is called the “anergy”. There have been just a few publications on exergy analyses for magnetocaloric energy conversion [21, 51–54].

The exergy efficiency of the magnetic refrigerator represents the ratio between the exergy of the cooling energy and the work (i.e. electric energy, which is pure exergy).

$$\xi = \frac{|E_c|}{W} = \frac{\Delta e_R}{w} \tag{1.30}$$

The specific exergy of the cooling energy can be defined by the following relation (see also Eq. (1.12)):

$$\mathrm{d}e = (\mathrm{d}h - T_{\mathrm{amb}}\,\mathrm{d}s) \tag{1.31}$$

In the case of a constant magnetic field (i.e. the Brayton process, where the temperature T_R is not constant), the enthalpy is equal to the heat (exergy of heat), and therefore:

$$\mathrm{d}e = (T_R - T_{\mathrm{amb}})\,\mathrm{d}s = \frac{(T_R - T_{\mathrm{amb}})}{T_R}\,\mathrm{d}q \tag{1.32}$$

and the integration of the exergy between the states a and b (which, e.g. correspond to the cooling process) corresponds to:

$$\Delta e_R = \int_a^b \mathrm{d}e = \int_a^b \left(1 - \frac{T_{\mathrm{amb}}}{T_R}\right)\mathrm{d}q = q_R - T_{\mathrm{amb}}\,\Delta s_R \tag{1.33}$$

where the second part of Eq. (1.33) represents the anergy. The exergy efficiency can be defined as:

$$\xi = \frac{q_R - T_{\mathrm{amb}}\,\Delta s_R}{q_H - q_R} \tag{1.34}$$

In the Carnot cycle, the Ericsson cycle or the Stirling cycle, in the case of an **isothermal demagnetization**, the specific exergy of the cooling energy can be defined as:

$$\Delta e_R = (T_R - T_{\mathrm{amb}})\,\Delta s_R = \frac{(T_R - T_{\mathrm{amb}})}{T_R}\,q_R \tag{1.35}$$

Now we can use Eqs. (1.29) and (1.35) to define the exergy efficiency for cycles that perform refrigeration by isothermal magnetization:

$$\Delta e_R = \frac{q_R}{\mathrm{COP}_C} \quad \rightarrow \quad \xi = \frac{\mathrm{COP}}{\mathrm{COP}_C} \tag{1.36}$$

In cases where the refrigeration is not performed isothermally (e.g. in the case of isofield cooling in the Brayton process), the exergy efficiency can be defined as follows:

$$\Delta e_R = \frac{q_R}{\overline{\mathrm{COP}_C}} \quad \rightarrow \quad \xi = \frac{\mathrm{COP}}{\overline{\mathrm{COP}_C}} \tag{1.37}$$

Note that in the cases shown here, the work or power correspond only to the magnetic work. However, one should be aware of other exergy inputs, which should also be taken into account (e.g. pump work, transmission work or losses) for the definition of the exergy efficiency or the COP of the real device.

1.3.2 Overview of the Basic Thermodynamic Cycles

A comprehensive theoretical description of the different thermodynamic cycles is given by Kitanovski et al. [55]. Here, we will present the basic magnetic thermodynamic cycles in *T–s* and *H–h* diagrams. The best-known magnetic thermodynamic cycles are the Brayton, the Stirling, the Ericsson, the Hybrid (Ericsson–Brayton) cycle and the Carnot cycle.

1.3.2.1 The AMR Thermodynamic Cycle

Most of the applications of magnetic refrigeration at room temperature use permanent magnets as the magnetic field source. Because these are very cost intensive and limited in their magnetic energy, researchers tend to apply moderate magnetic fields, mostly between 0.8 and 1.5 T. Such magnetic fields lead to adiabatic temperature changes in current magnetocaloric materials of up to about 5 K. However, the required temperature span between the heat source and the heat sink is usually much larger. Therefore, the most common way to increase the temperature span is to apply a magnetocaloric thermodynamic cycle that includes regeneration (note that a cascade system may be applied as well, but this needs to take account of the irreversible losses of heat transfer [37]).

There have been a number of studies performed for the different magnetic refrigeration cycles. For instance, analyses of the Ericsson magnetic refrigeration cycles have been reported in the following references: Hakuraku [56], who considered Ericsson cycle without regeneration; He et al. [57], Lucia [58], Wei et al. [59], Xia et al. [60] and Ye et al. [61], who considered a magnetic Ericsson cycle with passive regeneration. Magnetocaloric Carnot cycles without regeneration, i.e. basic Carnot cycles, were evaluated by, e.g. Sasso et al. [33], Steyert (for cryo-cooling without regeneration) [62]. Analyses of the Stirling magnetic refrigeration cycles (two isothermal processes and two iso-magnetization processes) have been performed by Steyert, who analysed the Stirling AMR cycle [24]; and by Chen [28], who carried out analyses for passive regeneration.

In most cases, the magnetocaloric material represents an active refrigerant, and also acts as the regenerator. This kind of regeneration is called an active magnetic regeneration (AMR). Therefore, it should be distinguished from the passive regenerators that are common in, e.g. conventional Stirling devices. The AMR cycle usually performs a kind of regenerative Brayton-like cycle (see, for example [63–69]).

None of the studies systematically focused on the evaluation of different AMR thermodynamic cycles. However, this was done recently by Kitanovski et al. [66] and Plaznik et al. [67].

As described in Chap. 4 for AMR thermodynamic cycles, the main difference between a cascade and an AMR cycle is that in the latter all the parts of the AMR simultaneously accept or reject heat to the heat transfer fluid, which further transfers heat between the neighbouring parts of the AMR (Tishin and Spichkin [32]). The regenerative process is established due to the oscillatory (counter current) fluid flow. In the AMR each infinitesimally small part of the magnetocaloric material performs its own thermodynamic cycle.

Figure 1.4 shows a simple schematic of the Brayton-type cycle based on the AMR (or so-called AMR cycle). The Brayton-like AMR cycle is the most commonly applied thermodynamic cycle in magnetic refrigeration at room temperature. Its basic operation can be described with the following processes. First, during the magnetization process, the magnetocaloric effect leads to an increase in the temperature of the magnetocaloric material. The working fluid, which leaves the heat source heat exchanger (CHEX), enters the voids of the porous magnetocaloric material in the AMR, when this is subjected to a magnetic field. Passing through the porous structure of the magnetocaloric material, the working fluid is heated and leaves the material. Then it enters the heat sink heat exchanger (HHEX), where it rejects the heat to the ambient. The same fluid, cooled by the ambient, again enters the magnetocaloric material, which is not subjected to the magnetic field (and thus cooled down due to the magnetocaloric effect), in the counter-flow direction. The working fluid cools, exits the magnetocaloric material structure (AMR), and enters the CHEX.

From Fig. 1.4 it is clear that the maximum temperature span at the side of the chilled fluid (cooling), which leaves the AMR and enters the heat source heat exchanger (CHEX), cannot exceed the adiabatic temperature change during the

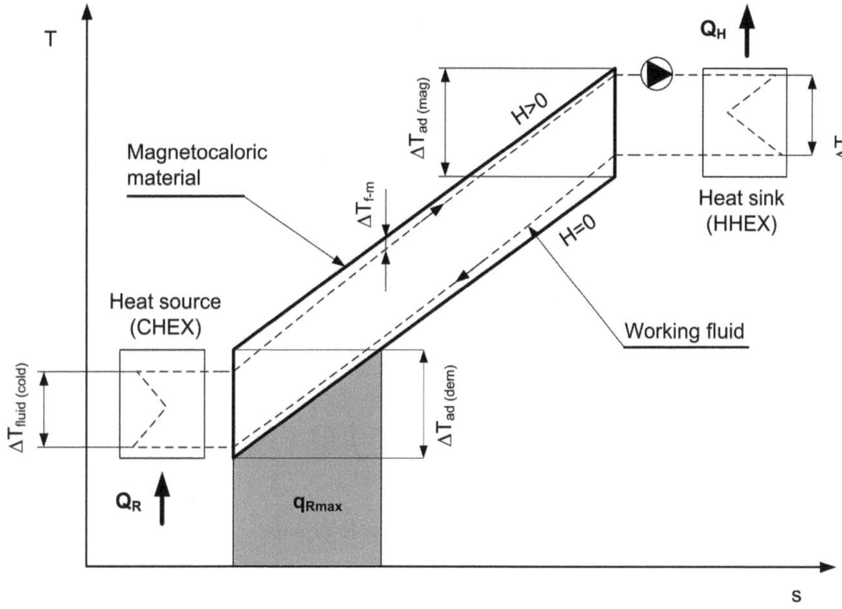

Fig. 1.4 A schematic example of the total AMR Brayton-like cycle in a *T*–*s* diagram

demagnetization. The adiabatic temperature span during the demagnetization defines the maximum temperature span of the working fluid that exits and enters the magnetocaloric material at the cold side. This is limited by the heat transfer between the working fluid and the magnetocaloric material. This also holds true for the warm side of the AMR. From this, one is also able to define the maximum refrigeration capacity $q_{R\max}$ of the magnetocaloric material. This issue is discussed in more detail in Chap. 4.

In order to perform other types of AMR thermodynamic cycles, one has to deal with the manipulation of the magnetic field distribution and the time variation of the fluid flow through the AMR, as explained in Kitanovski et al. [66] and Plaznik et al. [67]. Each of these cycles has a corresponding variation of the magnetic field, as well as a variation of the time that the fluid moves through the AMR. This further influences the internal thermodynamic cycles that are performed by each of the parts of the magnetocaloric material along the AMR, as explained in Chap. 4.

1.3.2.2 Magnetic Thermodynamic Cycles Without Regeneration

Despite of the fact that a regenerative process is required for the present stage of developments in magnetic refrigeration at room temperature, a basic knowledge of the different magnetic thermodynamic cycles is a necessity for a better understanding. Therefore, in the following text we describe the magnetic Brayton, the

Ericsson, the Carnot, and the magnetic Stirling cycles. All these thermodynamic cycles, or at least their approximation, can be obtained from the variations in the magnetic field and the fluid flow. We will not focus here on the passive regenerative process, since active regeneration represents a more efficient solution.

Magnetic Brayton Thermodynamic Cycle

Figure 1.5 shows the magnetic Brayton thermodynamic cycle, which is operated between two constant magnetic fields and two isentropic (adiabatic) processes of magnetization and demagnetization. For this particular cycle, the specific cooling capacity can be defined as follows:

$$q_R = \int_d^a T\,\mathrm{d}s = \int_d^a c_H\,\mathrm{d}T = \int_d^a \mathrm{d}h = h_a - h_d \tag{1.38}$$

The specific work, performed within the cycle equals:

$$\begin{aligned} w = \oint \mathrm{d}q = \int_b^c T\,\mathrm{d}s + \int_d^a T\,\mathrm{d}s = \int_b^c c_H\,\mathrm{d}T + \int_d^a c_H\,\mathrm{d}T \\ = \int_b^c \mathrm{d}h + \int_d^a \mathrm{d}h = (h_c - h_d) + (h_a - h_b) \end{aligned} \tag{1.39}$$

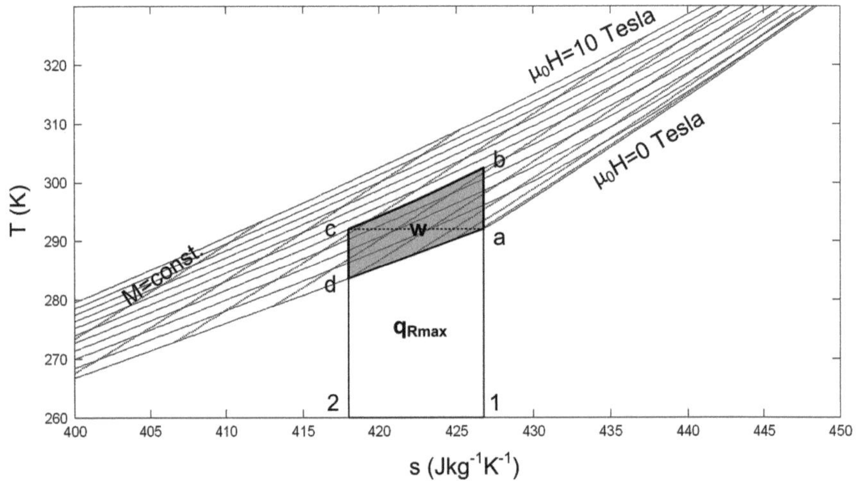

Fig. 1.5 The magnetic Brayton refrigeration cycle in a *T*–*s* diagram (with gadolinium as the refrigerant)

Now the COP can be defined as follows:

$$\mathrm{COP} = \frac{q_R}{|w|} = \frac{h_a - h_d}{|(h_c - h_d) + (h_a - h_b)|} \tag{1.40}$$

Now, the exergy efficiency can by defined using Eq. (1.37). The maximum refrigeration capacity of the magnetic Brayton refrigeration cycle is denoted by the surface 1–a–d–2 (note that points 1 and 2 should correspond to a temperature of 0 K). The specific magnetic work is denoted by the surface a–b–c–d.

Magnetic Ericsson Thermodynamic Cycle

Figure 1.6 shows the magnetic Ericsson thermodynamic cycle in a *T–s* diagram. During this cycle, the processes of magnetization and demagnetization are performed by an isothermal process. This can be provided by the simultaneous fluid flow (i.e. heat transfer) and demagnetization.

The cooling process and the related specific cooling capacity can be defined as follows:

$$q_R = T_R \int_d^a \mathrm{d}s = T_R \Delta s_{a-d} = T_R \cdot \int_d^a \mu_0 \left(\frac{\partial M}{\partial T}\right)_H \mathrm{d}H \tag{1.41}$$

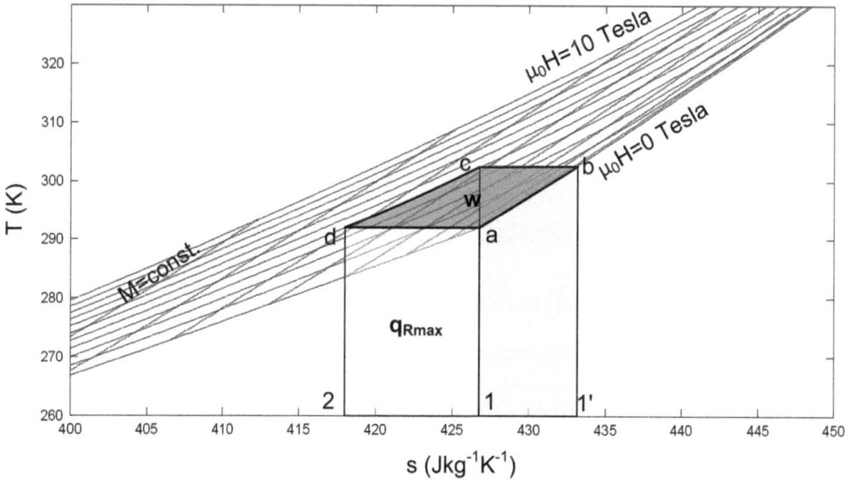

Fig. 1.6 The magnetic Ericsson refrigeration cycle in a *T–s* diagram (with gadolinium as the refrigerant)

where T_R denotes the temperature at which the refrigeration is performed. The work performed during the cycle is equal to:

$$w = \oint \mu_0 M \mathrm{d}H = \mu_0 \int_a^b M\mathrm{d}H + \mu_0 \int_c^d M\mathrm{d}H \tag{1.42}$$

The magnetic Ericsson cycle requires a regenerative process. Without it, Eq. (1.42) can also be written in other terms:

$$\begin{aligned} w = \oint \mathrm{d}q &= T_R \cdot \int_a^d \mathrm{d}s + T_H \cdot \int_b^c \mathrm{d}s + \int_a^b T\,\mathrm{d}s + \int_c^d T\,\mathrm{d}s \\ &= T_R \cdot \Delta s_{a-d} + T_H \cdot \Delta s_{c-b} + (h_d - h_c) + (h_b - h_a) \end{aligned} \tag{1.43}$$

Now the COP can be defined as:

$$\mathrm{COP} = \frac{q_R}{|w|} = \frac{T_R \cdot \Delta s_{a-d}}{|T_R \cdot \Delta s_{a-d} + T_H \cdot \Delta s_{c-b} + (h_d - h_c) + (h_b - h_a)|} \tag{1.44}$$

If we assume that the regeneration is performed without irreversible heat transfer losses, then the two enthalpy differences would have to be equal $(h_d - h_c) = (h_b - h_a)$. Furthermore, in the ideal Ericsson regenerative cycle, the entropy difference during refrigeration and heat rejection would be equal $\Delta s_{a-d} = -\Delta s_{c-b}$. Then the COP of the ideal regenerative Ericsson cycle has the same value as that of the Carnot cycle:

$$\mathrm{COP} = \frac{q_R}{|w|} = \frac{T_R \cdot \Delta s_{a-d}}{|T_R \cdot \Delta s_{a-d} - T_H \cdot \Delta s_{a-d}|} = \frac{T_R}{T_H - T_R} \tag{1.45}$$

Magnetic Carnot Thermodynamic Cycle

The magnetic Carnot refrigeration cycle (Fig. 1.7) is only useful for a comparison with the other refrigeration cycles. In practice its low cooling capacity limits any applicability in real devices. This well-known cycle operates between two isentropic processes of magnetization and demagnetization and two isothermal processes of magnetization and demagnetization. The last process is related to refrigeration, where the cooling capacity is defined as:

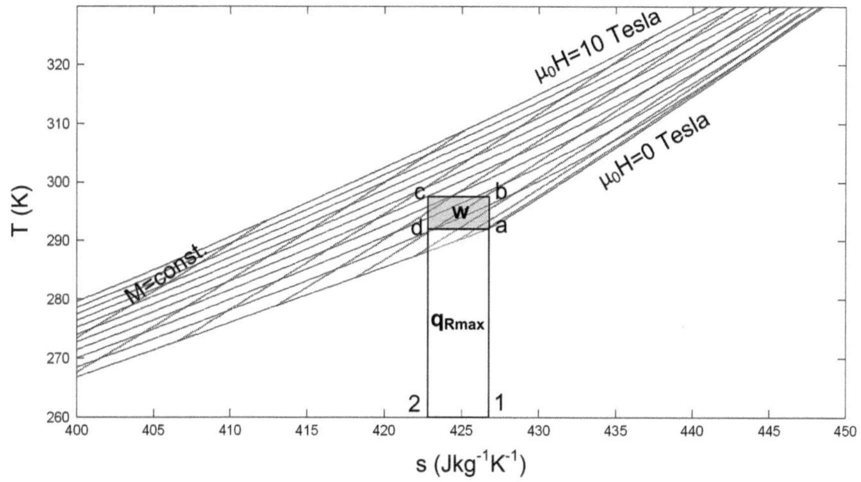

Fig. 1.7 The magnetic Carnot refrigeration cycle in a *T–s* diagram (with gadolinium as the refrigerant)

$$q_R = T_R \int_d^a \mathrm{d}s = T_R \Delta s_{a-d} = T_R \int_d^a \mu_0 \left(\frac{\partial M}{\partial T} \right)_H \mathrm{d}H \tag{1.46}$$

The specific work can be defined with the following relation:

$$w = \oint \mathrm{d}q = T_R \cdot \Delta s_{a-d} + T_H \cdot \Delta s_{c-b} \tag{1.47}$$

where $\Delta s_{a-d} = -\Delta s_{c-b.}$ Therefore, the COP can be defined as:

$$\mathrm{COP} = \frac{q_R}{|w|} = \frac{T_R}{T_H - T_R} \tag{1.48}$$

Magnetic Stirling Thermodynamic Cycle

Such a cycle (Fig. 1.8) is very difficult to operate in real situations. It is necessary to keep the iso-magnetization and iso-demagnetization processes (constant magnetization), which require a particular magnetic field variation with a simultaneous fluid flow. The specific cooling power for this cycle can be defined as:

$$q_R = T_R \int_d^a \mathrm{d}s = T_R \Delta s_{a-d} = T_R \cdot \int_d^a \mu_0 \left(\frac{\partial M}{\partial T} \right)_H \mathrm{d}H \tag{1.49}$$

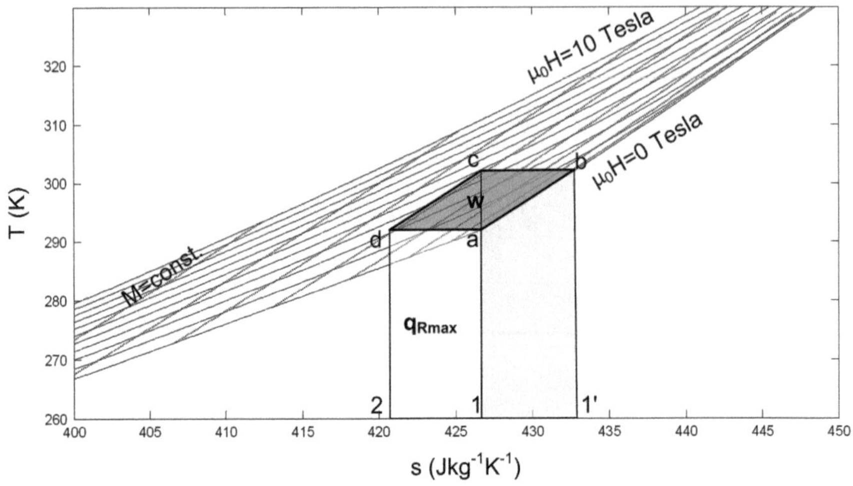

Fig. 1.8 The magnetic Stirling refrigeration cycle in a *T*–*s* diagram (with gadolinium as the refrigerant)

where T_R denotes the temperature at which the refrigeration is performed. The work performed during the cycle equals:

$$w = -\oint \mu_0 H \mathrm{d}M = \mu_0 \int_a^b H \mathrm{d}M + \mu_0 \int_c^d H \mathrm{d}M \tag{1.50}$$

Like with the Ericsson cycle, in the Stirling cycle it is also necessary to operate with regeneration. Without this regeneration, Eq. (1.50) can also be written in other terms:

$$w = \oint \mathrm{d}q = T_R \cdot \int_a^d \mathrm{d}s + T_H \cdot \int_b^c \mathrm{d}s + \Delta s_{c-b} + \int_a^b T\,\mathrm{d}s + \int_c^d T\,\mathrm{d}s \tag{1.51}$$

In the ideal Stirling regenerative cycle, the heat during iso-magnetization processes would be regenerated without any irreversible losses. If the entropy differences during the refrigeration and the heat rejection were to be equal $\Delta s_{a-d} = -\Delta s_{c-b}$, then the COP of the ideal regenerative Stirling cycle has the same value as that of the Carnot cycle:

$$\mathrm{COP} = \frac{q_R}{|w|} = \frac{T_R \cdot \Delta s_{a-d}}{|T_R \cdot \Delta s_{a-d} - T_H \cdot \Delta s_{a-d}|} = \frac{T_R}{T_H - T_R} \tag{1.52}$$

Despite the fact that the real Stirling cycle is difficult to perform in reality, it makes sense to use similar approaches in the design of a thermodynamic cycle for magnetic refrigeration. The reason for this is that the low-field and high-field iso (de)magnetization processes may follow the regenerative process with lower irreversible losses than the case with the Ericsson or Brayton regenerative cycles, which operate between two constant magnetic fields. This is due to the similar temperature–entropy variation in both iso-magnetization processes.

References

1. Tesla N (1890) Pyromagneto-electric generator. US Patent 428.057
2. Edison TA (1892) Pyromagnetic generator. US Patent 476.983
3. Brillouin L, Iskenderian HP (1948) Thermomagnetic generator. Federal Telecommunication Laboratories, Nutley, pp 300–313
4. Van der Voort E (1969) Ideal magnetocaloric conversion. Appl Sci Res 20:98–114
5. Chilowsky C (1952) Thermomagnetic generator and refrigerator. US Patent 2.619.603
6. GB 674284 (1952) Technical assets. Improvements in and relating to thermomagnetic generators and refrigerators
7. Elliott JF (1959) Thermomagnetic generator. J Appl Phys 11(30):1774–1777
8. Resler EL Jr, Rosensweig RE (1964) Magnetocaloric power. AIAA J 8(2):1418–1422
9. Resler EL Jr, Rosensweig RE (1967) Regenerative thermomagnetic power. J Eng Power 89:399–406
10. Kirol LD, Mills JI (1984) Thermomagnetic generator. In: Proceedings of the 19th intersociety energy conversion engineering conference, vol 3, pp 1361–1368
11. Kirol LD, Mills JI (1984) Numerical analysis of thermomagnetic generators. J Appl Phys 56:824–828
12. Salomon D (1989) Thermomagnetic mechanical heat engines. J Appl Phys 65:3687–3693
13. Pecharsky VK, Gschneidner KA Jr (1997) Effect of alloying on the giant magnetocaloric effect of $Gd_5(Si_2Ge_2)$. J Magn Magn Mater 167:179–184
14. Palmy C (2007) A thermo-magnetic wheel. Europhys News 38(3):32–34
15. Palmy C, Egolf PW (2007) Levitation and acceleration of a thermo-magnetic wheel. In: Proceedings of the 2nd international conference on magnetic refrigeration of the international institute refrigeration, pp 299–307
16. Diebold M, Kitanovski A, Vuarnoz D et al (2007) Force, torque and energy of machines with porous magnetisable wheels. In: 2nd Workshop on magnetostrictive materials and magnetic refrigeration, pp 15–16
17. Egolf PW, Kitanovski A, Diebold M et al (2009) Magnetic power conversion with full or porous wheel heat exchangers. J Magn Magn Mater 321(7):758–762
18. Kitanovski A, Diebold M, Vuarnoz D et al (2008) Applications of magnetic "power production" and its assessment. A feasibility study—final report. SFOE
19. Gama S (2010) Tesla type thermomagnetic motor powered by water heated using solar energy. In: Presentation at 4th international IIR conference on magnetic refrigeration, PR China Inner Mongolia, 2010
20. Brück E, Dung NH, Ou ZQ et al (2011) Magnetocaloric Materials—not only for cooling applications. In: Presentation at delft days on magnetocalorics. Delft University (NL)
21. Vuarnoz D, Kitanovski A, Gonin C et al (2012) Quantitative feasibility study of magnetocaloric energy conversion utilizing industrial waste heat. Appl Energy 100:229–237
22. Christiaanse T, Brück E (2014) Proof-of-concept static thermomagnetic generator experimental device. Metall Mater Trans 1E:36–40

23. Brown GV (1976) Magnetic heat pumping near room temperature. J Appl Phys 47:3673–3680
24. Steyert W (1978) Stirling cycle rotating magnetic refrigerators and heat engines for use near room temperature. J Appl Phys 49(3):1216–1226
25. Kirol LD, Mills JI (1984) Numerical analysis of thermomagnetic generators. J Appl Phys 56 (3):824–828
26. Rosensweig R (1985) Ferrohydrodynamics. Cambridge University Press, Cambridge
27. Barclay J (1990) A review of magnetic heat pump technology. Proceedings of the 25th intersociety energy conversion engineering conference IEEE, pp 222–227
28. Chen F, Murphy R, Mei V et al (1992) Thermodynamic analysis of four magnetic heat-pump cycles. Trans ASME J Eng Gas Turbines Power 114:715–720
29. Bisio G, Rubatto G, Schiapparelli P (1999) Magnetic systems depending on three or two variables; thermodynamic analysis and some existing and possible applications. Energy Convers Manage 40:1267–1286
30. Castellano G (2003) Thermodynamic potentials for simple magnetic systems. J Magn Magn Mater 260:146–150
31. Moos G (1997) On the thermodynamic framework for the description of reversible magnetomechanical and magnetocaloric phenomena. J Magn Magn Mater 172(3):247–253
32. Tishin AM, Spichkin YI (2003) The magnetocaloric effect and its applications. Series in condensed matter physics. Institute of Physics Publishing, Bristol and Philadelphia
33. Sasso CP, Basso V, LoBue M et al (2006) Carnot cycle for magnetic materials: the role of hysteresis. Phys B 372(1):9–12
34. Shir F, Mavriplis C, Bennett LH et al (2005) Analysis of room temperature magnetic regenerative refrigeration. Int J Refrig 28(4):616–627
35. Yu BF, Gao Q, Zhang B et al (2003) Review on research of room temperature magnetic refrigeration. Int J Refrig 26:1–15
36. Yu B, Yan Z, Gao Q et al (2006) Research on performance of regenerative room temperature magnetic refrigeration cycle. Int J Refrig 29:1348–1357
37. Kitanovski A, Egolf PW (2006) Thermodynamics of magnetic refrigeration. Int J Refrig 29:3–21
38. Pecharsky VK, Gschneidner KA Jr (1999) Magnetocaloric effect and magnetic refrigeration. J Magn Magn Mater 200:44–56
39. Kittel C (1958) Elementary statistical physics. Wiley, New York, p 228
40. Odenbach S, Liu M (2001) Invalidation of the kelvin force in ferrofluids. Phys Rev Lett 86:328–331
41. Baehr HD (1989) Thermodynamik. Sibente Auflage. Springer, Heidelberg, p 460
42. Wasserman AL (2011) Thermal physics concepts and practice. Cambridge University Press, New York, p 309
43. Plischke M, Bergersen B (1994) Equilibrium statistical physics, 2nd edn. World Scientific Publishing Co Inc, London, p 520
44. Buschow KJH, de Boer FR (2004) Physics of magnetism and magnetic materials. Kluwer Academic, New York, p 182
45. Morrish AH (2001) The physical principles of magnetism. IEEE Classic Reissue, original edn. Wiley, New York, p 696
46. Pecharsky VK, Gschneidner KA Jr (1999) Magnetocaloric effect from indirect measurements: magnetization and heat capacity. J Appl Phys 86:565–575
47. Morrish AH (1965) The physical principles of magnetism. Wiley, New York
48. Ashcroft NW, Mermin D (1976) Solid state physics. Saunders College, New York
49. Petersen TF, Pryds N, Smith A et al (2008) Two-dimensional mathematical model of a reciprocating room temperature active magnetic regenerator. Int J Refrig 31:432–443

50. Rant Z (1956) Exergie, ein neues Wort fur "Technische Arbeitsfahigkeit" (Exergy, a New Word for "Technical Available Work"). Forschung auf dem Gebiete des Ingenieurwesens 22:36–37
51. Vuarnoz D, Kitanovski A, Gonin C et al (2012) Thermodynamic and exergy efficiencies of magnetocaloric energy conversion utilising industrial waste heat. Int J Exergy 10(4):365–378
52. Kotani Y, Kansha Y, Tsutsumi A (2013) Active magnetic regenerative heat circulator for energy saving in thermal process. Chem Eng Trans 35:229–234
53. Kitanovski A, Gonin C, Vuarnoz D et al (2009) A standardization of the coefficient of performance and cost determination for magnetic refrigerators, heat pumps and power conversion machines. In: 3rd International conference on magnetic refrigeration at room temperature. Des Moines. Iowa. USA, pp 229–237
54. Rowe A (2011) Configuration and performance analysis of magnetic refrigerators. Int J Refrig 34:169–177
55. Kitanovski A, Tušek J, Poredoš A (2014) Magnetic cooling: from fundamentals to high efficiency refrigeration. In: Sandeman K, Gutfleisch O (eds) Magnetic thermodynamic cycles. Wiley, New York (submitted)
56. Hakuraku Y (1987) Thermodynamic simulation of a rotating Ericsson-cycle magnetic refrigerator without a regenerator. J Appl Phys 62:1560–1563
57. He J, Chen J, Wu C (2002) Heat-transfer effect on the performance of a magnetic ericsson refrigerator. J Non-Equilibr Thermodyn 27:57–69
58. Lucia U (2011) Second law analysis of the ideal Ericsson magnetic refrigeration. Renew Sustain Energy Rev 15:2872–2875
59. Wei F, Ling G, Chen J et al (2010) Performance characteristics and parametric optimization of an irreversible magnetic ericsson heat-pump. Phys B: Condens Matter 406:633–639
60. Xia ZR, Ye XM, Lin GX et al (2006) Optimization of the performance characteristics in an irreversible magnetic Ericsson refrigeration cycle. Phys B: Condens Matter 381:246–255
61. Ye XM, Lin GX, Chen JC et al (2007) Parametric optimization of an irreversible magnetic ericsson refrigerator with finite heat reservoirs. Phys B: Condens Matter 391:350–356
62. Steyert WA (1978) Rotating carnot-cycle magnetic refrigerators for use near 2 K. J Appl Phys 49:1227–1231
63. Xia Z, Zhang Y, Chen J et al (2008) Performance analysis and parametric optimal criteria of an irreversible magnetic Brayton-refrigerator. Appl Energy 85:159–170
64. Diguet G, Lin G, Chen J (2012) Performance characteristics of an irreversible regenerative magnetic Brayton refrigeration cycle using $Gd_{0.74}Tb_{0.26}$ as the working substance. Cryogenics 52:500–504
65. He JZ, Xin Wuand X, Deng XF (2008) Performance characteristics of an irreversible magnetic Brayton refrigeration cycle. Int J Refrig 31(1):138–144
66. Kitanovski A, Plaznik U, Tušek J et al (2014) New thermodynamic cycles for magnetic refrigeration. Int J Refrig 37:28–35
67. Plaznik U, Tušek J, Kitanovski A et al (2013) Numerical and experimental analyses of different magnetic thermodynamic cycles with an active magnetic regenerator. Appl Therm Eng 59(1–2):52–59
68. Zhang X, Chen J, Lin G et al (2010) Influence of irreversible losses on the performance of a two-stage magnetic Brayton refrigeration cycle. Phys B: Condens Matter 405:1632–1637
69. He JZ, Xin Wuand X, Deng XF (2008) Performance characteristics of an irreversible magnetic Brayton refrigeration cycle. Int J Refrig 31:138–144

Chapter 2
Magnetocaloric Materials for Freezing, Cooling, and Heat-Pump Applications

Magnetocaloric materials (MCM) are the 'heart' of every magnetic refrigeration or heat-pump application. Apart from having a crucial role in the heat-regeneration process, they also exhibit a special and vital phenomenon for magnetic refrigeration called the magnetocaloric effect. As mentioned in the previous chapter on the Thermodynamics of magnetocaloric energy conversion and as described later in the book in the Chap. 7 (*Overview of existing magnetocaloric prototype devices*) the first real magnetic devices working near room temperature were not built until the middle of the 1970s. However, the discovery of the magnetocaloric effect (MCE) in ferromagnetic materials dates back more than 90 years to 1917.

The MCE was discovered by the French and Swiss physicists Weiss and Piccard [1, 2]. This is an important historical fact that needs to be emphasized, since in the past 15 years a misconception has arisen in the magnetocaloric research community, wrongly attributing the discovery of the MCE to the work of Warburg [3]. At this point we have to address the recent paper of Smith [4], who did a thorough and interesting review of the research on thermodynamics that led to the discovery of the MCE. The next few lines are a brief summary of the historical events according to Smith's findings. He reviewed the original works of scientists that date back to the nineteenth century, starting with Joule in 1843 [5]. Joule observed that heat was evolved from iron samples when they were subjected to a magnetic field. Later, in 1860, Thomson (Lord Kelvin) [6] was already aware of the fact that ferromagnetic materials lose their magnetic properties when heated above a certain temperature (now known as the Curie temperature). Thomson correctly predicted that ferromagnetic materials would experience a heating effect when magnetized and a cooling effect when demagnetized, and that these effects would be the largest around the temperature where they lose their magnetization. However, he did not associate these predictions with the magnetocaloric effect. Then, in 1881, Warburg published a paper [3], which is nowadays wrongly cited when referencing the discovery of the MCE. Nevertheless, the work of Warburg was of great significance, since he was the first to explain magnetic hysteresis. He correctly predicted that the magnetization of a material is larger when the magnetic field is decreasing than when it is increasing. One year later, in 1882, Ewing [7] discovered the same phenomenon and was the first to name it "hysteresis". It was not until the works of

A. Kitanovski et al., *Magnetocaloric Energy Conversion*,
Green Energy and Technology, DOI 10.1007/978-3-319-08741-2_2

Weiss and Piccard were published in 1917 and 1918 [1, 2], where they discovered a reversible heating of a nickel sample near its Curie temperature (354 °C) when a magnetic field was applied. They found that the nickel sample increased its temperature by 0.7 K when a magnetic field of 1.5 T was applied. Furthermore, they also stated that the reversibility of the effect and also its larger order of magnitude could distinguish it from the heat that emerges from the hysteresis. Finally, they called their discovery a "novel magnetocaloric phenomenon", thereby coining the word "magnetocaloric".

The discovery of Weiss and Piccard was undoubtedly acknowledged and well known in the scientific community until the end of twentieth century, when the sudden misconception arose, attributing the discovery of the MCE to Warburg. The reasons for this misconception will not be discussed here. However, we encourage the reader of this book to investigate the paper of Smith [4], where this is explained in detail.

The first ideas that ferromagnetic materials could be usefully applied in power generation, refrigeration or heat pumping emerged with the works of the Slovenian physicist Stefan in the last quarter of the nineteenth century [8, 9]. Stefan explained how a thermomagnetic motor should work by exploiting the transition from the ferromagnetic to the paramagnetic state of the material by heating it above its Curie temperature. Edison [10, 11] and Tesla [12, 13] then patented their versions of thermomagnetic generators at the end of nineteenth century. In 1926 Debye [14] and in 1927 Giauque [15] independently discussed that if paramagnetic salts are adiabatically demagnetized, extremely low temperatures (under 1 K) could be achieved. This was experimentally proven in 1933 by Giauque and MacDougall [16]. In 1935, Urbain et al. [17] discovered ferromagnetism in gadolinium. This was the first ferromagnetic material discovered that has a Curie temperature near room temperature. However, it was not until the middle of the 1960s that the MCE of gadolinium was investigated [18, 19] by researchers from West Virginia University. This opened up the possibility of magnetic refrigeration devices operating near room temperature. In this manner Brown showed in his paper from 1976 [20] that gadolinium could be a possible MCM to be used in magnetic refrigeration. He built and experimentally tested the first-ever magnetic refrigeration prototype working near room temperature. From that point on the amount of research in magnetic refrigeration near room temperature started to increase. For example, Barclay and Steyert presented and patented the idea of an active magnetic regenerator in 1982 [21]. Active magnetic regeneration is an important invention in magnetic refrigeration. Active magnetic regeneration also implies that not only are the magnetocaloric properties of a material important, but also its thermal properties, as well as the manufacturability and processing properties to enhance the heat-transfer characteristics.

Another important milestone in magnetic refrigeration happened in 1997 with the discovery of the so-called giant MCE close to room temperature in a first-order transition material $Gd_5Si_2Ge_2$ by Pecharsky and Gschneidner [22]. The giant MCE observed at a transition temperature of 276 K was much higher (in terms of magnetic entropy change) than that of other known MCMs at that time. This discovery further increased the research on magnetic refrigeration near room temperature. Nowadays, a lot of effort is put into the research and design of magnetocaloric devices; however, even more effort is directed at the research of new MCMs that would be suitable for use in near-room-temperature applications.

There are a number of different MCMs available for use near room temperature; these are thoroughly described in various reviews in the literature [23–28]. Furthermore, there is also a book written by Tishin and Spichkin [29] which describes numerous different MCMs in details.

In general, MCMs can be divided into two groups based on the order of their phase transition from the ferromagnetic to paramagnetic state, thus calling them second-order or first-order materials [30]. The phase transition happens at the certain temperature, referred to as the Curie temperature. Above the Curie temperature the spontaneous magnetization disappears and the material becomes paramagnetic. Furthermore, the MCE is most noticeable at this phase transition. The difference between first-order and second-order materials is how this transition takes place (Fig. 2.1).

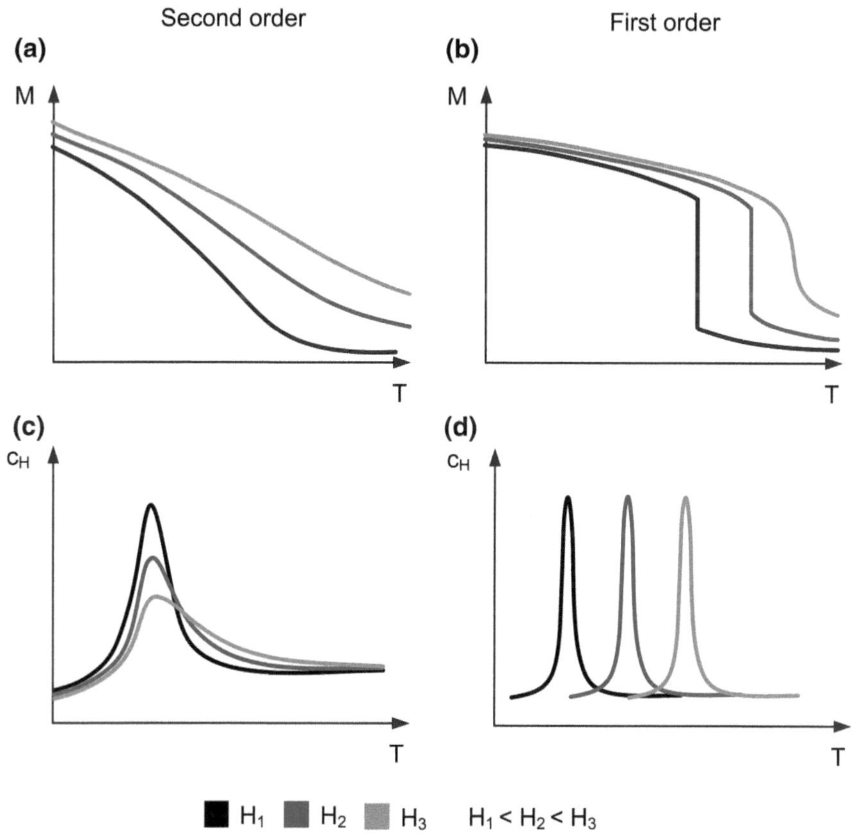

Fig. 2.1 Schematic general distinctions between second-order and first-order materials via magnetization (**a** and **b**) and specific heat (**c** and **d**) in relation to temperature and magnetic field

The characteristic of a second-order phase transition is the continuous change of the magnetization around the Curie temperature (Fig. 2.1a), while in the first-order phase transition the magnetization changes discontinuously at some temperature (Fig. 2.1b) with associated structural-deformation.

Regarding the magnetic entropy change, in second-order materials the magnetic entropy change increases with a larger magnetic field. In the first-order magnetic materials the entropy change only increases drastically to a certain value of magnetic field. However, with a larger field the magnetic entropy change becomes considerable over a wider temperature range.

Adiabatic temperature changes for both phase transitions increase their values with increasing field. However, in second-order materials the peak is broader than in first-order materials. Another important distinction between second- and first-order is in the specific heat (Fig. 2.1c, d). In second-order materials the specific heat is sharply peaked with a lower field and then decreases and broadens the peak without any significant shift in the peak temperature, while in first-order materials the specific heat significantly changes its peak-temperature position with larger magnetic fields, whilst not changing the peak values drastically.

Since this book is focussed more on an engineering approach to research and the design of magnetocaloric devices, the next pages of this chapter will present different MCMs from the engineering point of view rather than that of a material scientist. Some of the MCMs that are, at least at the moment, the most promising, will be presented. In this way, an engineer reading this book could get some initial impression about which direction she or he could focus her or his research and design of magnetic devices. Furthermore, some important issues and aspects regarding other characteristics (e.g. thermal, mechanical, chemical properties) will also be discussed.

2.1 General Criteria for the Selection of the Magnetocaloric Material

The MCMs as the coolants and the regeneration materials represent the most crucial elements of the magnetic refrigerator. Therefore, it is very important to apply the best material possible for a particular application. In general, they should have the following properties (see also e.g. [31]):

2.1.1 Suitable Curie Temperature of the Material

A precondition for the application of a MCM for a particular application is the suitability of its Curie temperature. With this we ensure that the MCE occurs at the required temperature or temperature range. The Curie temperature represents the temperature of the phase transition of the magnetic material between the ferromagnetic and paramagnetic phases, which is related to the most pronounced magnetocaloric effect. It should be noted that the magnetocaloric effects at temperatures that are relatively far away from the Curie temperature are practically negligible (depending on the width of the temperature range of the magnetocaloric effect).

2.1.2 The Intensity of the Magnetocaloric Effect

The most important criterion for the selection of a MCM is the intensity of its magnetocaloric effect. The MCE manifests itself as the adiabatic temperature change and/or isothermal entropy change, which are related through the specific heat of the material (see Eqs. (1.14) and (1.19)). It should be noted that for applications of the MCM in the AMR its adiabatic temperature change is more important than the isothermal entropy change. The material is, therefore, more suitable for an application if it has a greater adiabatic temperature change on account of the smaller isothermal entropy change. This is strongly related to the heat-transfer characteristics between the material and the heat-transfer medium, since the heat-transfer irreversibility losses can strongly reduce the device's performance in the case of a small adiabatic temperature change, as is also explained in Sect. 4.5. A detailed analysis of the impact of the adiabatic temperature change and the isothermal entropy change on the AMR's performance is presented in [32].

2.1.3 The Wide Temperature Range of the Magnetocaloric Effect

It is a great advantage for the MCM to have a (large) MCE over as wide a temperature range as possible. This is especially important in an AMR where the temperature span is established over the material. With a wide temperature range for the MCE we ensure that the intense MCE occurs over the entire material, even in the parts of the material that are temperature-wise away from its Curie temperature. Since the great majority of currently known MCMs exhibit a MCE over a relatively narrow temperature range, a layering of different MCMs with different Curie temperatures along the length of the AMR (in a direction of the temperature gradient) is required. As also explained and shown in Sect. 4.2, the layering also ensures an intense MCE over the entire length of the AMR (with the established temperature profile). It should also be noted that the MCE of second-order phase-transition materials like Gd occurs over a relatively wide temperature range compared to the first-order phase-transition materials, e.g. Mn–Fe–P and La–Fe–Si alloys (see Sect. 4.2), where layering is therefore more important.

2.1.4 Near-Zero Hysteresis of the Magnetocaloric Effect

The MCMs should have as small a hysteresis as possible. The hysteresis occurs as the magnetic hysteresis (during an alternating magnetic field) and the thermal hysteresis (during heating and cooling). It should be noted that the hysteresis is, in general, related with the first-order phase transition and its structural changes, and in general does not occur in a second-order phase transition materials (e.g. Gd), which is a great advantage. However, both hystereses result in an energy loss and therefore, an increase in the input work of the thermodynamic cycle as the result of the entropy generation [33]. This can drastically reduce the MCE during the cycling operation as well as the efficiency of the magnetocaloric device. The impact of the hysteresis on the performance of the magnetic refrigerator can be found in [34, 35].

2.1.5 High Thermal Conductivity and Diffusivity

In general, the thermal conductivity and thermal diffusivity of the MCM should be as high as possible, since it ensures a faster temperature response and a more intense heat transfer between the material and the heat-transfer fluid. However, the high thermal conductivity of the MCM can also reduce the AMR's performance due to the heat flux along the direction of the temperature gradient in material, parallel to the fluid flow. This is especially pronounced in the case of a shorter AMR with an ordered geometry (where the material in AMR is continuous along its length) and a large temperature span. As shown in Nielsen and Engelbrecht [36], the optimal thermal conductivity of the MCM applied in a parallel-plate AMR strongly depends on the length of the AMR and the operating frequency. They showed that in the case of a long AMR (200 mm) the thermal conductivity should be as high as possible (up to 30 $Wm^{-1}K^{-1}$), regardless of the operating frequency (up to 4 Hz) and the temperature span, while in the case of a shorter AMR (50 mm) there is an optimal thermal conductivity for each operating frequency (the higher the frequency the higher the optimal thermal conductivity will be: around 10 $Wm^{-1}K^{-1}$ at 1 Hz and 30 $Wm^{-1}K^{-1}$ at 4 Hz). For example, Gd and its alloys with Er and Tb have a thermal conductivity around 10 $Wm^{-1}K^{-1}$, La–Fe–Co–Si alloys around 8 $Wm^{-1}K^{-1}$ and La–Ca–Sr–MnO_3 ceramics around 1 $Wm^{-1}K^{-1}$.

2.1.6 Good Manufacturing Properties

It is desirable for the MCMs to have good manufacturing, casting, mechanical and processing properties, which allow them to be fabricated into the desired shape, suitable for use in an efficient AMR. The impact of the geometrical properties of the

AMR on its performance is presented in Sect. 4.4, while the review of the different applied fabrication techniques for the AMRs is given in Sect. 4.7.

2.1.7 High Electrical Resistivity

The high electrical resistivity of the MCM prevents the generation of eddy currents (which results in energy dissipation and heating of the material) under the influence of the changing of the external magnetic field. However, in a typical AMR, operating with frequencies up to 10 Hz, the impact of the eddy currents are, in general, negligible, but for applications at higher operating frequencies this might play an important role. For details of the energy dissipation due to the eddy currents in magnetic materials see, e.g. [37].

2.1.8 Good Corrosion Properties

It is preferable that the MCM does not corrode when in contact with water (or other heat-transfer fluids). From this point of view the ceramic manganite MCMs (e.g. La–Ca–Sr–MnO_3) have a certain advantage as they are non-corrosive. However, as explained in Sect. 4.6 the corrosion of other MCMs can be prevented by adding the proper inhibitors to the heat-transfer fluid.

2.2 Most Common Magnetocaloric Materials with a Near-Room-Temperature MCE

The subsequent subsections are intended to present groups of different MCMs that are currently the most promising in the field of magnetic refrigeration near room temperature. Only a brief description of the different materials are given to show the design engineer of the magnetocaloric prototypes basic idea of how to approach MCMs so as to apply them in the AMR. Detailed descriptions, reviews and studies of MCMs are already well covered in the known literature and are also more of the domain of material scientists.

Note that the majority of studies on different MCMs report their MCEs in the form of magnetic entropy change. This is, of course, a fundamental physical property for defining the MCE; however, in terms of system design and heat transfer it would be more useful to also have the data for the adiabatic temperature change and the specific heat for given MCMs. In this way, one could quickly consider the different MCMs to be suitable for the AMR design, at least during the initial design stages.

2.2.1 Gd and Its Alloys

Gadolinium (Gd) is definitely the most common MCM for magnetic refrigeration near room temperature. It is the only pure element that exhibits a MCE near room temperature (~293 K). Furthermore, its magnetocaloric properties are fairly good ($\Delta T_{ad} = 3.3$ K, $c_H = 300$ Jkg^{-1}K^{-1}, $\Delta s_M = 3.1$ Jkg^{-1}K^{-1} at magnetic field change of 1 T [38]), making it a strong candidate for use in magnetic refrigeration. Actually, gadolinium has already been thoroughly investigated and characterized for use as a constituent of various AMRs in different magnetic refrigeration devices as it is reviewed later in Chap. 7 on magnetic prototypes. As a result, Gd is known as a kind of reference material when considering different MCM candidates for an AMR design. However, one of the most important factors when choosing Gd is its purity. As was shown by Dan'kov et al. [39], different impurities in Gd may significantly alter its magnetocaloric properties.

However, introducing different amounts of other elements to make alloys with Gd can also have positive effects. Especially in terms of designing layered AMRs. For instance, different ratios of Gd and Mn in Gd–Mn alloys can lower the T_C to 278 K without any drastic changes in the MCE, as was shown by Jayaraman et al. [40, 41]. Furthermore, for example, in Gd–R alloys, where R is some other rare-earth element (Tb, Dy, Ho, Er) the T_C may also be shifted to lower temperatures [42]; however, without any drastic changes in their MCEs. For instance, Kaštil et al. [43] presented the MCE in Gd–Tb alloys. By changing the Tb content they could shift the Curie temperature of Gd–Tb alloys in the temperature range from 269 to 294 K, with an average adiabatic temperature change of approximately 2.5 K for a 1 T magnetic field change.

On the other hand, the Curie temperature T_C may also be shifted above that of pure Gd, which can be suitable in magnetic heat pumping or magnetic power generation. For instance, Couillaud et al. [44] presented the magnetocaloric properties of two MCMs, Gd–Sc–Ge and Gd–Sc–Si. The former has a T_C of 348 K and a magnetic entropy change Δs_M of 2.5 Jkg^{-1}K^{-1} from 0 to 1.5 T, while the latter has a T_C of 252 K and Δs_M of 2 Jkg^{-1}K^{-1} when changing the magnetic field from 0 to 1.5 T. Furthermore, Law et al. [45] showed that different ratios of elements in Fe–Gd–Cr–B alloys can lead to an increase in T_C above 400 K. However, the magnetic entropy change for the materials with a Curie temperature around such high temperatures can decrease to approximately 1 Jkg^{-1}K^{-1} for magnetic field change from 0 T to 1.5 T.

However, there is a group of Gd-based alloys that exhibit a remarkable MCE, noticeably higher than that of pure Gd and the above-mentioned alloys. These are the first-order Gd–Si–Ge alloys. These alloys exhibit the so-called giant magnetocaloric effect. The giant MCE was discovered in 1997 in $Gd_5Si_2Ge_2$ by Pecharsky and Gschneidner [22]. The same researchers later showed that by varying the Si-to-Ge ratio and by introducing small amounts of Ga into the Gd–Si–Ge the giant MCE may be tuned in the temperature range between approximately 20 and 305 K [42]. The MCE (in terms of magnetic entropy change) of Gd–Si–Ge is near room temperature at least two times higher than that of pure Gd. However, due to its first-order nature, Gd–Si–Ge alloys display a high magnetic hysteresis, which can drastically contribute to the parasitic losses in the magnetocaloric device.

However, the main drawback of Gd (and its second-order transition alloys) is its price, which limits its practical application. However, its magnetocaloric, thermal and manufacturing properties and the absence of hysteresis make it currently the best MCM for room-temperature magnetic refrigeration.

2.2.2 La–Fe–Si-Based MCMs

La–Fe–Si-based MCMs are well represented in magnetic refrigeration and are considered to be one of the possible alternatives to the expensive Gd-based MCMs.

The basis of La–Fe–Si materials is a hypothetical compound $LaFe_{13}$, which does not exist. However, by substituting a certain proportion of the Fe for Si or Al one can make a stable compound. For instance, in 2001 Hu et al. [46] discovered a first-order transition at 208 K in the compound $LaFe_{11.4}Si_{1.6}$. Later researchers discovered that the Curie temperature can be tuned by adding H to the structure of La–Fe–Si, as, for example, was presented by Fujita et al. in 2003 [47, 48]. Furthermore, researchers also found that the T_C may also be tuned by partially substituting Fe with Al, Co or Mn. This was presented by several authors, such as Katter et al. [49], Hansen et al. [50] and Bjørk et al. [38]. These kinds of substitutions may also alter a material's transition from first to second order.

Nowadays, there is a substantial number of different La–Fe–Si-based MCMs, which were thoroughly reviewed by Shen et al. [24]. One of the major issues regarding such materials is their long-term stability. However, this can be avoided by properly processing the material [51]. La–Fe–Si-based MCMs have a great potential to be used in layered AMRs, since their T_C may be tuned in a temperature range from approximately 200 to 340 K. Regarding their magnetocaloric properties, La–Fe–Si-based materials exhibit a larger magnetic entropy change than that of Gd. It may vary from approximately 5 to 12 $Jkg^{-1}K^{-1}$ (regarding magnetic field change of 1.6 T) [49, 51], depending on the material. The adiabatic temperature change is, for a magnetic field change from 0 to 1.4 T, in the range of 2.8 K [38]. La–Fe–Si-based materials have a substantially higher specific heat than Gd (from approx. 1,200 $Jkg^{-1}K^{-1}$ at 0 T to 700 $Jkg^{-1}K^{-1}$ at 1.4 T [38, 51]).

The reasons why La–Fe–Si-based MCMs are so appealing for use in magnetocaloric devices lie mostly in their low cost (in comparison to Gd). Some of the materials also exhibit no or low magnetic hysteresis, which is positive from the device-performance point of view. Moreover, the technology for producing such materials and then processing them is available for large-scale industrial production [49].

2.2.3 Mn-Based MCMs

Compounds with Mn are another large group of MCMs with the potential to be used in magnetic refrigeration. A comprehensive review on Mn-based MCMs was recently presented by Brück et al. [25]. In 2001, Wada et al. [52] presented a giant MCE in the compound Mn-As. It has a Curie temperature around 317 K with a magnetic entropy change of approximately 40 $Jkg^{-1}K^{-1}$ (when changing the magnetic field from 0 to 2 T), which is substantially larger than that of $Gd_5Si_2Ge_2$. Wada et al. [52] also showed that increasing the magnetic field above 2 T does not contribute much more to the increase in Δs_M. One of the issues associated with Mn–As is that it exhibits a large hysteresis due to its first-order nature. However, the hysteretic behaviour as well as the Curie temperature may be adjusted, to some extent, by substituting a certain proportion of As with Sb [52], making a Mn–As–Sb compound. Later in 2002, Tegus et al. [53] presented a new Mn-based compound Mn–Fe–P–As. By adjusting the P/As ratio, their T_C may be significantly tuned in a large temperature range from 150 to 335 K. For example, the compound $MnFeP_{0.5}As_{0.5}$ has a T_C at 280 K with a magnetic entropy change of 25 $Jkg^{-1}K^{-1}$ (2 T) [54]. A slight change of the P/As ratio in the compound $MnFeP_{0.45}As_{0.55}$ shifts its T_C to 306 K, while the magnetic entropy change decreases to 13 $Jkg^{-1}K^{-1}$ (2 T) [25, 55], which is still quite substantial. In 2011 Dung et al. [56] discovered that the hysteresis in Mn–Fe–P–As may be tuned by changing the Mn/Fe ratio. Furthermore, adjusting the Mn/Fe ratio may also lead to a change from a first- to second-order transition.

Another interesting group of Mn-based MCMs is the Mn–Fe–P–Si–Ge alloys [57]. Their main advantage is that they do not contain toxic As. Mn–Fe–P–Si–Ge materials exhibit a similar MCE to Mn–Fe–P–As. However, Mn–Fe–P–Si–Ge materials have a large hysteresis, which can also be tuned by changing the concentrations of Fe and Mn. In this manner, the Curie temperature can also be varied to some extent [58].

2.2.4 Manganites

Another group of materials that show a potential for use in magnetic refrigeration near room temperature are the perovskite manganites, which are basically ceramic materials. Their general formula may be expressed as $R_{1-x}M_xMnO_3$, where R = La, Nd or Pr and M = Ca, Sr or Ba. There is a large number of different manganite MCMs, which were comprehensively reviewed by Phan and Yu [26]. Manganites are second-order materials, thus exhibiting a low MCE (lower than Gd). However, their Curie temperature may be tuned over a large temperature range. For example, the manganites $La_{0.67}Ca_{0.33-x}Sr_xMnO_3$ (LCSM) can be tuned for their Curie temperature in the range from 267 to 369 K by changing the x value from 0 to 0.33 [59]. The LCSM compound ($x = 0$) with a Curie temperature of 267 K has a magnetic entropy change of 5.9 $Jkg^{-1}K^{-1}$ and an adiabatic temperature change of 2 K (from 0 to 1.2 T). However, the MCE decreases with increasing x value. Therefore, the LCSM compound ($x = 0.055$) with a Curie temperature of 285 K (which is relevant for near-room-temperature magnetic applications) has a magnetic entropy change of 2.8 $Jkg^{-1}K^{-1}$ and an adiabatic temperature change of 1 K for a magnetic field change from 0 to 1.2 T. In spite of the LCSMs having a rather low MCE, they are still a promising group of MCMs that could be used in magnetic refrigeration. This is mostly due to their low price, good corrosion resistance, the easy tunability of the T_C and the ease of processing [30].

2.2.5 Layered MCMs

As stated in Sect. 2.1 (*General criteria for the selection of the MCM*), one of the important characteristics of MCMs is to have a large MCE over as wide a temperature range as possible, since the AMR should operate with a large temperature span. Since the MCMs exhibit their largest MCE around their Curie temperature, the idea has been developed to build the AMR from different MCMs along the regenerator. Each material should have its Curie temperature (and therefore the largest MCE) in a different temperature range. In this way, the AMR could have a significant MCE across its full operating temperature span. As was presented in the previous sections on different MCMs, the tuning of Curie temperatures is of course possible by changing the concentrations of the certain elements in magnetocaloric compounds. Building the AMR from several MCMs with different Curie temperatures will lead to a step-wise change in the Curie temperatures along the AMR's length (Fig. 2.2a).

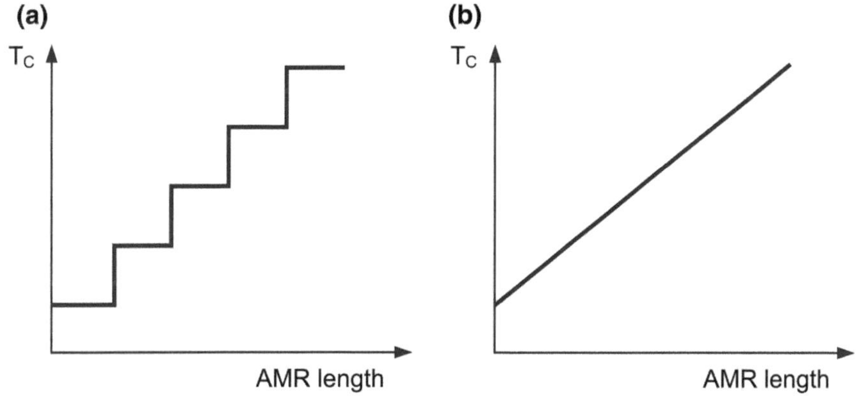

Fig. 2.2 **a** A schematic diagram of a step-wise T_C; **b** Linearly continuously T_C layered AMR

Table 2.1 Some MCMs with their magnetocaloric properties near room temperature

Material	T_C (K)	$-\Delta s_M$ ($Jkg^{-1}K^{-1}$)	ΔT_{ad} (K)	ΔB (T)	$q_{R,max}$ (Jkg^{-1})	References
Gd	~293	3.1	3.3	1	913	[38]
$Gd_{0.9}Tb_{0.1}$	~286	2.3	1.9	1	1,148	[43]
$Gd_5Si_2Ge_2$	~278	14	7.3	2	3,943	[22]
$LaFe_{11.06}Co_{0.86}Si_{1.08}$	~276	6.1	2.3	1	1,690	[38]
$LaFe_{11.05}Co_{0.94}Si_{1.01}$	~287	5.1	2.1	1	1,469	[38]
$LaFe_{10.96}\ Co_{0.97}Si_{1.07}$	~289	5.3	2.2	1	1,537	[38]
$La(Fe_{0.88}Si_{0.12})_{13}H$	~274	19	6.2	2	5,264	[48]
$La(Fe_{0.89}Si_{0.11})_{13}H_{1.3}$	~291	24	6.9	2	7,066	[48]
$La(Fe_{0.88}Si_{0.12})_{13}H_{1.5}$	~323	19	6.8	2	6,201	[48]
MnAs	~318	31	4.7	2	9,930	[52]
$MnFeP_{0.45}As_{0.55}$	~306	12.5	2.8	1	3,842	[55]
$Mn_{1.1}Fe_{0.9}P_{0.47}As_{0.53}$	~292	11	2.8	1	3,227	[55]
LCSM ($x = 0$)	~267	5.9	2.0	1.2	1,581	[59]
LCSM ($x = 0.055$)	~285	2.8	1	1.2	800	[59]
LCSM ($x = 0.165$)	~332	1.8	0.93	1.2	598	[59]

An overview of a numerical and experimental analysis of layered AMRs is presented in Sect. 4.2. However, recently, a new material was presented by Barcza et al. [60]. They presented a layered $LaFe_{13-x-y}Co_xSi_y$ material in which the Curie temperature changes continuously along the length. They managed to produce such a layered material by pressing several powders with different Curie temperatures on top of each other with a subsequent sintering and diffusion treatment [60]. In this manner gradients of the Curie temperature between 0.3 and 10 Kmm^{-1} were obtained. The general idea is to make such a material in which the Curie temperature gradient would linearly and continuously change along the length (Fig. 2.2b).

2.2.6 Conclusions

In the above sections a brief review of some of the most common MCMs that exhibit a MCE close to room temperature were presented. In this way the reader can obtain a general impression about which direction to search in the study of magnetic refrigeration, heat pumping or power generation. There are a number of extensive and thorough reviews on MCMs of different sorts already published in the literature, if the reader needs to study MCMs in more detail.

In conclusion, Table 2.1 shows some of the most interesting types of MCMs with regards to their MCE.

Only materials for which the data regarding their magnetic entropy change and adiabatic temperature change were available are shown. In this manner the maximum specific cooling energy $q_{R,\max}$ can also be given. The maximum specific cooling energy $q_{R,\max}$ is explained in more details in Chap. 4 (Active magnetic regeneration, Sect. 4.1.4). However, in general, it can be expressed using the following equation:

$$q_{R,\max} = \frac{(2T_R + \Delta T_{\mathrm{ad}})\Delta s_M}{2} \tag{2.1}$$

where for the examples in Table 2.1 T_R is the Curie temperature T_C with the corresponding magnetic entropy change $-\Delta s_M$ and adiabatic temperature change ΔT_{ad}. As is clear from Table 2.1 different MCMs have different Curie temperatures.

In this manner, the parameter $q_{R,\max}$ could be significant when designing layered AMRs. For example, when designing a magnetic refrigerator that would operate at a certain temperature span it would make sense to choose the material at the cold end of the AMR with the highest maximum specific cooling energy $q_{R,\max}$.

References

1. Weiss P, Piccard A (1917) Le phénoméne magnétocalorique. J Phys (Paris) 7:103–109
2. Weiss P, Piccard A (1918) Sur un nouveau phénoméne magnétocalorique. Comptes Rendus 166:352–354
3. Warburg E (1881) Magnetische Untersuchungen. Ueber einige Wirkungen der Coërcitivkraft. Ann Phys (Leipzig) 249:141–164
4. Smith A (2013) Who discovered the magnetocaloric effect? Warburg, Weiss and the connection between magnetism and heat. Eur Phys J H 38(4):507–517
5. Joule J (1843) On the calorific effects of magneto-electricity, and on the mechanical value of heat. Philos Mag 23:263–276
6. Nichol JP (ed) (1860) Cyclopedia of the physical sciences. Richard Green and Company, London
7. Ewing JA (1882) On effects of retentiveness in the magnetisation of iron and steel. Proc Roy Soc 24:39–45
8. Stefan J (1871) Ueber die Gesetze der electrodynamischen Induction. Wien Ber 64:193–224
9. Stefan J (1889) Ueber thermomagnetische Motoren. Ann Phys 274:427–440
10. Edison T (1888) Pyromagnetic motor. US Patent 380.100
11. Edison T (1892) Pyromagnetic generator. US Patent 476.983 A
12. Tesla N (1889) Thermo-magnetic motor. US Patent 396.121 A
13. Tesla N (1890) Pyromagneto-electric generator. US Patent 428.057 A
14. Debye P (1926) Einige Bemerkungen zur Magnetisierung bei tiefer Temperatur. Ann Phys (Leipzig) 386:1154–1160
15. Giauque WF (1927) A thermodynamic treatment of certain magnetic effects. A proposed method of producing temperatures considerably below 1 absolute. J Am Chem Soc 49:1864–1870
16. Giauque WF, MacDougall DP (1933) Attainment of temperatures below 1 absolute by demagnetization of $Gd_2(SO_4)_3 \cdot H_2O$. Phys Rev 43:768

17. Urbain G, Weiss P, Trombe F (1935) Un nouveau métal ferromagnétique, le gadolinium. Comptes Rendus 200:2132–2134
18. Marsh J (1963) MSc. Disertation, West Virginia University
19. Silars MP (1965) MSc. Disertation, West Virginia University
20. Brown GV (1976) Magnetic heat pumping near room temperature. J Appl Phys 47:3673–3680
21. Barclay JA, Steyert WA (1982) Active magnetic regenerator. US Patent 4.332.135 A
22. Pecharsky VK, Gschneidner KA Jr (1997) Giant magnetocaloric effect in $Gd_5(Si_2\,Ge_2)$. Phys Rev Lett 78:4494–4497
23. Gschneidner KA Jr, Pecharsky VK, Tsokol AO (2005) Recent developments in magnetocaloric materials. Rep Prog Phys 68:1479–1539
24. Shen BG, Sun JR, Hu FX et al (2009) Recent progress in exploring magnetocaloric materials. Adv Mater 21:4545–4564
25. Brück E, Tegus O, Cam Thanh DT et al (2008) A review on Mn based materials for magnetic refrigeration: structure and properties. Int J Refrig 31:763–770
26. Phan MH, Yu SC (2007) Review of the magnetocaloric effect in manganite materials. J Magn Magn Mater 308:325–340
27. Gutfleisch O, Willard MA, Brück E et al (2010) Magnetic materials and devices for the 21st century: stronger, lighter, and more energy efficient. Adv Mater 23(7):821–842
28. Sandeman KG (2012) Magnetocaloric materials: the search for new systems. Scripta Mater 67 (6):566–571
29. Tishin AM, Spichkin YI (2003) The magnetocaloric effect and its applications. Institute of Physics Publishing, Philadelphia
30. Smith A, Bahl CRH, Bjørk R et al (2012) Materials challenges for high performance magnetocaloric refrigeration devices. Adv Energy Mater 2:1288–1318
31. Yu B, Gao Q, Zhang B et al (2003) Review on research of room temperature magnetic refrigeration. Int J Refrig 26:622–636
32. Engelbrecht K, Bahl CRH (2010) Evaluating the effect of magnetocaloric properties on magnetic refrigeration performance. J Appl Phys 108:123918
33. Basso V, Sasso CP, Bertotti G et al (2006) Effect of material hysteresis in magnetic refrigeration cycles. Int J Refrig 29:1358–1365
34. Engelbrecht K, Nielsen KK, Bahl CRH et al (2013) Material properties and modeling characteristics for $MnFeP_{1-x}As_x$ materials for application in magnetic refrigeration. J Appl Phys 113(17):173510
35. von Moos L, Nielsen KK, Engelbrecht K et al (2014) Experimental investigation of the effect of thermal hysteresis in first order material MnFe(P, As) applied in an AMR device. Int J Refrig 37(1):303–306
36. Nielsen KK, Engelbrecht K (2012) The influence of the solid thermal conductivity on active magnetic regenerators. J Phys D Appl Phys 45:145001
37. Fiorillo F (2004) Measurement and characterization of magnetic materials. Elsevier Academic Press, Amsterdam
38. Bjørk R, Bahl CRH, Katter M (2010) Magnetocaloric properties of $LaFe_{13\text{-}x\text{-}y}Co_xSi_y$ and commercial grade Gd. J Magn Magn Mater 322:3882–3888
39. Dan'kov SY, Tishin AM (1998) Magnetic phase transition and the magnetothermal properties of gadolinium. Phys Rev B 57(6):3478
40. Jayaraman TV, Boone L, Shield JE (2011) Near room temperature magnetic entropy changes in as-cast $Gd_{100\text{-}x}Mn_x$ (x = 0, 5, 10, 15, and 20 at.%) alloys. J Alloy Compd 509:1411–1417
41. Jayaraman TV, Boone L, Shield JE (2013) Magnetocaloric effect and refrigerant capacity in melt-spun Gd-Mn alloys. J Magn Magn Mater 345:153–158
42. Pecharsky VK, Gschneidner KA Jr (1999) Magnetocaloric effect and magnetic refrigeration. J Magn Magn Mater 200:44–56
43. Kaštil J, Javorský P, Kamarád J (2011) Magnetocaloric effect of Gd-Tb alloys: influence of the sample shape anisotropy. Appl Phys A 104:205–209
44. Couillaud S, Gaudin E, Franco V et al (2011) The magnetocaloric properties of GdScSi and GdScGe. Intermetallics 19:1573–1578

45. Law JY, Ramanujan RV, Franco V (2010) Tunable Curie temperatures in Gd alloyed Fe-B-Cr magnetocaloric materials. J Alloy Compd 508:14–19
46. Hu F, Shen B, Sun J et al (2001) Influence of negative lattice expansion and metamagnetic transition on magnetic entropy change in the compound LaFeSi. Appl Phys Lett 78:3675–3677
47. Fujita A, Fujieda S, Fukamichi K (2003) Isotropic giant linear magnetostriction and large magnetocaloric effects in $La(Fe_xSi_{1-x})_{13}$ intinerant-electron metamagnetic compounds and their hydrides. Metal Mater Proc 15:273
48. Fujita A, Fujieda S, Hasegawa Y et al (2003) Intinerant-electron transition and large magnetocaloric effects in $La(Fe_xSi_{1-x})_{13}$ compounds and their hydrides. Phys Rev B 67:104416
49. Katter M, Zellmann V, Reppel GW et al (2008) Magnetocaloric properties of La(Fe Co, Si)13 bulk material prepared by powder metallurgy. IEEE Trans Magn 44:3044–3047
50. Hansen BR, Katter M, Kuhn LT et al (2009) Characterization study of a plate of the magnetocaloric material $La(Fe,Co,Si)_{13}$. Paper presented at the 3rd international conference of IIR on magnetic refrigeration at room temperature, Des Moines, Iowa, USA, 11–15 May 2009, pp 67–73
51. Barcza A, Katter M, Zellmann V et al (2011) Stability and magnetocaloric properties of sintered $La(Fe, Mn, Si)_{13}Hz$ alloys. IEEE Trans Magn 47:3391–3394
52. Wada H, Tanabe Y (2001) Giant magnetocaloric effect of $MnAs_{1-x}Sb_x$. Appl Phys Lett 79:3302
53. Tegus O, Brück E, Buschow KHJ et al (2002) Transition-metal-based magnetic refrigerants for room-temperature applications. Nature 415:150–152
54. Brück E, Tegus O, Li XW et al (2003) Magnetic refrigeration–towards room-temperature applications. Phys B 327:431–437
55. Brück E, Ilyn M, Tishin AM et al (2005) Magnetocaloric effects in $MnFeP_{1-x}As_x$–based compounds. J Magn Magn Mater 290–291:8–13
56. Dung NH, Zhang L, Ou ZQ et al (2011) From first-order magneto-elastic to magneto-structural transition in $(Mn, Fe)_{1.95}P_{0.50}Si_{0.50}$ compounds. Appl Phys Lett 99:092511
57. Songlin D, Tegus O, Fuquan B et al (2005) Magnetic entropy change in $Mn_{0.9}Fe_{1.1}P_{1-x}Ge_x$ compounds. IEEE Trans Magn 41(10):2778–2780
58. Cam Thanh DT, Brück E, Tegus O et al (2006) Magnetocaloric effect in MnFe(P, Si, Ge) compounds. J Appl Phys 99:08Q107
59. Dinesen AR, Linderoth S, Mørup S (2005) Direct and indirect measurements of the magnetocaloric effect in $La_{0.67}Ca_{0.33-x}Sr_xMnO_{3\pm\delta}$ ($x \in [0;0.33]$). J Phys Condens Matter 17:6257–6269
60. Barcza A, Zellmann V, Katter M (2012) Linearly, continuously graded transition temperatures in La-Fe-Co-Si parts for magnetic cooling applications. Paper presented at the 5th international conference of IIR on magnetic refrigeration at room temperature, Grenoble, France, 17–20 Sept 2012

Chapter 3
Magnetic Field Sources

In this chapter we describe some of the most important issues that relate to the sources for magnetic fields. We will try to provide information for engineers about the different possible magnetic field sources with respect to their application in magnetocaloric energy conversion. However, the main emphasis will be on permanent magnets and their assemblies. This is due to the fact that other kinds of magnetic field sources, for instance, electric resistive coils (electromagnets) operate with a rather low efficiency. The efficiency in electromagnets in terms of thermodynamics must be considered as the ratio of the energy output (magnetization energy) and the energy input to the coil. The current passing through the coiled wires in an electromagnet will induce Joule heating, due to the electric resistance. Not only will a loss of valuable energy occur, but the magnetic field source will, in most cases, also require cooling, especially in cases where a high magnetic flux density is required. In superconducting magnets, which can be treated as a special class of electromagnets, the current may be kept circulating through the wires of a coil without electrical resistance losses. However, in order to keep the superconducting magnet operating, a cryogenic system is required to provide cooling for the superconducting electrical coils. Namely, there is no practical application of a room-temperature superconducting material. There are, however, some indicators that this could be possible [1–5]. And if this happens in future, this chapter will unfortunately (or fortunately) not be of much use, but the application will open up an unbelievable range of applications, which will certainly change our way of life, including magnetocaloric energy conversion.

The reader should note that the magnetic field source is the most expensive part of a magnetic refrigerator or a heat pump, especially if those are based on permanent or superconducting magnets. Therefore, the optimal design of a magnetic field source is crucial for obtaining a cost-effective and energy-efficient device.

A. Kitanovski et al., *Magnetocaloric Energy Conversion*,
Green Energy and Technology, DOI 10.1007/978-3-319-08741-2_3

3.1 Introduction

Before starting, we will provide some basic definitions, which are necessary for an understanding of the magnetism.

3.1.1 Magnetic Field and Magnetic Induction

From the **Biot–Savart law** [6] we consider the generation of a magnetic field H due to the steady current flow I (constant over time) in a long conductor. Then the magnetic field $\mathrm{d}\overrightarrow{H}$ at some radial distance r from the elemental length of conductor $\mathrm{d}\overrightarrow{s}$ can be defined as: the vector product, which is perpendicular to the plane given by $\mathrm{d}\overrightarrow{s}$ and $\overrightarrow{r}$, and where $\overrightarrow{r} = r \cdot \overrightarrow{u}$, and therefore $\overrightarrow{u}$ represents a unit vector in the radial direction:

$$\mathrm{d}\overrightarrow{H} = \frac{1}{4\pi r^3} I\,\mathrm{d}\vec{s} \times \overrightarrow{r} = \frac{1}{4\pi r^2} I\,\mathrm{d}\vec{s} \times \overrightarrow{u} \tag{3.1}$$

In Fig. 3.1, a long electrical conductor carries the electric current. If we measure the magnetic field at the point P for the element dL, this can be, according to Eq. (3.1), defined as:

$$\mathrm{d}H = \frac{1}{4\pi r^2} I\,\mathrm{d}s\,u\,\sin(90^\circ - \alpha) = \frac{I\cos\alpha}{4\pi a}\mathrm{d}\alpha \tag{3.2}$$

where $s/a = tg\,\alpha \Rightarrow \mathrm{d}s = \frac{a\,\mathrm{d}\alpha}{\cos^2\alpha}$ and $a = r\cos\alpha$.

For the infinitely long electrical conductor at a distance of a, the magnetic field strength can be derived as:

$$H(a) = \int_{-\frac{\pi}{2}}^{\frac{\pi}{2}} \frac{I\cos\alpha\,\mathrm{d}\alpha}{4\pi a} = \frac{I}{2\pi a} \tag{3.3}$$

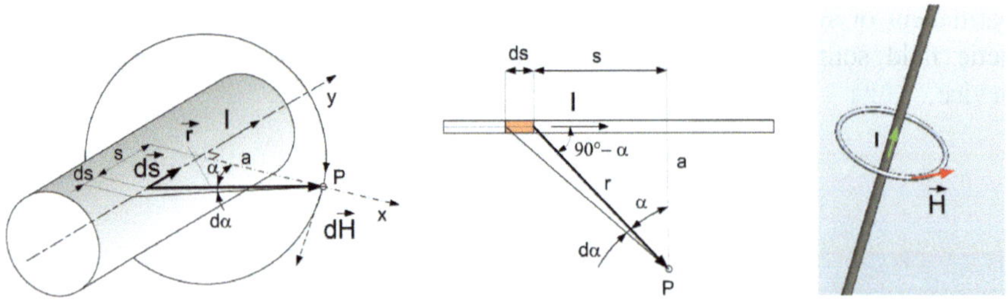

Fig. 3.1 The magnetic field induced around a long conductive wire

Now, let us consider **Ampere's** law [7], which states that the magnetic field intensity, measured at a distance r from current I, is proportional to:

$$\left|\vec{H}\right| \propto \frac{I}{r} \tag{3.4}$$

For a circular conduit with the closed path Γ, and with radius r, Eq. (3.3) can also be written as:

$$H \cdot (2\pi r) \propto I \tag{3.5}$$

The magnetic field that is generated in this case can be expressed using Ampere's law as:

$$\oint_{\Gamma} \vec{H} \cdot \mathrm{d}\vec{l} = I \tag{3.6}$$

For any closed-loop path, the closed line integral of the magnetic field is equal to the electric current encircled by such a loop.

Figure 3.2a shows a coiled wire wound in several turns (a solenoid) in which the circulating current produces an almost uniform magnetic field. In such a case the magnetic field can also be treated as being parallel with the coil axis (Fig. 3.2b). For this particular case we will consider the path given by A–B–C–D in Fig. 3.2b. Since

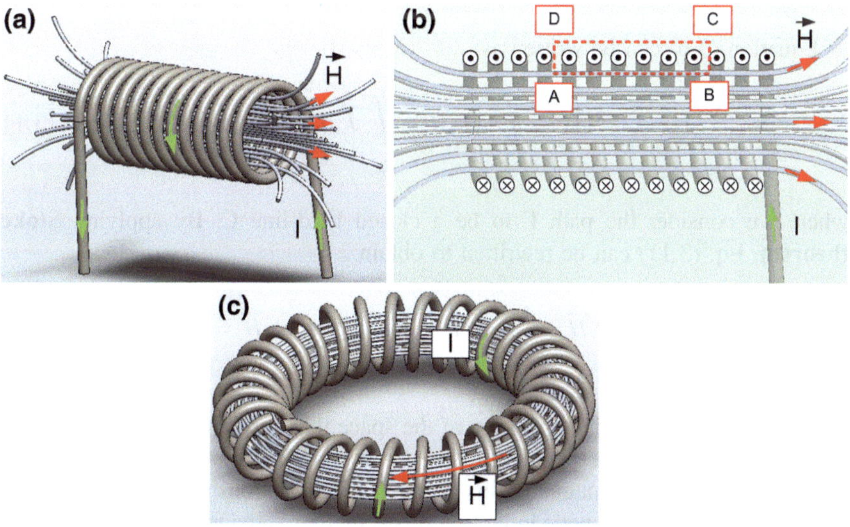

Fig. 3.2 **a** The current-field relation in a coil and, **b** a uniform solenoid, where Ampere's law is applied to the rectangular path *A–B–C–D–A*, **c** a uniform toroid

the magnetic field in the case of B–C and A–D can be considered as perpendicular to the path of the electric current, it therefore follows:

$$\vec{H} \perp \mathrm{d}\vec{l} \quad \Rightarrow \quad \vec{H} \cdot \mathrm{d}\vec{l} = 0 \tag{3.7}$$

Along the path CD the magnetic field equals zero (this will also be discussed later in the text where we deal with static magnetic fields, related to permanent magnets). Therefore, only the path AB contributes to the magnetic field:

$$H\, l_{AB} = I\, N_{AB} \tag{3.8}$$

where l_{AB} represents the length of the path A–B and N_{AB} represents the number of turns.

If the solenoid represents a torus-like shape (Fig. 3.2c), it is called a toroid or toroidal coil. All the magnetic flux is held inside such a coil. In a toroid with the number of turns N, the relation in Eq. (3.8) can be written as:

$$H = \frac{IN}{2\pi r} \tag{3.9}$$

where r is the radius of the toroid.

The electric current can be defined by the current density, which passes a certain area A as:

$$I = \int_A \vec{J}\, \mathrm{d}\vec{A} \tag{3.10}$$

Equation (3.6) can be written as:

$$\oint_{\Gamma = C} \vec{H}\, \mathrm{d}\vec{l} = \int_A \vec{J}\, \mathrm{d}\vec{A} \tag{3.11}$$

where we consider the path Γ to be a closed field-line C. By applying **Stokes theorem**, Eq. (3.11) can be rewritten to obtain:

$$\oint_A \mathrm{rot}\, \vec{H} \cdot \mathrm{d}\vec{A} = \int_A \vec{J}\, \mathrm{d}\vec{A} \quad \Rightarrow \mathrm{rot}\, \vec{H} = \vec{J} \tag{3.12}$$

If the magnetic field is a property of the space that surrounds an electric current and a magnet, then the response of the material to the external magnetic field is the magnetic induction B (magnetic flux density). The relationship between the magnetic field H and the magnetic induction B is actually related to the characteristics of the material. In a vacuum, these characteristics follow the linear relation as:

$$B = \mu_0 \cdot H \tag{3.13}$$

where

$$\mu_0 = 1.256 \times 10^{-6} \frac{V \cdot s}{Am} = 4 \cdot \pi \times 10^{-7} \frac{V \cdot s}{Am} = \frac{H}{m} \tag{3.14}$$

is the empirically obtained permeability of free space (vacuum).

Here, there may be some confusion, since sometimes in publications we can see B as the magnetic field. Note that in this particular case the authors tend to refer to the applied magnetic field in air (similar to vacuum).

The magnetic flux must be preserved through any closed volume (if not, this would require magnetic monopoles, which were hypothesized by Paul Dirac, and have been recently been receiving attention again [8, 9]). Therefore, the integral of the magnetic flux density through a certain volume, which is performed over the surface (envelope) of that volume, is:

$$\oint \vec{B} \cdot \mathrm{d}\vec{A} = 0 \tag{3.15}$$

This equation is equivalent to one of **Maxwell's equations**:

$$\nabla \cdot \vec{B} = 0 \tag{3.16}$$

3.1.2 Magnetic Moment

The force acting on the conducting wire carries the current I and is placed in a uniform magnetic field H (Fig. 3.3) that is equal to:

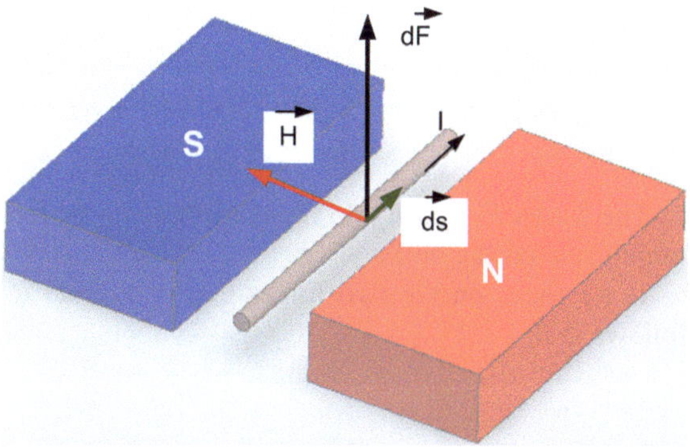

Fig. 3.3 Force acting on a conductor in a uniform magnetic field

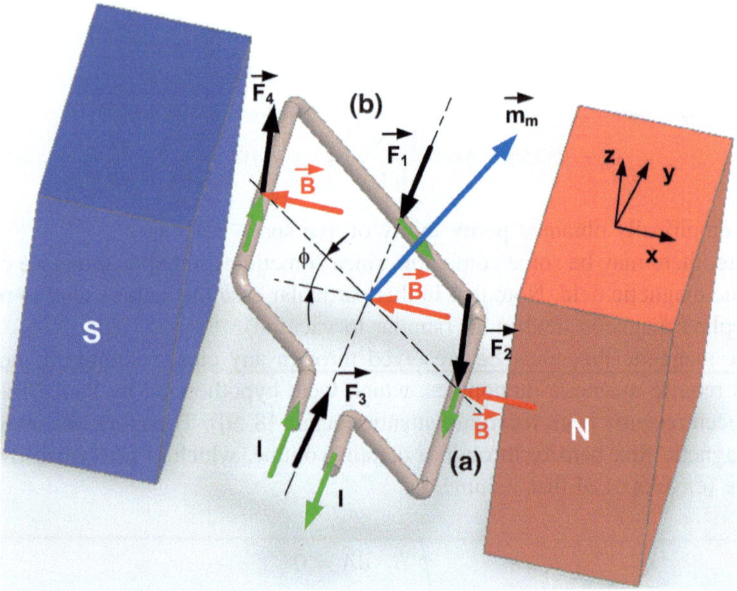

Fig. 3.4 **a** An electric circular loop carrying a current *I* and placed in the uniform magnetic field, **b** forces acting on the electric loop in the uniform magnetic field

$$\mathrm{d}\vec{F} = I\,\mathrm{d}\vec{s} \times \mu_0\vec{H} \tag{3.17}$$

If we now consider an electric circular loop carrying a current I and placed in a uniform magnetic field, the net force on such a loop will be zero (Fig. 3.4). However, this will not hold true for the magnetic torque:

$$\mathrm{d}\vec{T} = \vec{r} \times \mathrm{d}\vec{F} = \vec{r} \times \left(I\mathrm{d}\vec{s} \times \mu_0\vec{H}\right) \tag{3.18}$$

In Fig. 3.4, four forces are acting on the coils, two of which are equal in magnitude but opposite in direction ($\boldsymbol{F}_1$ and $\boldsymbol{F}_3$, and $\boldsymbol{F}_2$ and $\boldsymbol{F}_4$ respectively). It follows that:

$$\sum_1^4 \vec{F}_i = \vec{F}_1 + \vec{F}_2 + \vec{F}_3 + \vec{F}_4 = 0 \tag{3.19}$$

The forces $\boldsymbol{F}_1$ and $\boldsymbol{F}_3$ do not contribute to the torque, however, by $\boldsymbol{F}_2$ and $\boldsymbol{F}_4$, a torque acts on the coil.

If we focus on the two forces that contribute to the torque (Fig. 3.5), then it follows that:

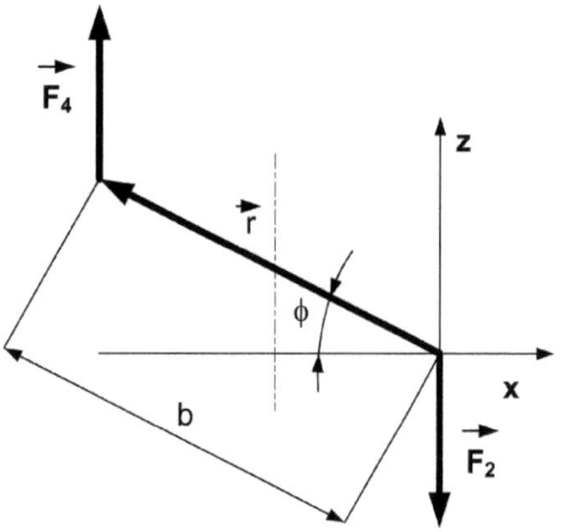

Fig. 3.5 The torque on a electric circular loop carrying a current I and placed in a uniform magnetic field

$$\vec{T} = \vec{r} \times \vec{F}_4 = \left(b\ \cos\ \varphi \vec{i} + b\ \sin\ \varphi \vec{k}\right) \times \left(-I\, a B \vec{k}\right) = a\, b\ \cos \varphi\, I\, B \vec{j} \quad (3.20)$$

$$T = IB \cdot ab \cos \varphi \quad (3.21)$$

where φ is the angle of inclination of the coil's plane, and the $A = a \cdot b$ represents the area surrounded by the coil.

The surface A can be represented by the vector:

$$\vec{A} = A \cdot \vec{n} \quad (3.22)$$

where $\vec{n}$ stands for the unit vector normal to the surface A. Now the torque, written as the vector, is:

$$\vec{T} = I\vec{A} \times \mu_0 \vec{H} \quad (3.23)$$

The area A (which constitutes the complete loop) is coincident with current I. The product of the two is defined as the magnetic dipole moment, which is considered to be the elementary magnetic quantity:

$$\vec{m}_m = I\vec{A} \quad (3.24)$$

3.1.3 Magnetization

The magnetic moment $\vec{m}_m$ per unit volume of a solid can be defined as the magnetization:

$$\mu_0 \vec{M} = \lim_{\Delta V \to 0} \frac{\sum \vec{m}_m}{\Delta V} \tag{3.25}$$

The magnetization can be considered as a macroscopic property of the material. It may also represent the spontaneous magnetization M within a ferromagnetic material, or the uniform magnetization induced by the applied field in a paramagnetic or a diamagnetic material.

3.1.4 Magnetic Field and Magnetic Induction Related to Magnetic Materials

In a permanent-magnet bar, each end of this bar will represent a pole. If two such bars are placed in the same plane as each other and separated by a vacuum, the like poles (e.g. p_1 and p_2, respectively) will repel and the unlike poles will attract with the force F, which is proportional to the product of the pole strengths and the inverse of the square of the distance (which can be referred to Coloumb's law [10, 11]):

$$F = \frac{p_1 p_2}{4\pi\mu_0 r^2} \tag{3.26}$$

The pole can be associated with the magnetic charge, similar to the case of electricity. For a single pole p_1, (not to be misinterpreted as a monopole), the magnetic field, as a vector, surrounding the pole can be defined [11, 12]:

$$\vec{H} = \frac{p_1 \vec{r}}{4\pi\mu_0 r^3} = k\frac{p_1 \vec{r}}{r^3} \tag{3.27}$$

where k can be defined as the value of the proportionality constant in a similar way as this is done for electric fields.

Now, lets us look at Eq. (3.28). If it is related to free space (vacuum), then the current density corresponds to Eq. (3.12). However, for material media, Eq. (3.12) can be written for the magnetic flux density as [11, 13]:

$$\text{rot } \vec{B} = \mu_0 \left(\vec{J_c} + \vec{J_m} \right) \tag{3.28}$$

In Eq. (3.28) the J_c relates to the electric current density due to the conduction. However, an additional term J_m represents the magnetization current density. The magnetization currents of the material are associated with quantum mechanics in the magnetized material. The integral of the magnetization current density through a certain volume, which is performed over the surface (envelope) of that volume, is:

$$\oint_A \vec{J}_m \cdot d\vec{A} = 0 \quad (3.29)$$

If we express the magnetization current as the rotor (curl) of the magnetization vector, then it follows by Stokes theorem:

$$\oint_\Gamma \vec{M} \cdot d\vec{s} = \oint_A \text{rot}\vec{M} \cdot d\vec{A} \quad (3.30)$$

If the path lies entirely outside the magnetized body it follows for any surface [13]:

$$\oint_A \text{rot } \vec{M} \cdot d\vec{A} = 0 \quad (3.31)$$

with [10, 13]:

$$\overrightarrow{J_m} = \text{rot } \vec{M} \quad (3.32)$$

Following Eq. (3.12), the conducting current density equals:

$$\vec{J_c} = \text{rot } \vec{H} \quad (3.33)$$

By following Eq. (3.32), this can be now expressed as:

$$\text{rot } \vec{B} = \mu_0 \left(\text{rot } \vec{H} + \text{rot } \vec{M}\right) \quad (3.34)$$

From Eqs. (3.16) and (3.34) it follows:

$$\nabla \cdot \vec{H} = -\nabla \cdot \vec{M} \quad (3.35)$$

We can also define the following general relation.

$$\vec{B} = \mu_0 \left(\vec{H} + \vec{M}\right) \quad (3.36)$$

This is known as the Sommerfeld convention. Equation (3.36) will reduce to Eq. (3.13) in the case of a vacuum with $M = 0$.

3.1.5 External, Internal Magnetic Field and the Demagnetization

Note also that Eq. (3.36) does not correspond only to the creation of the magnetic field H by conduction currents. Namely, in magnetic materials, the field will be produced around it as well as within its volume [13]. Therefore:

$$\vec{H} = \vec{H}_c + \vec{H}_m \tag{3.37}$$

The H_c represents the field created by conduction currents and H_m is the magnetic field due to the magnetization distribution of other magnetic field sources or the magnet itself. H_m can be denoted as the demagnetization field (stray field outside a magnet). Namely, the magnet creates "free poles" on the surface of the material that creates the demagnetization field H_{dem}, which acts in an opposite direction to magnetization M inside the magnet. The magnetic field also acts in a different direction than the magnetic flux density B. Figure 3.6 shows the case of the uniformly magnetized material with no external magnetic field sources. The magnetic field H can be named as the internal field H_{in}, and the H_c relates to the external magnetic field H_{out}, which is produced by steady electric currents or the stray field outside the sample volume. H_{out} is also named the applied field. Therefore, it can be written as [13]:

$$\vec{H}_{\text{in}} = \vec{H}_{\text{out}} + \vec{H}_d \tag{3.38}$$

The demagnetization field H_{dem} is related to the magnetization M as:

$$\vec{H}_{\text{dem}} = -N_{\text{dem}} \cdot \vec{M} \tag{3.39}$$

where N_{dem} represents the demagnetization factor—a tensor that is usually represented by a symmetric 3 × 3 matrix and is dependent on the geometry of the

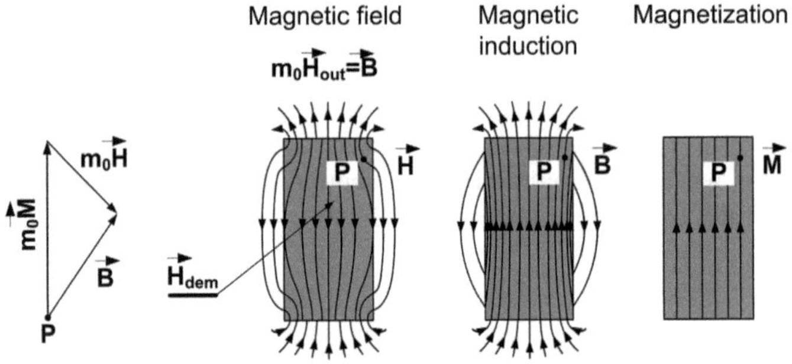

Fig. 3.6 Magnetic field outside and inside a bar magnet, magnetization inside the bar magnet and magnetic field induction (magnetic flux density *inside* and *outside* the magnet) and their relation

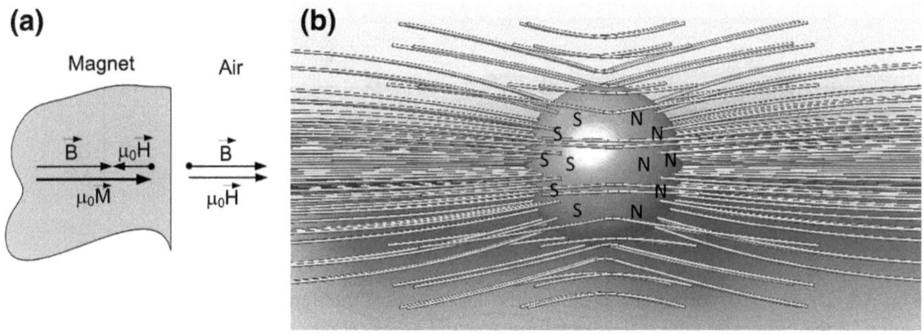

Fig. 3.7 **a** Demagnetization field and the material-air boundary, **b** a ferromagnetic sphere in a uniform external magnetic field

material. Therefore, the demagnetization field is associated with the shape anisotropy of the magnetic material.

For the case in Fig. 3.7-left, which shows a material-air boundary, the external magnetic flux density in the air gap ($B_{out} = \mu_0 H_{out}$) will equal the internal magnetic flux density:

$$\vec{B}_{out} = \mu_0 \vec{H}_{out} = \vec{B}_{in} = \mu_0 \left(\vec{H}_{in} + \vec{M}\right) \tag{3.40}$$

In the case of Fig. 3.7-right, the ferromagnetic sphere is placed in a uniform external magnetic field. This, because of the sphere's shape, induces a uniform magnetic field inside the sphere. However, in this case we will consider the entire volume. In contrast to the case above, with Eq. (3.39) the magnetic flux density in the sphere will be higher than the external magnetic flux density:

$$\left|\vec{B}_{out}\right| < \left|\vec{B}_{in}\right| \tag{3.41}$$

Therefore:

$$\vec{H}_{out} < \left(\vec{H}_{in} + \vec{M}\right) \tag{3.42}$$

The relationship in Eq. (3.42) can now be described by using of the definition of the demagnetization field:

$$\vec{H}_{out} \approx \left(\vec{H}_{in} + N_{dem}\vec{M}\right) \tag{3.43}$$

Equation (3.43) is actually an approximation, since the internal field and consequently the magnetization are not usually uniform (although they would be in the case of an ellipsoid). For the case of spontaneous magnetization with no external magnetic field, Eqs. (3.36) and (3.43) will lead to the following definition of the internal magnetic flux density:

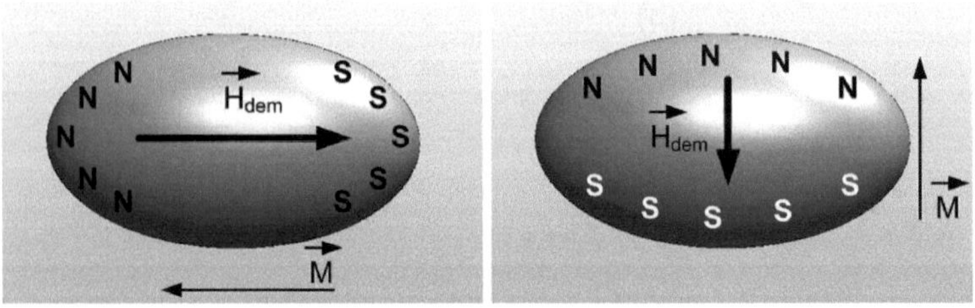

Fig. 3.8 Demagnetization field in ellipsoids. The value of the demagnetization field H_{dem} will be higher in the case when the poles are closer to each other (*right case*), and smaller when they are distant (*left case*)

$$\vec{B}_{\text{in}} = \mu_0 \vec{M}(1 - N_{\text{dem}}) \tag{3.44}$$

In order to have the smallest demagnetization field, the N_{dem} should limit to zero. This is the case when the material is thin, very long and magnetized along its major axis. In Fig. 3.8, two ellipsoids are shown. Along the principal axes of the ellipsoid, H_{dem} and M will be collinear. In the first case the demagnetization factor, due to distant "free poles", will be smaller than in the case when the "free poles" are closer.

3.1.6 Magnetic Susceptibility and Permeability

In order to define the response of a material to an applied magnetic field, a dimensionless property, the susceptibility, can be defined as:

$$\chi = \frac{M}{H} \tag{3.45}$$

Since the magnetization is not always a linear function of the magnetic field (internal magnetic field), the susceptibility or the magnetic permeability are also not constants. The susceptibility may also be expressed in a differential way:

$$\chi' = \frac{\mathrm{d}M}{\mathrm{d}H} \tag{3.46}$$

Note also that the external susceptibility can be defined. In this particular case, instead of the internal, the external (applied magnetic field) is used:

$$\chi_{\text{out}} = \frac{M}{H_{\text{out}}} \tag{3.47}$$

The magnetic permeability defines the ratio between the magnetic flux density and the internal magnetic field.

$$\mu = \frac{B}{H} \tag{3.48}$$

Analogous to susceptibility, the magnetic permeability can be also expressed in differential terms:

$$\mu' = \frac{\mathrm{d}B}{\mathrm{d}H} \tag{3.49}$$

Sometimes the term relative permeability is applied. This represents the ratio between the permeability of the material divided by the permeability of free space (vacuum, see Eq. 3.14):

$$\mu_r = \frac{\mu}{\mu_0} \tag{3.50}$$

It also follows:

$$\mu_r = \chi + 1 \tag{3.51}$$

3.1.7 Magnetic Force and Torque on a Dipolar Material

Let us now consider a magnetic material (a soft or hard ferromagnet) that is under an applied magnetic field $\vec{H}_{\text{out}}$. The magnetization vector $\vec{M}$ is aligned along the material, as shown in Fig. 3.9. At each end of the material there exists poles p, defined with the intensity of M as [12]:

$$p = M \cdot A \cdot \mu_0 = \rho_A \cdot A \tag{3.52}$$

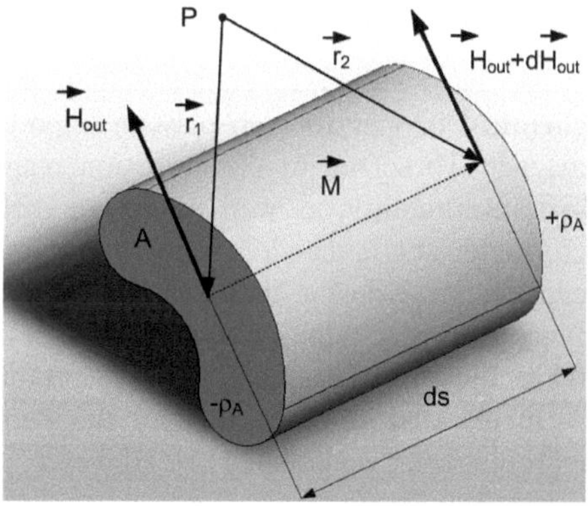

Fig. 3.9 The gradient magnetic field over the magnetic material

where M is the intensity of the magnetization and is related to the state of the polarization of the magnetized material, ρ_A represents the surface density of magnetic poles, and A represents the surface area of the pole with the elementary volume $A\ \mathrm{d}s = \mathrm{d}V$.

The Kelvin force density on such a material will be:

$$\vec{F} = \mu_0 \left(\vec{M} \cdot \nabla \right) \vec{H}_{\mathrm{out}} \tag{3.53}$$

where $\mu_0 \vec{M}$ represents the magnetic moment per unit volume of material and where the dipole moment is defined as (see Eq. 3.25):

$$\vec{m} = \vec{M} \cdot A \cdot \mu_0 \cdot \mathrm{d}s = \mu_0 \cdot \vec{M} \cdot \mathrm{d}V \tag{3.54}$$

In the soft ferromagnetic material, the magnetization $\vec{M}$ is parallel to the external field $\vec{H}_{\mathrm{out}}$. Therefore, Eq. (3.53) reduces to [12]:

$$\vec{F} = \mu_0 \vec{M}\, \nabla \vec{H}_{\mathrm{out}} \tag{3.55}$$

The torque density is defined as (see Eq. 3.23):

$$\vec{t} = \mu_0 \vec{M} \times \vec{H}_{\mathrm{out}} \tag{3.56}$$

3.2 Permanent Magnets

Permanent-magnet materials are “hard” magnetic materials that retain their magnetism after the removal of the applied magnetic field. The most important characteristics of permanent magnets can be divided into the following: **Magnetic properties,** which concern remanent magnetic flux density B_r, maximum energy product $(BH)_{\mathrm{max}}$, resistance to demagnetization—coercivity, demagnetization curve (the second quadrant), recoil permeability (which should be as low as possible). Among the **thermal properties** one should pay attention to the temperature range of the magnet’s operation and to the magnetization as a function of the temperature, which is associated with the reversible temperature coefficient. **Mechanical, geometric and chemical properties** also play an important role in the selection of a permanent magnet. These regard the corrosion resistance, the mechanical strength and the manufacturability, as well as the geometry (size, shape). An important **electric characteristic** is the electric resistivity, which is also associated with the potential eddy currents during the magnet’s operation. This should of course be avoided. Finally, the cost of the material, the cost of the production of the magnet assembly and the availability of material on the market play an important **economic** role.

In the case of the design of the permanent magnet, Eqs. (3.11) and (3.15) represent basic equations [14]. However, in the case of a permanent magnet alone,

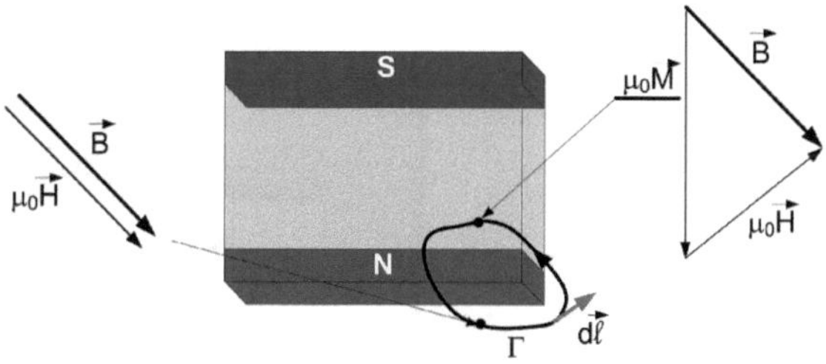

Fig. 3.10 Vectors on the path Γ on both sides of the magnet-air boundary

surrounded only by air, since there are no currents, Eq. (3.11) takes the following form:

$$\int_{\Gamma} \vec{H}\,\mathrm{d}\vec{l} = 0. \tag{3.57}$$

The term in Eq. (3.57) can also be considered as the magnetomotive force [14]. Let us consider a path Γ that passes the boundary between the bar magnet and the air (Fig. 3.10). If we apply Eq. (3.57) for this particular case, then we obtain:

$$\int_{\Gamma\ \mathrm{magnet}} \vec{H}\,\mathrm{d}\vec{l} + \int_{\Gamma\ \mathrm{air}} \vec{H}\,\mathrm{dl} = 0 \tag{3.58}$$

The negative magnetic field in the magnet will lead to a positive magnetic field in the air gap. Another important relation regards Eq. (3.15), which shows that the magnetic flux density is preserved over the considered surface of the material. Together with Eq. (3.58), these two equations can be used for the design of the magnetic circuit. Let us consider an example of a simple permanent-magnet assembly, which consists of soft iron (which is used to guide the magnetic flux) and the permanent magnet (Fig. 3.11).

In Fig. 3.10 we consider an infinitely permeable soft iron material; therefore, no heat-flux leakage will occur and the magnetic field outside the magnet assembly will equal to zero. The air gap between the two poles of soft iron has a cross-section area A_{gap} and its length is defined to be L_{gap}. In Fig. 3.11, the permanent magnet, attached to the soft iron, has a cross-section of A_{mag} and its length is L_{mag}. By using Eq. (3.15) and considering no leakage into the surroundings, Eq. (3.15) can be rewritten as:

$$B_{\mathrm{mag}}A_{\mathrm{mag}} = B_{\mathrm{gap}}A_{\mathrm{gap}} \tag{3.59}$$

Now, by applying Eq. (3.59), this can be rewritten for our particular case as:

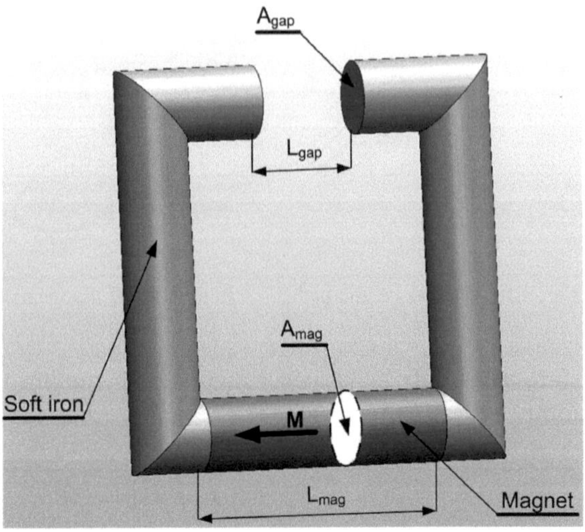

Fig. 3.11 A simple permanent-magnet assembly consisting of a magnet and soft iron with the air gap

$$H_{mag}L_{mag} + H_{gap}L_{gap} = 0 \tag{3.60}$$

Because there is no magnetization in the air gap, the magnetic flux density in the air gap can be simply written as (see Eq. 3.36):

$$B_{gap} = \mu_0 H_{gap} \tag{3.61}$$

By rearranging Eqs. (3.59)–(3.61) we obtain [14]:

$$\frac{B_{mag}}{\mu_0 H_{mag}} = -\frac{A_{gap}L_{mag}}{A_{mag}L_{gap}} \tag{3.62}$$

By using Eqs. (3.59)–(3.61) we can also obtain:

$$B_{mag}H_{mag}V_{mag} = -\frac{B_{gap}^2 V_{gap}}{\mu_0} \tag{3.63}$$

where V_{mag}, V_{gap} represent the volume of the magnet and the air gap, respectively. The maximum magnetic flux density in the air gap will be achieved with the product $B_{mag}H_{mag}$ being a maximum $(BH)_{max}$. This is why this product also represents a figure of merit for permanent magnets.

The working point of a permanent magnet is defined by the intersection of the load line and the *BH* loop (Fig. 3.12). The maximum $(BH)_{max}$ point for an isolated magnet with an ideal square magnetization loop (Fig. 3.12) can be calculated by maximizing the product of $\mu_0(H_{mag} + M)\, H_{mag}$. This depends further on the shape of the magnet, and the demagnetization factor N_{dem}:

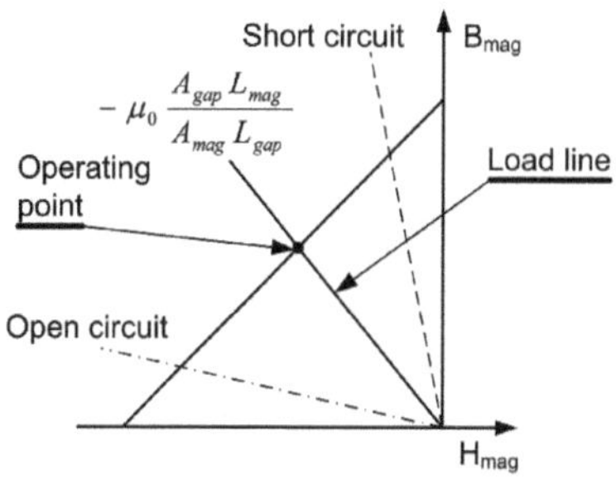

Fig. 3.12 The working point and the load line for a permanent magnet

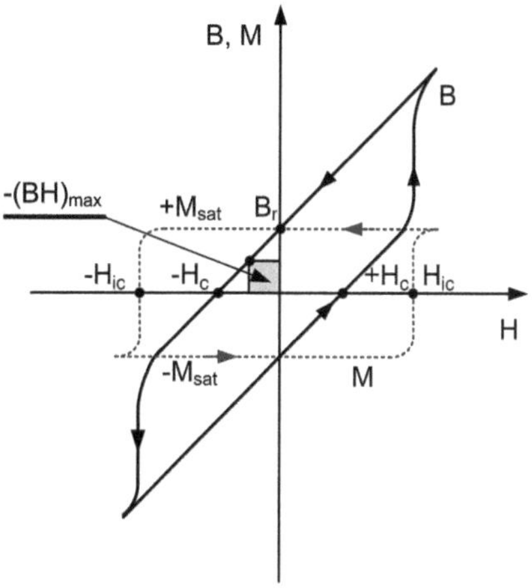

Fig. 3.13 Intrinsic magnetization M and magnetic flux density B as the function of the magnetic field H for a permanent magnet

$$H_{\mathrm{mag}} = -N_{\mathrm{dem}} M \tag{3.64}$$

Typical characteristics of a permanent magnet are shown in Fig. 3.13. When the permanent-magnet material is exposed to a high magnetic field it retains a high magnetization. This is associated with the remanence flux density B_r. Another important characteristic of a permanent magnet is the coercivity H_c. This represents the magnetic field strength that is needed to reduce the magnetic flux density to zero. The intrinsic coercivity H_{ci} is the magnetic field that usually represents higher values than the coercivity H_c. At this magnetic field the saturation magnetization M_{sat} suddenly reverses. The intrinsic coercivity H_{ci} actually represents the magnet's resistance to demagnetization (see Fig. 3.13, dashed line). If the applied magnetic field has the value of $+H_{\mathrm{ci}}$, the magnetization will suddenly change from $-M_{\mathrm{sat}}$ to $+M_{\mathrm{sat}}$. The opposite occurs when the magnetic field is applied with $-H_{\mathrm{ci}}$. If the applied field is further increased, the saturation magnetization $-M_{\mathrm{sat}}$ will remain constant. Another characteristic, which is very important for the engineering of magnets, is given by the B–H relation.

An ideal square $M(H)$ loop is defined by a linear second quadrant B_{mag} (H) where:

$$-(BH)_{\mathrm{mag}} = \mu_0 \left(\frac{M_{\mathrm{sat}}}{2}\right)^2 \quad (3.65)$$

Note that real magnets do not exhibit such perfect loops as in the case of Eq. (3.65). Their energy product is always smaller than the upper ideal limit. Namely, the soft iron (used to guide the magnetic flux), has a finite magnetic permeability. Furthermore, there will always be some magnetic flux leakage into the environment of the magnet (or magnet assembly). Therefore, for a real magnet assembly certain correction factors can be applied. These can be defined as the leakage coefficient K_1 (Eq. 3.66) and the loss factor K_2 (Eq. 3.67), respectively [14].

$$K_1 = \frac{\text{magnet flux}}{\text{useful flux}} = \frac{B_{\mathrm{mag}}A_{\mathrm{mag}}}{B_{\mathrm{gap}}A_{\mathrm{gap}}} > 1 \quad (3.66)$$

$$K_2 = \frac{\text{magnet magnetomotive force}}{\text{useful magnetomotive force}} = -\frac{H_{\mathrm{mag}}L_{\mathrm{mag}}}{H_{\mathrm{gap}}L_{\mathrm{gap}}} > 1 \quad (3.67)$$

By following the expressions in Eqs. (3.66), (3.67) and (3.62), which represents the load line, can now be written as:

$$\frac{B_{\mathrm{mag}}}{\mu_0 H_{\mathrm{mag}}} = -\left(\frac{K_1}{K_2}\right)\frac{A_{\mathrm{gap}}l_{\mathrm{mag}}}{A_{\mathrm{mag}}l_{\mathrm{gap}}} \quad (3.68)$$

Let us consider two different characteristics of magnets, as presented in Buschow and de Boer [15] (Fig. 3.14a, b).

In the first case (Fig. 3.14a) the usual characteristics for permanent magnets based on rare-earth materials (e.g. Nd–Fe–B) are shown. In this particular case, the intrinsic coercivity H_{ci} can be much larger than the remanence and exceeds the field that corresponds to the maximum energy product $(BH)_{\max}$ and the coercivity H_c. This kind of magnet will be able to resist very high magnetic fields, which can be as high as triple value of the field at the $(BH)_{\max}$.

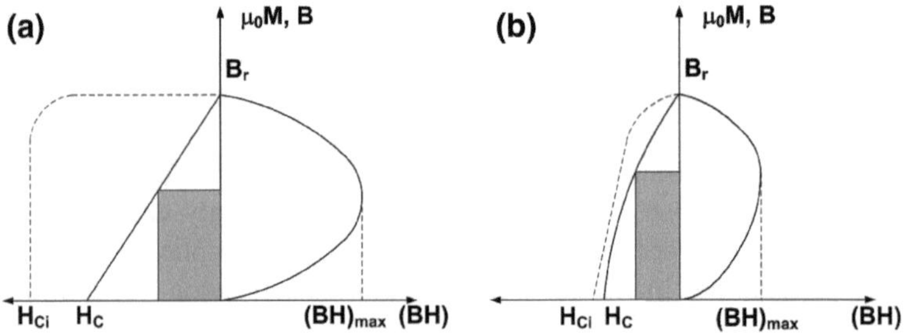

Fig. 3.14 The magnetic flux density and the magnetization curve as a function of the demagnetization field strength. The maximum energy product as a function of the magnetic flux density. **a** Rare-earth permanent magnets (i.e. Nd–Fe–B), **b** Al–Ni–Co magnets

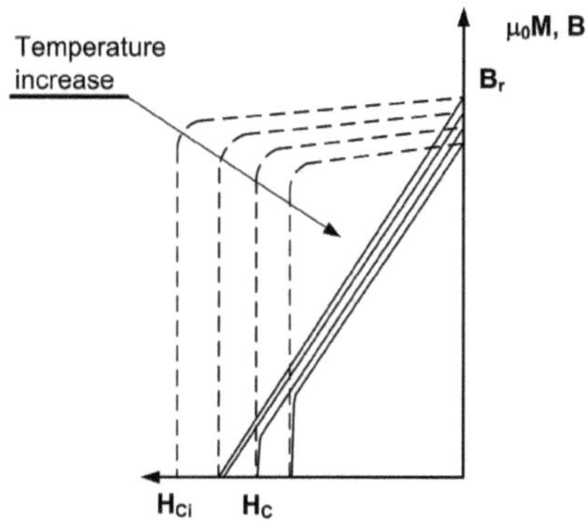

Fig. 3.15 The temperature dependence of the magnetic characteristics (magnetization —*dashed lines* and magnetic flux density) of a permanent magnet

In the second case (Fig. 3.14b), the coercivity is smaller than the remanence of the permanent-magnet material (e.g. AL–NI–CO) and it does not differ much from the intrinsic coercivity. Such a magnet will resist a relatively small field of demagnetization. If magnetic fields that are higher than those at double the value of the $(BH)_{max}$ are applied, this will lead to a full demagnetization of the magnet. Therefore, it is very important in a magnet's design that the demagnetization of a magnet does not occur.

A high intrinsic coercivity usually means magnets with rare-earth materials and magnets with an intrinsic magnetocrystalline anisotropy. In materials that are based on shape anisotropy (such as AL–NI–CO materials), these will not possess a high coercivity.

As was noted at the beginning of this section, the maximum energy product $(BH)_{max}$ is not the only criterion on which the selection of an appropriate magnet material should be made. Additionally, one should also consider the magnitude of the reverse magnetic fields that do not harm the properties of the magnet. Therefore, the recoil line is one of the important criteria, especially in structures where very high changes in the demagnetization magnetic field in the "air gap" can occur (for more information see Buschow and de Boer [15] and Campbell [14]).

In certain cases magnets can be exposed to high temperatures in the environment (note also that eddy currents may cause the heating of a magnet assembly). In this particular case one should consider the temperature dependence of the magnet material in terms of the remanence and the coercivity (see Fig. 3.15) as an example.

3.2.1 Permanent Magnet Materials

Permanent-magnet materials can generally be divided into groups of ceramic materials, Al–Ni–Co materials, rare-earth materials and polymer-bonded materials. The last group at present does not provide appropriate characteristics to be successfully applied in magnetocaloric energy conversion. We will briefly present the first three groups of permanent-magnet materials.

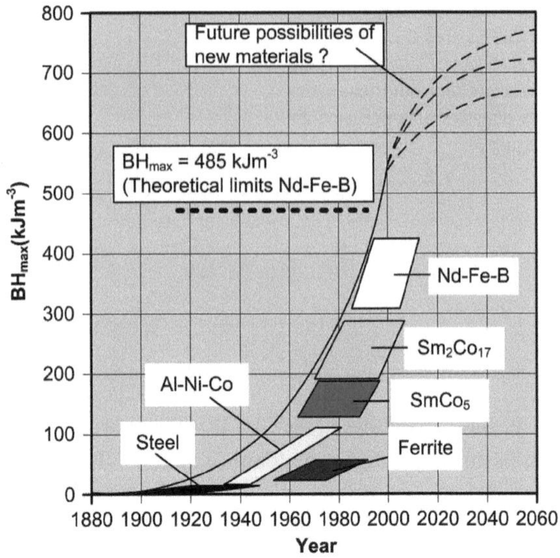

Fig. 3.16 Historical overview of the maximum energy density of permanent magnets. Modified figure, originally published in [16]; published with kind permission of © [Vacuumschmelze GmbH 2009]. All Rights Reserved

Figures 3.16 and 3.17 show the historical overview of the maximum energy density of permanent magnets and different magnetic materials and alloys, respectively. They have been reproduced from documents available from a European producer of soft and hard magnetic materials [16], i.e. Vacuumschmelze GmbH. This company is also involved in the production of La-based magnetocaloric materials. Despite the fact that the trend for the increase in the energy density was similar to that of Moore's law, we can see from the Fig. 3.16 that developments in new permanent-magnet materials have not led to any substantial breakthrough in the past 15 years.

3.2.1.1 Ceramic Materials

These materials are also called "ferrites". They are manufactured from a composite of iron oxide combined with $BaCO_3$ or $SrCO_3$. The manufacturing process normally involves pressing and sintering. These magnets are brittle and have a low energy product. On the other hand, they are inexpensive, and with the relatively high coercivity, resistant to corrosion and can be used at higher temperatures. They can be manufactured as isotropic (the same magnetic properties in all directions) or anisotropic magnets (have a preferred direction of magnetization as a result of particle alignment during the processing). These magnets are the most widely applied magnets on the market. Their energy product is in the range from 10 to 40 kJ m^{-3}.

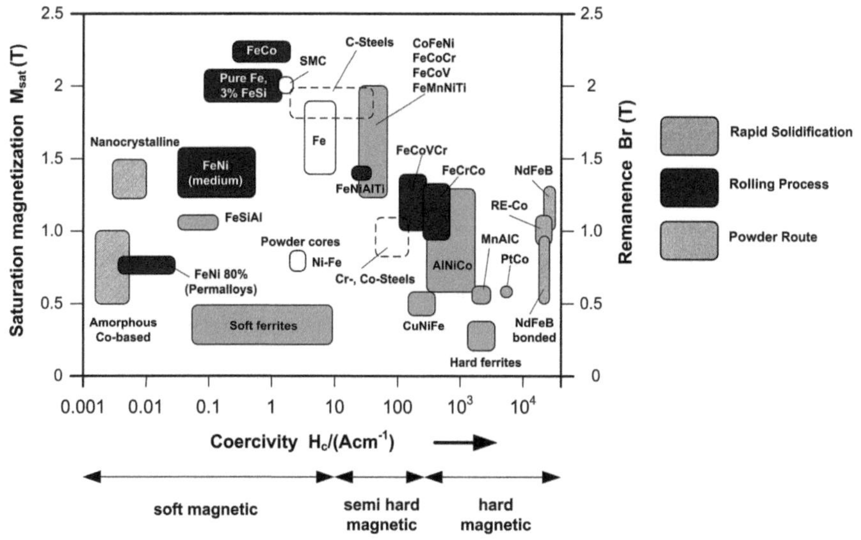

Fig. 3.17 Magnetic materials and alloys. Modified figure, originally published in [16]; published with kind permission of © [Vacuumschmelze GmbH 2009]. All Rights Reserved

3.2.1.2 Al–Ni–Co

These magnets are made by alloying Al–Ni–Co with Fe. They have a very good thermal stability, good corrosion resistance, relatively high remanence and a reasonable cost. However, they have a low coercivity compared to rare-earth magnets and a relatively low energy product: 10–80 kJ m^{-3}. This limit is given by the nonlinear characteristics of the B–H curve, which strongly limits the design of the device and its dynamic operation [15]. They are prone to demagnetization due to shock, and should be relatively long (for example, rod shape) in order to resist demagnetization in an open magnetic circuit. Some other elements, e.g., Cu or Ti, are used in order to improve the magnetic properties.

3.2.1.3 Rare-Earth Magnets

These materials are available in both, sintered and bonded forms. The last require a polymer matrix which is used for keeping the alloy powder in a form. Two classes of permanent-magnet materials represent most of the commercial applications, neodymium-iron-boron (Nd–Fe–B) and samarium cobalt (Sm–Co).

3.2.1.4 Nd–Fe–B Magnets

These magnets are the only group of permanent magnets that have been applied in magnetic refrigeration at room temperature. Nd–Fe–B magnets are the most powerful, having an energy product $(BH)_{max}$ from 200 to about 400 kJ m^{-3}. They also possess a high remanence Br, a relatively high coercivity and they are less expensive than Sm-Co magnets; however, their cost is much higher than that of other types of magnets.

3.2.1.5 Sm–Co Magnets

Their main characteristics are a high remanence Br, a high coercive field H_c and a relatively high energy product, which varies between 140 and 250 kJ m^{-3}. They are the most expensive permanent magnets. They are less temperature sensitive than Nd–Fe–B magnets, but very high temperature ranges are not of interest in magnetic refrigeration or heat pumping. However, it may be very important in power-generation applications. These are the reasons why Sm–Co magnets have not been applied in any of the prototype magnetic refrigeration devices.

More information about permanent-magnet materials can be found in Cullity and Graham [11] and Campbell [14].

3.3 Electromagnetic Coils

These types of magnets require an electric power source for their operation. They can be divided into electromagnets, superconducting magnets, and the special domain of electric pulsed magnets. In this section, we will briefly describe the first two types.

3.3.1 The Electromagnetic Coil

An **electromagnet** consists of a soft iron core surrounded by a number of turns of an electric wire (see also Fig. 3.2a). Figure 3.18 shows the case of an empty "gap", surrounded by a single electric wire.

In the case of Eq. (3.3) we have defined the magnetic field at the centre of the coil. Based on Fig. 3.18 and Eq. (3.1) we can define the magnetic flux density in the axis of the circular coil with the steady current.

First, we will define the position vectors to be in the following relation:

$$\vec{r} = \vec{z} - \vec{a} = z \cdot \vec{k} - a \cdot \cos\varphi\vec{i} - a \cdot \sin\varphi\vec{j} \tag{3.69}$$

where:

$$|\vec{r}| = r = \left(a^2 + z^2\right)^{\frac{1}{2}} \tag{3.70}$$

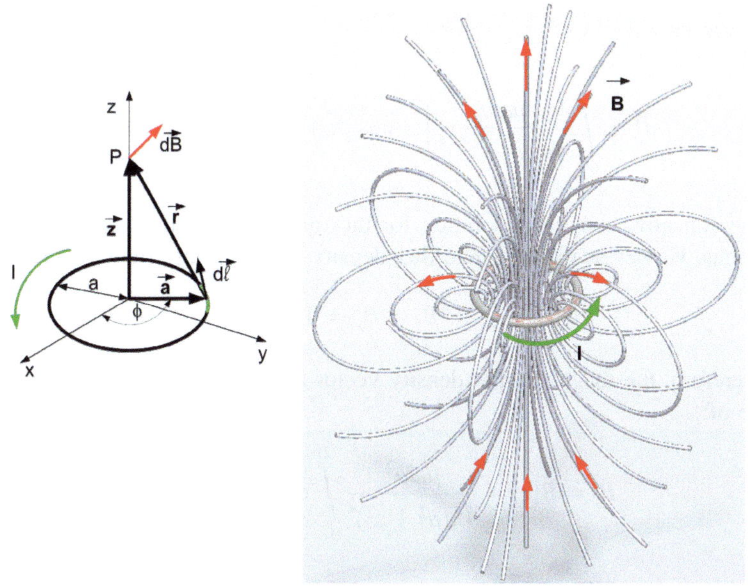

Fig. 3.18 Magnetic flux density as a result of a circular loop that carries a steady current

The vector product between $\mathrm{d}\vec{l}$ and $\vec{r}$ equals:

$$\mathrm{d}\vec{l} \times \vec{r} = \left[a\mathrm{d}\varphi \cdot \left(-\sin\varphi\vec{i} + \cos\varphi\vec{j}\right)\right] \times \left[z \cdot \vec{k} - a \cdot \cos\varphi\vec{i} - a \cdot \sin\varphi\vec{j}\right] \tag{3.71}$$

where:

$$\mathrm{d}\vec{l} = \left(\frac{\mathrm{d}\vec{a}}{\mathrm{d}\varphi}\right)\mathrm{d}\varphi = \left(-a\sin\varphi\vec{i} + a\cos\varphi\vec{j}\right)\mathrm{d}\varphi \tag{3.72}$$

It follows from Eq. (3.71):

$$\mathrm{d}\vec{l} \times \vec{r} = a\left[z \cdot \cos\varphi\vec{i} + z \cdot \sin\varphi\vec{j} + a \cdot \vec{k}\right]\mathrm{d}\varphi \tag{3.73}$$

The magnetic flux density can now be defined as:

$$\mathrm{d}\vec{B} = \frac{\mu_0}{4\pi r^3} I\mathrm{d}\vec{l} \times \vec{r} = \frac{\mu_0}{4\pi r^3} Ia\left[z \cdot \cos\varphi\vec{i} + z \cdot \sin\varphi\vec{j} + a \cdot \vec{k}\right]\mathrm{d}\varphi \tag{3.74}$$

$$\mathrm{d}\vec{B} = \frac{\mu_0 Ia}{4\pi(a^2 + z^2)^{\frac{3}{2}}}\left[z \cdot \cos\varphi\vec{i} + z \cdot \sin\varphi\vec{j} + a \cdot \vec{k}\right]\mathrm{d}\varphi \tag{3.75}$$

By the integration of the vector:

$$\vec{B} = \int_0^{2\pi} \frac{\mu_0 I a}{4\pi(a^2 + z^2)^{\frac{3}{2}}} \left[z \cdot \cos\varphi \vec{i} + z \cdot \sin\varphi \vec{j} + a \cdot \vec{k} \right] \mathrm{d}\varphi \tag{3.76}$$

One can now obtain the solution for the magnetic flux density in each of the directions. However, the magnetic flux density in directions x and y equals zero.

$$B_x = B_y = 0 \tag{3.77}$$

Therefore, the magnetic flux density vector acts in the direction z with a magnitude of:

$$B_z = B(z) = \frac{\mu_0 I a^2}{4\pi(a^2 + z^2)^{\frac{3}{2}}} \int_0^{2\pi} \mathrm{d}\varphi = \frac{\mu_0 I a^2}{2(a^2 + z^2)^{\frac{3}{2}}} \tag{3.78}$$

In the centre of the ring, when $z = 0$, we obtain:

$$B = \frac{\mu_0 I}{2a} \tag{3.79}$$

Let us focus now on the number of single N turns of an electric wire (Fig. 3.19). We will consider, in contrast to the case in Eqs. (3.7)–(3.9), an example of a finite solenoid.

By considering the previous example in Fig. 3.18, we will first define the number of turns per unit length as:

$$n = \frac{N}{L} \tag{3.80}$$

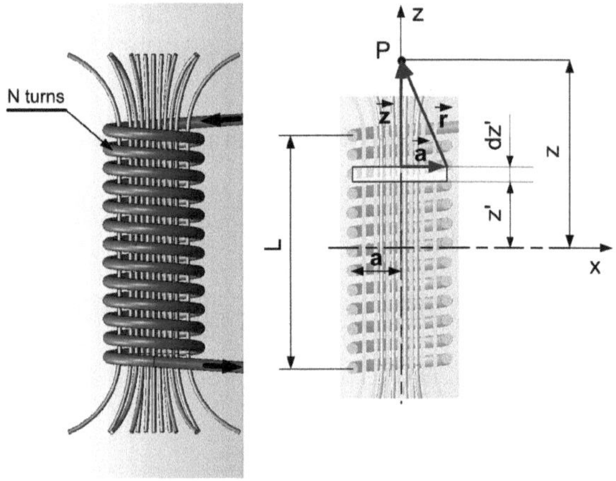

Fig. 3.19 A finite circular coil (solenoid) with N turns of wire

Furthermore, the amount of current flowing through is proportional to the thickness of the cross-section:

$$dI = \frac{N}{L} I dz' = nI dz' \tag{3.81}$$

The magnetic field at point P can now be defined as:

$$dB = \frac{\mu_0 \ a^2}{2\left(a^2 + (z - z')^2\right)^{\frac{3}{2}}} dI \tag{3.82}$$

By taking into account Eqs. (3.81) and (3.82), the integral over the whole length of the solenoid will lead to the following magnetic flux density at the point P.

$$\begin{aligned} B(z) &= \frac{\mu_0 \ n I a^2}{2} \int_{-\frac{L}{2}}^{\frac{L}{2}} \frac{dz}{\left(a^2 + (z - z')^2\right)^{\frac{3}{2}}} \\ &= \frac{\mu_0 \ n I}{2} \left[\frac{\frac{L}{2} - z}{\sqrt{\left(z - \frac{L}{2}\right)^2 + a^2}} + \frac{\frac{L}{2} + z}{\sqrt{\left(z + \frac{L}{2}\right)^2 + a^2}} \right] \end{aligned} \tag{3.83}$$

A single layer of wire will, in practice, not provide sufficient magnetic flux density. Therefore, in practice the coil will consist of several layers (Fig. 3.20) of wires (or a layer of a run with a certain thickness, such as for instance in so-called Bitter magnets, see Fig. 3.21).

By taking Eq. (3.82), we can rearrange it for this particular case. Now, we deal with the 2D problem, by adding the length a as the variable, by taking values from a_1 to a_2.

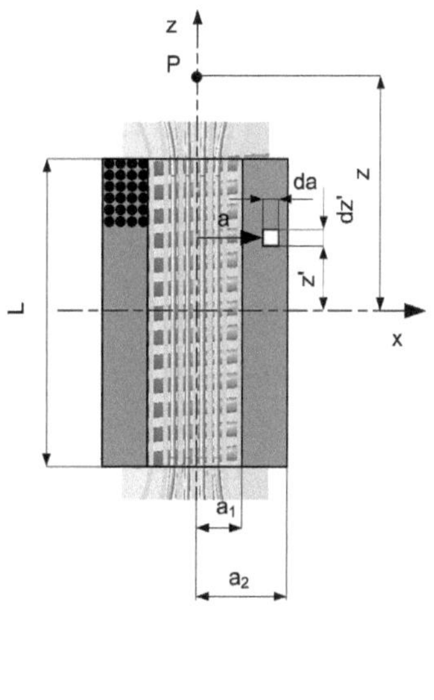

Fig. 3.20 The solenoid comprising several layers of turns

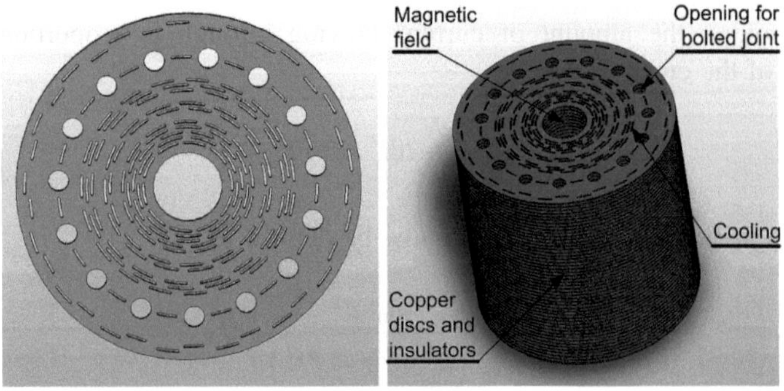

Fig. 3.21 Simple scheme of a Bitter magnet

If δ represents the diameter of the wire, then the change of the electric current can be defined by using the following proportion:

$$dI = I\frac{N_1}{L}\frac{N_2}{(a_2 - a_1)}\,da\,dz' = I\frac{L}{\delta L}\frac{(a_2 - a_1)}{\delta(a_2 - a_1)}\,da\,dz' = I\frac{da\,dz'}{\delta^2} \tag{3.84}$$

where N_1 denotes the number of turns of the electric wires with respect to the length of the solenoid L, and the N_2 represents the number of turns of the electric wire with respect to the width of the solenoid in the range from a_1 to a_2.

$$N_1 = \frac{L}{\delta};\quad N_2 = \frac{a_2 - a_1}{\delta} \tag{3.85}$$

The change of the magnetic flux density can now be expressed as:

$$dB = \frac{\mu_0\ a^2}{2\left(a^2 + (z - z')^2\right)^{\frac{3}{2}}}\frac{I}{\delta^2}\,da\,dz' \tag{3.86}$$

It follows:

$$B(a,z) = \int_{a_1}^{a_2}\int_{-\frac{L}{2}}^{\frac{L}{2}} \frac{\mu_0\ a^2}{2\left(a^2 + (z - z')^2\right)^{\frac{3}{2}}}\frac{I}{\delta^2}\,dz'\,da \tag{3.87}$$

$$B(a,z) = \frac{\mu_0 I}{2\delta^2}\left[\begin{array}{c}\left(\frac{L}{2} - z\right)\cdot\ln\left[\dfrac{a_2 + \sqrt{\left(\frac{L}{2} - z\right)^2 + a_2^2}}{a_1 + \sqrt{\left(\frac{L}{2} - z\right)^2 + a_1^2}}\right]\\ +\left(\frac{L}{2} + z\right)\cdot\ln\left[\dfrac{a_2 + \sqrt{\left(\frac{L}{2} + z\right)^2 + a_2^2}}{a_1 + \sqrt{\left(\frac{L}{2} + z\right)^2 + a_1^2}}\right]\end{array}\right] \tag{3.88}$$

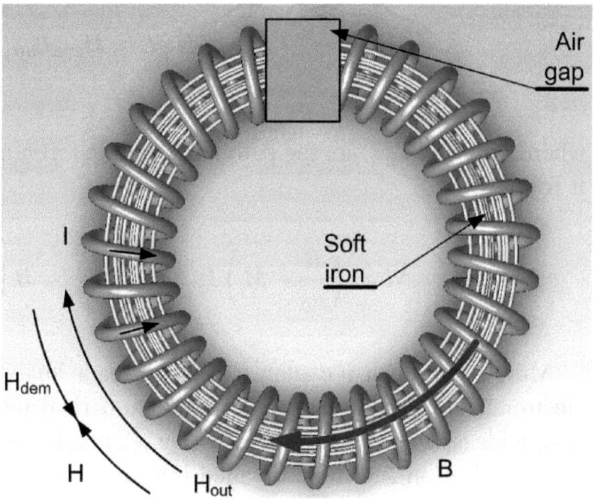

Fig. 3.22 A toroid with an opened "gap"

For the case of the magnetic flux in the middle of the solenoid ($z = 0$), Eq. (3.88) will reduce to:

$$B(a,0) = \frac{\mu_0 I L}{2\delta^2} \ln \left[\frac{a_2 + \sqrt{\left(\frac{L}{2}\right)^2 + a_2^2}}{a_1 + \sqrt{\left(\frac{L}{2}\right)^2 + a_1^2}} \right] \tag{3.89}$$

Now, if a soft magnetic permeable iron is placed inside such a gap, the magnetic flux density B will be substantially increased as the sum of the magnetic field and the magnetization. Therefore, the iron will multiply the magnetic field due to the current. If we consider a simple iron toroid (Fig. 3.22) with the wires around it, then the magnetic field (internal magnetic field) can be defined by Eq. (3.9):

$$B = \frac{IN}{2\pi a} = \frac{IN}{L} \tag{3.90}$$

The magnetic flux density can then be considered as:

$$B = \mu_0 \left(\frac{NI}{L} + M \right) \tag{3.91}$$

and:

$$NI = \left(\frac{B}{\mu_0} - M \right) L = \frac{BL}{\mu_{\text{iron}}} \tag{3.92}$$

If we consider the air gap to be very small, so no magnetic flux leakage occurs in the air gap and we ignore the demagnetization effect, then it follows:

$$NI = \int_{\text{closed path}} H\,dL = H_{\text{iron}}L_{\text{iron}} + H_{\text{gap}}L_{\text{gap}} \quad (3.93)$$

where L_{iron} and L_{gap} are the path lengths in the iron and in the air gap, respectively. It follows:

$$NI = \left(\frac{B}{\mu_0} - M\right) L_{\text{iron}} + \frac{B\,L_{\text{gap}}}{\mu_{\text{gap}}} = B\left(\frac{L_{\text{iron}}}{\mu_{\text{iron}}} + \frac{L_{\text{gap}}}{\mu_{\text{gap}}}\right) \quad (3.94)$$

More comprehensive analyses can be made analytically; however, for particular electromagnet designs normally a 2D numerical tool is used for solving the first design ideas, whereas 3D software (e.g. Comsol, Ansys Multiphysics) can serve for the optimization and more detailed information on the operation of a magnet. This also holds for permanent-magnet assemblies and their design.

The magnetic field in electromagnets can vary between a few mT up to 2 T, or even more. Of course, at high magnetic fields the electromagnets become very large in volume and mass. Furthermore, in almost all cases, high magnetic field electromagnets require additional cooling of the coil, which heats up due to the Joule heating caused by the electric resistance of the wire. High magnetic field electric magnets also do not apply an iron core, and at very high magnetic fields, these magnets operate as pulsed magnets. Moreover, the ratio between the useful magnetizing energy and the energy lost by Joule heating is rather small, indicating that electromagnets cannot by successfully applied in magnetic refrigeration devices. However, these kinds of magnets can be applied in testing devices, either for magnetocaloric materials or for the characterization of magnetocaloric regenerators. In such electromagnets the size of the air gap (between poles) can usually be adjusted by a screw mechanism, thus producing an alternating magnetic field.

For additional reading on the basic engineering of magnetism, besides the already cited books, we recommend Purcell and Morin [17].

A much more interesting application is that of **superconducting magnets**. However, their potential market application in magnetic refrigeration is restricted to rather large units, since this would affect the cost of the device. This issue is also discussed in the Chap. 9. However, one has to be aware that the operating costs of superconducting magnets can represent less than 5 % of that of an equivalent resistive magnet. An additional advantage compared to equivalent resistive magnets is in their compactness and durability. However, since the superconductivity of materials is strongly restricted by the critical magnetic field (see subsequent text), the strongest superconducting magnets cannot perform such a high magnetic flux density as counter, resistive Bitter magnets. But most probably, such high magnetic fields (e.g. far above 10 T) will not be applied in any market application of the magnetic refrigeration or heat pumping near room temperature.

Since the costs of the R&D in this special domain of magnetic refrigeration can be at least an order higher compared to permanent-magnet-based magnetic refrigerators, this will be rather difficult to perform at a particular research institute. More likely, industry has the potential for such developments. Therefore, we would like to encourage the industry involved in superconducting power applications to invest in the development of large-scale superconducting magnetic chillers or heat pumps.

3.3.2 Superconducting Magnets

The phenomenon of superconductivity has been known since the discovery of Onnes in 1911 [18]. He observed (in a sample of mercury), that certain electrical conductors display zero DC electrical resistance. Since then, a number of metals and alloys have been found to possess superconductivity at different temperatures from one degree of Kelvin up to much higher temperatures. In 1933 Meissner and Ochsenfeld [19] discovered extraordinary magnetic properties in superconductors. These, when cooled below their critical (superconducting transition) temperature and under an applied magnetic field, acted as a magnetic shield by expelling the magnetic flux around the sample. This led to the definition of the Meissner effect (Fig. 3.23), which states that in the superconductor, when this is in its superconducting state, the magnetic flux density is zero [20]. This effect is not related to the zero electrical resistance, but represents an additional property of superconductors.

If we consider the equation for the magnetic flux density of a magnetic material (Eq. 3.36) $B = \mu_0(H + M)$, then it follows for $B = 0$ and $H > 0$ that the magnetization M should be negative. Therefore, a superconductor under an applied magnetic field will possess a negative magnetization. This results from the electric current, flowing without resistance around the outer envelope (surface) of the superconductor.

There are type-I and type-II superconductors. In type-II superconductors, the surface energy of the superconducting/normal interface is negative. Therefore, under the applied magnetic field, in a type-II superconductor, a negative magnetization will be produced by the surface current. The same happens with type-I superconductors, but only until the lower critical field at which the magnetic flux

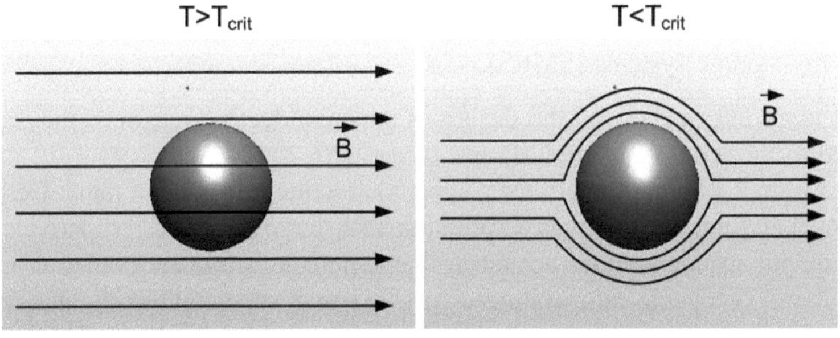

Fig. 3.23 Two different magnetic states of superconductors due to the Meissner effect

enters the sample and forms individual flux lines (fluxoids) [21, 22]. Most basic materials and some alloy superconductors show type-I behaviour. These will loose their superconductivity above the critical magnetic field and will behave as normal electric conductors. Therefore, type-II superconductors are of interest for engineering applications, since they are able to carry high current densities and large magnetic fields.

Figure 3.24 (top) has been produced based on data from Refs. [23–25]. It shows the historical development of superconducting materials according to their critical temperature. Since these temperatures relate to cryogenics, accordingly, superconducting magnets can be divided into different groups:

- **Low-temperature superconducting magnets (LTS Magnets)**, which apply liquid helium or are cooled by a cryocooler (please see subsequent text relating to different technologies for cryocoolers),
- **High-temperature superconducting magnets (HTS Magnets)**, which apply liquid helium, a cryocooler or liquid nitrogen,
- **Hybrid magnets,** which combine a copper magnet in an inner section with the superconducting magnetic in an outer section.

In Fig. 3.24 (bottom) a review on HTS materials with respect to their critical temperature is shown. This figure has been produced based on data from Refs. [26, 27]. According to the design, most superconducting magnets fall into the following groupings:

- **Solenoids**: Represent cylindrical structures and are most broadly applied (Fig. 3.25),
- **Dipoles**: These magnets generate a uniform field transverse to their longer axis and can be found in particle accelerators and magnetohydrodynamic (MHD) applications,
- **Quadrupoles**: Generate a linear gradient field transverse to their axis over the central region of their bore and can also be seen in particle accelerators,
- **Racetracks**: Racetracks are wound in a plane where each turn consists of two parallel sides and two semi-circles at each end, where a pair is assembled to approximate the field of a dipole. These magnets can be found in superconducting motors, generators, as well as in train applications (Maglev),
- **Toroids**: Generate magnetic fields in the azimuth direction along the toroid. They can be found in fusion reactors (Tokamak) or in superconducting magnetic energy-storage systems (SMES).

For more information on the design of superconducting magnets, the reader is referred to the work of Yuan [20] and Iwasa [28].

A cryogenic cooling system for a superconducting magnet is a must. Despite of the fact that some may use an expression like “cryogenic-free” superconducting magnets, this of course does not mean that superconductors are cooled at temperature above cryogenic temperatures. Any existing superconducting magnet will require cooling of its coils to cryogenic temperatures, whether this is applied using liquid (cryogenic) refrigerants or a cryogenic cooler (cryocooler). If we focus on the

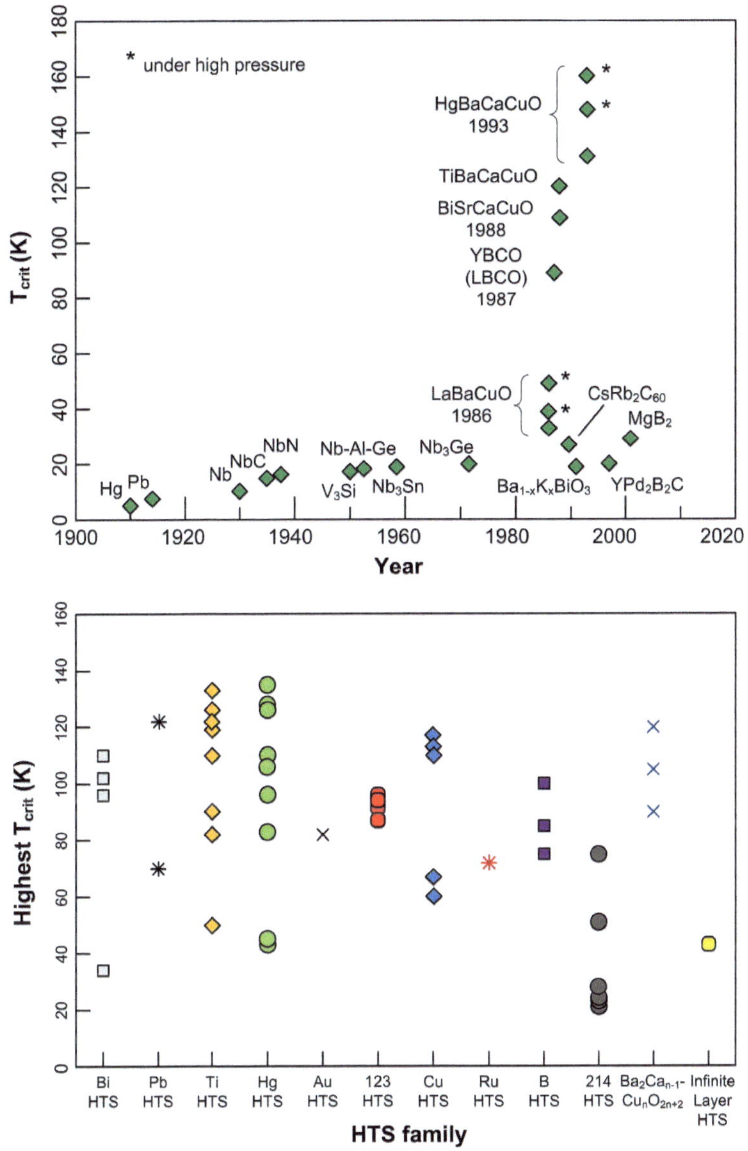

Fig. 3.24 A historical development of superconducting magnets (*upper*) and a review on HTS materials (*lower*)

latter, these can be divided into the Joule–Thomson (JT), the Brayton, Gifford–McMahon (GM), the Stirling and the Pulse-tube cryocoolers (see Fig. 3.26, which has been reproduced according to the work of Radebaugh [29]). Other principles, e.g. a cryogenic AMR, are not used in applications for the cooling of superconducting magnets.

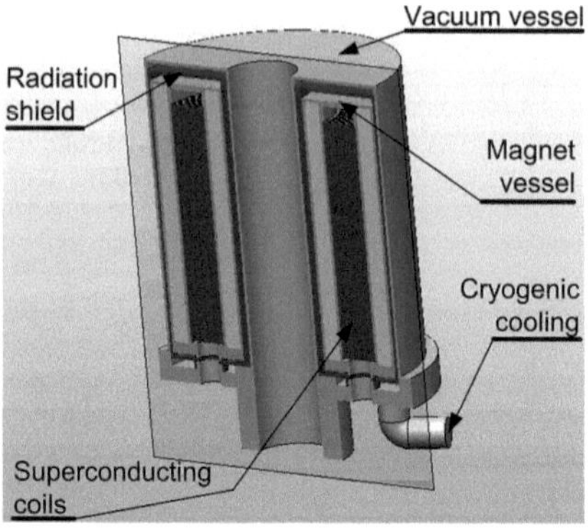

Fig. 3.25 An example of a superconducting solenoid

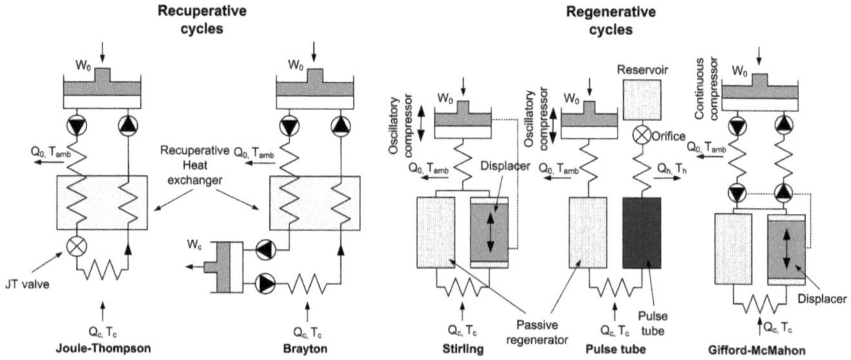

Fig. 3.26 Schematics of recuperative and regenerative cryocoolers. Modified figure, originally published in [29]; published with kind permission of © [IOP Publishing, J Phys: Condens Matter 2009]. All Rights Reserved

Generally, cryocoolers can be applied to superconducting devices in different ways, such as: open-cycle cooling with immersion, closed-cycle cooling by reducing pressure, closed-type cooled with immersion, forced-flow circulation cooling and direct cooling by refrigerator. Additional information on these methods can be found in Flynn [30] and Wang [31].

As can be seen from Fig. 3.26, cryocoolers can be either regenerative or recuperative [32]. The recuperative cryocoolers apply recuperative heat exchangers (e.g. counter-current) and operate with a steady flow of refrigerant through the system. The regenerative cryocoolers comprise at least one regenerative heat exchanger, i.e. a passive regenerator. They operate under oscillating flow and pressure. In a passive regenerator the incoming warm gas refrigerant transfers heat to the material of the regenerator during a certain thermodynamic process (e.g. isochoric in Stirling device) and it absorbs heat from the regenerator before the thermodynamic cycle is repeated. In steady-state conditions, like in an active regenerator (see the chapter on

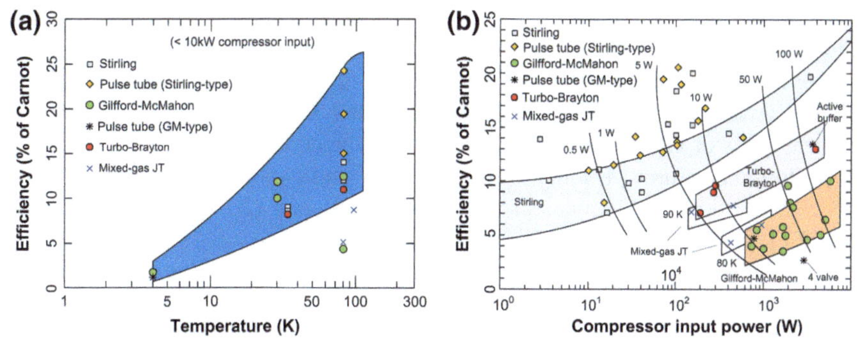

Fig. 3.27 Comparison of different cryocoolers **a** comparison of the Carnot efficiency for different cryocoolers depending on the temperature of cooling; **b** carnot efficiencies of different cryocoolers at a cooling temperature of 80 K. Modified figure, originally published in [29]; published with kind permission of © [IOP Publishing, J Phys: Condens Matter 2009]. All Rights Reserved

active magnetic regeneration), there will be a temperature gradient established over the matrix of the passive regenerator.

Figure 3.27 shows a comparison of the Carnot efficiency (exergy efficiency) of different cryocooler technologies. These are also briefly described in the following text.

The temperature range of **Brayton (recuperative) cryocoolers** is from 4.2 to 120 K. The maximum refrigerating capacity of a single refrigerator is 20 kW at a temperature of 4.5 K [31]. These kinds of refrigerators are broadly applied in LTS magnets. Their advantages are that they have a steady operation, long durability, good efficiency for large sizes; however, they are expensive, require a large heat exchanger, and therefore they are not so compact, and have strict operating requirements [29–31].

The **Joule-Thomson (JT) (recuperative)** refrigerator represents a steady operation and applies heat recuperation. This kind of refrigerator applies gases that have a strong dependence of enthalpy versus pressure. It mainly consists of four parts: compressor, counterflow heat exchanger, JT valve and a reservoir. The refrigerant gas is compressed at room temperature and the generated heat is rejected to the ambient. Gas then enters the warm part of the counter heat exchanger, where it is precooled to a certain temperature. Then it enters the JT valve and expands during the adiabatic-isenthalpic process where it is liquefied at a cryogenic temperature and at the initial pressure. The liquid refrigerant leaves the system and is used for refrigeration, whereas the remaining gas flows through the counter current heat exchanger towards the compressor. Since there are no "cold" moving parts in the Joule–Thomson cryocoolers, these kinds of devices can be scaled to microsizes [29–31].

Another type of cryocooler relates to the **Gilford–McMahon (GM) (regenerative)** refrigeration cycle. This kind of thermodynamic cycle comprises the following four processes: adiabatic charging (compression), isobaric cooling, adiabatic discharging and isobaric heating. The GM cryocoolers apply a passive regenerator and a displacer. Their frequency of operation is rather low (1 Hz). The advantages of these cryocoolers, despite their rather low exergy efficiency, are in the simplicity of the device, compactness, small vibration and the reliability. There are two general types of GM cryocoolers. The single-stage cryocoolers, which can operate down to temperatures of liquid nitrogen (77 K), and two-stage cryocoolers, which can operate down to temperatures of liquid hydrogen (20 K) or liquid helium (4.2 K) [29–31].

The **Pulse-Tube (regenerative) refrigerator** is another type of the cryocooler, which generally consists of a compressor, regenerator, gas piston, orifice, gas reservoir and other elements. A Carnot efficiency of about 20 % at 80 K has been reported by Radebaugh [32]. Pulse-tube cryocoolers apply the pressure oscillating flow through the hot end orifice (similar to a displacer in a GM or Stirling device). In the first type of pulse-tube cryocoolers, the compressor and the rotary valve are applied. Pressure oscillators (valve-less compressor) are applied in another type of pulsed tubes. Therefore, one can distinguish between the GM and Stirling type of pulse-tube refrigerator, respectively [32]. The GM pulse-tube refrigerator operates at very low frequency (1–2 Hz), and is ideal for small cooling powers and low temperatures (e.g., down to 2 K). The Stirling type operates at high frequency (50–60 Hz) and with the lowest operating temperature of about 10 K. Compared to the Stirling and Gifford-McMahon refrigerators, there is no need to apply a displacer in a pulse-tube refrigerator. This brings the advantage of the latter because of substantially reduced vibration, increased durability and reduced axial heat conduction [29–31].

Stirling (regenerative) cryocoolers apply the Stirling cycle, and they can be applied for the lowest temperature at about 12 K. The frequency of the operation can be as high as 60 Hz. In these types of cryocoolers the heating of the gas occurs during the compression, and cooling during the expansion of a gas refrigerant. Stirling cryocoolers (similar to GM) apply a displacer in order to move the gas refrigerant to the hot end during the compression (or to the cold end during the expansion). These cryocoolers are suitable for applications of the HTS magnets (e.g. in the range from 20 to 77 K and from the cooling power in the range from 50 to 500 W. Stirling cryocoolers are more efficient than Brayton cryocoolers, less sophisticated, more compact and they do not require strict operation conditions, which are necessary in Brayton cryocoolers [29–31].

3.4 Permanent-Magnet Designs in Magnetic Refrigeration

The first magnetic refrigerator based on permanent magnets was developed in 2001 and presented by Lee et al. [33], and later by Zimm et al. [34, 35] (note there are certain indicators of earlier prototypes—see Chap. 7, however, the information is not supported by a photography). Since then, most of the magnetic refrigeration prototypes that have been developed, whether for experimental purposes or as potential future market applications, have used permanent magnets. Superconducting magnets have been applied as well, although these were not constructed specifically for the application in magnetic refrigeration at room temperature (see also the Chap. 7). Namely, researchers simply took advantage of existing superconducting magnets and since these solenoids consisted of a dewar vessel with a bore for the magnetic field in the middle, prototypes based on such solutions were linear with the reciprocate movement of the magnetocaloric material (AMR) in and out of the magnetic field. No special design of superconducting magnet for magnetic refrigeration has been reported to date, despite the fact that this kind of technology could bring (large-scale) market applications. This issue is addressed in the Chap. 9. In this section, however, we will focus on some of the design concepts that have been used for permanent-magnet structures in magnetic refrigeration. More information about particular devices can be found in the Chap. 8.

We will divide the permanent-magnet designs into the following categories:

- **Static or moving simple (2D) magnet assemblies** (which mostly relate to different types of "horseshoe" magnets).
  These kinds of magnets are applied in linear or rotary machines in which the magnetocaloric material is the rotating or linearly moving part, and where the magnetic field source is static, or the magnetic field source (magnet) is moved linearly or rotated over the static magnetocaloric material in the form of an AMR. The magnetic flux can be approximated to flow in two dimensions (2D).
- **Static Halbach (2D) magnet assemblies**
  These magnets are in an arrangement that was first proposed by Halbach in 1980 [36]. Most of the prototypes that apply the (linear or rotary) motion of the magnetocaloric material are based on such a solution. Figure 3.28 shows a basic Halbach type of magnet assembly, consisting of a number of permanent magnets, with the magnetization arranged in a such a way that the magnetic flux is concentrated in the middle of the magnet assembly.
- **Rotary Halbach (2D) magnet assemblies**
  By taking the principle of a Halbach array, different approaches have been used in order to perform a rotation of the magnet assembly or the rotation of a part of the magnet assembly over the static magnetocaloric material. This kind of principle, at least according to our knowledge and experiences, represents one

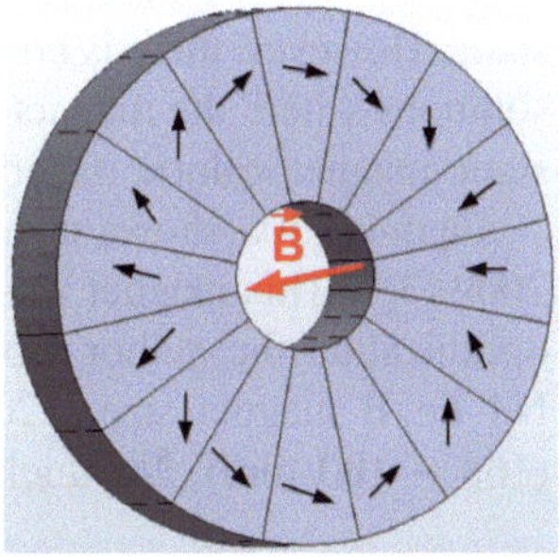

Fig. 3.28 A typical Halbach array of permanent magnets

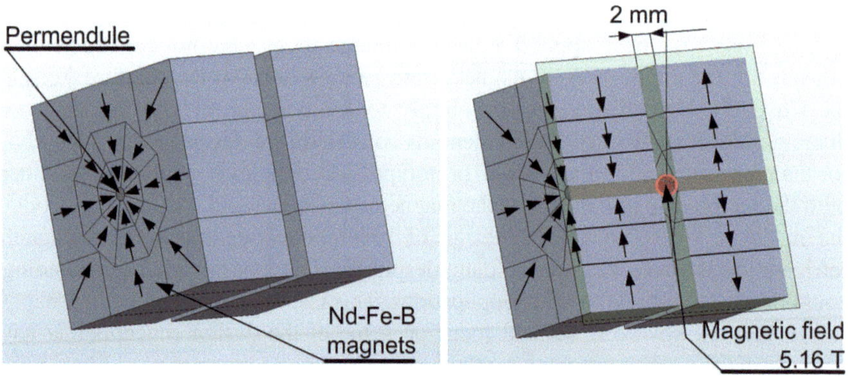

Fig. 3.29 A 5 T three-dimensional (3D) magnet assembly-note the air gap is not shown in the proportion the dimensions of the magnet assembly (see also Kumada et al. [37])

of the best solutions, especially due to the fact that different friction losses and fluid leakages are associated with the sealing of the rotary magnetocaloric material (usually in the form of a rotating disc). This, however, does not represent the same problem for solutions where the linear motion of the magnetocaloric material is applied.

- **Halbach (3D) magnet assemblies**
  A three-dimensional guidance of the magnetic flux can lead to a high magnetic flux density (Fig. 3.29). Of course, such a magnet assembly becomes rather complicated and in most cases too expensive for any future application in magnetic refrigeration near or at room temperature. Figure 3.29 shows an example of a 5 T magnet assembly, which has not been developed for magnetic refrigeration, but shows the possibility of reaching very high magnetic fields.

3.4.1 Static or Moving Simple (2D) Magnet Assemblies

Such magnet assemblies, especially because of the simplicity of shapes, can represent an interesting solution, especially for experimental devices or demonstrators. In all the cases represented below (Figs. 3.30 and 3.31), the magnet assembly is static. However, there is no reason why such an approach could not also be used for solutions where the magnet assembly or a part of it represents a rotating part (with static magnetocaloric material).

Figure 3.30a shows a magnet assembly that was presented by Zheng et al. in 2009 [38]. This magnet was designed using the Ansys multiphysics tool. After the optimization the author reported the following features: gap size (20 × 40 mm), Nd–Fe–B magnet size (120 × 80 mm), and magnet assembly outer dimensions (168 × 180 mm). No thickness of the magnet was reported. The magnetic flux

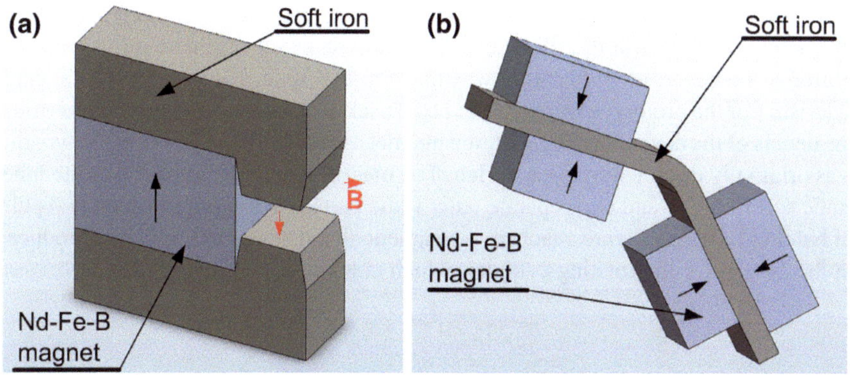

Fig. 3.30 **a** A horseshoe type of magnet constructed by Zheng et al. [38], **b** half of the magnet assembly constructed by Bohigas et al. [40]

Fig. 3.31 Static magnet assembly designed by Šarlah in 2007 (reported in Tušek et al. [41])

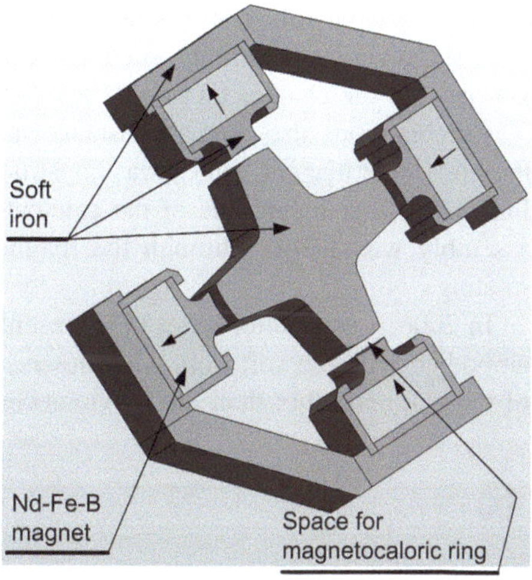

density in the gap varied from 0 T (at the wall of the Nd–Fe–B magnet) to approximately 1.1 T at a distance of 48 mm from the wall of the Nd–Fe–B magnet, with a maximum midpoint magnetic flux density of 1.5 T. The mass of the Nd–Fe–B magnet (without iron) was reported to be 7.2 kg. In the experimental device, the magnet assembly was moved over two static AMR beds, by means of mechanical subsystem, composed of a step motor and a ball screw [38]. A very similar magnet assembly was shown in a patent of Zimm et al. [39].

Another example is shown in Fig. 3.30b. In this particular case we show half of the magnet assembly that was used to provide a magnetic flux density of 0.9 T in the rotating disc containing magnetocaloric material. More information about this particular device can be found in the Chap. 7 in the reference to the work of Bohigas et al. [40].

In 2007, the team from the University of Ljubljana presented their first rotary magnetic refrigerator at the IIR Thermag 2 conference on magnetic refrigeration in Portorož; however, no publication was made at that time. Later, in 2010, the characteristics of the prototype were reported by Tušek et al. [41], showing and describing the details of the device. In Fig. 3.31, the magnet assembly of the device is shown; this was originally designed by Alen Šarlah. The magnet assembly consisted of the inner static soft iron part and two separate outer static parts of soft iron, to which two pairs of Nd–Fe–B magnets were attached. A magnetic flux density of 0.98 T was produced in the air gap for the rotating cylinder, which contained parallel Gd plates.

3.4.2 Static Halbach (2D) Magnet Assemblies

The Halbach principle has been applied in many different magnetic refrigerator prototypes. In some cases the magnet assembly consisted of magnets only, whereas in other cases, a soft iron was additionally applied to guide the magnetic flux.

A static magnet assembly, such is that shown in Fig. 3.28, was applied by many groups (see the chapter on prototypes for more details).

Combinations also exist, for instance, a hybrid between the Halbach structure and the horseshoe magnet, which was constructed by Lee et al. (presented in [42]—see Fig. 3.32). The magnitude of the calculated magnetic flux density of this magnet assembly was 1.9 T. Through the magnetic field, a disc containing AMRs was rotating.

In 2006, Vasile and Muller [43] presented a paper in which a Halbach structure, embodied with the soft iron, was shown. Such a principle enables better shielding of the magnetic flux than the previous case. However, in both cases the magnetic

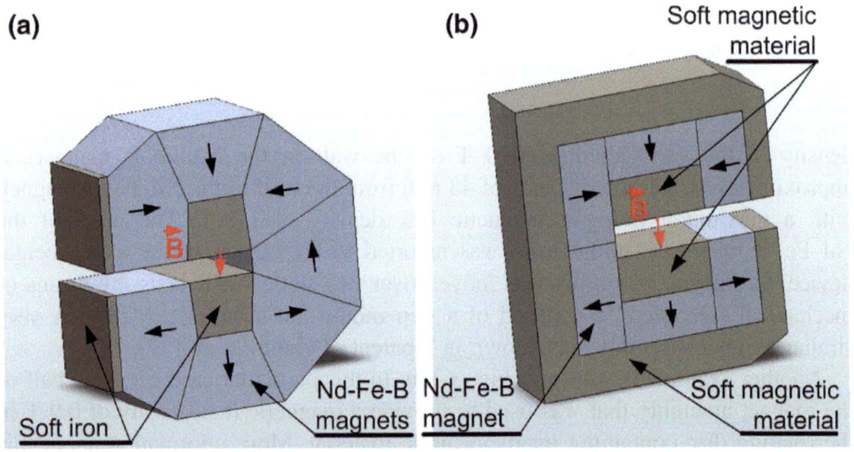

Fig. 3.32 **a** An example of a static magnet assembly developed by Lee et al. [42], **b** an example of a part of the magnet assembly developed by Vasile and Muller [43]

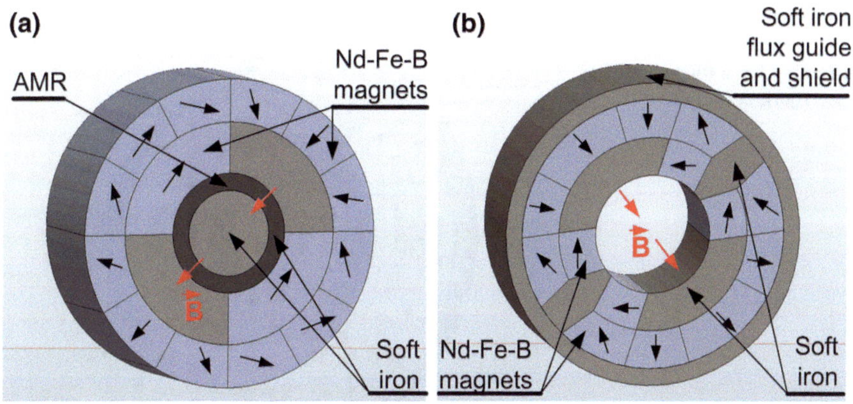

Fig. 3.33 Design of static coaxial permanent-magnet assemblies for a rotating AMR, **a** a solution without an outer soft iron shield (see Egolf et al. [44], Vuarnoz et al. [45]), **b** a solution with an outer shield, without showing the inner static part (see Egolf et al. [46])

shielding is not provided in a direction that is perpendicular to the surface plane of the magnet (direction of observer of Fig. 3.32).

Another class of static permanent magnets is shown in Figs. 3.33 and 3.34. In this case, the magnetocaloric material, in the form of a rotating cylinder, is positioned between two static coaxial cylinders, which represent a soft magnetic material for the guidance of the magnetic flux. The magnet assembly in Fig. 3.33a was designed by the team from the University of Applied Sciences of Western Switzerland (UASWS); it was first presented in a study for the Swiss Federal Office of Energy (SFOE) in 2008 [44], and later in the article of Vuarnoz et al. [45].

In the solution in Fig. 3.33a, the outer ring made of soft magnetic material can be added with a slight modification of the direction of the magnetization of permanent magnets. A similar design was presented by the same team (UASWS) in 2009 in another study for the SFOE (Fig. 3.32b) [46].

Later in 2010, the team from the DTU—Risoe presented a similar solution to the previous two, with the difference that both the inner and the outer static cylinders would represent a magnet assembly [47, 48]. This magnet assembly is shown in Fig. 3.34. The angles of the magnetization of the magnets were assumed and approximated by the authors of this book. A maximum magnetic flux density of 1.24 T was obtained in the regions shown by the red arrow in Fig. 3.34.

Note that all the magnet assemblies presented in Figs. 3.33 and 3.34 can have different magnetic flux densities, which can be controlled in the design by the thickness of the magnetocaloric cylinder, as well as the diameters of the outer and the inner static parts of the magnet assembly. Therefore, such an approach enables scaling of device from a few watts to kilowatts of cooling power. For instance, three different designs of the magnet assembly in Fig. 3.33b were performed in the study of Egolf et al. [46], i.e. for 1, 1.5 and 2 T, respectively. The dimensions of the outer diameter of the magnet assembly in this case were taken to be 400 mm for all the three cases, whereas the inner static soft iron part (not shown in Fig. 3.33b would have a diameter

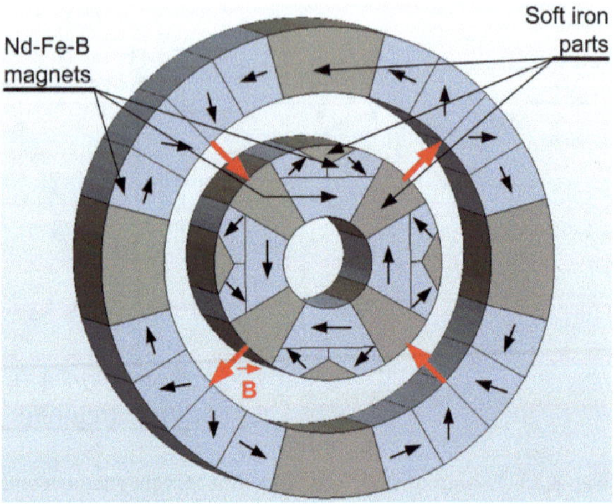

Fig. 3.34 Design of a static coaxial permanent-magnet assembly for a rotating AMR (see Bjørk et al. [47, 48])

of 180 mm (thickness—gap—of the magnetocaloric ring 34 mm), 166 (thickness—gap—of the magnetocaloric ring 18 mm) and 150 mm (thickness—gap—of the magnetocaloric ring 6.2 mm), respectively.

3.4.3 Rotary Halbach (2D) and Simple (2D) Magnet Assemblies

A static AMR in a magnetic refrigerator or a magnetic heat pump offers the possibility of a substantial reduction of losses compared to a rotating AMR. These losses in the latter relate to the friction of dynamic seals (and related heating), internal (or even external if the device is not well designed) leakage of the working fluid, which occurs between the static piping part and the rotating AMR.

Therefore, the rotation of a magnetic field over a static magnetocaloric material represents a more efficient solution. The simplest way to perform this in rotary devices is simply by taking a double Halbach magnet array, as shown in Fig. 3.35. This was actually done by Tura and Rowe [49, 50], who applied a double Halbach as the 1.34 T magnetic field source for a magnetic refrigerator. This kind of approach was also studied by Bjørk et al. [51] as well as by Bouchekara and Nahas [52].

According to Fig. 3.35, the inner array rotates with respect to the stationary outer array. When the magnetic flux density vector within the space between the two arrays is aligned, this will provide the ON-operation of the magnetic field source, by summing the individual inductions of each array (Fig. 3.35a). When the inner array is rotated by 180° the magnetic flux density vectors of the outer and inner ring cancel each other, thus, resulting in the minimum low-field state (Fig. 3.35a).

In 2014, Arnold et al. [53] presented a triple Halbach array (Fig. 3.36), which represented a further optimization of the previously [49, 50] constructed double

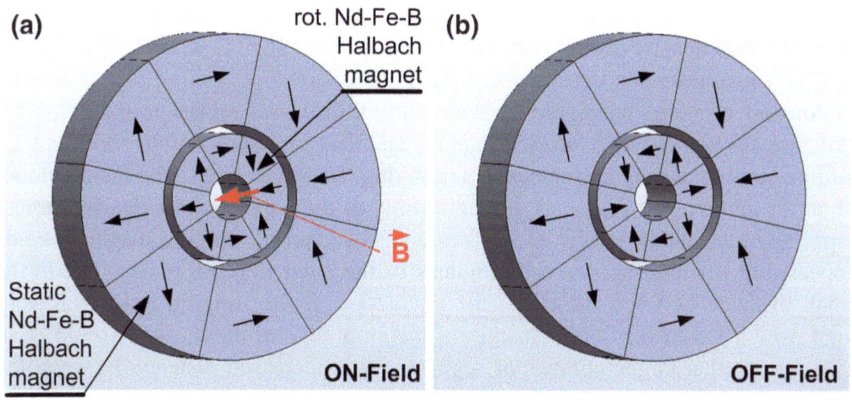

Fig. 3.35 A double Halbach array, which was for instance applied in magnetic refrigeration by Tura and Rowe [49, 50] (**a** ON field position, **b** OFF field position)

Halbach array (presented in Fig. 3.35). The problem that characterized the double Halbach solution was in the actual magnetic flux density at low field, which was not zero (having an average high-field magnetic flux density of 1.34 T and an average low-field magnetic flux density of 0.57 T [53]). This, of course, is not desired. As stated further by Arnold et al. [53], the total field vector orientation inside the AMR volume rotated, which could induce additional rotational forces and eddy currents.

The new magnet design of a triple Halbach array with the increased number of magnet segments in each ring (each ring comprised 12 segments of magnets—not shown in Fig. 3.36) improved the homogeneity and the magnetic flux density. The two outer magnetic rings rotated in the counter direction with respect to each other, while the inner magnetic ring was static. In this way a sinusoidal field waveform was produced. In analogy with the previous solution, the three rings, when their vector of the magnetic flux density was aligned, lead to the maximum magnetic flux density in the bore. When the two outer rings were rotated by 180 °C with respect to the inner static ring, the magnetic flux density vectors cancelled each other, thus

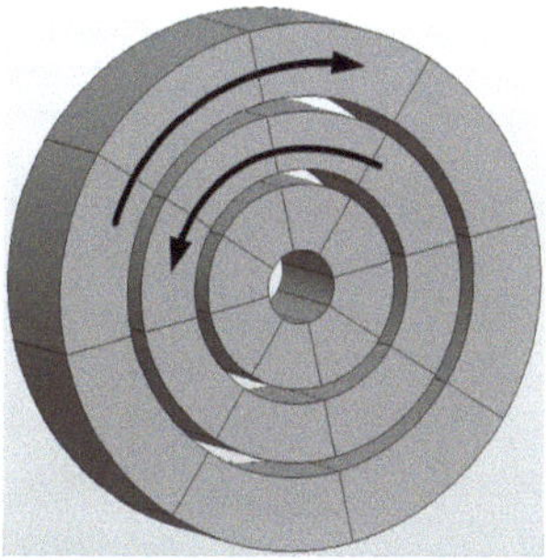

Fig. 3.36 A triple Halbach array with two outer magnetic rings rotating in the counter direction and with the inner static magnetic ring (see also Arnold et al. [53])

providing a low magnetic field in the bore of the magnet assembly. The average magnetic flux density during high-field operation was 1.25 T, whereas the average low-field magnetic flux density was 0.29 T [53].

Another problem, addressed by Arnold et al. [53], was related to the non-ideal structure of the Halbach array. Namely, an ideal structure would represent an infinite number of magnet segments in each ring. However, due to the finite number of permanent magnet segments, the uniformity of the magnetic flux density varied, especially near the interfaces of two segments. Therefore, fringing magnetic fields appeared at the interfaces of the segments of the outer ring. Moreover, due to the small distance between the rings, the interaction of the fringing fields was very high, which led to very high changes of the amplitude of the torque. Arnold et al. [53] proposed a larger number of segments or a increased distance between the magnet rings to solve that problem.

In 2007 at the third International Conference on Magnetic Refrigeration at Room Temperature, which was held in Portorož in Slovenia, Okamura et al. [54] presented an optimized magnet assembly, which was serving a prototype for the Chubu Electric Power Co. This magnet assembly represented an improvement over a previous solution in which two magnet bars (with the maximum magnetic flux density being 0.77 T) [55], attached to a soft iron core, were rotating and thus providing a magnetic field in the four beds of the AMRs. The AMRs were positioned between the rotating magnet bars and the static outer ring made of soft iron. As denoted by Okamura et al. [54], the problem related to the magnet assembly was in the eddy currents, which were produced along the outer soft iron ring. As a solution the authors cut the soft iron ring on a number of thinner rings, attached to each other with an electrical insulator in between. This measure drastically improved the performance. Furthermore, the authors have introduced a new inner rotating part, which provided a higher magnetic flux density with a magnitude of 1.1 T. The cross-section of the new magnet assembly is shown in Fig. 3.37.

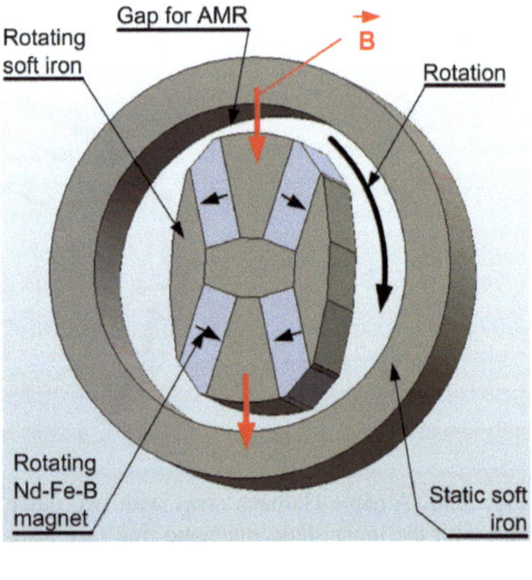

Fig. 3.37 The cross-section of the magnet assembly designed and constructed by Okamura et al. [54]

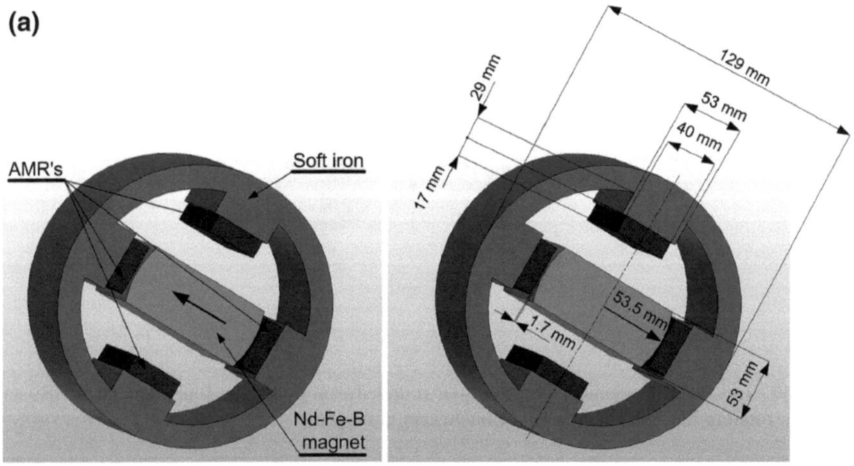

Fig. 3.38 The reference magnet assembly with the *four rectangular* poles and AMRs and with the rotating bar magnet (see also Bouchekara et al. [56])

A comprehensive numerical study of different solutions that relate to the rotation of a bar magnet (similar to the solution of Okamura et al. [54, 55]) was performed by Bouchekara et al. [56]. These solutions did not represent a Halbach structure, but a design in which the rotor was represented by a single permanent bar magnet and where the static parts (stator) were the AMRs and the outer iron ring. For each of the solution the authors evaluated their electromagnetic performance, i.e. the magnetic flux density, the forces and the torque. In the study, multipole stators and rotors with 4, 6 and 8 AMR beds were analyzed.

For the purposes of the study, a simple magnet assembly was used as the reference and this is shown in Fig. 3.38. The outer static iron ring in this magnet assembly consisted of four poles to which four AMR beds would be attached. The rotation of the inner bar magnet provided the magnetization/demagnetization for each of the AMR beds. For the purposes of the study, the reference dimensions of the magnet assembly were defined, and these are shown in Fig. 3.38. For all the other evaluated configurations of the magnet assemblies, the dimensions of the magnet, the AMR and the air gap were fixed with the values given in Fig. 3.38.

Figure 3.39 shows the magnet assembly denoted by (b). In this magnet assembly, compared to (a) (Fig. 3.38), only the AMR shape and the shape of the outer iron ring poles were slightly modified into rounded shapes.

In Fig. 3.40, two magnet assemblies are shown, denoted by (c) and (d). In the magnet assembly (c), the structure has the same shape and the same number of poles as the reference magnet assembly (c). Since the goal was to analyse the influence of the iron poles, the magnet assembly (c) is without poles, leaving the AMRs to be attached directly to the smooth iron ring. In the magnet assembly (d) (Fig. 3.40), compared to magnet assembly (c), additional steel parts with a thickness

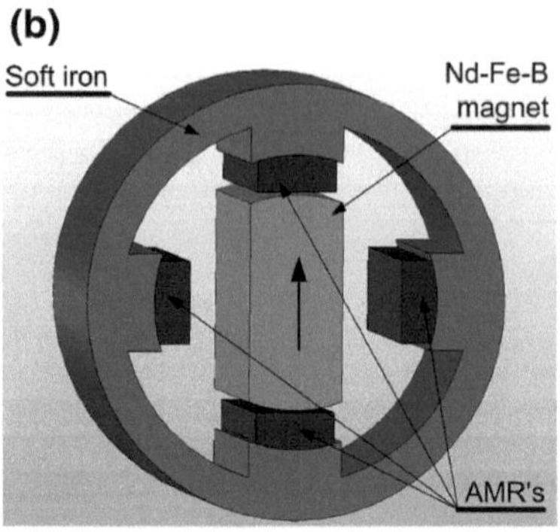

Fig. 3.39 The magnet assembly with the four rounded iron poles and four rounded AMRs and with the rotating bar magnet (see also Bouchekara et al. [56])

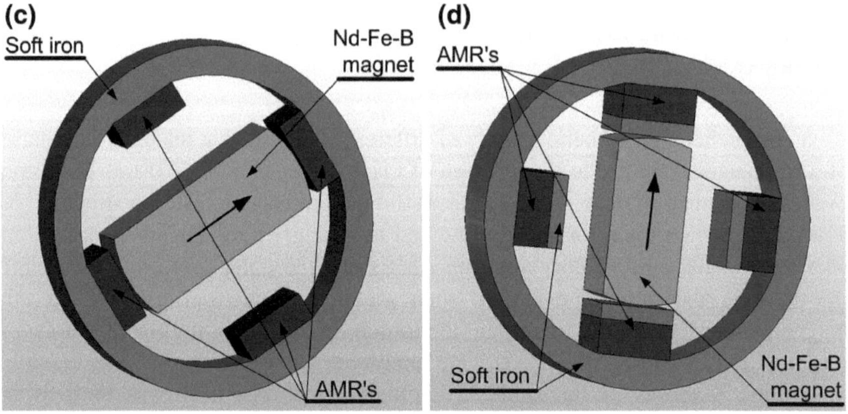

Fig. 3.40 The magnet assembly without iron poles, **c** the *four rectangular* AMRs are attached to the smooth static iron ring, **d** the *four rectangular* AMRs are attached to the smooth static iron ring and 10 mm steel is attached to AMRs (see also Bouchekara et al. [56])

of 10 mm were attached to the AMR beds. The idea here was to bring support that would sustain tangential forces instead of the AMRs.

The magnet assemblies (e) and (f) in Fig. 3.41 comprise six poles each. The idea behind the increased number of poles was that a larger number of poles reduce the magnetic torque of the system. Therefore, the magnet assembly (e) had the same shape as the reference magnet assembly (a); however, it comprised six poles of soft iron to which six AMRs were attached. The design of the magnet assembly denoted by (f) is similar. This magnet assembly did not comprise iron poles, similar to the magnet assembly (c) (in the Fig. 3.40); however, it comprised six beds of AMRs.

An additional change was made in the magnet assembly denoted by (g) (Fig. 3.42). This magnet assembly was similar to the magnet assembly (f) (from Fig. 3.41); however, with the difference being that the AMRs were embodied into the smooth iron ring, thus having no iron poles.

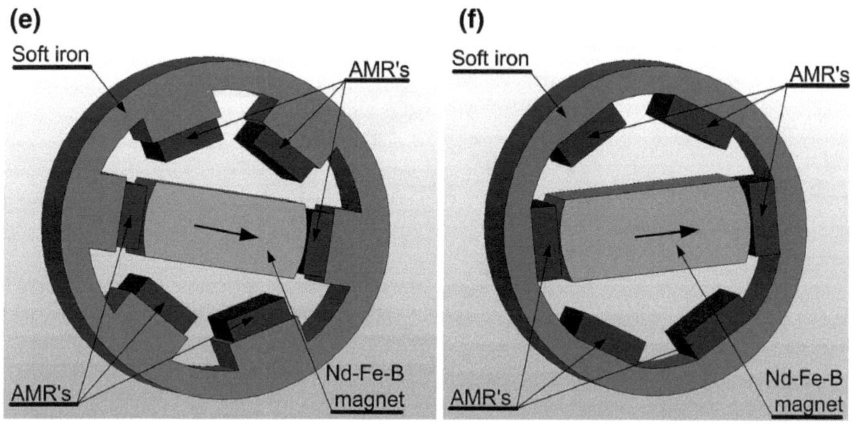

Fig. 3.41 The magnet assembly, **e** with the *six rectangular* poles and six AMRs and with the rotating bar magnet, **f** without six rectangular poles and with six AMRs attached to the smooth iron ring and with the rotating bar magnet (see also Bouchekara et al. [56])

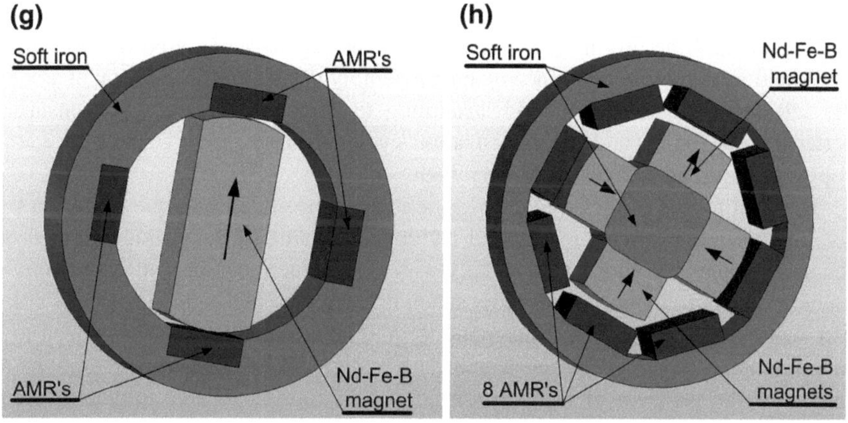

Fig. 3.42 **g** The magnet assembly without the rectangular poles and with four AMRs embodied into the smooth iron ring and with the rotating bar magnet, **h** the magnet assembly comprises four rotating magnet poles, three of which are use to concentrate the magnetic flux into the fourth one. It also comprises 8 AMR beds, which are attached to the smooth static iron ring. (see also Bouchekara et al. [56])

The magnet assembly (h) in Fig. 3.42 is different to all other 7 structures. Namely, it comprises four rotating magnet poles, three of which are used to concentrate the magnetic flux into the fourth one. It also comprised eight AMR beds, which are attached to the smooth static iron ring.

According to Bouchekara et al. [56], the reference magnet assembly (a) was also chosen because of its simple construction and it was optimized in order to provide the maximum magnetic flux density (the magnetic flux density in the centre of each

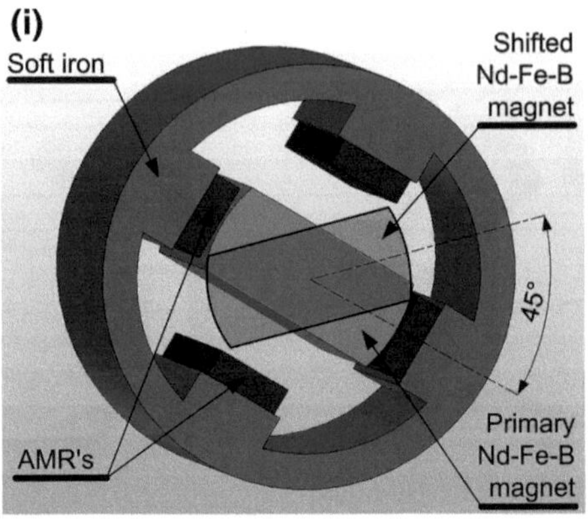

Fig. 3.43 The magnet assembly, **i** with two halves of the magnet system in which the rotor or the static part is shifted by 45° (only the shifted magnet is shown in the figure) (see also Bouchekara et al. [56])

AMR bed was varying from 0.05 to 0.94 T). The results, however, revealed that the reference magnet assembly is characterized by high mechanical stresses on both the AMR beds and the driving actuator [56]. The authors also reported that the magnetic field profile of the reference magnet assembly was the most trapezoidal among the selected cases, which should, according to Bouchekara et al. [56], ensure a better magnetization and demagnetization cycle. As a negative consequence, the cogging torque and forces were very high.

The total torque of the structure was defined as the cogging torque acting on the rotating bar magnet. This torque had the opposite value of the torque obtained on the stator (poles). It varied sinusoidally with the magnet position and its maximum value was about 20 Nm. The authors stated that the increase in the number of poles can significantly decrease the magnetic torque; however, this would also lead to a more complex device. It was also indicated by the authors that one of solutions to overcome the problem of cogging torque was to apply a passive torque-compensation system that represented the eight poles synchronous magnetic coupling, designed to produce the same torque as the reference magnets, but with the opposite phase. With this solution a reduction of 10 Nm was achieved.

Figure 3.43 shows another solution for the magnet assembly in which this consists of two partial magnet systems (as the cascade) and in which the rotor (magnet bar) or the static part are shifted by 45°. These solutions were evaluated in order to minimize the magnetic torque.

Based on numerical simulations, the following results were obtained and are discussed in the subsequent text (see also the results in Table 3.1):

- The solution represented by the magnet assembly (c) with the smooth iron ring led to a decrease of the total torque, and the reluctance variation was left due to the MCE magnetic beds, which acted as additional poles. The maximum magnetic flux density, compared to the reference magnets, was slightly lower.

Table 3.1 The magnetic flux density, the magnetic torque, and forces on magnet assemblies (see also Bouchekara et al. [56])

	Magnet assembly (a)	Magnet assembly (b)	Magnet assembly (c)
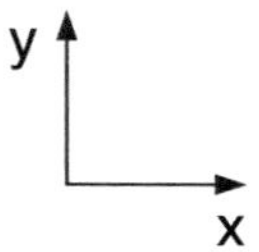			
B_{max} (T)	0.94	0.87	0.91
T_{max} (Nm)	19.6	15.6	12.6
$F_{x,\ max}$ (N)	78.7	73.5	64.5
$F_{y,\ max}$ (N)	93.5	159.6	107.5
	Magnet assembly (d)	Magnet assembly (e)	Magnet assembly (f)
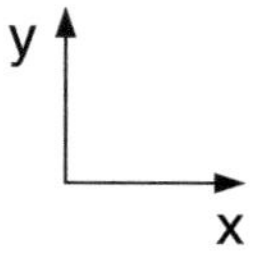			

(continued)

Table 3.1 (continued)

B_{max} (T)	0.7	0.9	0.9
T_{max} (Nm)	12.4	3.6	3.3
$F_{x,\ max}$ (N)	0	99.4	100.3
$F_{y,\ max}$ (N)	65.5	113.7	120
	Magnet assembly (g)	Magnet assembly (h)	Magnet assembly (i)
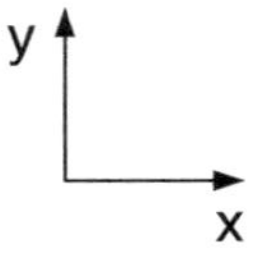	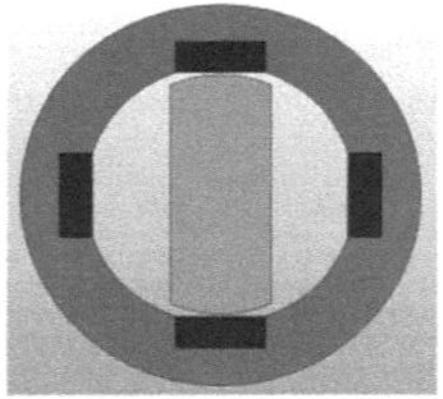	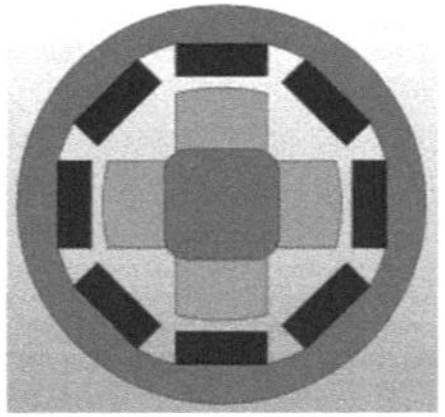	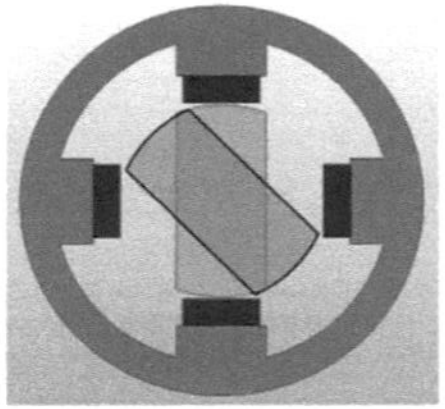
B_{max} (T)	0.72	0.75	0.91
T_{max} (Nm)	6	3	1.9
$F_{x,\ max}$ (N)	122	34.3	64.5
$F_{y,\ max}$ (N)	633	378.5	107.5

- Increasing the number of beds from 4 to 6, strongly decreased the magnetic torque. This was observed for the magnet assembly (e) versus magnet assembly (a), where the torque was reduced by a factor of 5. In the case when the magnet assemblies without the iron poles are compared, and with a different number of AMRs (i.e. magnet assembly (f) versus magnet assembly (c), the torque was reduced by a factor of 4. However, the magnetic flux density was also reduced in both cases. The distribution of the magnetic flux density was not trapezoidal any more.
- The embodied AMRs in the soft iron ring do not represent a good solution. The torque is reduced, and the magnetic flux density also substantially decreases. Moreover, the force along the magnet bar is substantially increased, compared to the reference magnet assembly.
- The results of the simulations showed that the soft magnetic parts attached to the AMRs in the magnet assembly (d) can significantly overtake the force applied on the AMR, by cancelling the force in the direction X (perpendicular to the magnet bar) and by reducing the force in the Y direction by 40 %. This solution, however, substantially decrease the magnetic flux density.
- In the magnet assembly, denoted by (h), the torque, compared to the reference magnet assembly (a), was substantially reduced. However, the magnetic flux density was decreased as well.
- The special cascade arrangement of magnet assembly (i) substantially decreased the magnetic torque compared to the reference magnet (a). Also, the magnetic flux density remained almost the same. According to Bouchekara et al. [56] such a system with several shifted blocks could be used for the higher cooling power or the cascade use of different magnetocaloric materials in order to increase the temperature span (similar to the layering of magnetocaloric materials—see also the chapter on AMRs).

As can be seen from solutions presented in this section, we can learn much about the design of rotating magnet structures by studying another domain, which regard permanent magnet motors (see also Hanselman [57], Gieras [58]). For instance, the cogging torque (or the reluctance torque) between the permanent magnets in a rotor and the slot openings in a stators occurs in such motors.

When each magnet in the rotor of such a motor rotates, the reluctance is experienced by magnets of the rotor passing the sloth opening between the stator teeth, which are elongated into so-called shoes (Fig. 3.44a).

The slot openings create a varying reluctance for the magnet flux, and therefore, the cogging torque. Without the "shoes" and only with stator teeth, the reluctance variance and the consequent cogging torque are much larger. Therefore, the shoes can drastically reduce the cogging torque. Also, the smaller the slot opening is, smaller will be the cogging torque, so without slot openings, the cogging torque should become zero. One should not misinterpret this situation with the magnet assembly (g) in the Fig. 3.42; however, in both cases, the magnetic flux density (induction) will be smaller, as will the torque.

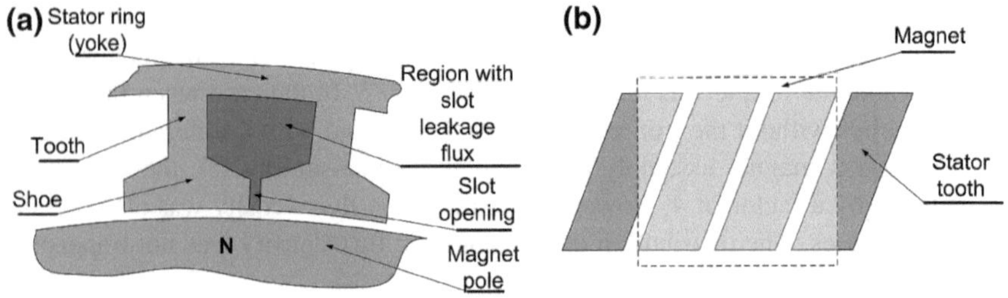

Fig. 3.44 *Left* Slot leakage flux, *Right* Skewed stator slots

Other measures in permanent-magnet motors in order to reduce the cogging torque are related to the radial dimensions of the shoe, the length of the teeth, the distance between the teeth, the reduction of the variation of the magnetic flux length from the magnet to the stator, and the relationship between the number of magnet poles and the number of stator slots. According to the last of these, if each magnet of the rotor appears in the same position relative to the stator slots (e.g. in a four-pole rotor with 12 slots in a stator), then the cogging torques of each of the magnets are in a phase with each other. Therefore, the total torque will represent the addition of each magnet.

If the motor, however, consists of, for instance, four poles and a stator with 15 slots, this will lead to different positions of the magnet poles relative to the slots. Then the cogging torques will not fully add, since they will be out of phase with each other.

The net change in the reluctance and, consequently, the cogging torque can also be reduced if the slots are skewed, as shown in Fig. 3.44b. In this case, each magnet experiences a net reluctance that stays almost unchanged when the slots are passing by. Therefore, changes in the axial direction are used to decrease the effect of changes along the circumferential dimension (see Hanselman [57]). Note that if the magnets are skewed, this effect will not have a role.

In magnet design one should also pay attention to the shape of the magnets, since this is also related to the cost of the production, as well as the cost related to the assembly of such magnets. Simple shapes are therefore of great interest. An interesting solution was presented by Bouchekara et al. [59]. In this particular case (see Fig. 3.45), simple bar-shaped magnets were applied. The authors performed a comprehensive analysis based on finite elements in order to establish which configuration of bar magnets (the magnetic field of 1 T) could provide the best solution.

With the rotation of the bar magnets, the ON and OFF field operation was provided to a pair of AMR beds (because of the need for a continuous operation of the device). However, as an advanced solution, two such pairs of AMRs have been considered. The dimensions of the two parts in Fig. 3.45 are identical. These should be assembled in such a way that a 180° angle shift should be provided to magnets with odd numbers in the second part, compared to the first part. Like this the total torque of a device could be decreased (the torque of the first part has a different sign to the torque in the second part) [59].

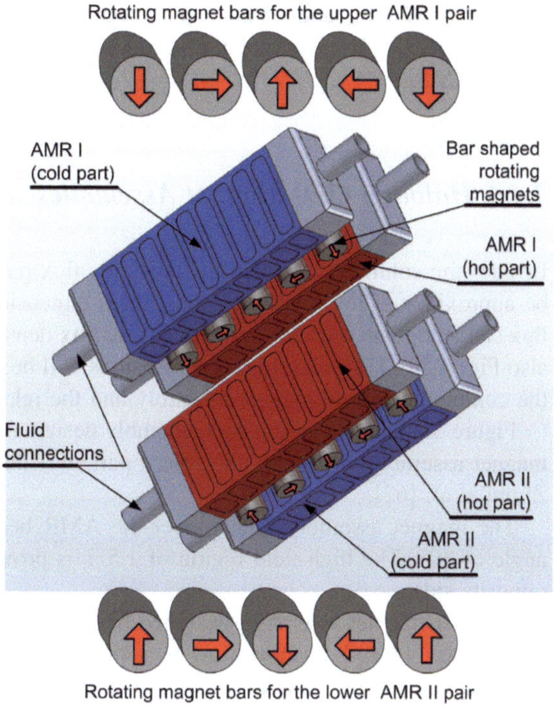

Fig. 3.45 An example of *bar-shaped* magnets providing an ON and OFF magnetic field to two pairs of AMR beds (see also Bouchekara et al. [59])

An example of a magnet assembly for a linear (reciprocate) device is shown in Fig. 3.46 [60]. This magnet assembly consists of four Nd–Fe–B magnets, which, together with the body of soft iron, provided 1.15 T in the gap. The magnet assembly is attached to a mechanical system, and therefore, it moves over the static AMR. A better solution would be to use a pair of AMRs in order to provide

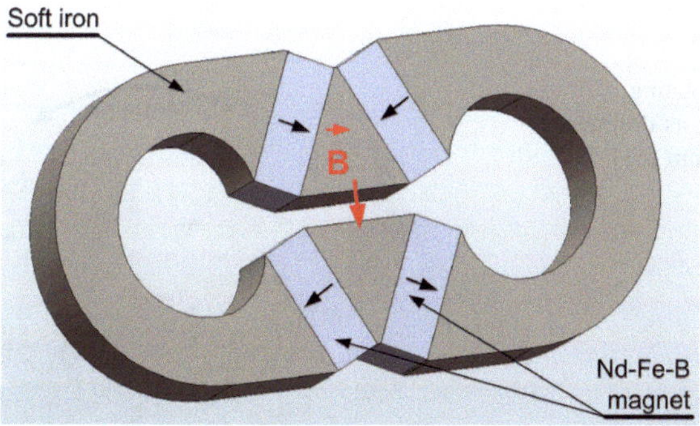

Fig. 3.46 The magnet assembly consisted of four permanent magnets with a soft iron guide for the magnetic flux. This magnet assembly was used for a linear experimental device by Tušek et al. [60]

continuous cooling and balance the magnetic forces; however, the device was constructed only for experimental purposes for the investigation and characterization of AMRs.

3.4.4 Halbach (3D) Magnet Assemblies

Besides the solutions in which the magnetic flux path in the magnet assembly can be approximated by two dimensions, three-dimensional guidance of the magnetic flux can lead to an increase in the magnetic flux density in the desired direction (see also Fig. 3.29). However, higher dimensions will be, in most cases, associated with the complexity of the magnet assembly and the related costs for such a structure.

Figure 3.47 shows the magnet assembly designed by Chell and Zimm [61]. This magnet assembly represents the rotary part of a device, which was presented by Zimm et al. [62].

The magnet assembly rotates over the AMR beds, which are separated by an angle of 30°. The high-field region of 1.5 T is provided at an angle of 60° on its opposite sides.

Another magnet assembly, designed by Kitanovski et al. [63] and presented by Egolf et al. [44, 64], is shown in Fig. 3.48. This design was made for a rotary ring with magnetocaloric material beds. Two high and two low regions of magnetic field were provided, each on the opposite side of the magnet assembly. In Fig. 3.48b, the cross-section of the magnet assembly shows only the high magnetic field regions. For these, the Ansys simulations showed a magnetic flux density of 2.1 T. The magnetic flux passes the magnetocaloric material in the radial direction. A soft iron body envelopes the magnet assembly, thus, providing guidance of the magnetic flux as well as magnetic shielding. It contains openings for the introduction of the fluid connection (shown in Fig. 3.48a).

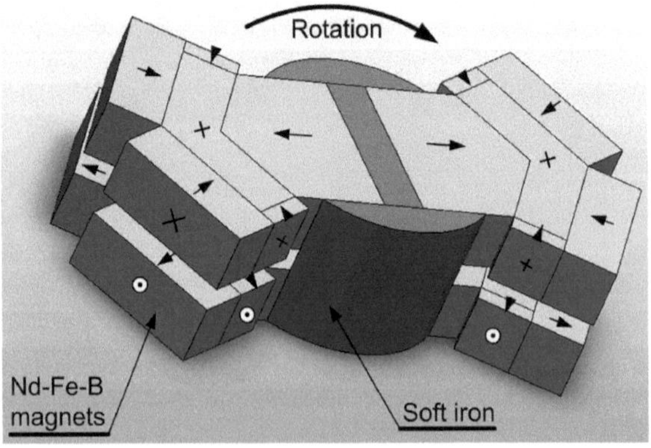

Fig. 3.47 The rotating magnet assembly as designed by Chell and Zimm [61]

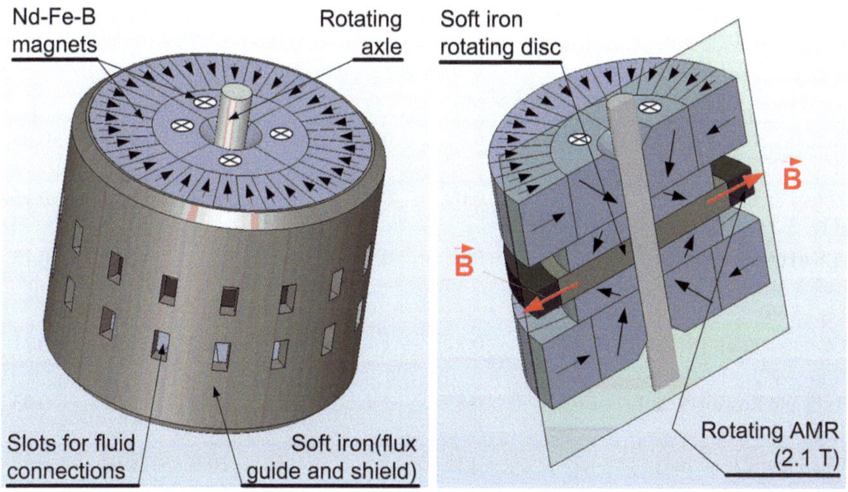

Fig. 3.48 A 3D Halbach array designed at the University of Applied Sciences of Western Switzerland [44, 63, 64]

3.5 Evaluation of Different Magnet Assemblies Designed or Constructed for Magnetic Refrigeration

The following section provides information about the figure of merit for the design of a permanent-magnet assembly. This can also be considered as the "efficiency" of the design of magnets for magnetic refrigeration. Most of the work that regards this issue was conducted for magnetic refrigeration at room temperature by Bjørk [65], Bjørk et al. [66] and Roudaut et al. [67].

Bjørk also proposed the most recognized figure of merit for a magnetic field source in magnetic refrigeration. Namely, he introduced the so-called Λ_{cool}, which was defined using the following relation [65, 66]:

$$\Lambda_{\text{cool}} = \left((\mu_0 H)^{\frac{2}{3}}_{\text{high}} - (\mu_0 H)^{\frac{2}{3}}_{\text{low}} \right) \frac{V_{\text{field}}}{V_{\text{magnets}}} P_{\text{field}} \tag{3.95}$$

In Eq. (3.95), the V_{field} represents the volume where a high field is generated at some moment. The volume as the sum of all the permanent magnets in the magnet assembly is represented by the V_{magnets}. The terms $(\mu_0 H)^{2/3}_{\text{high}}$ and $(\mu_0 H)^{2/3}_{\text{low}}$ represent the volume average of the applied magnetic field in the high-field region (magnetization) and in the low-field region (demagnetization), respectively. The P_{field} parameter represents the fraction of time when the magnet is in use.

This figure of merit can be applied especially for the design of devices that are based on the Brayton-like AMR cycle. Namely, in other thermodynamic cycles, the transition magnetic field (from high to low field) should be considered as well and should not be averaged (see the chapter on AMRs and different thermodynamic

Table 3.2 Characteristics of some designs of permanent-magnet assemblies in magnetic refrigeration (reproduced from Bjørk [65])

Name	$V_{magnets}$ (dm^3)	V_{field} (dm^3)	B_{high} (T)	B_{low} (T)	P_{field}	Λ_{cool}/P_{field}	Λ_{cool}
Bohigas et al. [40] (Fig. 3.29)	0.38	0.02	0.9	0[a]	1	0.05	0.05
Zheng et al. [38] (Fig. 3.29)	0.5	0.09	0.93	0[a]	0.9[a]	0.17	0.15
A.Šarlah (reported by Tušek et al. [41] Fig. 3.30)	0.11	0.65	0.97	0.1	1	0.13	0.13
Okamura et al. [54] (Fig. 3.36)	3.38	0.8	1	0	0.9[a]	0.24	0.21
Tura and Rowe [49, 50] (Fig. 3.34)	1.03	0.05	1.4	0.1	0.5	0.05	0.03
Chell and Zimm [61] (Fig. 3.46)	4.7	0.15	1.5	0.1[a]	0.9[a]	0.04	0.03

[a] estimation

cycles). Furthermore, the Λ_{cool} does not give information about torque issues, eddy currents and finally, the cost of the device, which strongly depends on the costs of the elements of the magnet assembly as well as the costs related to different shapes of permanent magnets. Moreover, the Λ_{cool} parameter strongly depends on the type of operation of the AMR, for which as could be seen, different operating regimes can be taken into account, which strongly depend on the AMR's effectiveness (which also defines the time required for the fluid flow) as well as the method or mechanism that is used for the motion of the magnet or the AMRs. Table 3.2 shows some of the results from the analysis that was performed by Bjørk et al. [66].

The most important way to define the effectiveness of the magnet assembly is simply through the economics of a device in terms of investment or total costs (investment, maintenance and operation costs) versus the cooling power of the device. Note that the permanent-magnet assembly in almost all cases represents the highest costs among all the elements of a device (see also the Chap. 9 on economics of magnetic refrigeration). On the other hand, a full economic analysis requires a full knowledge of the geometry, the operation and other characteristics of a device. Since there was no general standardisation method implemented, we are left with more simple figures of merit, such as the one from Bjørk [63]. This can serve for a first and very fast evaluation of the different magnet designs.

In the article of Bjørk et al. [66], the authors evaluated different designs and operation characteristics of permanent-magnet assemblies used in prototypes of magnetic refrigerators. Besides the Λ_{cool} parameter, the authors also defined the parameter Λ_{cool}/P_{field}, since this takes into account only the magnet design and not the ratio of the time period during which the magnet assembly is actively used. Detailed information about the calculation of particular parameters and some other designs can be found in Bjørk [65] and Bjørk et al. [66].

References

1. Ashcroft NW (1968) Metallic hydrogen: a high-temperature superconductor? Phys Rev Lett 21(26):1748
2. Scheike T, Böhlmann W, Esquinazi P et al (2012) Can doping graphite trigger room temperature superconductivity? Evidence for granular high-temperature superconductivity in water-treated graphite powder. Adv Mater 24(43):5826–5831
3. Avci S, Chmaissem O, Allred JM et al (2013) Magnetically driven suppression of nematic order in an iron-based superconductor. Nat Commun 5:3845
4. Shi X, Lin PV, Sasagawa T et al (2014) Two-stage magnetic-field-tuned superconductor-insulator transition in underdoped $La_{2-x}Sr_xCuO_4$. Nat Phys 10:437–443
5. Smolyaninov II, Smolyaninova VN (2013) Is there a metamaterial route to high temperature superconductivity?. eprint arXiv:1311.3277, p 17
6. Jiles DC (1998) Introduction to magnetism and magnetic materials, 2nd edn. CRC Press, Boca Raton, p 568
7. Fujimoto M (2007) Physics of classical electromagnetism. Springer, New York, p 325
8. Ray MW, Ruokokoski E, Kandel S et al (2014) Observation of Dirac monopoles in a synthetic magnetic field. Nature 505:657–660
9. Pietilä V, Möttönen M (2009) Creation of Dirac Monopoles in Spinor Bose-Einstein Condensates. Phys Rev Lett 103:030401
10. Feynman RP, Leighton RB, Sands M (2011) The Feynman lectures on physics, boxed set: the New Millennium Edition, vol II. Basic Books, New York
11. Cullity BD, Graham CD (2009) Introduction to magnetic materials. Institute of electrical and electronics engineers. Wiley, Hoboken, p 543
12. Rosensweig R (1985) Ferrohydrodynamics. Cambridge University Press, Cambridge
13. Coey JMD (2010) Magnetism and magnetic materials. Cambridge University Press, Cambridge, p 625
14. Campbell P (1996) Permanent magnet materials and their application. Cambridge University Press, Cambridge, p 220
15. Buschow KJH, de Boer FR (2004) Physics of magnetism and magnetic materials. Kluwer Academic, New York, p 182
16. Weickhmann M (2009) Nd–Fe–B magnets, properties and applications. VACUUM SCHMELZE GmbH, Hanau, p 10
17. Purcell EM, Morin DJ (2013) Electricity and magnetism, 3rd edn. Cambridge University Press, Cambridge, p 853
18. Onnes HK (1911) Further experiments with liquid helium. On the change of electric resistance of pure metals at very low temperatures etc. IV: The resistance of pure mercury at helium temperatures. Comm Phys Lab Univ Leiden 120b, Reprinted in Proc K Ned Akad Wet 13:1274
19. Meissner W, Ochsenfeld R (1933) Naturwissenschaften 21:787
20. Yuan W (2011) Second-generation high-temperature superconducting coils and their applications for energy storage. Doctoral thesis, University of Cambridge, Springer, p 145
21. Fossheim K, Sudbř A (2004) Superconductivity, physics and applications. Wiley, London, p 427
22. Dougherty R, Kimel JD (2013) Superconductivity revisited. Taylor & Francis Group, Boca Raton, p 172
23. Jérome D, Pasquier CR (2005) One dimensional organic superconductors. In: Narlikar AV (ed) Frontiers in superconducting materials. Springer, New York, pp 183–230
24. Muranaka T, Zenitani Y, Shimoyama J et al (2005) Superconductivity in MgB_2. In: Narlikar AV (ed) Frontiers in superconducting materials. Springer, New York, p 937–981
25. Buckel W, Kleiner R (2013) Supraleitung: Grundlagen Und Anwendungen. Wiley, Auflage, p 7

26. Hott R, Kleiner R, Wolf T et al (2014) Review on superconducting materials. In: Seidel P (ed) Chapter in applied superconductivity, handbook on devices and applications encyclopedia of applied physics. Wiley, Weinheim
27. Hott R, Kleiner R, Wolf T et al (2005) Superconducting materials–a topical overview. In: Narlikar AV (ed) Frontiers in superconducting materials. Springer, Berlin
28. Iwasa Y (2009) Case studies in superconducting magnets, design and operational issues, 2nd edn. Springer, Berlin, p 682
29. Radebaugh R (2009) Cryocoolers: the state of the art and recent developments. J Phys Condens Matt 21(16):164219
30. Flynn T (2004) Cryogenic engineering, 2nd edn. CRC Press, Boca Raton, p 912
31. Wang Y (2013) Fundamental elements of applied superconductivity in electrical engineering. Wiley, New York, p 444
32. Radebaugh R (2000) Development of the pulse tube refrigerator as an efficient and reliable cryocooler. Proc Inst Refrig 96:11–29
33. Lee SJ, Kenkel JM, Pecharsky VK et al (2002) Permanent magnet array for the magnetic refrigerator. J Appl Phys 91(10):8894–8896
34. Zimm C, Boeder A, Chell J et al (2005) Design and performance of a permanent magnet rotary refrigerator. In: Proceedings of the first international conference on magnetic refrigeration at room temperature, Montreux, Switzerland, 2005, pp 367–373
35. Zimm C, Boeder A, Chell J et al (2006) Design and performance of a permanent-magnet rotary refrigerator. Int J Refrig 29(8):1302–1306
36. Halbach K (1980) Design of permanent multipole magnets with oriented rare earth cobalt material. Nucl Instrum Methods 169(1):1–10
37. Kumada M, Iwashita Y, Aoki M et al (2003) The strongest permanent dipole magnet. In: Proceedings of the 2003 particle accelerator conference, pp 1993–1995
38. Zheng ZG, Yu HY, Zhong XC et al (2009) Design and performance study of the active magnetic refrigerator for room-temperature application. Int J Refrig 32(1):78–86
39. Zimm C, Sternberg A, Jastrab AG et al (2002) Rotating bed magnetic refrigeration apparatus. US 20020053209
40. Bohigas X, Molins E, Roig A et al (2000) Room-temperature magnetic refrigerator using permanent magnets. IEEE Trans Magn 36(3):538–544
41. Tušek J, Zupan S, Šarlah A et al (2010) Development of a rotary magnetic refrigerator. Int J Refrig 33:294–300
42. Lee SJ, Kenkel JM, Jiles DC (2002) Design of permanent-magnet field source for rotary-magnetic refrigeration systems. IEEE Trans Magn 38(5):2991–2993
43. Vasile C, Muller C (2006) Innovative design of a magnetocaloric system. Int J Refrig 29 (8):1318–1326
44. Egolf PW, Kitanovski A, Vuarnoz D et al (2008) Magnetische Wärmepumpe mit Erdwaerme-Quelle, BFE Project Number 100873, Annual Report, p 12
45. Vuarnoz D, Kitanovski A, Gonin C et al (2009) Modeling of a two-stage magnetic refrigerator with wavy-structure Gadolinium heat exchangers. In: Third IIR international conference on magnetic refrigeration at room temperature, Des Moins, Iowa, 11–15 May, pp 311–319
46. Egolf PW, Kitanovski A, Gonin C (2009) Central magnetic cooling and refrigeration machines (Chiller) and their assessment. Annual report, BFE Project Number 153691, p 13
47. Bjørk R, Bahl CRH, Smith A et al (2010) An optimized magnet for magnetic refrigeration. J Magn Magn Mater 322:3324–3328
48. Bjørk R, Smith A, Bahl CRH (2010) Analysis of the magnetic field, force, and torque for two-dimensional Halbach cylinders. J Magn Magn Mater 322:133–141
49. Tura A, Rowe A (2007) Design and testing of a permanent magnet magnetic refrigerator. In: Proceedings of the 2nd international conference of magnetic refrigeration at room temperature, Portoroz, Slovenia, pp 363–370
50. Tura A, Rowe A (2011) Permanent magnet magnetic refrigerator design and experimental characterization. Int J Refrig 34(3):628–639

51. Bjørk R, Bahl CRH, Smith A et al (2010) Comparison of adjustable permanent magnetic field sources. J Magn Magn Mater 322:3664–3671
52. Bouchekara HREH, Nahas M (2012) Magnetic refrigeration technology at room temperature. In: Barsan V (ed), Trends in electromagnetism–from fundamentals to applications. InTech, Austria, pp 225–250
53. Arnold DS, Tura A, Ruebsaat-Trott A et al (2014) Design improvements of a permanent magnet active magnetic refrigerator. Int J Refrig 37:99–105
54. Okamura T, Rachi R, Hirano N et al (2007) Improvement of 100 W class room temperature magnetic refrigerator. In: Proceedings of the 2nd international conference of magnetic refrigeration at room temperature, Portoroz, Slovenia, pp 377–382
55. Okamura T, Yamada K, Hirano N et al (2006) Performance of a room-temperature rotary magnetic refrigerator. Int J Refrig 29:1327–1331
56. Bouchekara HREH, Kedous-Lebouc A, Yonnet JP et al (2011) Electromagnetic design of a magnetic field source for a magnetocaloric refrigerator. Prog Electromagnet Res M 19:251–263
57. Hanselman D (2006) Brushless permanent magnet motor design, 2nd edn. Magna Physics Publishing, Hillsboro, p 392
58. Gieras JF (2009) Permanent magnet motor technology: design and applications. Electrical and computer engineering, 3rd edn. CRC Press, London, p 608
59. Bouchekara HREH, Kedous-Lebouc A, Yonnet JP (2012) Design of a new magnetic refrigeration field source running with rotating bar-shaped magnets. Int J Refrig 35:115–121
60. Tušek J, Kitanovski A, Zupan S et al (2013) A comprehensive experimental analysis of gadolinium active magnetic regenerators. Appl Therm Eng 53:57–66
61. Chell J, Zimm CB (2006) Permanent magnet assembly. US patent 7.148.777
62. Zimm C, Auringer J, Boeder A et al (2007) Design and initial performance of a magnetic refrigerator with a rotating permanent magnet. In: Proceedings of the 2nd International conference of magnetic refrigeration at room temperature, Portoroz, Slovenia, pp 341–347
63. Kitanovski A, Vuarnoz D, Egolf PW (2009) A magnetocaloric device, especially a magnetic refrigerator, a heat pump or a power generator. European Patent Application, EP 2.108.904, p 54
64. Egolf PW, Gravier L, Croci M (2012–2013) Nanotechnology in new magnetic heating, refrigeration and energy conversion devices. In: Industria & informazione, international special issue 2012–2013, Refigeration and air conditioning, pp 54–61
65. Bjørk R (2010) Designing a magnet for magnetic refrigeration. DTU Risoe, Phd thesis, 2010, p 343
66. Bjørk R, Bahl CRH, Smith A et al (2010) Review and comparison of magnet designs for magnetic refrigeration. Int J Refrig 33:437–448
67. Roudaut J, Yonnet JP, Kedous-Lebouc A et al (2010) Permanent magnet sources for magnetic refrigeration. In: REPM'10-Proceedings of the 21st workshop on rare-earth permanent magnets and their applications, pp 86–89

Chapter 4
Active Magnetic Regeneration

It is well known that the magnetocaloric effect of most magnetocaloric materials at moderate magnetic fields (up to 1.5 T) is limited to a maximum adiabatic temperature change of 5 K [1, 2]. This value is not sufficient for such materials to be directly implemented into a practical cooling or heating device where temperature span over 30 K is required. Therefore, in order to increase the temperature span, one and so far the best option is for a heat regenerator to be included in the magnetic thermodynamic cycle.

A heat regenerator is a type of indirect heat exchanger where the heat is periodically stored and transferred from/to a thermal storage medium (regenerative material) by a working (heat-transfer) fluid. The regenerative material usually has a porous structure, through which a working fluid is pumped in an oscillatory, counter-flow manner (which is more efficient than a parallel flow system). During the 'hot period', a warmer fluid flows through the regenerative material, which cools down, while the material heats up. As a result, heat is stored in the material. During the 'cold period', a cooler fluid flows through previously heated regenerative material, so the fluid heats up, while the material cools down. The heat is therefore transferred back to the fluid (the same fluid or a different one) at a different phase of the thermodynamic cycle. After a certain number of such steps, a periodic steady state is reached and, as a result, a temperature profile can be established along the length of the regenerator [3].

The need to apply heat regenerators in a magnetic refrigerator was already realized by Brown [4] in the first prototype of a magnetic refrigerator, built in 1976. He applied a regenerative Stirling-like thermodynamic cycle (very similar to AMR), which significantly increased the temperature span of the device [4, 5]. A few years later Steyert [6] and Barclay and Steyert [7] presented and explained the active magnetic regenerator, which remains the most applied method for the exploitation of the magnetocaloric effect at room temperature. Furthermore, all prototypes of magnetic refrigerators built since then have been based on the AMR process [5]. An AMR, unlike a passive (regular) heat regenerator, contains a magnetocaloric material as the regenerative material. It has a double function in a magnetic regenerator, i.e. it works as a regenerator and enables an increase in the temperature span as well as working as a coolant and generating/absorbing heat between the particular phases of the

A. Kitanovski et al., *Magnetocaloric Energy Conversion*,
Green Energy and Technology, DOI 10.1007/978-3-319-08741-2_4

thermodynamic cycle (hence the name active). It is therefore more compact than a system with a separated passive regenerator as well as being more efficient. The latter is mostly due to the smaller heat transfer losses, which in the case of a passive regenerator occur during the heat transfer between the regenerative and the magnetocaloric materials (directly or through the external fluid flow) [8]. As discussed later in this chapter, the AMR performs its own unique thermodynamic cycle, while the passive regenerator performs only a particular process of the thermodynamic cycle. Note the passive regenerative material is not a thermodynamically working substance like the magnetocaloric material is in the AMR.

Figure 4.1 shows schematics of four different AMRs in different geometrical forms (perforated-plates AMR (a); parallel-plate AMR (b); wires-like AMR (c) and packed-bed AMR (d)). The parallel-plate and packed-bed AMRs are the most widely applied to date [5].

In this chapter, the principle of the AMR operation is explained and discussed. Different thermodynamic cycles with the AMR and its characteristics are shown and the layered-bed AMR principle is introduced. Furthermore, numerical modelling of the AMR is described. Theoretical (numerical) as well as some experimental results of the AMR operation are shown. These are based on simulations and optimization of different operating conditions (utilization factor and operating frequency), different geometries of packed-bed and parallel-plate AMRs, different AMR thermodynamic cycles and application of different heat-transfer fluids. At the

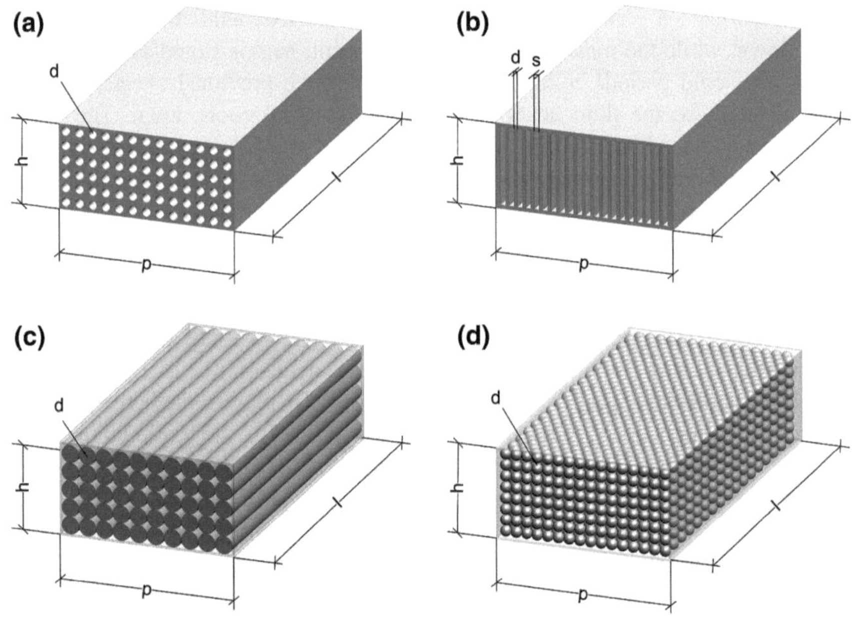

Fig. 4.1 Schematic example of four different AMR geometries

end of the chapter, a brief review on different processing techniques of AMRs is presented and discussed.

4.1 Operation of an Active Magnetic Regenerator (Different Thermodynamic Cycles with an AMR)

The operation of an AMR is generally based on four operational steps, as shown in Fig. 4.2. It should be noted that these four steps can also overlap and can be performed simultaneously, which can lead to a number of different thermodynamic cycles [9, 10].

However, the most basic, and by far the most widely applied, thermodynamic cycle of the AMR is the Brayton-like cycle, which is based on the following four operational phases (see Fig. 4.2):

- Magnetization—each part of the magnetocaloric material along the AMR is heated up due to the magnetocaloric effect (Fig. 4.2a)
- Fluid flow from the 'cold side' (cold heat exchanger—CHEX) through the heated magnetocaloric material to the 'hot side' (hot heat exchanger—HHEX) in the magnetic field. During this process of heat transfer, the fluid heats up (while the material cools down) and in the HHEX transfers the heat to the surroundings (Fig. 4.2b)
- Demagnetization—each part of the magnetocaloric material along the AMR is cooled down due to the magnetocaloric effect (Fig. 4.2c)

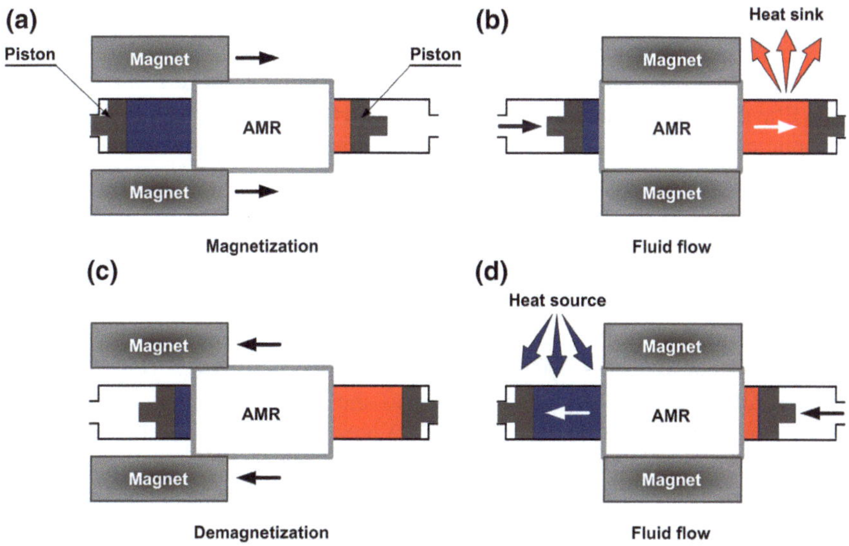

Fig. 4.2 Schematics showing the four basic operational phases of the AMR process for the Brayton-like AMR cycle in a linear (reciprocating) device

- Fluid flow from the 'hot side' (HHEX) through the heated magnetocaloric material to the 'cold side' (CHEX) in the absence of the magnetic field (the same fluid as in first phase (Fig. 4.2a), but in the counter-flow direction and therefore displacing it). During this process of heat transfer, the fluid cools down (while the material heats up) and in the CHEX it absorbs heat from the surroundings (Fig. 4.2d).

We can, in general, distinguish between rotary and linear (reciprocating) AMRs that further result in rotary and linear magnetic refrigeration devices, respectively (see Chaps. 7 and 8). The difference (with regard to AMR) is only in the kinetics of the movement between the AMR and the magnetic field source. The operational phases and the performed thermodynamic cycles of a particular AMR can be treated equally for both types.

It should be noted that the term Brayton-like (and later also Ericsson-like and Carnot-like) is used since the AMR cycle differs from the conventional Brayton or even the Brayton regenerative cycle, as explained later in this chapter. Furthermore, the conventional Brayton magnetic thermodynamic cycle is based on an adiabatic magnetization and demagnetization. However, in a practical device, regardless of the magnetic field source, it is rather difficult to ensure an instantaneous increase and decrease of the magnetic field. Since there is an unavoidable heat transfer between the magnetocaloric material and the working fluid (although static) situated in the AMR during the time needed to increase or decrease the magnetic field from the minimum to the maximum value, or vice versa, we can only talk about quasi-adiabatic (de) magnetization. One should note that a quasi-adiabatic (de)magnetization process should be as fast as possible and much faster than that of the convective heat transfer between the magnetocaloric material and the working fluid during the fluid flow period in order to reduce the heat transfer losses (see Sect. 4.4).

An AMR can also be applied for other thermodynamic cycles, for example the Ericsson-like, Carnot-like, Stirling-like or some hybrid cycles (e.g. a combination of a Brayton-like and an Ericsson-like cycle). However, different AMR thermodynamic cycles will be explained later in this chapter, while the basic operation of the AMR, establishing the temperature profile along its length and its unique thermodynamic cycle will be explained in detail for the most widely applied Brayton-like AMR cycle (see also Sect. 1.3).

Figure 4.3 shows a Brayton-like AMR cycle, schematically represented in a *T–s* diagram for periodic steady-state conditions (Fig. 4.3a, b) and the required magnetic field profile and fluid flow regime (Fig. 4.3d), together with an example of a magnet assembly that is suitable for a Brayton-like AMR cycle (Fig. 4.3c).

Generally, the AMR cycle is similar to the cascade system, where several thermodynamic cycles (connected in series) are used to increase the temperature span. As explained in Hall et al. [11], the major difference between an AMR and a cascade cycle is in the fact that the AMR cycle does not pump heat directly between the next-neighbour particles, but all the particles accept or reject heat to the heat-transfer fluid at the same time and are coupled indirectly through the working fluid. So, in the AMR cycle, there is no overlapping of the internal cycles at the same

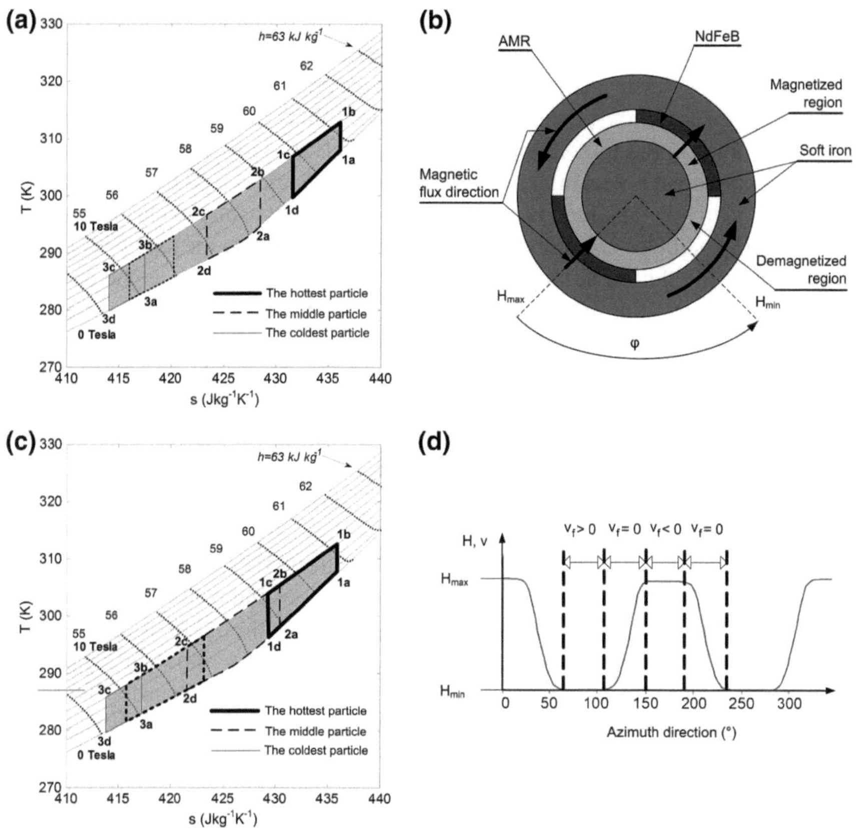

Fig. 4.3 The Brayton-like AMR cycle and its characteristics

time, as is the case in cascade systems. However, their thermodynamic paths do overlap, as can be seen in Fig. 4.3a, b, but for different time sequences. Due to the oscillating (counter) fluid flow, the internal (local) cycles interact with the neighbouring ones through the heat-transfer fluid and thus perform the regenerative process, which enables an increase of the temperature span along the AMR [12].

As each infinitesimally small particle of magnetocaloric material along the AMR performs its own thermodynamic cycle, it is difficult to show clearly the entire AMR thermodynamic cycle. For the purpose of the presentation (see Fig. 4.3a, b), we have chosen the hottest part of the AMR, the middle part, the coldest part and a part that is in between the coldest and the middle parts of the AMR.

Each of the selected parts performs the thermodynamic cycles as follows: (a–b) adiabatic magnetization (with no fluid flow); (b–c) fluid flow from CHEX to HHEX (isofield cooling of the magnetocaloric material); (c–d) adiabatic demagnetization (with no fluid flow); and (d–a) fluid flow from HHEX to CHEX (isofield heating of the magnetocaloric material). The entire AMR performs the thermodynamic cycle denoted by (1a–1b–3c–3d).

The difference between Fig. 4.3a, b corresponds to different utilization factors of the fluid flow. The utilization factor, which gives us information about the displaced fluid through the AMR and gives us the ratio between the heat capacity of the fluid and the magnetocaloric material during a single fluid flow period, is defined in Eq. (4.26) later in the text. Figure 4.3a corresponds to a low utilization factor and Fig. 4.3b corresponds to a high utilization factor. As can be seen from both figures, there is a relatively smaller overlapping of the thermodynamic cycles between the neighbouring particles at a low utilization factor, compared to the case with a higher utilization of the fluid. The impact of operational conditions such as the utilization factor and the operating frequency on the AMR's performance is discussed further in Sect. 4.4.

In Fig. 4.3c, d, magnet assembly is shown together with the corresponding variation of the magnetic field and the fluid flow profile through the AMR. The magnet assembly is an example of a rotary type of magnetic refrigerator. This kind of magnet assembly (see also [13–15]) provides the magnetic flux variation, which can be applied for the Brayton, Ericsson or Hybrid (Brayton–Ericsson) thermodynamic cycles.

In order to understand in detail the operation of the AMR, Figs. 4.4 and 4.5 show how the temperature span is established between the hot and cold sides of the AMR from the initial state with constant temperatures along the AMR to the periodic, steady state (for the Brayton-like AMR cycle). Figure 4.6 shows a temperature profile of the magnetocaloric material (black line) and the heat-transfer fluid (grey line) along the AMR during all four operational phases in the periodic steady state.

Figure 4.4 shows the initial temperature response of the magnetocaloric material at the hot and cold sides of the AMR based on the magnetic field and the fluid flow profile shown in Fig. 4.3d. It should be noted that in this case, the hot side of the AMR is assumed to be maintained at a constant (i.e. room) temperature.

It is evident from Fig. 4.4 that after just two complete cycles, there is a slight temperature difference between the magnetocaloric material at the hot and cold sides.

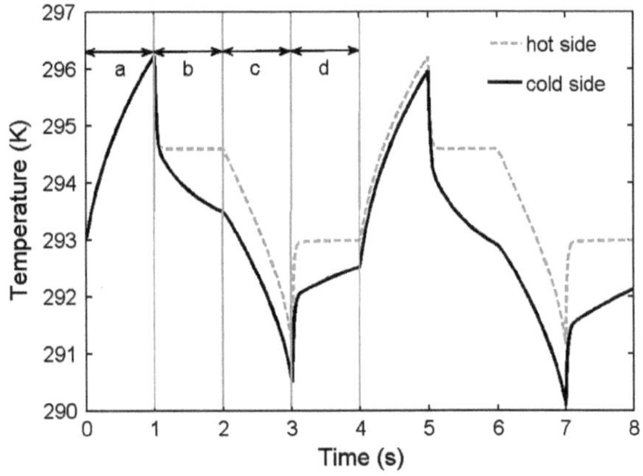

Fig. 4.4 The initial temperature response of the magnetocaloric material at the hot and cold sides of the AMR, *a* Magnetization, *b* Fluid flow, *c* Demagnetization, *d* Fluid flow

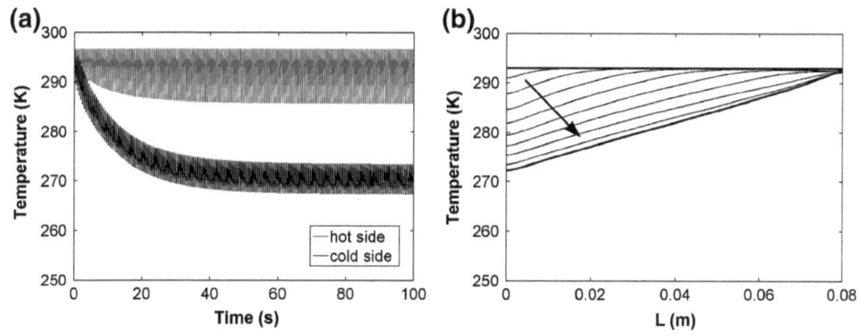

Fig. 4.5 **a** The establishment of the temperature profile between the magnetocaloric material at the hot and cold sides of the AMR, **b** The establishment of the temperature profile along the AMR from the initial state to the periodic steady state

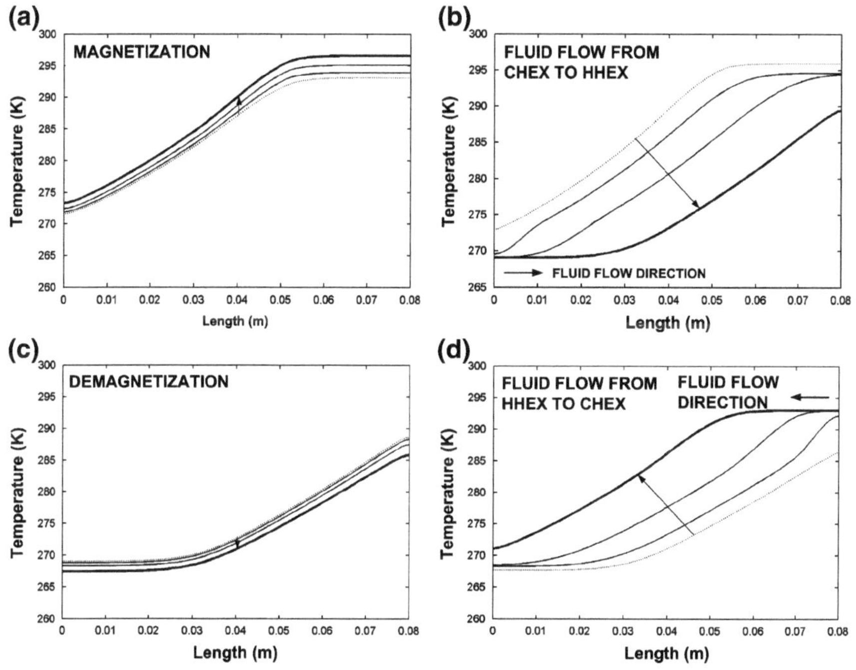

Fig. 4.6 Temperature profile along the AMR during all four operational phases (for a Brayton-like AMR cycle)

After a certain number of cycles (the number strongly depends on the operating conditions), a periodic steady state is reached. This happens when the temperature of the AMR between two successive cycles does not change (in this case decrease) anymore. The entire response of the magnetocaloric material at the hot and cold sides of the AMR, from the initial temperatures to the periodic steady state, is shown in Fig. 4.5a, while Fig. 4.5b shows the temperature profile along the AMR during the establishment of the temperature span from the initial state to the point where the periodic steady state is reached.

The dotted lines in Fig. 4.6 show the temperature profile at the beginning of a particular phase, while the full lines show the profile at its end. It should be noted that these phases are the same as schematically shown in Fig. 4.2. Figures 4.4, 4.5 and 4.6 are reproduced using a numerical model of the AMR (with Gd as the magnetocaloric material), which is presented in detail in Tušek et al. [16, 17].

4.1.1 Characteristics of an Ericsson-like AMR Cycle

An AMR can also be applied with some other thermodynamic cycles. This is achieved by using different combinations of the (de)magnetization and fluid flow process simultaneously.

The Ericsson-like AMR cycle is, unlike the Brayton-like AMR cycle, based on an isothermal (de)magnetization rather than adiabatic. In Fig. 4.7 the Ericsson-like AMR cycle is schematically presented in a *T*–*s* diagram in the periodic steady state (Fig. 4.7a), together with the required magnetic field profile and fluid flow regime (Fig. 4.7b). In this case, each part of the magnetocaloric material along the AMR performs its own small Ericsson thermodynamic cycle. Even though the distribution of the magnetic field can be similar to the case of the Brayton-like AMR cycle, the Ericsson-like AMR cycle, in order to perform the isothermal (de)magnetization,

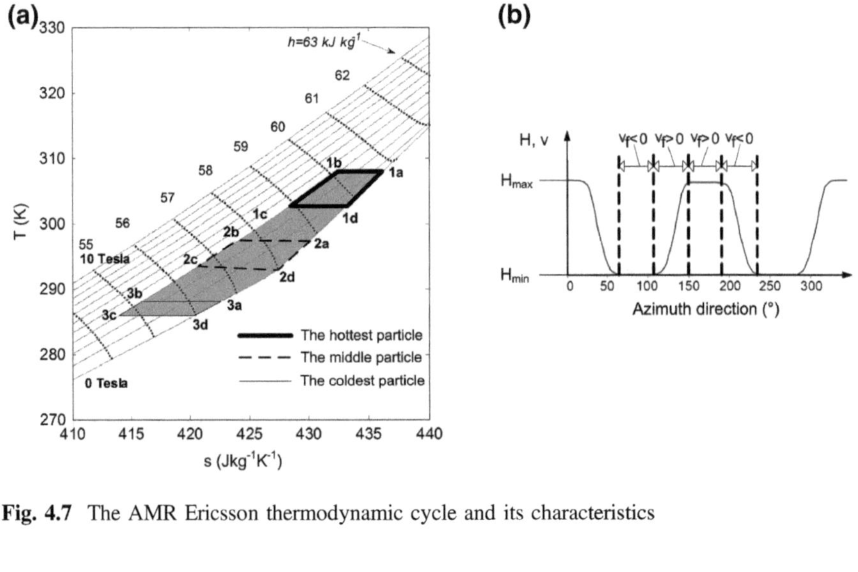

Fig. 4.7 The AMR Ericsson thermodynamic cycle and its characteristics

also requires a simultaneous fluid flow during the (de)magnetization process. This means that the fluid flows during all four processes, i.e. isothermal magnetization (a–b), isofield fluid flow at a higher magnetic field (b–c), isothermal demagnetization (c–d) and isofield fluid flow at a low magnetic field (d–a), see Fig. 4.7. It should be noted that during the demagnetization and isofield fluid flow, the fluid flow is in the counter-flow direction ($v_f < 0$) compared to the magnetization and isofield fluid flow ($v_f > 0$), as is also shown in Fig. 4.7b. A similar idea of combining the (de)magnetization and the fluid flow process, despite not being exactly related to the study of thermodynamic cycles, has been presented and analysed by Bjørk and Engelbrecht [18].

In order to ensure isothermal (de)magnetization (or at least to be close to it), the mass flow rate of the fluid flow during that time must be appropriate for the magnetocaloric effect. However, since the magnetocaloric effect along the AMR and its properties (mostly the specific heat) are not constant, but rather strongly temperature dependent, it is practically impossible to ensure a truly isothermal (de) magnetization process with a spatially constant mass flow rate (which is unavoidable in real devices) along the entire length of the AMR. Therefore, we can only talk about quasi-isothermal (de)magnetization. In order to ensure the best possible conditions for the isothermal (de)magnetization, the required fluid flow during the (de)magnetization process can differ from the optimum fluid flow during the isofield process.

4.1.2 Characteristics of a Hybrid Brayton–Ericsson-like AMR Cycle

By combining the best features of the Brayton-like and Ericsson-like AMR cycles, we are led to the development of a Hybrid Brayton–Ericsson-like AMR cycle (Fig. 4.8) [9, 10]. In this particular case, the thermodynamic cycle consists of six processes, as follows: adiabatic magnetization (a–b), isothermal magnetization (b–c), isofield fluid flow at a high magnetic field (c–d), adiabatic demagnetization (d–e), isothermal demagnetization (e–f), and isofield fluid flow at a low magnetic field (f–a). The Hybrid Brayton–Ericsson-like AMR cycle is, in the periodic steady state, schematically represented in the *T–s* diagram in Fig. 4.8a, together with the required magnetic field profile and fluid flow regime (Fig. 4.8b).

In the case of such a cycle, the (de)magnetization process is divided into the adiabatic and the isothermal process (like in the case of the Carnot-like AMR cycle, as shown later). The AMR is first magnetized to certain but not the maximum, magnetic field in a quasi-adiabatic process (without fluid flow). A subsequent process of magnetization is then performed quasi-isothermally with a simultaneous fluid flow. The process of demagnetization is analogous (first adiabatic and then isothermal with the counter-flow direction of the fluid flow). The ratio of the adiabatic and isothermal (de)magnetization should be optimized for each particular

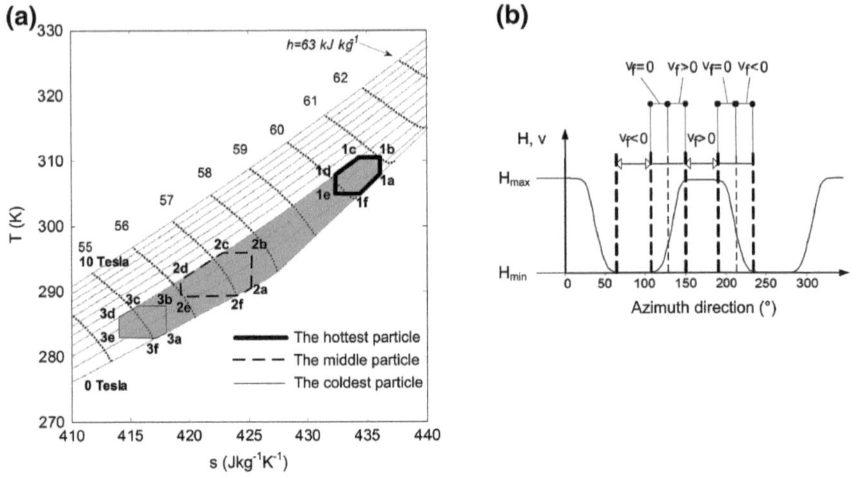

Fig. 4.8 The AMR hybrid Brayton–Ericsson-like AMR cycle and its characteristics

system, separately; however, better performances are to be expected if the AMR is magnetized adiabatically, at least to the middle value of the magnetic field or more [18, 19].

The combination of the (de)magnetization and the fluid flow (heat transfer) process is especially interesting in the case where the magnetic field is generated by permanent magnets. There is always a volume with a gradient of the magnetic flux density close to the magnetized air gap, which is due to the unavoidable leakage of the magnetic flux out of the desired magnetized volume in such systems. It is therefore not possible to ensure a steep Dirac function of the magnetic field variation, as would be ideal for the Brayton thermodynamic cycle. This makes the Ericsson-like and the Hybrid Brayton–Ericsson-like AMR cycles very interesting solutions where permanent magnets are used to magnetize the AMR (which is the case for all prototypes of magnetic refrigerators built in the past 10 years).

4.1.3 Characteristics of a Carnot-like AMR Cycle

In the case that the AMR cycle performs adiabatic magnetization with no fluid flow (a–b), followed by isothermal magnetization with a fluid flow (b–c), adiabatic demagnetization with no fluid flow (c–d) and finally isothermal demagnetization with a fluid flow (d–a), this will lead to a Carnot-like AMR cycle. It is schematically shown in Fig. 4.9a in the periodic steady state, together with the required magnetic field profile and the fluid flow regime (Fig. 4.9c).

Again, the fluid flow during the demagnetization is in the counter-flow direction ($v_f < 0$) compared to the magnetization process ($v_f > 0$), as noted in Fig. 4.9c.

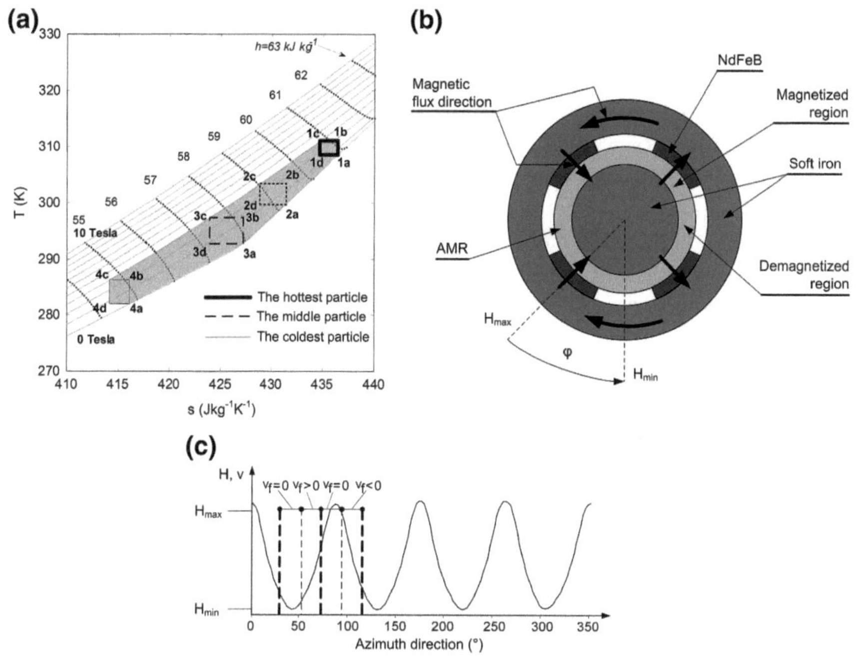

Fig. 4.9 The AMR Carnot thermodynamic cycle and its characteristics

It should be pointed out that the Carnot-like AMR cycle looking at the overall surface in a *T–s* diagram (the entire AMR) does not look like a Carnot thermodynamic cycle, but similar to a Hybrid Brayton–Ericsson-like AMR cycle (see Fig. 4.8a). We should therefore not expect such a cycle to be, a priori, the most efficient. The well-known Carnot thermodynamic cycle from classic thermodynamics is not even regenerative. However, if each infinitesimally small particle of magnetocaloric material is treated separately, it can be seen from Fig. 4.9a that each of them performs its own local Carnot thermodynamic cycle. It is based on adiabatic and isothermal (de)magnetization, which for the same reason as described above cannot be ideal, but rather quasi-adiabatic and quasi-isothermal.

Figure 4.9b represents a simple schematic of a magnet assembly, which could be applied to provide the magnetic field distribution that is required for the operation of the Carnot-like AMR cycle.

It is evident that this kind of AMR cycle does not require a homogeneous magnetic field, but utilizes only its increase and decrease. Such a magnetic field can be generated with a smaller input energy or a smaller mass of permanent magnets compared to the homogenous magnetic field required for the Brayton-like and the Ericsson-like AMR cycles. This is one of the advantages of the Carnot-like AMR cycle.

Note the application of thermodynamic diagrams, for instance a *T–s* diagram, is not sufficient to theoretically define the performance of a particular AMR cycle. The heat transfer losses to the working fluid, the heat gains and losses to the surroundings, the heat generation (friction of the fluid, eddy currents, valve friction, pump losses) and the viscous losses have to be taken into account as well. Additionally, one should estimate the losses related to valve systems, fittings and the pump's operation, as well as the efficiency of the heat source heat exchanger (CHEX) and the heat sink heat exchanger (HHEX). These losses can be included in the dynamic numerical model of the AMR (see Sect. 4.3). However, a theoretical investigation with thermodynamic diagrams (steady-state models) can give us basic information, which is still very important for the understanding and the design of a potential prototype device. With the optimization of thermodynamic cycles, we can substantially improve the efficiency or the cooling power of a potential device, the compactness and the corresponding cost. A direct numerical and experimental comparison of the different, above presented, AMR thermodynamic cycles is shown in Sect. 4.5.

4.1.4 Maximum Specific Cooling Power in the AMR Cycle

It is important to know the maximum cooling power of a particular magnetocaloric material. If regeneration is performed, the cooling power corresponds only to a part of the whole cooling energy, which is available within a particular thermodynamic cycle. Namely, a large part of this power is transferred in the processes of regeneration, and therefore it does not contribute to the cooling power. Figure 4.10 shows an example of a maximum temperature span that is restricted by the adiabatic temperature changes of the magnetization and demagnetization process without regeneration. Usually, this is not sufficient for a real application. Therefore, regeneration should be applied. Figure 4.10 also shows the maximum specific

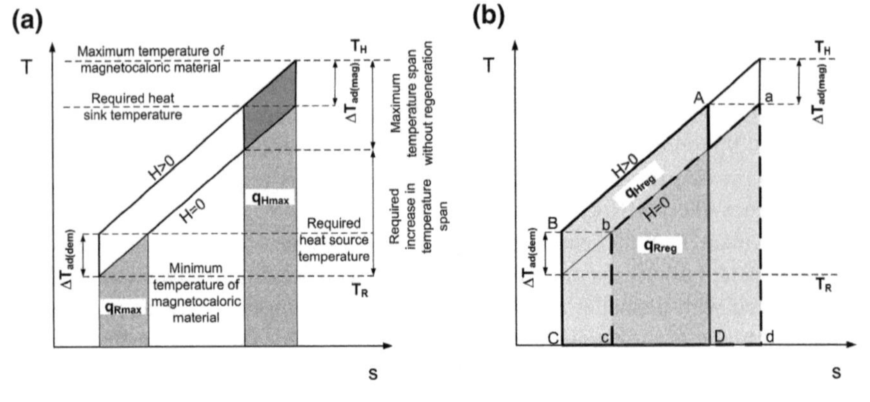

Fig. 4.10 **a** The maximum temperature span and the maximum specific cooling and heating energy in a magnetic refrigeration cycle. **b** The ideally regenerated heat $q_{Rreg} = q_{Hreg}$, which corresponds to the surfaces ABCD = abcd

cooling energy $q_{R\max}$ and the maximum specific heating energy $q_{H\max}$ that can be obtained in the magnetic thermodynamic cycle (this case is shown for a Brayton-like cycle). Despite the fact that the regenerative process can be applied to increase the temperature span, this must not be misunderstood as the maximum specific cooling power (because of the heat regeneration).

In the case that all the irreversible losses are neglected, the theoretical specific cooling energy per mass or volume of magnetocaloric material for the case of the Brayton-like AMR cycle can be defined using the following equation:

$$q_R(H=0) = \int_{s(T_R)}^{s\left(T_R+\Delta T_{\mathrm{ad(dem)}}\right)} T\,\mathrm{d}s \tag{4.1}$$

where T_R denotes the lowest temperature of the magnetocaloric material in the cycle. In the case of a heat pump, the theoretical specific heating energy per mass or volume of magnetocaloric material for the case of the Brayton-like AMR cycle is equal to:

$$q_H(H>0) = \int_{s\left(T_H-\Delta T_{\mathrm{ad(mag)}}\right)}^{s(T_H)} T\,\mathrm{d}s \tag{4.2}$$

The difference in both capacities defined in Eqs. (4.1–4.2) is equal to the theoretical work of the AMR cycle:

$$w = \int_{s\left(T_R+\Delta T_{ad(dem)},\,H>0\right)}^{s(T_H,\,H>0)} T\,ds - \int_{s(T_R,\,H=0)}^{s\left(T_H-\Delta T_{ad(mag)},\,H=0\right)} T\,ds \tag{4.3}$$

where T_H represents the highest temperature of the magnetocaloric material within the cycle.

Note that Eqs. (4.1–4.3) are also valid for the case where regeneration is performed, but it does not account for the overlapping of the internal cycles. They could be generally applied for the characterization of magnetocaloric materials in AMRs if the overlapping is neglected or included afterwards.

For a rapid (engineering) estimation, it is convenient to simplify (linearize) Eq. (4.1) as:

$$q_{R\max}(H=0) = \int_{s(T_R)}^{s\left(T_R+\Delta T_{\mathrm{ad(dem)}}\right)} T\,\mathrm{d}s \approx \frac{\left(2\,T_R+\Delta T_{\mathrm{ad(dem)}}\right)\Delta s_{(T_R)}}{2} \tag{4.4}$$

4.2 Layered AMR

In the Fig. 4.11, an AMR with layered magnetocaloric materials is shown. Why layering of magnetocaloric materials is important?

One of the disadvantages of the known magnetocaloric materials (especially those with a first-order phase transition) is the relatively narrow temperature range over which the magnetocaloric effect occurs (see Chap. 2 for details).

This is especially important when the magnetocaloric material is used in an AMR and/or when large temperature spans between the heat source and heat sink are required.

Since during the operation of the AMR a temperature profile is established along the length of the magnetocaloric material, this implies that according to the temperature profile, some parts of the magnetocaloric material can be, temperature-wise, relatively far from the material's Curie temperature. The magnetocaloric effect for this part of the magnetocaloric material is therefore smaller compared to the parts that are closer to the Curie temperature.

It has been shown experimentally and numerically in various studies [20–32] that layering magnetocaloric materials with different Curie temperatures along the AMR (in the fluid flow direction) increases the AMR's performance (see Fig. 4.11a) due to the larger average magnetocaloric effect along its length.

The layering of magnetocaloric materials along the length of the AMR is especially important when one applies first-order phase transition materials (e.g., Mn–Fe–P; La–Fe–Si), since their magnetocaloric effect is limited to a narrower temperature range compared to the second-order phase transition materials like Gd.

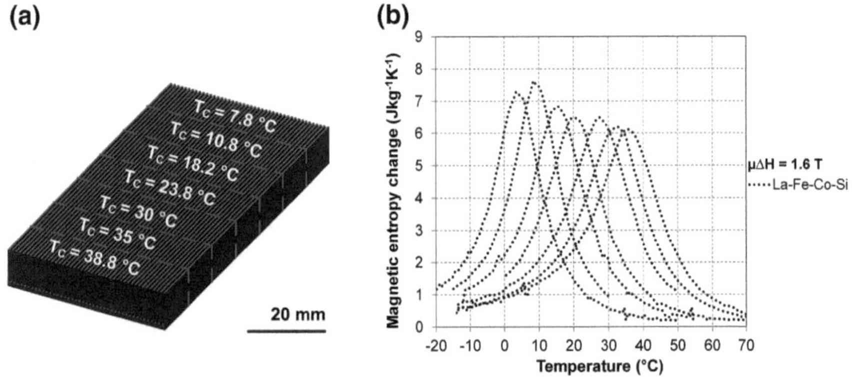

Fig. 4.11 **a** Example of a AMR with seven layers. **b** The magnetocaloric effect (magnetic entropy change) of La–Fe–Co–Si with different Curie temperatures (reproduced from data obtained from Vacuumschmelze GmbH)

4.3 Numerical Modelling of an Active Magnetic Regenerator

With the aim of analysing the operation of a magnetic refrigerator in detail, a number of theoretical (numerical) models were developed. Since for more than 30 years, the principle of the AMR has been considered as the most efficient way to exploit the magnetocaloric effect, the vast majority of these models are actually AMR models. The modelling approach is especially important since the AMR process has a highly multidisciplinary nature and involves thermodynamics, heat transfer, fluid dynamics, solid-state physics and magnetism problems. Its performance strongly depends on various operational conditions (utilization factor, operational frequency, magnetic field change and its profile, heat-transfer fluid, performed thermodynamic cycle) as well as on the magnetocaloric and geometrical characteristics of the AMR (the type and geometry of the magnetocaloric materials used and the related losses). It is crucial to understand the fundamental loss mechanisms, their relations and the performance limits in order to develop and design an efficient AMR and a subsequent high-performance magnetic refrigerator.

4.3.1 A Brief Review of AMR Numerical Models

A detailed review of the AMR numerical models developed up to 2010 is presented in Nielsen et al. [33]. In general, there are two established approaches to AMR modelling: steady-state, time-independent modelling and dynamic, time-dependent modelling. However, the overall goal of all AMR models is to predict the performance of a particular AMR in terms of cooling power, temperature span, efficiency, input magnetic work, etc.

The steady-state models are, in general, simpler, with the aim being to estimate the AMR's performance without a detailed knowledge of its dynamic characteristics during operation. The basic assumption of such a model is an ideal thermodynamic cycle for the magnetocaloric material, while the corresponding losses, like the heat transfer losses between the heat-transfer fluid and the magnetocaloric material, the viscous losses, the losses due to longitudinal thermal conductivity, etc. are subsequently taken into account, through the appropriate correlations and correction factors. The major benefit of these models is their computational efficiency. However, the predictive capabilities of steady-state models are limited, as they are unable to include the interactions between the loss mechanisms, which are the main disadvantage if this model is to be used for the development and optimization of a real AMR device [33]. Some of the steady-state models can be found in, e.g. [27, 34–37].

With the aim of a detailed understanding of the AMR cycle and its properties, a number of time-dependent numerical models were developed. They are based on the energy equations of the magnetocaloric material and the heat-transfer fluid

(Eqs. 4.5 and 4.6). Regarding the number of addressed dimensions in the energy equation, we can distinguish between one-dimensional (1-D), two-dimensional (2-D) and three-dimensional (3-D) time-dependent AMR models. The 1-D approach is the most widely applied, mostly due to its higher computational efficiency. In the 1-D models, it is assumed that the fluid flow and the thermal conductivity (if included) only occur in the direction of the fluid flow. A crucial parameter for the 1-D models is the heat transfer coefficient (e.g. the Nusselt number), which defines the heat transfer rate between the magnetocaloric material and the heat-transfer fluid. The accuracy of the 1-D models, therefore, often very much depends on the suitability and the accuracy of the correlation of the heat transfer coefficient being used. In recent years, several 1-D time-dependent models were developed. They can be found in, e.g. [16, 38–48]. Recently, also a few 2-D models have been applied [49–52]. The 2-D model generally assumes a two-dimensional fluid velocity profile and includes the longitudinal and transversal thermal conductivity of the magnetocaloric material and the heat-transfer fluid (parallel and perpendicular to the direction of the fluid flow). The governing equations for the 2-D model are not directly coupled through the heat transfer coefficient, as in the case of 1-D models, but through the additional boundary condition, which makes them physically more consistent. However, its application is, in general, limited by the ordered AMR geometries (e.g. parallel-plates, wire-like and perforated structures or similar). Random geometries, like packed-bed structures, cannot be fully addressed in a two-dimensional space. The 1-D models, in contrast, do not have this limitation, as the impact of the geometry is considered through the use of suitable thermohydraulic correlations. However, Li et al. [53] developed a partial 2-D model of a packed-bed AMR with spheres, where the thermal conduction of the spheres is considered in 2-D, while the fluid equation is applied in 1-D and the required thermohydraulic properties are included through the appropriate correlations. A similar approach was presented for a honeycomb AMR by Šarlah et al. [54].

4.3.2 Mathematical (Physical) Model of an AMR (Basic Energy Balance Equations)

The mathematical model of an AMR is based on the well-established passive regenerator model. The major difference between passive regenerator models and AMR models is the implementation of the magnetocaloric effect and the timing between the magnetic field's profile and the fluid flow's profile. The basic model of the passive heat regenerator was first developed by Anzelius [55]. A few years later, Nusselt [56] and Hausen [57] separately described the general operation of the heat regenerator and its mathematical model. The applications and research on heat regenerators were greatly expanded in later years due to the increasing need for heat regeneration and storage in various sectors. The mathematical model of the heat regenerator can, therefore, be found in the literature, e.g. [58–64].

The mathematical model of an AMR is based on the governing energy equations of the magnetocaloric material (solid) and the heat-transfer fluid, which are derived from the energy equation [64]. In general, an AMR model is based on similar assumptions to those usually applied for passive heat regeneration [33, 64]:

- Parasitic heat transfer losses (gains) to the surroundings are neglected (see Sect. 4.3.4 for details)
- The only heat source and heat sink in the AMR are due to the magnetocaloric effect
- No flow leakage or flow bypassing around the AMR occurs
- The magnetocaloric material is equally distributed; the porosity is homogenous and thus no flow maldistribution and edge effects occur (see Sect. 4.3.4 for details)
- No phase change occurs in the fluid
- The fluid is incompressible and thus no compression/expansion of the fluid and no pressure oscillations occur during the flow periods (this is valid liquids, which are usually applied)
- No mixing of the 'hot' and 'cold' fluid flows occurs
- Any dead volume (the volume between the entry/exit of the AMR and valve/ flow divider) is neglected
- The heat transfer caused by radiation within the regenerator is negligible compared to the convective and conductive heat transfer
- The physical properties of the heat-transfer fluid and the magnetocaloric material (except its specific heat, as shown later in this section) are defined based on its average (local) temperature and pressure
- The applied magnetic field (in the empty air gap of the magnetic field source) is equal to the internal magnetic field in the magnetocaloric material (the demagnetization field is neglected—see Sect. 4.3.4 for details).

In its most general form the governing energy equation of the magnetocaloric material (solid) and the heat-transfer fluid can be written as follows [33]:

$$\rho_s c_s \frac{\partial T_s}{\partial t} = \nabla \cdot (\lambda_s \nabla T_s) + \dot{q}_{\mathrm{MCE}} + \dot{q}_{\mathrm{HT}} \tag{4.5}$$

$$\rho_f c_f \left(\frac{\partial T_f}{\partial t} + (v \cdot \nabla) T_f \right) = \nabla \cdot (\lambda_f \nabla T_f) + \dot{q}_{\mathrm{HT}} + \dot{q}_{\mathrm{vis}} \tag{4.6}$$

where ρ, c, λ, v, T, t are the density, specific heat, thermal conductivity vector, velocity vector, temperature and time, respectively. The subscript s is for solid and f is for fluid.

The terms on the left-hand side of both equations describe the temperature variations of the fluid and solid with time. The fluid equation also contains the convection fluid transport term. Since the 1-D model applies only to the fluid flow direction (x-direction), this term is reduced to the one-dimensional form, where the fluid velocity (as a scalar) is input data. It also usually assumes that the fluid flow has a uniform temperature at each cross section of the fluid channel and the velocity profile is uniform along the entire length of the AMR (fully developed flow). In the 2-D and 3-D models, a velocity profile (vector) is calculated using the well-known Navier–Stokes equations, which can be simplified into an analytical expression, as showed by Nielsen et al. [50] for a parallel-plate AMR or solved numerically at the same time as the governing energy equations (Eqs. 4.5 and 4.6) [49]. Since the fluid flow in the AMR, especially at high operating frequencies, is oscillating and thus not fully developed, the correct application of this effect can play an important role, mostly with respect to the heat transfer characteristics.

The first terms on the right-hand side of Eqs. (4.7) and (4.8) describe the thermal conductivity through the borders of the differential control volume. In the 1-D model the thermal conductivity is applied only in the flow direction (x-direction). In order to simplify the 1-D model and increase the computational efficiency, some models, e.g. [16, 41, 42] apply the effective thermal conductivity in the flow direction. In doing so, they combine the thermal conductivity of the solid and the fluid into the effective thermal conductivity applied only in the solid equation. This is defined as [65]:

$$\lambda_{\text{eff}} = \lambda_{\text{stat}} + \lambda_f D^d \tag{4.7}$$

The effective thermal conductivity can be divided into the static thermal conductivity, with no fluid flow λ_{stat}, and the thermal conductivity due to the fluid dispersion (D^d) of the fluid flow. The static thermal conductivity depends on the solid and fluid thermal conductivity and the porosity of the regenerator. The correlations for the static thermal conductivity can be found in the literature, e.g. [65–69] and can be, in general, separated into the ordered structures (parallel-plate) and the packed-bed structures (spheres, cylinders, irregular particles). The thermal dispersion of the fluid flow reflects the thermal conduction due to the hydrodynamic mixing in the fluid flow through the porous structure. It occurs due to the velocity fluctuations in the fluid flow and the separation and reunification of the fluid along its path [70]. In the literature, e.g. [65, 71–73] various correlations of the fluid's thermal dispersion can be found. Again, one can distinguish between the correlations for the ordered and the packed-bed structures.

On the other hand, the 2-D model usually applies the thermal conductivity (of the fluid and the solid) in the flow direction as well as perpendicular to it (x- and y-directions). This is, of course, physically more correct, but computationally less efficient.

The term $\dot{q}_{\text{HT}}$ applies to the heat transfer between the magnetocaloric material (solid) and the heat-transfer fluid. In the 1-D model the heat transfer term is defined using the well-known Newton's law of cooling:

$$\dot{q}_{\mathrm{HT}}(x,t) = \alpha \frac{A_{\mathrm{HT}}}{V}\left(T_s(x,t) - T_f(x,t)\right) \tag{4.8}$$

where α is a heat transfer coefficient usually obtained through Nusselt number correlations (see Section Heat transfer and Fanning friction factor correlations for details), A_{HT} is the total heat transfer area of the AMR and V is the volume of the solid (for Eq. 4.5) or fluid (for Eq. 4.6). Since the 1-D model neglects the thermal conductivity and thus the temperature distribution in the solid perpendicular to the fluid flow, many models, e.g. [16, 41, 43] apply a correction factor for the heat transfer coefficient. In this way, we can deal with the effective (lumped) heat transfer coefficient (α_{eff}) [74]. Using such an approach, the effect of the non-uniform temperature distribution in the magnetocaloric material is taken into account to a certain extent. The effective heat transfer coefficient is defined as:

$$\alpha_{\mathrm{eff}} = \frac{\alpha}{\left(1 + \frac{Bi}{a_0}\right)} \tag{4.9}$$

The factor a_0 depends on the geometry of the magnetocaloric material in the AMR and has a value of 3 for spheres, 4 for cylinders and 5 for plates [74]. The Biot number is defined as:

$$Bi = \frac{\alpha d}{2\lambda_s} \tag{4.10}$$

where d is the sphere diameter or plate thickness. It should be noted that a well acceptable limit is established, i.e. if the Biot number is less than 0.1, the thermal conductivity and the temperature gradient in the material perpendicular to the fluid flow can be neglected (due to there being at least ten times higher convective heat transfer at the surface) [74]. Since the Biot number in the AMR is not a priori less than 0.1, an effective heat transfer coefficient should be considered in the 1-D models. Furthermore, Eq. (4.9) is fully valid only for the steady-state heat transfer conditions [75]. Since in an AMR, the heat transfer is transient (the temperature difference between the solid and fluid is not constant at any time in the process). Engelbrecht et al. [75] developed a correction factor for transient conditions that should be included in Eq. (4.9).

However, the 2-D and 3-D models may apply the heat transfer term (Eq. 4.8) as well, but it is more consistent to apply an additional boundary condition that describes the heat transfer and temperature gradients at the surface instead (the term $\dot{q}_{\mathrm{HT}}$ is thus not applied in Eqs. (4.5) and (4.6)):

$$\lambda_s \left.\frac{\partial T_s}{\partial y}\right|_{y=h} = \lambda_f \left.\frac{\partial T_f}{\partial y}\right|_{y=h} \tag{4.11}$$

The viscous losses term ($\dot{q}_{\text{vis}}$) in Eq. (4.6) on the macroscopic scale represents the pressure drop of the fluid flow in the AMR. Since the hydraulic diameters of AMRs are usually very small (in the micro-heat exchanger range) the viscous losses are mostly affected by the viscous dissipation caused by friction in the core of the AMR (the entry and exit effects are thus usually neglected), which further causes a degradation of the mechanical energy into heat. Viscous losses can play an important role, especially in the packed-bed AMRs, which suffer from order higher pressure drops compared to an ordered structure, e.g. parallel-plate AMRs [76]. Most 1-D models include a viscous losses term in the governing energy equations, while most 2-D models, which are in general limited to the ordered-structure AMRs, do not [33]. In the 1-D models, the viscous losses term is usually applied as a pressure drop along the AMR (which is assumed to be constant) calculated through the friction factor—Reynolds number correlations (see the Section Heat transfer and Fanning friction factor correlations for details). It can be written as:

$$\dot{q}_{\text{vis}} = v\frac{\partial p}{\partial x} = v\frac{\Delta p}{L} = 2f_F\frac{v^3\rho_f}{d_h} \tag{4.12}$$

where v, Δp, L, ρ_f, d_h, f_F are the average velocity of the fluid, pressure drop, AMR length, density of the fluid, hydraulic diameter and Fanning friction factor. Besides the impact on the AMR's energy state, the viscous losses strongly affect the efficiency of the process, since high viscous losses require a larger input work to pump the fluid (see Eq. 4.17).

The $\dot{q}_{\text{MCE}}$ term in Eq. (4.5) applies the magnetocaloric effect. In general, AMR models implement the magnetocaloric effect using different approaches. A more straightforward and simple way to include the magnetocaloric effect is to apply the adiabatic temperature change of the magnetocaloric material during the magnetization and demagnetization processes, as was done by, e.g. [43, 49, 52]. In doing so, the term $\dot{q}_{\text{MCE}}$ is not included in Eq. (4.5) directly, but the following equations are applied (solely) during the magnetization and demagnetization processes instead:

$$T_{s,fi} = T_{s,i} + \Delta T_{\text{ad,mag}}\left(T_{s,i}; \mu_0 H_{fi}; \mu_0 H_i\right) \tag{4.13a}$$

$$T_{s,fi} = T_{s,i} - \Delta T_{\text{ad,dem}}\left(\left(T_{s,i} + \Delta T_{\text{ad,mag}}\right); \mu_0 H_{fi}; \mu_0 H_i\right) \tag{4.13b}$$

where *fi* and *i* denote the initial and final temperatures (T) and magnetic fields ($\mu_0 H$), respectively, and *mag* relates to the magnetization, while *dem* relates to the demagnetization. The adiabatic temperature change (ΔT_{ad}) depends on the material's temperature and the magnetic field change. An example of an adiabatic temperature change for gadolinium (as a function of the magnetic field change and temperature) calculated using the Mean Field Theory and the well-known Maxwell relation $((\partial s_m/\partial \mu_0 H) = (\partial M_0/\partial T))$ is shown in Fig. 4.12a for both the magnetization and demagnetization processes. The Mean Field Theory model is a

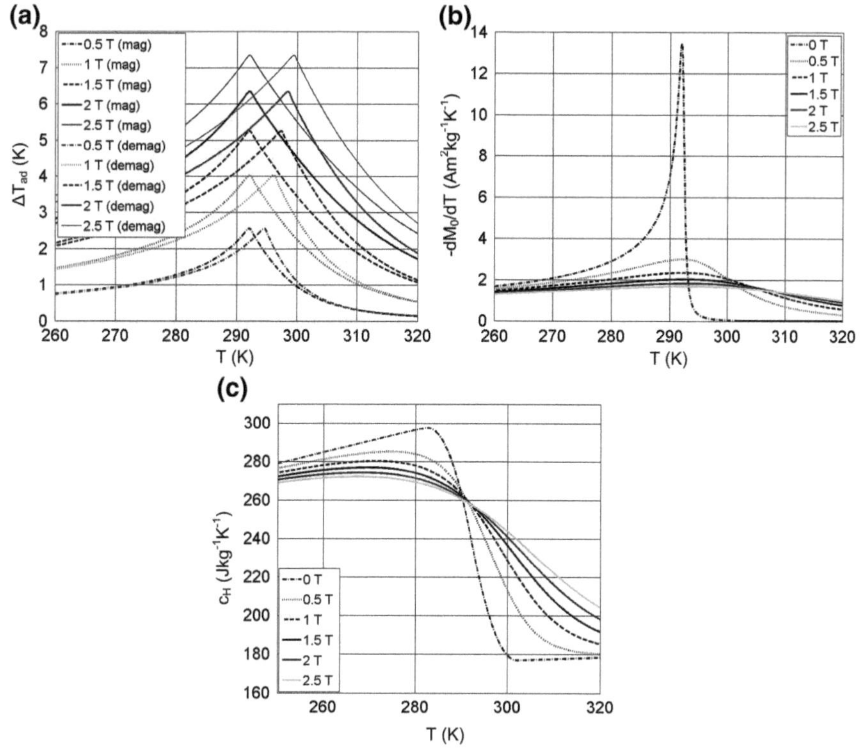

Fig. 4.12 Magnetocaloric and thermal properties (of Gd obtained by the mean field theory) applied to the AMR model for different magnetic fields between 0 and 2.5 T. **a** Adiabatic temperature change for magnetization and demagnetization. **b** Derivative of the specific magnetization over the temperature. **c** Specific heat

combination of the Weiss mean field model for the calculation of the specific magnetization and the magnetic contribution to the total specific heat, the Debye model for the calculation of the lattice contribution to the total specific heat and the Sommerfeld model for the calculation of the electron contribution to the total specific heat [49, 77, 78]. It should be noted that the demagnetization curve is shifted for the corresponding value of the adiabatic temperature change at each temperature in order to ensure the thermodynamic consistency of the model (see Fig. 4.12a). The inclusion of the magnetocaloric effect directly through the adiabatic temperature change is associated with the assumption of a discrete, 'on-off' magnetic field change during the (de)magnetization process. The models that apply the magnetocaloric effect directly, but not through the governing equations, are further limited by the Brayton-like AMR cycle, since other AMR cycles are based on a time-dependent (de)magnetization with a simultaneous heat transfer.

Another approach to include the magnetocaloric effect into the AMR model is to apply it directly in the governing equations through the $\dot{q}_{\mathrm{MCE}}$ term. It is defined as (the so-called built-in method):

$$\dot{q}_{\mathrm{MCE}} = -\rho T_s \frac{\partial M_0}{\partial T}(T_s; \mu_0 H)\frac{\partial \mu_0 H}{\partial t} = -\rho T_s \frac{\partial s_m}{\partial \mu_0 H}(T_s; \mu_0 H)\frac{\partial \mu_0 H}{\partial t} \quad (4.14)$$

where M_0 is the specific magnetization (per mass of magnetocaloric material), s_m is the magnetic entropy and ρ is the density. The term $\frac{\partial \mu_0 H}{\partial t}$ describes the magnetic field profile defined by the magnet assembly and the operating conditions. Equation (4.14) applies the well-known Maxwell relation $\partial M_0/\partial T = \partial s_m/\partial \mu_0 H$, normally used to calculate the magnetic entropy change from the measured values of the specific magnetization.

An example of a derivative of the specific magnetization over temperature for gadolinium is shown in Fig. 4.12b. Such an inclusion of the released energy in the magnetocaloric material during the (de)magnetization process over a period of time is more appropriate as a more realistic magnetic field profile and time-dependent (de)magnetization with simultaneous heat transfer can be applied. Various models, e.g. [41, 42, 51] include the magnetocaloric effect in such a way. However, this method requires a detailed data set (or input functions) of the magnetization or magnetic entropy and the specific heat at different temperatures and magnetic fields, which may not always be available. It should also be noted that the specific heat of magnetocaloric materials around the transition temperatures strongly depends on the temperature and magnetic field and cannot be taken as a constant value, as shown for gadolinium in Fig. 4.12c.

The developed AMR numerical models, in general, rely on the experimentally or theoretically obtained magnetocaloric properties of a particular magnetocaloric material. Especially in the case of gadolinium, which is a kind of reference magnetocaloric material for the AMR models, the Mean Field Theory became a well-established tool for estimating its magnetocaloric and thermal properties. As a result, it was applied by various authors of AMR models, e.g. [16, 49, 51]. The Mean Field Theory (together with the applied Maxwell relation) is able to generate a data set for the required magnetocaloric properties that can be included in the AMR model using both above-presented methods (the direct or built-in method), even though due to the applied assumptions it over predicts the magnetocaloric effect. However, the experimentally obtained data are usually measured for an insufficient number of different temperatures, and especially magnetic fields, to be correctly and consistently used via the built-in method. An example of sufficiently detailed, measured magnetocaloric properties for gadolinium and La–Fe–Co–Si magnetocaloric materials, which allows its implementation into a built-in AMR model, was presented by Bjørk et al. [32].

As was also noted by some authors of AMR models [79], it is extremely important to obtain all the magnetocaloric properties used in the model (adiabatic temperature change or specific magnetization and specific heat) from the same source. So, all should be calculated using the Mean Field Theory or experimentally measured (the latter can also suffer from a significant measuring uncertainty). Even a minimal mismatch between the applied magnetocaloric data can lead to a potential thermodynamic inconsistency and further to unrealistic results (especially for the COP values).

The AMR numerical models usually apply the following initial and boundary conditions (Eq. 4.15a, b, c), which are required in order for the above-presented mathematical model to be solved [33]:

- The inlet fluid temperatures at the hot (subscript h) and cold (subscript c) sides of the AMR are predefined (based on the predefined heat sink and source temperatures or on the temperatures of the fluid that exited the AMR during the previous fluid flow period)—Eq. (4.15a)
- The temperatures of the AMR at the beginning of a particular phase ($t = 0$) in the cycle are the same as at the end of previous phase ($t = \tau_f$)—Eq. (4.15b)
- At both sides of the AMR an adiabatic boundary condition is applied—Eq. (4.15c)

$$T_{f,h}(x = 0, t) = T_{f,h,\text{in}}; T_{f,h}(x = L, t) = T_{f,h,\text{in}} \tag{4.15a}$$

$$T_{s,h}\left(x, t = \tau_f\right) = T_{s,c}(x, t = 0); T_{s,h}(x, t = 0) = T_{s,c}\left(x, t = \tau_f\right) \tag{4.15b}$$

$$\left.\frac{\partial T_s}{\partial x}\right|_{x=0} = \left.\frac{\partial T_s}{\partial x}\right|_{x=L} = 0 \tag{4.15c}$$

As already described, most 2-D and 3-D models also apply additional boundary conditions describing the heat transfer between the magnetocaloric material and heat-transfer fluid at the heat transfer surface (Eq. 4.11).

As is evident from the governing equations, the operation and performance of the AMR depends on various independent parameters, like the operating conditions, the geometrical, thermal and magnetocaloric properties of the AMR and the properties of the heat-transfer fluid. As recently shown [42, 52, 80], the number of independent parameters can be reduced by the application of the dimensionless AMR model. It is based on a dimensionless length (and thickness in the 2-D model), time, temperature and magnetic field, as well as different dimensionless numbers, like NTU, utilization factor, thermal capacity ratio, Fourier number, Peclet number, etc. which are used directly in the governing equations. The resulting reduced number of key parameters simplifies the performance optimization procedure. At the same time, the dimensionless model improves the comparability of various numerically as well as experimentally analysed AMRs, which are otherwise hard to compare due to several parameters that crucially influence the operation of the AMR.

In general, the developed numerical models of the AMR and the above-presented mathematical model are solved using various numerical techniques. However, a majority of the models apply the finite-difference method [16, 41, 47, 48, 50], which guarantees total energy conservation across the boundaries at all times.

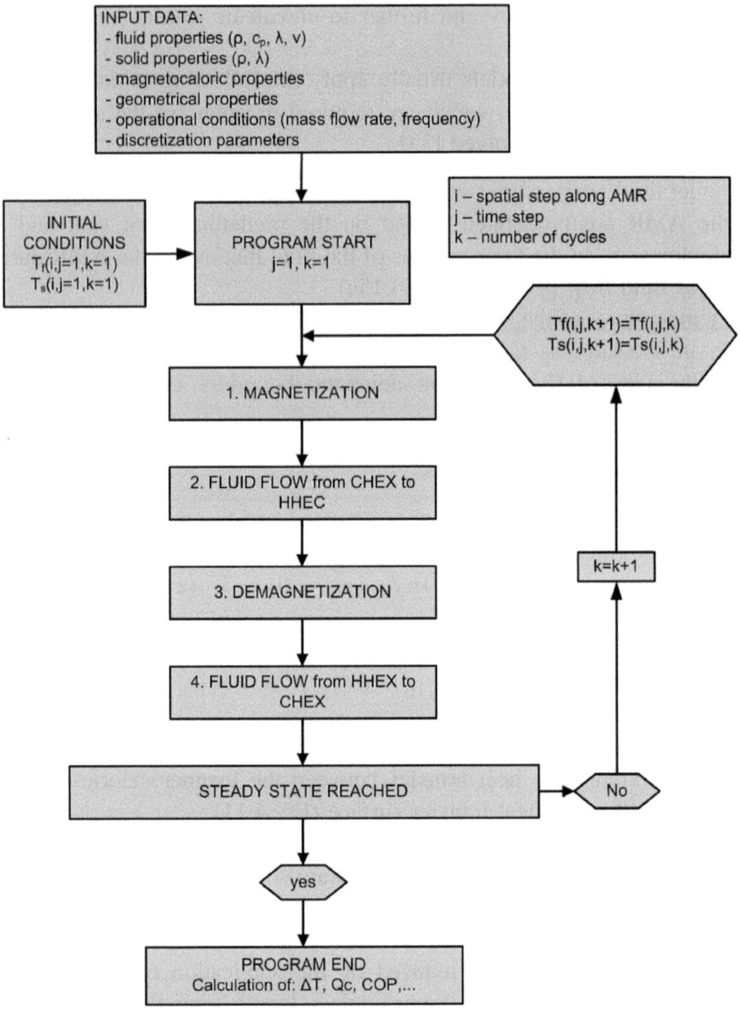

Fig. 4.13 General flow chart of the AMR model

Figure 4.13 shows a general flow chart of the AMR model for the Brayton-like AMR cycle [16, 49, 51].

The programme starts with the required input data (fluid, solid, magnetocaloric and geometrical properties, operational conditions and discretization parameters) and the initial conditions, which define the initial temperatures along the AMR. The programme then performs all the pre-defined operational phases and repeats these phases until the periodic steady state is reached. This occurs when the temperature of the fluid at the exit of the AMR between two consecutive cycles does not change any more (or by less than a predefined value).

In the steady-state condition, the programme is then able to calculate the cooling (or heating) characteristics of the AMR, like the temperature span, the cooling and heating power and the COP (or further efficiency). The temperature span of the AMR is defined as the average exit fluid temperature difference between the 'hot' and 'cold' sides of the AMR. The cooling and heating powers generated by the AMR are defined using the following equations:

$$\dot{Q}_R = f \int_0^{\tau_f} \dot{m}_f c_{p,f} \left(T_{f,c,\mathrm{in}}(t) - T_{f,c,\mathrm{out}}(t) \right) \mathrm{d}t \tag{4.16a}$$

$$\dot{Q}_H = f \int_0^{\tau_f} \dot{m}_f c_{p,f} \left(T_{f,h,\mathrm{out}}(t) - T_{f,h,\mathrm{in}}(t) \right) \mathrm{d}t \tag{4.16b}$$

where $T_{f,c,\mathrm{out}}$ and $T_{f,c,\mathrm{in}}$ are the fluid temperature as it exits and enters the AMR at the cold side over the fluid flow period (τ_f), while $T_{f,h,\mathrm{in}}$ and $T_{f,h,\mathrm{out}}$ are the fluid temperatures as it enters and exits the AMR at the cold side over the fluid flow period, respectively. The f is the operational frequency, $\dot{m}_f$ is the mass flow rate and $c_{p,f}$ is the specific heat of the fluid. The COP is defined as the ratio of the cooling power (in the case of the refrigeration cycle) and the input work that includes the magnetic work to run the cycle and the work needed to pump the fluid:

$$\mathrm{COP}_R = \frac{\dot{Q}_R}{P_m + P_{\mathrm{pump}}} \tag{4.17}$$

The work needed to pump the fluid can be calculated using the following equation (where η_{pump} is the pump's efficiency):

$$P_{\mathrm{pump}} = \frac{\dot{m}_f \Delta p}{\rho_f \eta_{\mathrm{pump}}} \tag{4.18}$$

For an ideal AMR (meaning that the efficiencies of auxiliary devices are not accounted for) operating under periodic steady-state conditions, the magnetic work input is equal to the difference between the cooling and heating powers. Furthermore, the magnetic work can also be calculated through the surface of the performed cycle (for each particle of the magnetocaloric material along the AMR, including the overlapping) in the T–s diagram per unit of time:

$$P_m = \dot{Q}_H - \dot{Q}_R = m_{\mathrm{mcm}} f \oint T_s \mathrm{d}s = m_{\mathrm{mcm}} f \int_0^L \int_0^\tau T_s(x,t) \frac{\mathrm{d}s(x,t)}{\mathrm{d}t} \mathrm{d}t \mathrm{d}x \tag{4.19}$$

As explained in Engelbrecht [41] and Šarlah [81], when both values of the magnetic work agree within a certain precision (excluding the heating of the system due to viscous losses), the model can be regarded as both thermodynamically consistent and correct.

In Petersen et al. [82], a direct comparison of 1-D and 2-D models for a parallel-plate AMR is shown. The authors concluded that both models show an excellent qualitative and quantitative agreement between the cooling and heating powers for thin regenerator channels. However, the results of the two models diverge as the thickness of the regenerator plates is increased. The comparison between the model results for the COP values did not show the same degree of agreement. The reason for this is that the magnetic work and the COP are derived from the cooling and heating powers and small differences in either result in large deviations between the estimated COP values of the models. The cause of the discrepancy between the models at larger channel thicknesses is due to the effect of the perpendicular temperature gradients, primarily in the fluid as well as in the solid. These are not accounted for by the 1-D model. They concluded that the 1-D model is valid (regarding the 2-D model) when the channels are smaller than about 0.5 mm, which is necessary for an efficient AMR with good heat transfer characteristics, as shown in Sect. 4.4.

4.3.3 Heat Transfer and Fanning Friction Factor Correlations

Most 1-D models rely on correlations for the convective heat transfer and friction factor coefficients. The use of the appropriate correlations is crucial for an accurate AMR model. As explained by Šarlah and Poredoš [42], a 10 % higher heat transfer coefficient can yield a 4.4 % higher temperature span of the AMR. In this section, the most widely applied correlations for the convective heat transfer and friction factor coefficients are reviewed and compared through the Nusselt number and the Fanning friction factor number correlations, respectively.

Here, the Reynolds number, the Nusselt number and Fanning friction factor are defined as:

$$Re = \frac{\rho v \epsilon d_h}{\eta} \tag{4.20}$$

$$Nu = \frac{\alpha d_h}{\lambda} \tag{4.21}$$

$$f_F = \frac{d_h}{2(v\varepsilon)^2 \rho} \frac{\Delta p}{L} \tag{4.22}$$

where ρ, v, ε, d_h, η, α, λ, Δp, L are the fluid density, internal (pore) velocity of the fluid, porosity, hydraulic diameter, dynamic viscosity, heat transfer coefficient, fluid thermal conductivity, pressure drop along the AMR and its length, respectively. The Reynolds number defined in Eq.(4.20) includes the external velocity (v ε), which is valid for the packed bed structures. However, the Reynolds number of the parallel-plate is usually defined with the internal velocity. For details see the original source of a particular correlation in Tables 4.1, 4.2, 4.3 and 4.4. It should be noted that for a packed-bed AMR, in most cases the correlations are based on the particle (sphere) diameter and not the hydraulic diameter. Therefore, this should be included in Eqs. (4.20)–(4.22) as well (see Tables 4.1, 4.2, 4.3 and 4.4 for details). The hydraulic diameter in Eqs. (4.20)–(4.22) is defined as (if not stated differently in Tables 4.1, 4.2, 4.3 and 4.4):

$$d_h = \frac{4V_{\mathrm{AMR}}\varepsilon}{A_{\mathrm{HT}}} \tag{4.23}$$

Tables 4.1, 4.2, 4.3 and 4.4 present the Nusselt number and the friction factor number correlations for the parallel-plate and packed-bed AMR, respectively. Tables 4.1, 4.2, 4.3 and 4.4 also show the conditions for which the particular correlation is valid, the AMR models that apply it and the source where it can be found.

Figures 4.14, 4.15 and 4.16 show a comparison of the Nusselt number and the Fanning friction factor number correlations presented in Tables 4.1, 4.2, 4.3 and 4.4 for two different AMRs (with outer dimensions of 80 mm (length) × 40 mm (width) × 10 mm (height)) and with water as the heat transfer fluid. The parallel-

Table 4.1 The Fanning friction factor number correlation used in the parallel-plate AMR models

Friction factor correlation	Conditions of validation	Used in the AMR model by	Source
$f_F = 24/Re$	For fully developed laminar flow	Tušek et al. [17], Nikkola et al. [47], Chiba et al. [48], Petersen et al. [82]	Keys and London [88]

Table 4.2 The Nusselt number correlations used in the packed-bed AMR models

Nusselt number correlation	Note	Conditions of validation	Used in the AMR model by	Source
$Nu = 2 + 1.1Re^{0.6}Pr^{1/3}$	Based on d and not d_h	For all Re numbers	Tušek et al. [16], Aprea et al. [29], Engelbrecht [41], Burdyny and Rowe [46], Li et al. [53]	Wakao and Keguei [71]
$Nu = 2 + 1.8Re^{1/2}Pr^{1/3}$	Based on d and not d_h	$Re > 50$	Engelbrecht [41]	Rohsenow and Hartnett [89]
$j = 0.23Re^{-0.3}$	j is Colburn factor defined as:	$Re > 20$	Dikeos and Rowe [90]	Keys and London [88]
$Nu = jRePr^{1/3} = 0.23Re^{0.7}Pr^{1/3}$	$j = StPr^{2/3} = \frac{NuPr^{1/3}}{Re}$			
$Nu = \left(0.5Re^{1/2} + 0.2Re^{2/3}\right)Pr^{1/3}$	$d_h = d\frac{\varepsilon}{(1-\varepsilon)}$	$Re > 20$	Shir et al. [40]	Whitaker et al. [91]

Table 4.3 The friction factor number correlations used in the packed-bed AMR models

Friction factor correlation	Note	Conditions of validation	Used in the AMR model by	Source
$f_F = 90\frac{(1-\varepsilon)^2}{Re\varepsilon^2} + 0.9\frac{1-\varepsilon}{\varepsilon^3}$	Based on d and not d_h	For smooth particles For all Re numbers	Engelbrecht [41]	Kaviany [65]
$f_F = 75\frac{(1-\varepsilon)^2}{Re\varepsilon^2} + 0.9375\frac{1-\varepsilon}{\varepsilon^3}$	Based on d and not d_h	For smooth particles For all Re numbers	Engelbrecht [41], Li et al. [53]	Rohsenow et al. [89]

plate AMR was constructed of plates with a thickness and spacing of 0.25 mm (porosity of 0.5), while the packed-bed AMR had spheres with a diameter of 0.25 mm (porosity of 0.39).

It is clear from Fig. 4.14 that the majority of the Nusselt number correlations for fully developed laminar flow do not depend on the Reynolds number, but only on the boundary conditions. The exception here is the correlation from Nickolay and Martin [84], which is also valid for developing flow (which usually occurs in the AMR); therefore, it is more correct than others and used by most authors of the (1-D) parallel-plate AMR model (see Table 4.4).

In the case of the packed-bed AMR, the difference between the different correlations is quite significant, as shown in Fig. 4.15. However, the most widely used by the authors of packed-bed AMR models is the correlation from Wakao and Keguei [71]. Figure 4.16 shows the Fanning friction factor correlation for a parallel-plate AMR and two very similar AMR correlations for the packed-bed AMR, which are the most widely used in the models.

It should be underlined that there are no well-established heat transfer correlations for the flow through porous media in the low Reynolds number flow regime and for high operating frequencies (oscillating flow), as is usually the case during the AMR's operation. This field has to be further investigated in order to develop more reliable correlations [43].

4.3.4 Improved Modelling of an AMR (Modelling of the Additional Loss Mechanisms in an AMR)

The above-presented mathematical model of the AMR is based on certain assumptions (neglected loss mechanisms) for which it was shown that it might have a significant influence on the AMR's operation and its cooling characteristics [92]. This section thus shortly discusses the impact and deals with the inclusion of the following additional AMR loss mechanisms: the **demagnetization field**; the **flow maldistribution**; the **parasitic heat losses to the surroundings** and the **hysteresis losses of the magnetocaloric material** into the AMR mathematical model.

Table 4.4 The Nusselt number correlations used in the parallel-plate AMR models

Nusselt number correlation	Note	Conditions of validation	Used in the AMR model by	References
$Nu = \left(7.541^{3.592} + \left(1.841 Gz^{(1/3)}\right)^{3.592}\right)^{(1/3.592)}$	$Gz = \frac{d_h}{L} RePr$	$0.1 < Gz < 10^6$ For fully developed laminar flow	Tušek et al. [17], Nielsen et al. [83]	Nickolay and Martin [84]
$Nu(x) = \left(7.541^{3.592} + \left(1.841 Gz(x)^{(1/3)}\right)^{3.592}\right)^{(1/3.592)}$	$Gz(x) = \frac{d_h}{x} RePr$	For developing laminar flow	Petersen et al. [82]	
$Nu = 7.54$		For a fully developed laminar flow and a constant heat transfer wall temperature (the latter is not actually the case in the AMR)	Chiba et al. [48]	Bejan et al. [85]
$Nu = 7.541\begin{pmatrix} 1 - 2.61R + 4.97R^2 \\ -5.119R^3 + 2.702R^4 \\ -0.548R^5 \end{pmatrix}$	R is the aspect ratio of the fluid channel	For fully developed laminar flow	Nikkola et al. [47]	Kandlikar et al. [86]
$Nu = \left(\left(0.41\left(\frac{x}{d_h Re}\frac{1}{Pr}\right)^{-1/2}\right)^2 + \Psi^2\right)^{1/2}$ $\Psi = 8.235\begin{pmatrix} 1 - 2.0421R + 3.0853R^2 \\ -2.4765R^3 + 1.0578R^4 \\ -0.1861R^5 \end{pmatrix}$	R is the aspect ratio of the fluid channel	For developing laminar flow	Risser et al. [44]	Bavie're et al. [87]

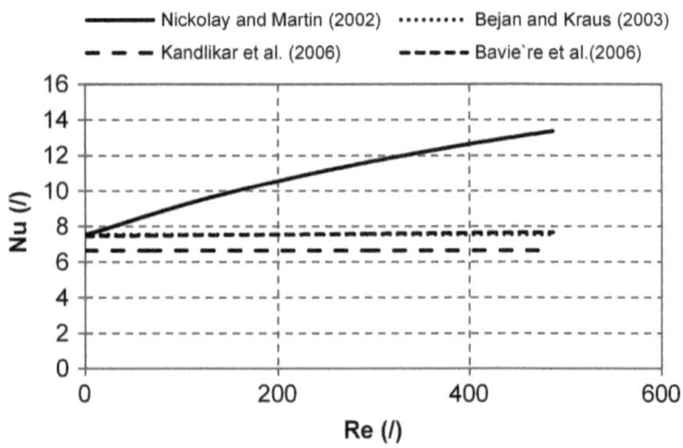

Fig. 4.14 Nusselt number correlations for a parallel-plate AMR (see Table 4.4)

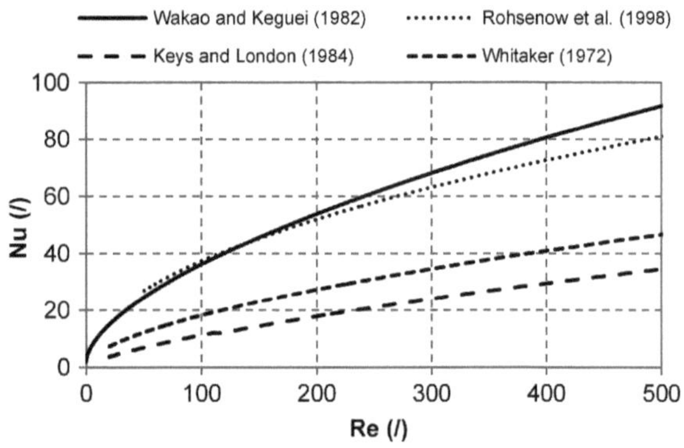

Fig. 4.15 Nusselt number correlations for a packed-bed AMR (see Table 4.3)

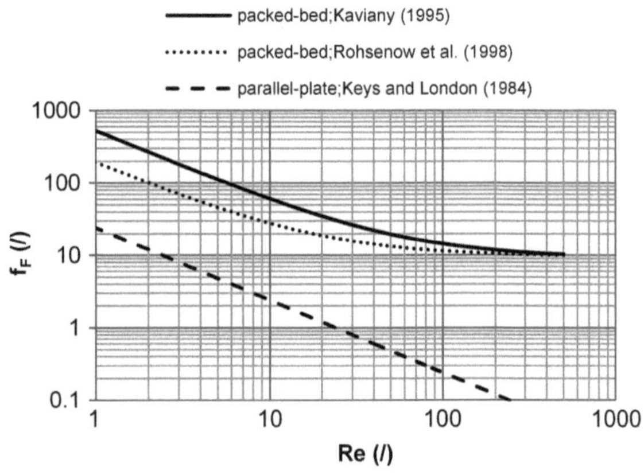

Fig. 4.16 Fanning friction factor number correlations for a packed-bed AMR and a parallel-plate AMR (see Tables 4.3 and 4.4)

4.3.4.1 The Demagnetization Field

It was shown that the internal magnetic field in the magnetic material (and subsequently in the AMR) is, in general, always lower than the applied magnetic field in the empty air gap of the magnetic field source [93, 94]. This is due to the demagnetizing field, which is a consequence of the magnetization of the magnetic body and opposes the magnetization inside the body, thus reducing the resulting internal field. In order to apply the correct values for the magnetocaloric effect into the numerical model of the AMR, the internal magnetic field in the magnetocaloric material must be used instead of the applied one. It can be written as:

$$H_{\mathrm{in}} = H_{\mathrm{appl}} - H_{\mathrm{dem}} = H_{\mathrm{appl}} - N(x)M(T,H) \tag{4.24}$$

where H_{appl}, H_{dem}, N and M are the applied magnetic field, the demagnetization field, the demagnetization tensor and the magnetization, respectively. Equation (4.24) is usually expressed as a scalar equation and the average demagnetization factor is used instead of the demagnetization tensor [95].

The internal demagnetization field is, in general, spatially dependent over the magnetized body. It depends on the geometry of the body, its temperature and the internal magnetic field. In the case of an AMR, which is a non-uniformly magnetized body, the demagnetization field therefore strongly depends on its geometry, the temperature span at which it operates, the magnetocaloric material(s) applied and the applied magnetic field. Its fully correct application is therefore non-trivial and cannot be solved analytically (analytical solutions are only available for an uniformly magnetized body, like ellipsoidal bodies or an infinite sheet and cylinder [95]). Recently, a 3-D magnetostatic numerical model for the calculation of the demagnetization tensor and subsequently the demagnetization field was presented for a non-uniformly magnetized rectangular prism [46] and a stack of rectangular prisms [96] (e.g., a parallel-plate AMR). Bjørk and Bahl [97] numerically analysed the demagnetization factor of non-uniformly magnetized, randomly packed, spherical particles (e.g. a packed-bed AMR). Furthermore, a coupled AMR model with the demagnetization model can be found in [92, 98, 99].

Various studies showed that the impact of the demagnetization field on the internal field in the AMR can be significant. In practice, it depends mostly on the AMR's outer dimensions and the distribution of the material inside (for a particular magnetic field) [93, 94, 96]. In the case of a typical AMR with a porosity of about 30 % subjected to a magnetic field of 1 T the internal magnetic field, in the two extreme cases (parallel-plate AMR with the plate's distribution parallel to the applied magnetic field and perpendicular to it) can be reduced from 10 % (parallel-distribution) and up to 70 % (perpendicular distribution) [96, 100]. It was also shown numerically as well as experimentally that a parallel-plate AMR with the plate's distribution parallel to the magnetic field can generate significantly higher cooling characteristics compared to the perpendicular distribution [92, 99, 101].

4.3.4.2 The Flow Maldistribution

The flow maldistribution, also referred to as flow channelling, results from a non-uniform porosity distribution along the fluid flow direction in the AMR. It mostly depends on the geometry and the fluid flow rate. For a packed-bed AMR, the porosity near the wall is typically greater than in the middle of the AMR. Due to the smaller pressure drop near the wall the velocity is increased and the resulting cold or hot bypasses decrease the regeneration efficiently [33, 102]. The flow maldistribution in a packed-bed AMR strongly depends on the ratio between the outer AMR dimensions and the particle diameter [103]. We do not know whether the impact of the flow maldistribution has yet been directly included in the numerical model of a packed-bed AMR. However, there are several models which enable the calculation of the radial porosity distribution [104] as well as the radial velocity distribution in the packed bed [105], but since the great majority of the packed-bed AMRs are 1-D, this cannot be directly applied. As explained previously in this chapter, in the 1-D models the thermohydraulic properties are included by using appropriate correlations (mostly Nusselt number and friction factor correlations). Since these correlations are also usually obtained through the experimental testing of a packed bed [81, 106] they, to the certain extent, already include the impact of the flow maldistribution. However, each AMR can have a different degree of flow maldistribution, depending on the ratio between the outer AMR dimensions and the particle diameter and the manufacturing technique applied.

A flow maldistribution is also observed for the parallel-plate AMR and results from non-uniformly distributed plates causing a non-equal distribution of the velocity and the fluid flow. Since the parallel-plate AMRs are usually well into the micro-channel region, the manufacturing tolerance might be significant. It was shown that it can have a very significant effect, especially for AMRs with a channel thickness below 0.3 mm [107]. Jensen et al. [108] and Nielsen et al. [109] developed a technique and a model that enables an estimation of the reduction of the Nusselt number due to the flow maldistribution for a particular AMR based on its plate distribution standard deviation. They suggested the application of the Nusselt scaling factor, which represents the ratio of the effective Nusselt number of the particular regenerator and the ideal Nusselt number of the uniformly distributed regenerator. This can later be directly included in the AMR model of a particular AMR by multiplying it with the Nusselt number based on the applied correlation (for which it can be assumed that it is obtained for the uniformly distributed plates).

4.3.4.3 Heat Losses to the Surroundings

As described in Nielsen et al. [33], most AMR models assume perfect insulation with respect to the ambient. Thermal interactions with the regenerator housing and parasitic losses to the surroundings are therefore ignored. However, it was shown

[92, 110] that they have a significant effect when modelling a real AMR device (and predicting its performance). There are two general approaches for including the thermal interactions with the surroundings into the AMR model: directly into the governing equations [47, 50] or through an iterative thermal analysis where the estimated loss of cooling power for particular operating conditions (through Eq. 4.25) is used as an input for the model [92, 110]. The direct approach applies additional terms in the AMR governing equations (Eqs. 4.5 and 4.6):

$$\dot{q}_{\text{loss}} = \frac{1}{R_{\text{loss}}} \frac{A_{\text{loss}}}{V} (T_{\text{AMR}} - T_{\text{surr}}) \qquad (4.25)$$

where R_{loss} is the lumped thermal resistance of the regenerator housing (and other isolations if applied), A_{loss} is the outer area of the AMR and T_{surr} is the ambient temperature. The parasitic loss term can be applied either to the solid or to the fluid equation (where T_{AMR} is the average temperature of the fluid and solid at a certain location and time) or to both simultaneously (where T_{AMR} is T_f for the fluid equation and T_s for the solid equation, respectively) as shown by Nikkola et al. [47].

Both approaches for the inclusion of the heat losses to the surroundings apply only a lumped thermal resistance and do not include a regenerator housing and its thermal mass into the numerical scheme through an additional domain and governing equation. To the best of our knowledge no such AMR model exists so far. However, Nielsen et al. [111] presented a heat regenerator model with an included regenerator wall (and its governing equation) in the numerical scheme. They analysed the impact of the housing wall on the packed-bed regenerator's performance under the conditions usually applied in the AMR and showed that its performance can be significantly reduced for a Reynolds number below 10 and a thermal conductivity of the housing material above 10 $\text{Wm}^{-1}\text{K}^{-1}$.

4.3.4.4 Hysteresis Losses

A hysteresis behaviour during the magnetization and demagnetization process (magnetic hysteresis) and during heating and cooling (thermal hysteresis) is observed in most of the known first-order magnetocaloric materials (e.g. Mn–Fe, La–Fe–Si). It was shown that hysteresis behaviour can drastically reduce the magnetocaloric effect under the cyclic conditions applied in the AMR as well as the AMR performance [112, 113]. To the best of our knowledge, there is no AMR model that directly applies the hysteresis of the magnetocaloric material. For details about the effect of the material hysteresis on the magnetic refrigeration cycle associated with the non-equilibrium thermodynamics and its modelling through the Preisach model approach see [114, 115].

4.4 The Impact of the Operational Parameters and Geometry on the Performance of the AMR

The operation and performance of the AMR have been the subject of various researchers, from the theoretical (numerical) [33] as well as the experimental [5, 116] points of view. Among others, those studies showed that the performance of the AMR (with a particular magnetocaloric material) strongly depends on the operational properties, e.g. [43, 45, 50, 117] and the geometrical characteristics, e.g. [15–17, 118]. In particular, the operating conditions (utilization factor and operational frequency) must be carefully chosen. If, for example the mass flow rate and the related utilization factor (Eq. 4.26) are too small or too high this can disable the heat regeneration process and prevent a temperature span from being established. The utilization factor is defined as:

$$U = \frac{\dot{m}_f c_f \tau_f}{m_{\mathrm{MCM}} \bar{c}_{\mathrm{MCM}}} \quad (4.26)$$

where $\dot{m}$, c_f, τ_f, m_{MCM}, $\bar{c}_{\mathrm{MCM}}$ represent the mass flow rate of the heat-transfer fluid, the specific heat of the fluid, the fluid flow period, the mass of magnetocaloric material in the AMR and its average specific heat, respectively. If the amount of fluid displaced (pumped) through the AMR (utilization factor) is too small (also regarding the operational frequency and the AMR geometry), the fluid does not have the capability to transfer or absorb the whole amount of energy generated during the magnetocaloric effect, as explained in [16, 119]. If, on the other hand, more than the optimum amount of fluid is pumped through the AMR, this can cause fluid from the hot end of the AMR (or even from the HHEX) to be transported into the CHEX. This of course reduces the temperature span as well as the total cooling power of the AMR [52]. Under such conditions, the temperature span between the heat source and heat sink will be small and the operation will be similar to one without regeneration (single-stage device). The impact of the utilization factor on the AMR's performance is clearly shown in Figs. 4.17, 4.18 and 4.19. It was shown by various researchers [17, 43, 45, 50, 117] that in most cases the optimum utilization factor is between 0.2 and 0.8, which means that in the optimum case the entire fluid in the AMR is not displaced during the fluid flow period. In other words, the fluid entering the AMR will, in most cases, only move until, e.g. the middle of the AMR, and displace the rest of the fluid out of the AMR at the other side. Furthermore, a higher utilization factor also results in a higher degree of overlapping of the local (internal) thermodynamic cycles (see Fig. 4.3a, b), which further increase the input magnetic work needed to run the cycle. This should, therefore, be carefully optimized.

Another crucial operational parameter of the AMR is the operating frequency, which is usually defined as the number of performed thermodynamic cycles per unit of time:

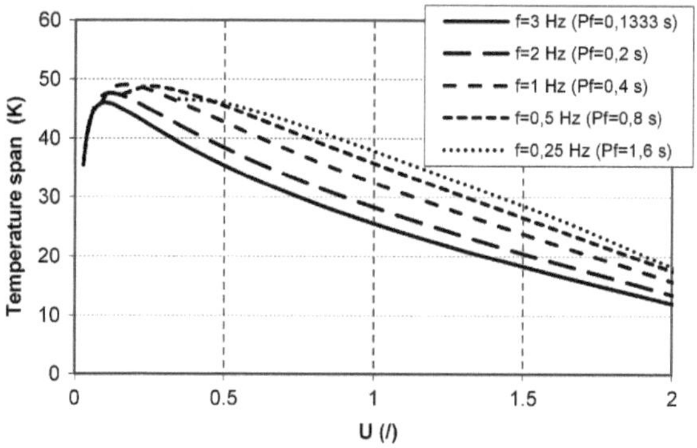

Fig. 4.17 The temperature span at zero cooling power as a function of the utilization factor for different operating frequencies in the case of a packed-bed AMR with Gd spheres of 0.5 mm for a 1 T magnetic field change

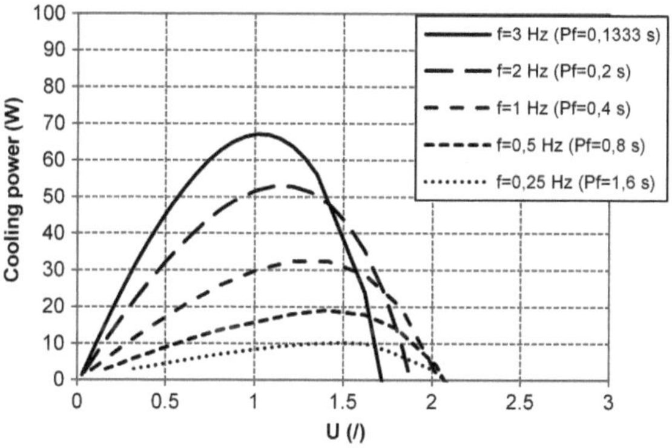

Fig. 4.18 The cooling power at 15 K of temperature span as a function of the utilization factor for different operating frequencies in the case of a packed-bed AMR with Gd spheres of 0.5 mm for a 1 T magnetic field change

$$\nu = \frac{1}{2\tau_{\text{mag}} + 2\tau_f} \tag{4.27}$$

where τ_{mag} is the duration of the magnetization and demagnetization process ((de) magnetization time) and τ_f is the duration of the fluid flow period. Here, we assume that both fluid flow periods as well as both the magnetization and demagnetization times are equal (symmetrical operation of regenerator), which is usually the case in an AMR cycle. Since the thermodynamic cycle of the AMR process contains two relatively different types of processes: (de)magnetization and fluid flow period, the impact of the operating frequency of such a cycle must therefore be more carefully analysed. The (de)magnetization process should be as short as possible, regardless of the duration of the fluid flow period and the performed AMR cycle, since a long

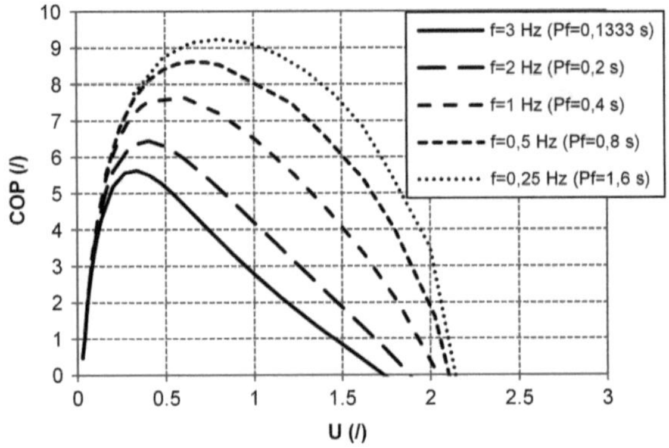

Fig. 4.19 The COP at 15 K of temperature span as a function of the utilization factor for different operating frequencies in the case of a packed-bed AMR with Gd spheres of 0.5 mm for a 1 T magnetic field change

(de)magnetization time, in general, increases the heat transfer losses. The time of the (de)magnetization is usually defined by the magnetic field source (magnet assembly) and the design of the device itself. On the other hand, the time of the duration of the fluid flow period should be carefully optimized. This is also strongly related to the geometry of the AMR. In the case of thicker walls or larger particles (spheres) in the AMR, a longer time is required for the fluid to transfer or absorb the total amount of energy generated by the magnetocaloric effect from the magnetocaloric material. This means that in this case, the optimum duration of the fluid flow period would be longer (and the operating frequency lower) compared to the case of an AMR with thinner plates or smaller particles.

Figures 4.17, 4.18 and 4.19 show the impact of the operational parameters (utilization factor and operating frequency) on the temperature span (at zero cooling load) and on the cooling power (Eq. 4.16a) and COP (Eq. 4.17) at 15 K of the temperature span for a packed-bed AMR with Gd spheres of 0.5 mm and a Brayton-like AMR cycle (with a magnetic field change of 1 T and water as a working fluid).

The total mass of gadolinium was assumed to be 0.15 kg, while the hot side temperature was maintained constant at 293 K. The magnetocaloric properties of the gadolinium were calculated using the Mean Field Theory (see [49, 77, 78] for details). By changing the operating frequency, both the fluid flow and the (de) magnetization period were changed, while keeping a constant ratio between those two periods (τ_{mag}:τ_f = 1:4). The results presented here are reproduced using the numerical model described in Tušek et al. [16, 17]. Here, and also in other numerical analyses presented in this chapter, the evaluated operating frequency was limited to 3 Hz. Higher frequencies would in some cases (very fine AMR geometry as explained later in the text) result in higher cooling powers (but in general in smaller efficiency). The limit of 3 Hz was selected since the great majority of so far built magnetic refrigerator prototypes operate with frequency below 3 Hz (see Chap. 7).

It is evident from Figs. 4.17, 4.18 and 4.19 that there is an optimum operating frequency and utilization factors that ensure the optimum operation of the AMR. However, both optimum operating parameters strongly depend on the optimization criterion. The question is which AMR cooling characteristic is the most crucial for a particular application: a high temperature span, a high cooling power or a high efficiency (COP)? In general, the cooling power increases with the operating frequency and the utilization factor, but only up to frequencies and utilization factors at which the required temperature span along the AMR can be established and exceeded. In the particular case shown in Fig. 4.18, the cooling power is increasing with the operating frequency for all the analysed frequencies, since also at the highest analysed frequency the required temperature span of 15 K can be exceeded (see Fig. 4.17). In general, this depends strongly on the AMR geometry and its heat transfer characteristics: the better they are the higher the optimum frequency of operation will be. On the other hand, the cooling power is not monotonically increasing with the utilization factor, since it starts to decrease for utilization factors between 1 and 1.5 (depending on the operating frequency) and reaches negative values at utilization factors of about 2, at which the required temperature span of 15 K cannot be exceeded anymore (see Fig. 4.17).

The largest temperature spans are, in general, obtained at significantly smaller operating frequencies and smaller utilization factors compared to the highest cooling powers. The situation is similar for the COP values, which are, in general, decreasing with the operating frequency (see Fig. 4.19). Higher frequencies are related to higher optimum mass flow rates and therefor larger viscous losses as well as a higher degree of overlapping of the local thermodynamic cycles in the *T–s* diagram and thus a higher input magnetic work, which reduces the overall efficiency. This is also explained in detail in, e.g. [16, 43, 45, 119]. However, an efficient AMR that is related to a fine AMR geometry would also be able to establish a large temperature span at higher frequencies, which further results in a higher cooling power as well as higher COP values.

Figures 4.18 and 4.19 show the AMR's performance at 15 K of temperature span. The reason that this analysis and the analysis later in this chapter are based on a relatively low temperature span (not applied in practical applications) is in the limited maximum temperature span that can be reached by some of the evaluated AMRs, their operating conditions and regimes (thermodynamic cycles). Therefore, a more comprehensive comparison can be performed for a lower temperature span. The dependence of the cooling power on the temperature span and further on the COP is shown in Sect. 4.5 for different AMR thermodynamic cycles. The theoretical results usually show a linear-like dependence of the cooling power and temperature span (for magnetocaloric materials with the second-order phase transition), while the experimentally measured dependence is more complex mostly due to losses to the surroundings [92, 110].

Another crucial parameter that strongly influences the AMR's performance is the geometry of the magnetocaloric material in the AMR. It is in fact beneficial to have as fine an AMR structure as possible (thin plates or small particles and a small hydraulic diameter) in order to be able to operate at a higher optimum operating frequency and with smaller heat transfer losses. However, the AMR geometry (hydraulic diameter) also influences the viscous losses of the fluid flow. On one hand, the small hydraulic diameter increases the convective heat transfer, but on the other hand, it increases the viscous losses and the related pressure drop as well. Therefore, the AMR geometry should be optimized with respect to both thermo-hydraulic properties (heat transfer and pressure drop) in order to obtain the best overall performance [17, 76].

Figures 4.20 and 4.21 show the impact of the AMR's geometry for different packed-bed AMRs with Gd spheres and different parallel-plate AMRs with flat Gd plates for a Brayton-like AMR cycle (with water as the working fluid and 1 T of magnetic field change). The figures show the maximum specific cooling power (per mass of Gd) and the maximum COP obtained for the optimum utilization factor for each analysed case for two different operating frequencies at 15 K of temperature span. For details, including the assumptions used for the geometrical analysis presented in Figs. 4.20, 4.21, see Tušek et al. [17]. Among other assumptions, the analysis neglects the flow maldistribution and assumes a constant porosity in the AMR as well as demagnetization effects in the material, which in general both decrease the AMR's performance.

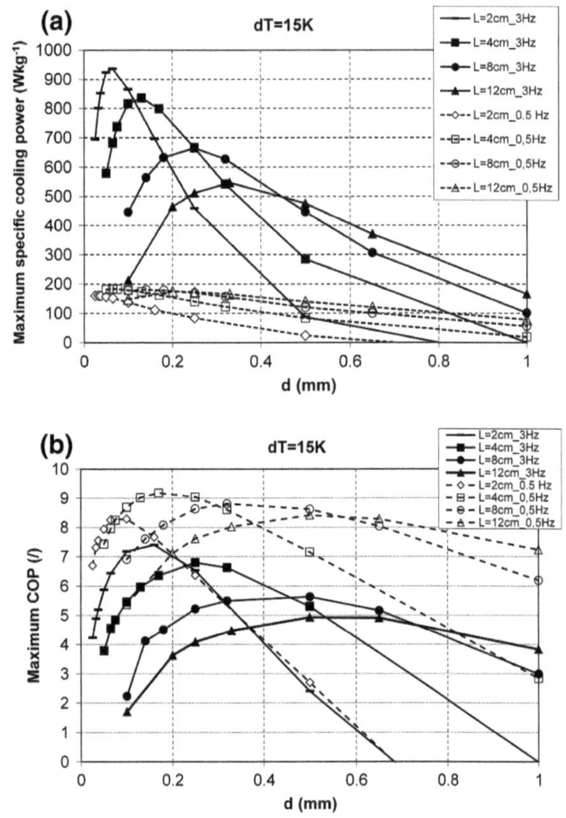

Fig. 4.20 **a** The maximum specific cooling power as a function of the sphere diameter for two different operating frequencies (*dotted lines* for 0.5 Hz and *full lines* for 3 Hz) and different lengths of the packed-bed AMR. **b** The maximum COP as a function of the sphere diameter for two different operating frequencies (*dotted lines* for 0.5 Hz and *full lines* for 3 Hz) and different lengths of the packed-bed AMR

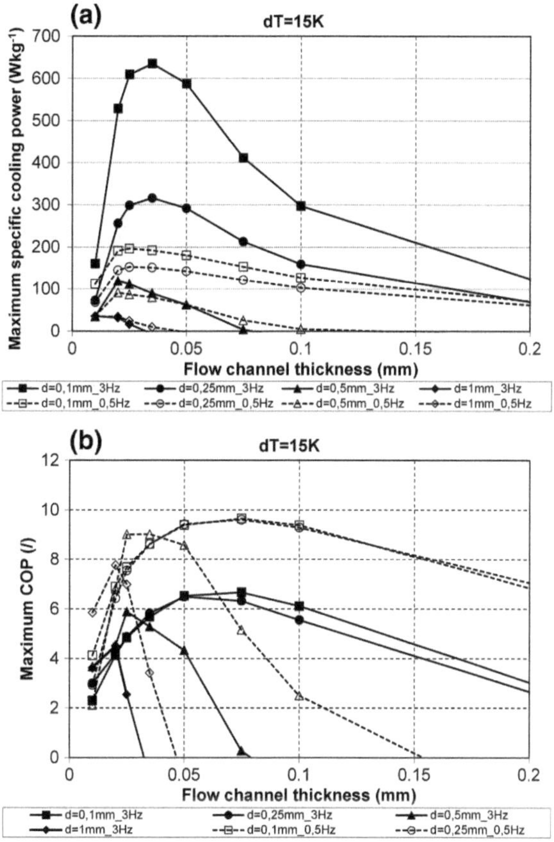

Fig. 4.21 **a** The maximum specific cooling power as a function of the channel thickness for the parallel-plate AMR at two different operating frequencies (*dotted lines* for 0.5 Hz and *full lines* for 3 Hz) and four plate thicknesses. **b** The maximum COP as a function of the channel thickness for the parallel-plate AMR at two different operating frequencies (*dotted lines* for 0.5 Hz and *full lines* for 3 Hz) and four plate thicknesses

However, even though it is to be expected that different magnetocaloric materials, different heat-transfer fluids, different magnetic field changes, etc. would probably lead to different optimum AMR geometries, as shown below, it is crucial to be aware of the impact of the thermohydraulic properties and the AMR geometry on its cooling characteristics.

Figure 4.15 shows the specific cooling power and the COP as a function of the sphere diameter for different lengths of a packed-bed AMR. It is evident that there is a well-defined optimum sphere diameter for each length of the packed-bed AMR from the point of view of the specific cooling power as well as the COP. In the case of a small sphere diameter (below the optimum value), the viscous losses dominate and are increasing with a decreasing sphere diameter and an increased length of the AMR. In the case of a greater sphere diameter (above the optimum value) the heat transfer losses dominate, since in this case the AMR is filled with larger spheres, which leads to a smaller heat transfer coefficient and a smaller total heat transfer area. It is further evident that at a high operating frequency, the packed-bed AMR shows the best performance for very short lengths (20 mm or even less, according to the presented trend of dependency), while at a low operating frequency the best performance is for a length of 40 mm from both the cooling power and COP points of view. In both cases, the optimum corresponding sphere diameter is around 0.1 mm or less (see Fig. 4.20).

In practice, a short, packed-bed AMR is related to a smaller mass of magnetocaloric material and thus a small (absolute) cooling power. In order to increase the cooling power, it is suggested to use a wider or higher AMR. However, the latter is further related to the height of the air gap in the magnet assembly, which is further directly related to a smaller magnetic field in the air gap.

Figure 4.21 shows the specific cooling power and the COP as a function of the channel flow thickness for different plate thickness of a parallel-plate AMR (with a constant length of 80 mm). It was shown by Tušek et al. [17] that due to the small viscous losses of the parallel-plate AMR, its length has only a minor effect on the specific cooling power and the COP values. Therefore, its impact is not shown here.

It is evident from Fig. 4.21a that a parallel-plate AMR shows the largest specific cooling power in the case of a plate that is as thin as possible and that there is an optimum fluid channel thickness of about 0.03 mm, regardless of the plate thickness and the operating frequency. Such a trend was somehow expected, since thin plates provide a larger total heat transfer area, while the fluid channel thickness influences the hydraulic diameter, which is closely related to the heat transfer coefficient as well as to the viscous losses. For small fluid channel thicknesses (below the optimum value) the viscous losses dominate, while at higher fluid channel thicknesses (above the optimum value) the heat transfer losses dominate. A similar trend is observed in the case of the COP values shown in Fig. 4.21b. However, the increase in the COP with the reducing plate thickness is limited by the plate thickness of 0.25 mm, while the optimum corresponding fluid channel thickness increases with the plate thickness and has a value of 0.075 mm at the optimum plate thickness (from the COP point of view).

It is further evident from Figs. 4.20 and 4.21 that there is a difference in the optimum AMR geometry for different operating frequencies. It occurs due to the fact that at low frequencies the optimum mass flow rate (not shown in Figs. 4.20 and 4.21) and consequently the viscous losses are, to some extent, smaller than in the case of a high operating frequency. As a result, at a lower operating frequency the optimum AMR geometry is different (e.g. a smaller optimum sphere diameter for a packed-bed AMR) as compared to a higher operating frequency. This trend is more pronounced in the case of the packed-bed AMR than in the parallel-plate AMR due to the smaller viscous losses of the latter. Furthermore, the viscous losses influence the COP to a much larger extent than the cooling power. In the case of the packed-bed AMR this causes a smaller optimum sphere diameter from the cooling power point of view than from the COP point of view.

Based on the results presented in Figs. 4.20 and 4.21, we can conclude that the packed-bed AMR is able to generate a higher specific cooling power (especially at higher operating frequencies), which is due to the better heat transfer characteristics (a larger total heat transfer area and a smaller hydraulic diameter). It is to be expected that a parallel-plate AMR with a plate thickness below 0.1 mm would generate specific cooling loads comparable to the packed-bed AMRs.

On the other hand, the parallel-plate AMR performs slightly better COP values, mostly due to the smaller viscous losses. However, shorter packed-bed AMRs (below 40 mm with an optimum corresponding sphere diameter) can operate with a similar efficiency as the parallel-plate AMR.

The development of a parallel-plate AMR with a plate thickness of 0.1 mm or less and a spacing of approximately 0.05 mm, and packed-bed AMRs with spheres below 0.1 mm (with short lengths), also from other magnetocaloric materials (not just Gd) is one of the major future challenges for magnetic refrigeration at room temperature. This is very important, especially for operations at higher frequencies, which enables higher cooling loads and high temperature spans (and consequently higher compactness). Furthermore, it is necessary that the AMRs are constructed with none or only a minor maldistribution, since this can drastically affect the AMR's performance (see Sect. 4.3.4).

4.5 The Analysis of Different AMR Thermodynamic Cycles

In this section, we present numerical and experimental analyses of thermodynamic cycles for an AMR. For this purpose, we have evaluated a Brayton-like, Ericsson-like, and a Hybrid Brayton–Ericsson-like AMR cycle. The Carnot-like AMR thermodynamic cycle was numerically investigated; however, it was not tested experimentally, since this was not possible with the experimental device and its magnetic field distribution applied.

4.5.1 Numerical Investigation and Comparison of Different AMR Thermodynamic Cycles

The numerical programme used for the comparison of the thermodynamic cycles with an AMR was based on a model developed by Tušek et al. [16, 17].

In the first analysis, the AMR was assumed to have the shape of a coaxial cylinder with a length of 8 cm, a porosity of 39 %, and a total mass of magnetocaloric material equal to 2.05 kg. This kind of ring can be, for instance, applied in a rotary magnetic refrigerator. It was assumed that the ring consists of packed beds of gadolinium spheres with a diameter of 0.5 mm. The magnetocaloric properties of the gadolinium were calculated using the Mean Field Theory [49, 77, 78]. The working fluid was assumed to be water. The simulations were performed for a 15 K temperature span between the heat source (293 K) and the heat sink.

The working regime of the magnetic cooling device for a particular AMR cycle was defined by the characteristics of the fluid flow profile and the magnetic field profile. The regimes for the Brayton-like, Ericsson-like, and Carnot-like AMR cycles are shown in Table 4.5.

Table 4.5 Fluid flow and (de)magnetization periods for the analysed thermodynamic cycles

Time periods/Magnetic field	Type and characteristics of the process				Magnetic field distribution
Brayton	Adiabatic magnetization	Isofield cooling of MCM	Adiabatic demagnetization	Isofield heating of MCM	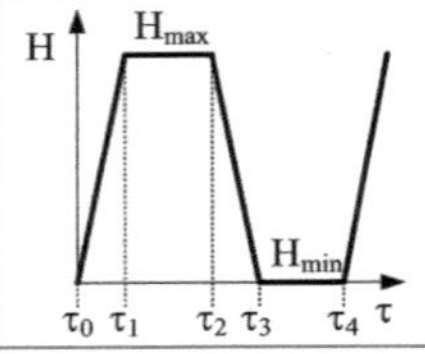
	$(\tau_1 - \tau_0)(v_f = 0)$	$(\tau_2 - \tau_1) = 4(\tau_1 - \tau_0)$ $(v_f > 0)$	$(\tau_3 - \tau_2)$ $(v_f = 0)$	$(\tau_4 - \tau_3) = 4(\tau_3 - \tau_2)$ $(v_f < 0)$	
Ericsson	Isothermal magnetization	Isofield cooling of MCM	Isothermal demagnetization	Isofield cooling of MCM	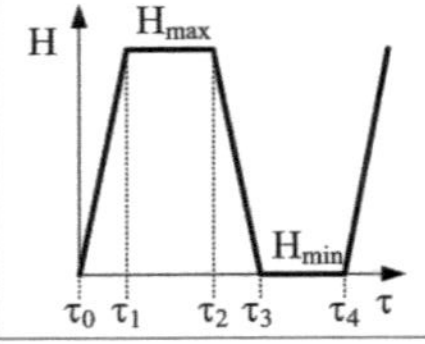
	$(\tau_1 - \tau_0)(v_f > 0)$	$(\tau_2 - \tau_1) = 4(\tau_1 - \tau_0)$ $(v_f > 0)$	$(\tau_3 - \tau_2)$ $(v_f < 0)$	$(\tau_4 - \tau_3) = 4(\tau_3 - \tau_2)$ $(v_f < 0)$	
Carnot	Adiabatic magnetization	Isothermal magnetization	Adiabatic demagnetization	Isothermal demagnetization	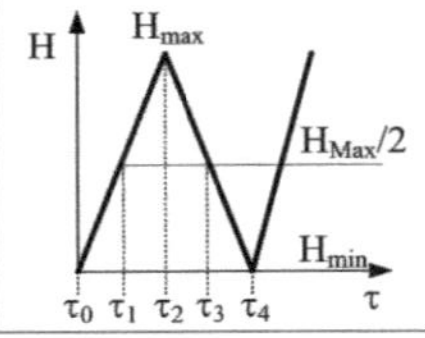
	$(\tau_1 - \tau_0)$ $H_1 = H_{max}/2$ $(v_f = 0)$	$(\tau_2 - \tau_1)$ $H_2 = H_{max}$ $(v_f > 0)$	$(\tau_3 - \tau_2)$ $H_3 = H_{max}/2$ $(v_f = 0)$	$(\tau_4 - \tau_3)$ $H_4 = H_0 = 0$ $(v_f < 0)$	

In Fig. 4.22, the maximum specific cooling power and the maximum COP (obtained at the optimum utilization factor—mass flow rate for each case) are presented as a function of the magnetic field. In this particular case the frequency of the device was held constant at 3 Hz.

In Fig. 4.23, the maximum specific cooling power and the maximum COP are presented as a function of the frequency of the operation (for a constant magnetic field of 1 T). Like for the results shown in Figs. 4.19, 4.20, 4.21 and 4.22, the utilization factor was varied in order to obtain the maximum values of the COP and the maximum values for the specific cooling power per mass of magnetocaloric material.

According to Fig. 4.22, the maximum specific cooling power can be obtained with a Brayton-like AMR cycle, which shows slightly better cooling performance than the Ericsson-like AMR cycle. However, the highest COP can be obtained with the Ericsson-like AMR cycle. When compared to a Brayton-like and an Ericsson-like AMR cycle, a substantially lower specific cooling power and a lower COP can be obtained with the Carnot-like AMR cycle. The increases in the cooling power and the COP with the magnetic field are more evident for the Carnot-like AMR cycle.

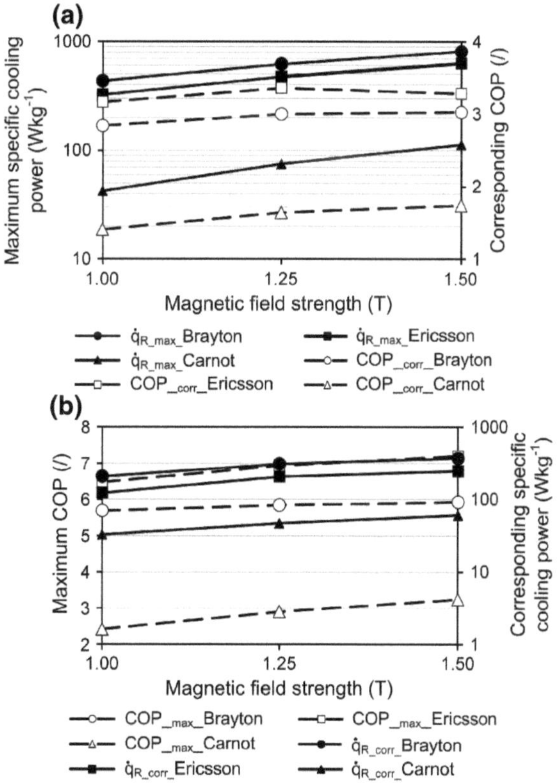

Fig. 4.22 **a** The maximum specific cooling power as a function of the magnetic field for three different magnetic refrigeration thermodynamic cycles. **b** The maximum COP as the function of the magnetic field for three different magnetic refrigeration thermodynamic cycles

Fig. 4.23 **a** The maximum specific cooling power as the function of the frequency of the operation for three different magnetic refrigeration thermodynamic cycles. **b** The maximum COP as a function of the frequency of the operation for three different magnetic refrigeration thermodynamic cycles

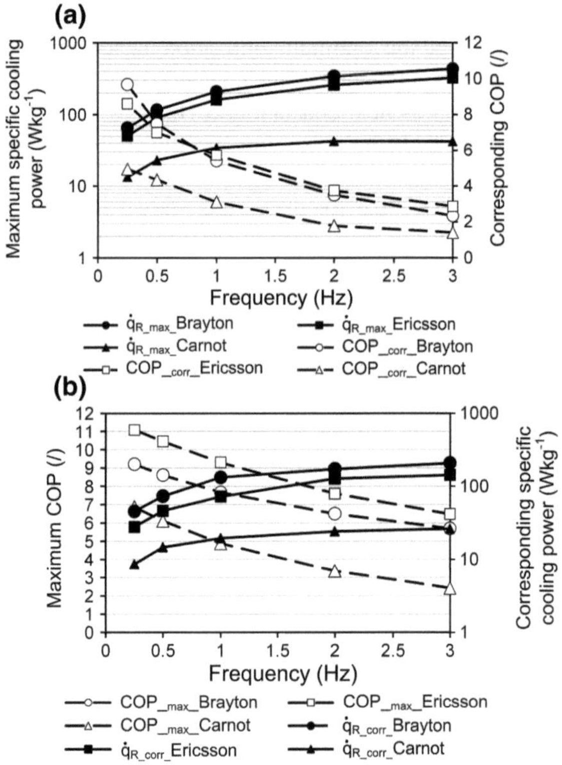

In all three cases, both the specific cooling power and the COP increase with an increase of the magnetic field. It is well known that a higher magnetic field change leads to a higher cooling power and also higher COP values (although to a smaller extent, since with a larger magnetic field change the performed magnetic work is increased) [37]. The reason for this is that a higher magnetic field change leads to a higher adiabatic temperature change, which compensates the irreversible heat losses between the magnetocaloric material and the working fluid. For example if the adiabatic temperature change during the magnetization equals, e.g. 4 K, this also represents theoretically the maximum possible temperature difference between the magnetocaloric material and the working fluid.

The temperature difference between the magnetocaloric material and the working fluid during the fluid flow period strongly depends on the heat transfer coefficient and the available heat transfer area. In the case that the heat transfer coefficient or the area is too small, then the temperature difference between the magnetocaloric material and the working fluid is too large (or the heat cannot be transferred for such a small temperature difference). Literally, the irreversible heat losses in terms of temperature difference will overcome the "generated" adiabatic temperature change, and the device will not be able to cool any more.

It can be seen from Fig. 4.23 that the maximum cooling power per mass of magnetocaloric material increases with the frequency of operation for all the analysed frequencies, except for the Carnot-like AMR cycle, where the peak specific cooling power is obtained at a frequency of around 2 Hz (this is also strongly related to the AMR's geometry, as explained in Sect. 4.4). Between the Brayton-like and Ericsson-like AMR cycles there is small difference in the cooling power, whereas the Carnot-like AMR cycle provides much less cooling power. The highest COP can be obtained with an Ericsson-like cycle, and the lowest COP can be obtained with a Carnot-like AMR cycle.

It is important to note that the permanent magnet assembly (in addition to the fluid flow regime) defines the type of thermodynamic cycle that can be performed. The mass of the magnet assembly for the Carnot-like cycle can be much smaller than the mass of the magnet assembly for the Brayton-like or Ericsson-like cycle (see [10] for details). Since the magnet assembly represents the major costs of a magnetic refrigerator [15], it also makes sense to represent results as a function of the magnet mass, as was presented by Kitanovski et al. [10].

Based on the results of the first analysis, we can conclude that the considered Carnot-like AMR cycle, compared to the Brayton-like or the Ericsson-like cycles, operates with a much lower specific cooling power and efficiency. This is mostly due to the increased irreversible heat transfer losses related to the non-ideal regeneration between the neighbouring material particles in the regenerator. Small increments in the material's temperature due to quasi-isothermal magnetization will also require a very small temperature difference between the magnetocaloric material and the working fluid, which are unfortunately strongly limited by the heat transfer coefficient and the available heat transfer area. Therefore, the selection of an appropriate cycle can drastically influence the regeneration process.

In both the Carnot-like and Ericsson-like AMR cycles, the performance is not as high as would be expected from just studying the *T–s* diagram. Therefore, the application of a *T–s* diagram is not the right method to study the AMR cycle's performance, especially because it does not account for the AMR's regeneration process and the corresponding heat transfer losses between the working fluid and the magnetocaloric material.

In the second analysis, which was performed by Plaznik et al. [9], three different thermodynamic cycles were analysed: the Brayton-like, the Ericsson-like and the Hybrid Brayton–Ericsson-like AMR cycles. The aim of the investigation was the same as in the first analysis, i.e. to investigate the performance of a magnetic cooling device with an AMR. However, in this particular case, the influence of the magnetization profile (defined by the magnet assembly) on the performance of the AMR was investigated. Furthermore, two different types of AMRs were evaluated in the simulation (Fig. 4.24). The parallel-plate and packed-bed AMR, respectively. In both AMRs, gadolinium was used as the magnetocaloric material and water was considered as the working fluid. In all cases, the magnetic field change was considered to be 1 T.

(a)

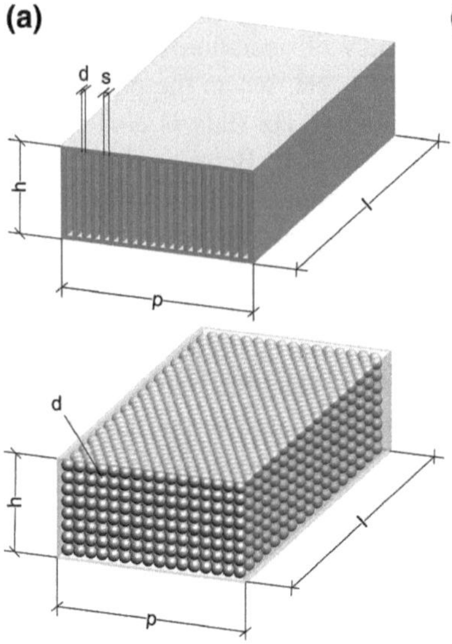

(b)

	Parallel plate AMR	Packed bed AMR
d [mm]	0.25	0.5
s [mm]	0.1	/
p [mm]	40	40
l [mm]	80	80
h [mm]	10	10
m [kg]	0.18	0.15

Fig. 4.24 **a** Schematic presentation of a parallel-plate AMR (*up*) and packed-bed AMR (*down*). **b** Table with geometry of the analysed AMRs

Since the Brayton and Ericsson-like AMR cycles and the corresponding time periods are already shown in Table 4.5, the Table 4.6 only shows the time periods for the Hybrid Brayton–Ericsson-like AMR cycle.

The first simulations were performed for different predefined temperature spans between the heat source and heat sink, i.e. 4, 8,12, 16 and 20 K (with a heat source temperature of 293 K). In this particular case, two operating frequencies (ν) were applied in the simulations, i.e. 0.5 and 3 Hz, respectively. The utilization factor (defined by Eq. 4.26) was fixed with the value U = 0.3. Note that the selected operating conditions have been often realized in the best magnetocaloric cooling devices built to date [5].

For the particular cases in Figs. 4.25 and 4.26, the ratios between the magnetization period and the period of constant magnetic field were $\tau_{mag}:\tau_{const}$ = 1:4 for both frequencies of operation; however, for different absolute time periods (matched to a particular cycle).

Figure 4.25 shows the dependence of the temperature span and the COP on the specific cooling power for three different types of magnetic refrigeration cycle for a packed-bed AMR.

Figure 4.26 shows the dependence of the temperature span and the COP on the specific cooling power for three different types of AMR cycles for a parallel-plate AMR.

From Figs. 4.25 and 4.26, it is evident that the Brayton-like cycle can achieve the highest specific cooling power, regardless of the temperature span and the AMR geometry. A slightly lower specific cooling power can be obtained with the Hybrid

Table 4.6 Fluid flow and (de)magnetization periods for the Hybrid Brayton–Ericsson-like AMR cycle

Time periods/magnetic field					
	Type and characteristics of the process				Magnetic field distribution
Hybrid	Adiabatic magnetization	Isofield cooling of MCM	Adiabatic demagnetization	Isofield heating of MCM	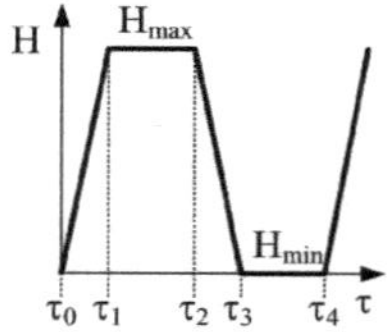
	$(\tau_{1/2} - \tau_0)(v_f = 0)$	$(\tau_2 - \tau_1) = 4(\tau_1 - \tau_0)$ $(v_f > 0)$	$(\tau_3 - \tau_2)/2$ $(v_f = 0)$	$(\tau_4 - \tau_3) = 4(\tau_3 - \tau_2)$ $(v_f < 0)$	
	Isothermal magnetization		Isothermal demagnetization		
	$(\tau_1 - \tau_{1/2})$ $(v_f > 0)$		$(\tau_3 + \tau_2)/2$ $(v_f < 0)$		

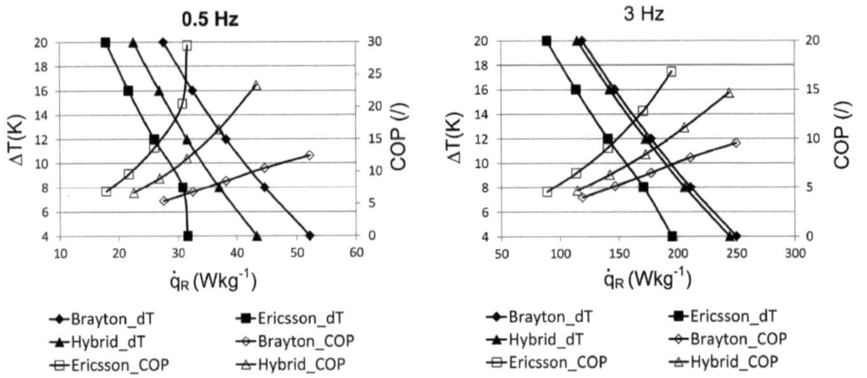

Fig. 4.25 The dependence of the temperature span and the COP on the specific cooling power for three different types of magnetic refrigeration cycles for a gadolinium packed-bed AMR in a magnetic field change of 1 T for two different frequencies of operation (0.5 and 3 Hz) at $U = 0.3$

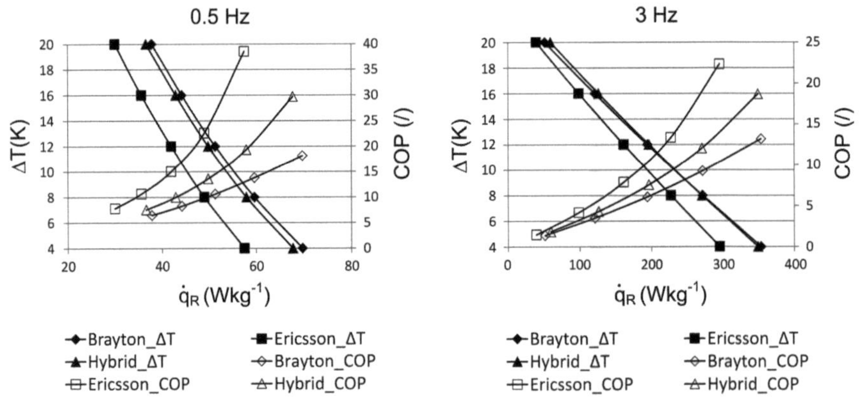

Fig. 4.26 The dependence of the temperature span and the COP on the specific cooling power for three different types of magnetic refrigeration cycles for a gadolinium parallel-plate AMR in a magnetic field change of 1 T for two different frequencies of operation (0.5 and 3 Hz) at $U = 0.3$

AMR cycle, and the smallest cooling power can be obtained with the Ericsson-like AMR cycle.

The highest COP can be obtained with the Ericsson-like, followed by the Hybrid AMR cycle, while the Brayton-like AMR cycle shows the lowest efficiency. There are two factors that make the Ericsson and Hybrid cycles more efficient than the Brayton cycle, regardless of the irreversible, higher heat transfer losses. The first reason is a smaller amount of magnetic work that is defined as the total surface area of each particle of magnetocaloric material in the *T–s* diagram. The other factor that has a positive effect on the COP of the Ericsson-like and Hybrid AMR cycles is the lower pressure drop. Since the fluid flow period in the case of the Ericsson-like and

Hybrid AMR cycles is longer (as it is also performed during the (de)magnetization process), smaller velocities are required for the same utilization factor. Consequently, a smaller pressure drop occurs.

It has been reported in Bjørk and Engelbrecht [18] that changing the magnetic field profile can have a strong impact on the performance of the AMR. A further investigation was performed in order to study the impact of the magnetization profile on the performance of a particular thermodynamic cycle. In this particular case, four different magnetic field profiles were considered (Fig. 4.27). These were defined as the ratio (τ_{mag}:τ_{const}) between the time periods of a variable magnetic field (equal to the magnetization period) and the time period during which the magnetic field remains constant. The simulation was carried out for τ_{mag}:τ_{const} ratios between 0.125 and 1. The temperature span was taken to be 15 K.

Figure 4.28 shows the results of a numerical analysis that was performed for a gadolinium packed-bed AMR with different magnetic field profiles and different AMR thermodynamic cycles. Note again that the utilization factor was varied in order to obtain the maximum values of the COP and the maximum values for the specific cooling power per mass of magnetocaloric material.

Based on the results in Fig. 4.28, we can conclude that for a particular geometry of the AMR (Fig. 4.24), a higher cooling power or COP can be obtained with a smaller ratio τ_{mag}:τ_{const}, since a long (de)magnetization time increases the irreversible heat transfer losses.

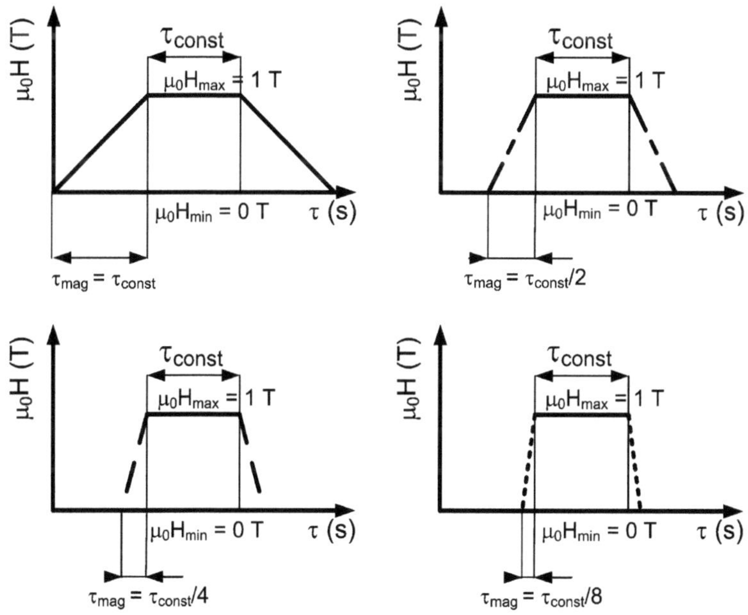

Fig. 4.27 The magnetic field profiles as a function of the time period

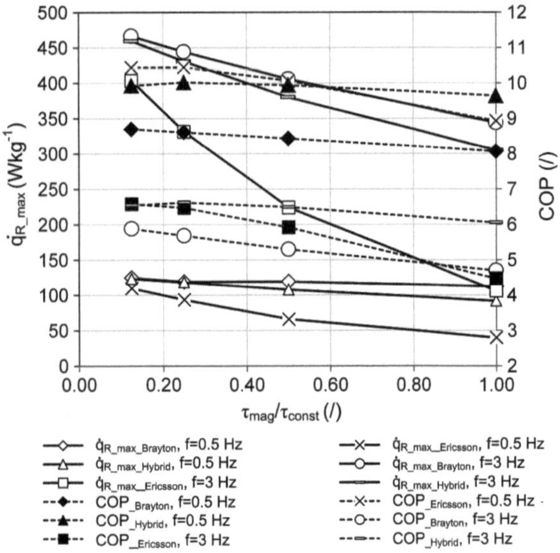

Fig. 4.28 The specific maximum cooling power per mass of magnetocaloric material and the maximum COP as a function of the magnetic field profile ($\Delta\mu_0 H = 1$ T, $\Delta T = 15$ K)

As can be seen from Fig. 4.28, the Brayton-like AMR cycle, regardless of the τ_{mag}:τ_{const} ratio, can exhibit greater cooling powers than the Hybrid and Ericsson-like AMR cycles. However, both the Hybrid and the Ericsson-like AMR cycles can operate with a higher COP than the Brayton-like AMR cycle.

4.5.2 Experimental Investigation and Comparison of Different AMR Thermodynamic Cycles

Tests were carried on an experimental device that was presented in detail in [101]. This experimental device is also presented in the Chap. 7.

The gadolinium parallel-plate AMR was applied in experiments (see Fig. 4.24 for the photographs and the geometry). The heat-transfer fluid (solution of water (70 %) and ethylene glycol (30 %)) is pumped through the AMR by means of two connected pistons that are driven by an electric actuator. Different mass flow rates can be achieved by varying the piston's offset distance and its velocity. The average magnetic field provided by the permanent magnet assembly was measured to be 1.15 T (magnetization area).

In the experiment, three different working regimes (AMR thermodynamic cycles) were investigated (Fig. 4.29). They represent the real measured time dependence between the magnetic field profile and the fluid flow profile to which the AMR was exposed.

Because of the technical characteristics of the experimental device, the (de) magnetization time was relatively long ($\tau_{mag} = 0.75$ s) and therefore the frequencies of the test device and consequentially the cooling powers were relatively low.

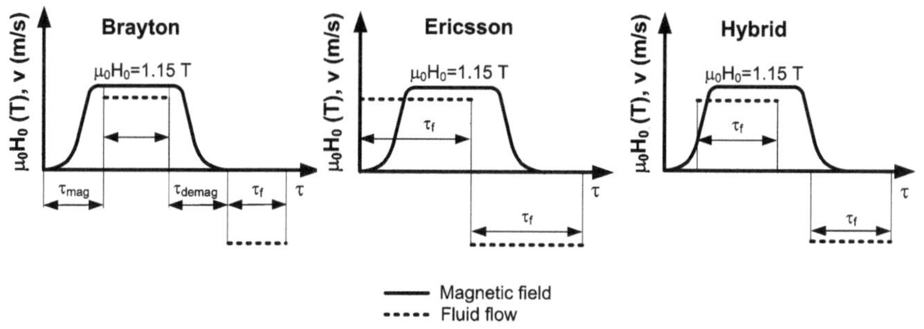

Fig. 4.29 The magnetic field profiles and the corresponding fluid flow periods for the AMR thermodynamic cycles investigated in the experiment

The $\tau_{mag} = 0.75$ s was taken in all the experimental analyses shown in this section. The τ_f was changed for a particular test of the AMR's thermodynamic cycle (information about this period is provided in the text below). In addition to τ_{mag} and τ_f, the response time of the data acquisition and control elements was, in all the cases, equal to $\tau_0 = 0.2$ s per cycle. The frequency of the operation was therefore defined by the duration of the total period. In the case of the Brayton-like AMR refrigeration cycles this equals to $1/(2\ \tau_{mag} + 2\ \tau_f + \tau_0)$. However, in the case of the Ericsson-like AMR cycle, the frequency of operation was defined as $1/(2\ \tau_f + \tau_0)$, since the fluid flow was performed during all the thermodynamic processes. In the Hybrid AMR cycle the frequency was defined as $1/(2(\tau_f + \tau_{mag/2}) + \tau_0)$.

The first tests were performed for no cooling load conditions in order to measure the maximum possible temperature span for a given AMR thermodynamic cycle. For this purpose, the utilization factor and the working frequency were varied.

Figure 4.30 shows the ratio between the maximum no-load temperature spans (obtained at U = optimum, which has been defined at the maximum temperature span) and the adiabatic temperature change ΔT_{ad} for different frequencies (we denote this as the regeneration factor). As can be seen from Fig. 4.30, the largest regeneration factor (6.6 at $f = 0.3$ Hz and $\tau_f = 0.85$ s) was obtained with the Hybrid

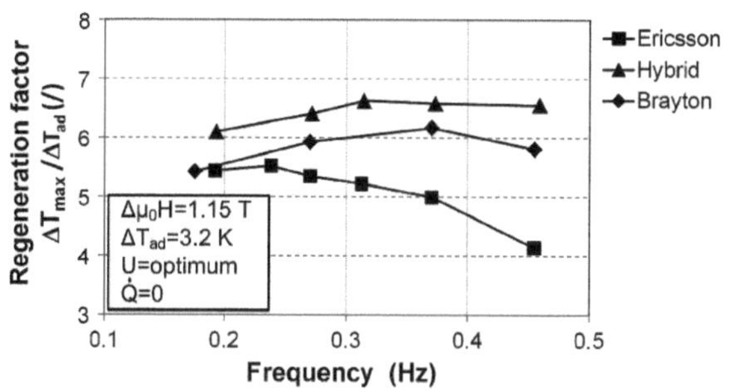

Fig. 4.30 The experimentally obtained no-load maximum temperature span as a function of the operating frequency for a parallel-plate gadolinium AMR

AMR cycle. Furthermore, in the Hybrid AMR cycle, the regeneration factor remains almost constant for a wider range of operating frequencies, whereas in a Brayton-like, and especially an Ericsson-like AMR cycle, the regeneration factor drastically decreases with the increased frequency. However, at low operating frequencies (i.e. a longer duration of the fluid flow period), the Hybrid AMR cycle approaches the regeneration factor of the Brayton-like and Ericsson-like AMR cycles.

By changing the working regime of the device and performing the Hybrid AMR cycle, the performance of the test device was improved (e.g. for the maximum no-load temperature span and consequently the regeneration factor) compared to the Brayton-like AMR cycle. A similar conclusion with respect to combining the process of (de)magnetization and fluid flow was also obtained by Bjørk and Engelbrecht [18].

Next, the experiments were performed at a fixed frequency of the operation ($f = 0.37$ Hz) and with a constant utilization factor ($U = 0.3$). The cooling load was varied in order to obtain different temperature spans, related to a particular AMR thermodynamic cycle (Fig. 4.31).

The experimentally predicted COP of the AMR cycle was calculated using the numerical programme [16, 17] as the ratio between the experimentally obtained cooling power and the sum of the numerically calculated input work, which can be further divided into the work required to pump the fluid and the magnetic work (in this particular case, the related mechanical losses were neglected). The magnetic work input was calculated as the integral of the performed thermodynamic cycles in the T–s diagram. A detailed description of the experimentally predicted COP can be found in Plaznik et al. [9]. For the particular case of Fig. 4.31, the ratios between the magnetization period and the fluid flow period were the following: Brayton-like AMR (τ_{mag}:τ_f = 0.75:0.5); Ericsson-like AMR (τ_{mag}:τ_f = 0.75:1.25); Hybrid (τ_{mag}:τ_f = 0.75:0.87).

Based on the results of Fig. 4.31, we can estimate that the cooling performances and the maximum temperature span (under no-load conditions) of the Brayton-like and Hybrid Brayton–Ericsson-like AMR cycles are similar. On the other hand, the

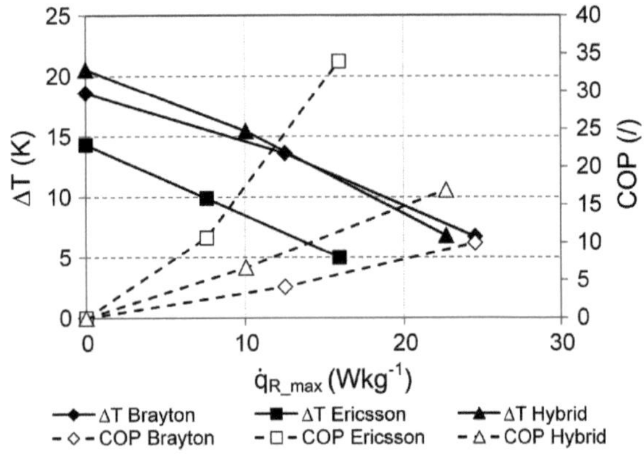

Fig. 4.31 The experimentally obtained maximum temperature span and COP as a function of the cooling load for a parallel-plate gadolinium AMR ($\Delta\mu_0 H = 1.15$ T, $U = 0.3, f = 0.37$ Hz)

Ericsson-like AMR cycle exhibits a significantly lower cooling power for all the temperature spans. However, it is predicted that the Ericsson-like AMR cycle can operate with much higher efficiency, especially when the thermodynamic cycles are compared at higher specific cooling power. Note again, that the frequency in this particular case as well as the utilization $U = 0.3$ were kept constant.

A direct comparison of the numerical results and the experimental ones under the same operating conditions is beyond the scope of this analysis. This is due to the fact that there are some factors (the demagnetization effect and flow maldistribution—see Sect. 4.3.4 for details) that have a strong influence on the performance of the AMR (see [92] for details), and have not been included in the particular numerical model. However, it can be concluded that both, the numerical and experimental analysis showed the same trend of dependency for all the analysed cycles.

4.5.3 Guidelines for Future Research on AMR Thermodynamic Cycles

Future magnetic refrigeration devices will have to efficiently operate at high temperature spans between the heat source and the heat sink. Furthermore, high frequencies are required, since these are related to high power densities (compactness and related cost). In order to perform an efficient operation, the thermodynamic cycles have to be carefully studied. We have proven that at present the mostly applied Brayton-like AMR thermodynamic cycle should be replaced by other types of thermodynamic cycles.

It has been shown in this chapter that the performance of the Carnot-like AMR cycle was significantly poorer compared to other analysed AMR cycles. Among all the evaluated thermodynamic cycles, the Ericsson-like AMR cycle is the most efficient, which can also be seen from the *T*–*s* diagram. It is evident that the thermodynamic cycles based on isothermal (de)magnetization require less magnetic work, e.g. compared to the Brayton-like AMR thermodynamic cycle. On the other hand, isothermal magnetization is related to higher heat transfer irreversibility losses due to a smaller average temperature difference between the fluid and the magnetocaloric material and therefore a less intense heat transfer. This directly results in a smaller cooling power (and the temperature span), which is the main disadvantage of an Ericsson-like AMR cycle, and most probably also the Stirling-like AMR cycle, although the latter was not the subject of an analysis.

With regard to high efficiency and the cooling power, the Hybrid Brayton –Ericsson-like AMR cycle represents a serious alternative to the Brayton cycle. The introduction of this kind of thermodynamic cycle will not only affect the efficiency and the power density of a device, but will also have an important impact on the design features of the magnet assembly. It has been shown by Kitanovski et al. [10], that the homogenization of the magnetic field, such is required for instance for the Brayton type of magnetic refrigeration cycle, may lead to a higher required mass of the magnets, compared to some other thermodynamic cycles. In a magnetic refrigerator or heat pump, two aspects—the energy efficiency and the specific cost (e.g. eurokW^{-1}) of a device can be strongly affected by the chosen thermodynamic cycle. Besides the AMR itself, this is also related to the corresponding magnet assembly and the fluid flow characteristics. In a certain magnet assembly, researchers usually tend to obtain a homogeneous magnetic field (in both high and low magnetic field regions), which is preferable, especially for the Brayton-like AMR cycle [120, 121]. However, such an optimization of the magnetic field also usually leads to more complex permanent magnet assemblies and sometimes also a reduced magnetic field change. Different AMR cycles can therefore lead to different, potentially simpler and less expensive, magnet assemblies.

By choosing the optimum cycle, the performance of the cooling device can be substantially improved. Therefore, the research community should invest greater efforts in improving magnetic refrigeration devices with new thermodynamic cycles.

4.6 The Impact of the Heat Transfer Fluid

The heat-transfer (working) fluid and its thermohydraulic properties have an important role in the performance of the AMR. In order to ensure good cooling characteristics for the AMR (at high frequency of operation) the applied working fluid should, in general, have high a thermal conductivity and thermal diffusivity, and a low viscosity.

The majority of the magnetic refrigerator prototypes use water-based heat-transfer fluids with different alcohol additives. Some earlier prototypes also applied gases, such as helium, nitrogen or even air [5]. Similarly, also the majority of numerical analyses of the AMR performance consider water as a heat-transfer fluid [33]. Water is often chosen due to its very good heat transfer properties, non-toxicity and simplicity of use. However, the majority of modern applied magnetocaloric materials in the AMR corrode when in direct contact with water (for details see [122–125] where different corrosion inhibitors are considered). It was shown [28, 101] that a mixture (e.g. ratio of 70:30) of distilled (or deionised) water with different alcohols (e.g. commercial automotive ethylene glycol) can prevent the corrosion of the most promising magnetocaloric materials (Gd, La–Fe–Co–Si). Furthermore, the alcohol additives decrease the freezing temperature of the mixture below 0 °C, which will be required in future magnetic refrigeration systems (but also reduce its thermal diffusivity).

Not many analyses on the impact of the heat-transfer fluid on the AMR performance have been performed to date. This subject was somehow neglected, although it is very important. Petersen [119] numerically and experimentally, while Kitanovski et al. [15] and Wu et al. [126] numerically, evaluated and compared the performance of the AMR with different working fluids. They evaluated water, liquid metals (mercury and Galinstan—a liquid alloy consisting mainly of gallium, indium and tin), different alcohols (ethanol, ethylene glycol, propylene glycol, and glycerol) and different mixtures of water and ethanol. They concluded that the AMR shows by far the best cooling characteristics if liquid metals are used as a working fluid (especially at higher operating frequencies (>2 Hz)), while among other fluids pure water showed the best performance. Similar conclusions were obtained also by Silva et al. [127], who theoretically compared the performance of different fluids (water, mercury, gallium and different alcohols) used in a high-frequency micro-size magnetocaloric refrigerator. They concluded that gallium shows even better performance than mercury, which is highly toxic and in any case cannot be used in a real magnetic refrigeration or heat pump device.

This section presents a numerical comparison of the AMR performance with six different heat-transfer fluids. The analysed fluids and their relevant thermohydraulic properties are shown in Table 4.7.

The analysis was performed using the AMR numerical model [16, 17] with gadolinium as the magnetocaloric material in two different geometries (packed-bed and parallel-plate) and 1 T of magnetic field change. The packed-bed AMR is constructed with spheres of 0.25 mm diameter, while the parallel-plate AMR with plates having a thickness of 0.25 mm and a spacing of 0.1 mm. The outer dimensions of both AMRs are: 80 mm (length) × 40 mm (width) × 10 mm (height).

The evaluated fluids were selected with respect to the current state of the art. Since the majority of the magnetic refrigerator prototypes apply water or a mixture of water with different alcohols, we chose water, two mixtures of water and ethanol and pure ethylene glycol. Silicone oil was selected as it is often used in various hydraulic and thermal applications. Since some earlier works [15, 119, 126, 127] showed a clear advantage when applying liquid metals as the working fluid in the magnetic refrigerator (mercury and gallium) we evaluated Galinstan (mainly consisting of gallium, indium, and tin), which has low toxicity. Lately, it has been used as a replacement for mercury and has the potential to be used in a future magnetic refrigerator or a heat pump.

Table 4.7 Thermohydraulic properties of the working fluids evaluated in the analysis

	Specific heat ($Jkg^{-1}K^{-1}$)	Density (kgm^{-3})	Thermal conductivity ($Wm^{-1}K^{-1}$)	Thermal diffusivity (m^2s^{-1})	Dynamic viscosity (Pas)
Water	4,180	998	0.599	1.44×10^{-7}	1.00×10^{-3}
Water + wt. 20 % ethanol	4,350	968	0.465	1.10×10^{-7}	2.23×10^{-3}
Water + wt. 50 % ethanol	3,470	922	0.342	1.07×10^{-7}	2.31×10^{-3}
Ethylene glycol	2,303	1,120	0.303	1.17×10^{-7}	3.00×10^{-2}
Silicone oil	1,620	855	0.163	1.18×10^{-7}	1.29×10^{-3}
Galinstan	370	6,440	16.05	6.74×10^{-4}	2.40×10^{-3}

Figures 4.32 and 4.33 show the maximum specific cooling power (per mass of magnetocaloric material) and the maximum COP (obtained at the optimum utilization factor—mass flow rate for each case) for all the analysed working fluids as a function of the operating frequency for packed-bed and parallel-plate AMRs, respectively. The temperature span between the heat source (293 K) and the heat sink was set at 15 K.

The trend of dependency for the specific cooling power and the COP with the operating frequency presented in Figs. 4.32 and 4.33 is in general expected and explained in more detail in Sects. 4.4 and 4.5 (for water as a heat-transfer fluid).

It was also expected that due to the better heat transfer geometry the packed-bed compared to parallel-plate AMR would have a higher specific cooling power and a higher optimum operating frequency (the exception here is Galinstan in the parallel-plate AMR, as explained below), but also smaller COP values due to higher viscous losses.

In the case of the packed-bed AMR the best cooling characteristics were obtained with water. The additives of ethanol significantly reduce the performance of a device, especially the cooling power at frequencies above 1 Hz (mostly due to the lower thermal diffusivity and thermal conductivity). Even though Galinstan has by far the highest thermal diffusivity of all the evaluated fluids, it does not perform as well as some might expect in the packed-bed AMR. This is mostly due to its very high density and the viscous losses, which prevent a better performance. A similar situation is also true for ethylene glycol, which suffers from a very high viscosity

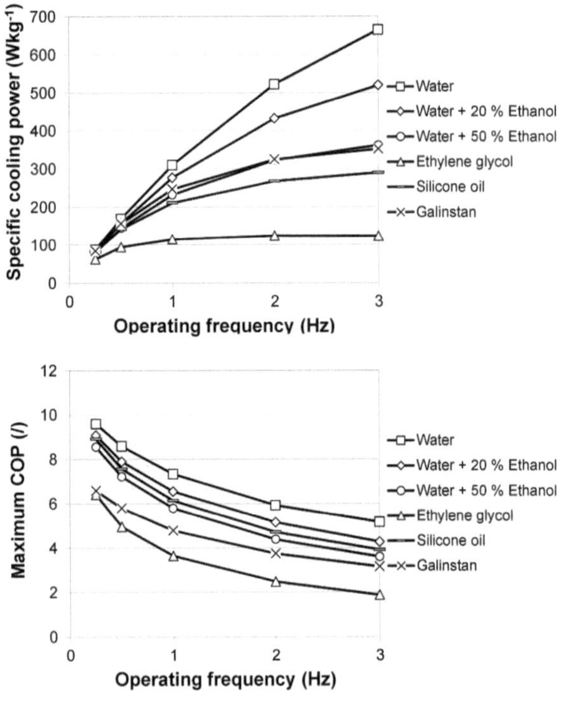

Fig. 4.32 The maximum specific cooling power and the maximum COP for the analysed working fluids as a function of the operating frequency for a packed-bed AMR at 1 T of magnetic field change and 15 K of temperature span

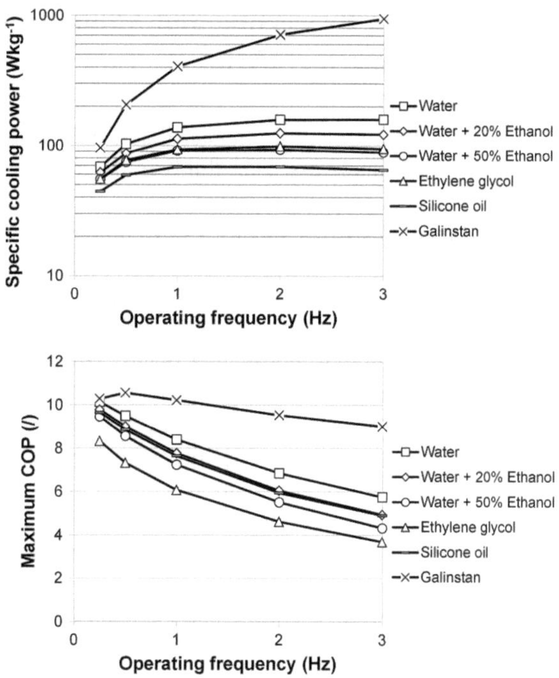

Fig. 4.33 The maximum specific cooling power and the maximum COP for the analysed working fluids as a function of the operating frequency for a parallel-plate AMR at 1 T of magnetic field change and 15 K of temperature span

and thus shows by far the lowest performance in the packed-bed AMR (from the COP and the cooling power points of view).

On the other hand, due to a smaller friction factor the parallel-plate AMR is much less effected by the hydraulic properties of the working fluid (density, viscosity), but more by the thermal diffusivity and thermal conductivity.

It is evident from Fig. 4.33 and Table 4.7 that there is almost a direct correlation between the thermal conductivity of the fluid and the cooling power. This is not entirely true for the COP values, which are more effected by the viscous losses and therefore ethylene glycol leads to the lowest COP. However, the AMR with Galinstan performs significantly higher specific cooling powers than with any other fluid (even in the packed-bed AMR) and the optimum frequency of operation is well above 3 Hz (which is not the case with other fluids in the parallel-plate AMR). This is mostly due to the superior thermal diffusivity and thermal conductivity.

It can be concluded that the fluids with a high thermal diffusivity and, especially, thermal conductivity would significantly increase the AMR's performance and enable operation at higher frequencies and higher COP values compared to water and especially other analysed fluids. This is clearly evident for Galinstan in the parallel-plate AMR. However, such fluids (e.g. liquid metals) often have high densities (and/or viscosity), which increase the viscous losses and prevent its efficient use in packed-bed or similar AMRs. This is why in the packed-bed AMRs water was found to be the optimum working fluid (among the analysed fluids).

4.7 Review of Processing and Manufacturing Techniques for AMRs

It was shown earlier in this chapter that a fine AMR geometry (the micro-channel range of wall thickness and fluid voids that are well below 1 mm) with a homogenous porosity is one of the crucial preconditions for efficient operation. The so far evaluated AMR geometries were limited to more or less simple solutions of packed beds of powders, grains, or spheres and parallel-plate structures [5]. The first type of AMR suffers from high viscous losses, while the second has relatively poor heat transfer properties [76]. The geometries that can have better thermohydraulic properties (e.g. corrugated plates, honeycomb structures, different foams) are difficult to manufacture with currently available magnetocaloric materials and conventional processing technologies. However, some recent achievements in powder metallurgy, e.g. [128] might enable the fabrication of different micro-channel AMRs, also with currently the most interesting La-based (i.e. La–Fe–Co–Si), Mn-based (i.e. Mn–Fe–P–As; Mn–Fe–P–Si) and perovskite-type manganese oxides based (i.e. La–Ca–Sr–MnO_3) magnetocaloric materials.

In this section, some of the most promising applied fabrication techniques for AMRs are reviewed. The presented methods are generally divided into two groups: the fabrication of Gd-based AMRs, which are currently limited to packed-bed and parallel-plate structures, and the fabrication of sintering-based AMRs, which in general makes possible more advanced structures.

4.7.1 Fabrication of Gd-based AMRs

Gadolinium is a metal with relatively good malleable and ductile properties. This means that it can be formed into different geometries, like thin plates (>50–100 μm), wires, spheres, and cylinders using standard forming technologies (Fig. 4.34). These are also the only four geometries (and powder) of Gd evaluated in an AMR to date. The thin Gd plates or wires that are needed for an efficient AMR are too brittle to be further formed into more efficient heat transfer geometries, like corrugated plates, honeycomb structures or packed wire screens. However, it is to be expected that further research in this field would lead to more advanced, Gd-based AMRs. By Gd-based, we mean pure gadolinium or different gadolinium alloys with other materials (mostly Er, Tb or Dy).

The fabrication of a packed-bed AMR is relatively straightforward. This holds for all packed-bed AMRs and not only those which are Gd-based, while packed-bed AMRs with any other magnetocaloric materials are limited to powder. However, special attention should be given to the infill of particles into the AMR housing. In order to achieve as homogenous porosity as possible, the AMR should be filled gradually and stacked in between, so the particles can find their optimum positions. Furthermore, the particles should be fixed in their position so their movement

Fig. 4.34 Photograph of Gd in different forms. **a** Spheres with a diameter of approximately 0.4 mm. **b** Cylinders with a length of 4 mm and a diameter of 2.5 mm. **c** Plates with a thickness of 0.25 mm. **d** Powder with an average size of 0.4 mm

should be avoided. Namely, during the (de)magnetization process the particles will, if not fixed, move, and cause friction. It should also be noted that the shape of the particles significantly affects the uniformity of the porosity. The powder, for example has, in addition to spheres, more than one spatial degree of freedom. Therefore, it can be randomly oriented inside the AMR and its porosity would be higher and less homogenous compared to the spheres.

There are a few techniques applied for the fabrication of parallel-plate AMRs based on thin Gd plates. In general, they can be divided into two groups: the AMRs with a spacing integrated into the housing and the jointed AMR. An example of the first one can be found in, e.g. [129–131] and is schematically shown in Fig. 4.35.

The magnetocaloric plates are inserted into the housing with integrated spacings (space dividers). The main disadvantage of this method is the limit in the spacing thickness, as it cannot be fabricated as thin as would be necessary for more efficient heat transfer (<0.1 mm).

The jointed AMRs are based on magnetocaloric plates and narrower plates (ribbon) that work as the spacing between the plates (see Fig. 4.35b). In this case, we are not limited by the spacing thickness. The magnetocaloric plates and the

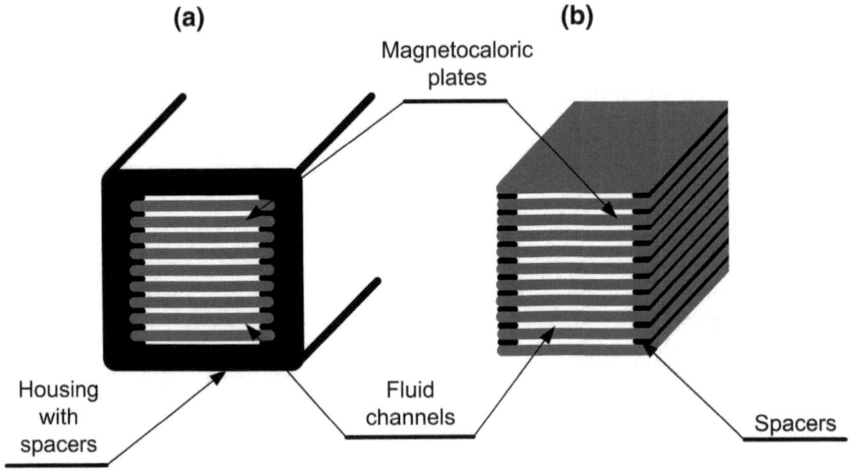

Fig. 4.35 **a** AMR with spacers integrated into the housing. **b** Jointed AMR

Fig. 4.36 A photograph of a laser-welded AMR with a plate thickness of 0.25 mm

spacings can be joined by gluing [132] or laser welding [101]—see Fig. 4.36. The laser welding of the AMR is especially interesting since it enables a precise and permanent joint of an arbitrary thin spacer and the magnetocaloric plate, and is further insensitive to the working fluid. The spacer must, of course, be weldable, with the magnetocaloric material used (for example gadolinium and stainless steel as a spacer are compatible).

4.7.2 Fabrication of Powder-Based (sintered) AMRs

The magnetocaloric materials produced by powder metallurgy, e.g. La-based and Mn-based materials, require special forming and processing techniques for the AMR's fabrication [133]. The great majority of these types of materials were tested in the form of a powder in packed-bed AMRs [5]. However, some fabrication methods for producing powder-based ordered AMRs were presented and evaluated as well.

Those methods enable the fabrication of parallel-plate AMRs or even some more advanced geometries. The largest effort in this field was placed on La-based (La–Fe–Co–Si) and perovskite-type manganese oxides-based (La–Ca–Sr–MnO_3) magnetocaloric materials, which in addition to Gd and some of its alloys show the greatest potential for application in a magnetic refrigerator in the near future. The most promising fabrication methods are briefly presented below.

Katter et al. [134] proposed a process termed thermally induced decomposition and recombination (TDR) for the fabrication of an AMR with La–Fe–Co–Si alloys. Fully dense, bulk samples of $La(Fe,Co,Si)_{13}$ were initially prepared by the reactive sintering of powder mixtures. Due to the high thermal expansion and low bending strength, the conventional machining caused cracks in the material, which prevents its machining to a fine element suitable for use in an AMR. After a heat treatment at about 1073 K the material is in the decomposed (magnetocalorically passive) state, and the material possesses enhanced mechanical properties (low thermal expansion) and is easier to machine. Then, after the material has been machined (using EDM cutting) in the decomposed state, it is recombined back into the magnetocalorically active state (recombination heat treatment at about 1323 K) and its magnetocaloric properties are fully recovered. This technique enables the production of relatively fine magnetocaloric elements suitable for use in an AMR. Figure 4.37 shows the AMR part produced by the wire EDM cutting of an initial block of magnetocaloric material using the TDR technique. The block was produced by Vacuumschmelze [31, 135].

In 2013, Moore et al. [136] proposed a selective laser melting (SLM) method for the production of a more sophisticated AMR geometry (compared to most widely applied, packed-bed and parallel-plate geometries). SLM is a rapid prototyping technique to form a variety of three-dimensional shapes. The same technique was applied already by Tura et al. [137] using bronze powder to produce passive regenerators with small (0.7 × 0.7 mm) channels suitable for the AMR's geometries. However, Moore et al. [136] used $La(Fe,Co,Si)_{13}$ as a starting powder and produced two types of AMRs: a wavy-channel block and the block of an arrays of fin-shaped rods. The first had a channel diameter of 0.8 mm, a high surface-to-volume ratio and excellent thermohydraulic properties [138, 139], while the

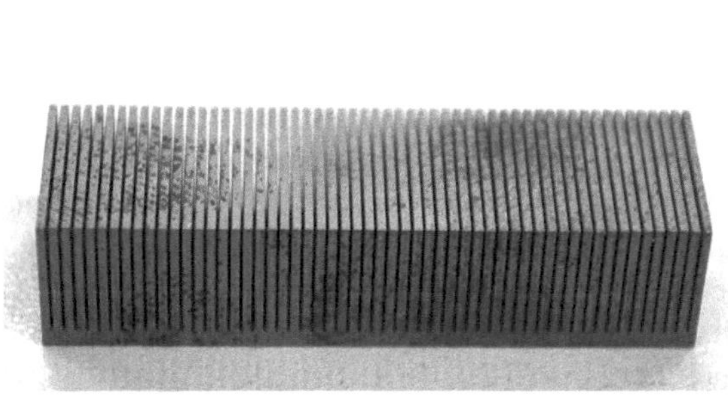

Fig. 4.37 A photograph of a La–Fe–Co–Si AMR part produced by wire EDM cutting using the TDR technique. The plate thickness is 0.5 mm and the spacing is 0.2 mm. The AMR was manufactured by the German company Vacuumschmelze

second had fin-like rod diameters of about 1 mm and a spacing between them of 0.2 mm.

Even smaller dimensions of magnetocaloric elements and fluid channels (down to 0.1 mm) are possible using the SLM technique with different starting powders. Since $La(Fe,Co,Si)_{13}$ is brittle and has irregular shaped particles a larger starting powder must be used (50–80 μm). It was shown that spherical particles of a few microns in diameter would be the best choice in this regard. However, the produced AMRs were annealed at 1,323 K for several days and further tested for their magnetocaloric properties and cycling stability. They concluded that the adiabatic temperature change, especially around its Curie temperature, was somewhat reduced due to the smaller amount of 1:13 phase present (the ratio between La and other materials). The AMRs were tested according to their cyclic stability. Therefore, a magnetization and demagnetization with cycling frequencies of 4 Hz for 106 cycles was performed. No degradation of the magnetocaloric effect occurred during the cycling, which makes it promising candidate for future applications in a magnetic refrigerator.

Tape casting is a promising method for producing thin and flat plates made of magnetocaloric materials, as recently shown for $La–Ca–SrMnO_3$ (LCSM) perovskites. As explained in Bahl et al. [140], the powders of LCSM material were suspended in slurry with additional azeotropic additives. Using a so-called doctor blade to control the thickness (in this case 0.3 mm), the slurries were applied from a vessel onto a moving substrate. The resulting tapes were further sintered at 1,473 K for 4 h. The produced plates were subsequently assembled into an AMR by using a gluing technique (see Fig. 4.35b) and applied in a magnetic refrigerator [140, 141]. The tape-casting technique can also be successfully applied for making layered (graded) magnetocaloric plates (and further layered AMRs) by using several powders (slurries) with different Curie temperatures side by side (see [142] for details).

Pryds et al. [143] applied a thermoplastic extrusion process to fabricate a monolithic squared channel (honeycomb) AMR with LCSM perovskites. The LCSM powder was mixed with stearic acid and thermoplastic (polyethylene) binder. The slurry was than extruded into a honeycomb structure (AMR) with a wall thickness of 0.5 mm and a channel thickness of 1 mm. The monolith was further sintered to remove the binder, which causes a reduction in the specific heat and an increase in the adiabatic temperature change. The fabricated regenerator was also tested in a passive regenerator experiment showing similar performance to a parallel-plate regenerator with the same porosity. As explained by the authors [143], the main advantages of such a fabrication method (compared to parallel-plate regenerators) are mainly the low cost, the low time consumption and the structural (stable thin-wall structure) benefits, which can be produced by a one-step processing technique.

In Pulko et al. [144] epoxy-bonded magnetocaloric plates and then the AMR were evaluated with the goal of fabricating an AMR with thinner plates and a smaller spacing compared to the state-of-the-art sintered AMR (produced by TDR method). As magnetocaloric material $LaFe_{13-x-y}Co_xSi_y$ powder (130 μm) was used.

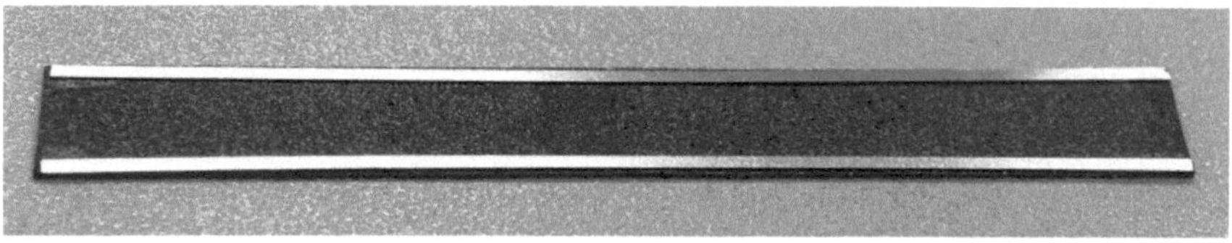

Fig. 4.38 Photograph of the epoxy-bonded plate with glued spacers

The plates were made in a special Teflon mould. The powder was mixed with epoxy resin and after vacuuming, the plates were cold pressed inside the Teflon mould. The composite plate stayed in the mould for 24 h, when it was removed and subsequently cured under ambient conditions. The thickness of the plates was approximately 0.4 mm. However, in order to study the impact of the epoxy and the plate composition in detail, different epoxy-bonded magnetocaloric plates were made and analysed regarding their magnetocaloric, thermal and cycling properties and further directly compared to the samples prepared by the TDR technique (as explained above). The samples had a volume fraction of magnetocaloric powder between 45 and 60 %. As was expected, the sintered plates have somewhat better magnetocaloric and especially thermal properties compared to the epoxy-bonded plates. The main reason for that is the impact of the epoxy, since it has a relatively high specific heat and a low thermal conductivity. The sample with the lowest amount of epoxy exhibited the highest adiabatic temperature change (but about 50 % smaller than the sintered samples). In the next step the AMR was made using epoxy-bonded magnetocaloric plates (see Fig. 4.38) and applied in the magnetic refrigerator. The AMR consists of plates with a thickness of 0.4 mm and a spacing of 0.1 mm and was produced by gluing, which is better than the current state of the art, which involves the TDR method (but its magnetocaloric properties are poorer).

A very similar idea was also applied by Skokov et al. [145] with $La(Fe,Si)_{13}$ powder. The authors used 5 wt% of silver epoxy and powder between 50 and 300 μm. The samples were then pressed into thin plates with up to 0.3 GPa. The results for the optimized epoxy-bonded samples (powder of 100 μm and compacted pressure of 0.1 GPa) show a very promising magnetocaloric effect (in some cases an even higher adiabatic temperature change compared to the bulk samples). They also fabricated an AMR based on that technique with plate thickness and a channel size of 0.6 mm (but this was not applied to the magnetic refrigerator).

It is not yet clear which fabrication method or even which magnetocaloric material is going to be successfully applied in any future (commercial) magnetic refrigerator. As many times pointed out in this chapter, the AMR geometry (besides the magnetocaloric effect) plays one of the most important roles in a highly efficient magnetic regenerator and therefore suitable fabrication methods must be applied. However, its time—and cost efficiency are required as well.

4.8 Where Is the Limit for Applying a Conventional AMR Cycle?

It is well known that large temperature differences between the fluid and the magnetocaloric material lead to large irreversible heat transfer losses. This is why researchers tend to improve the heat transfer characteristics of the AMR.

Therefore, the limit of the AMR cycle is given by the applied mechanism of heat transfer (together with the modest magnetocaloric effect in known magnetocaloric materials). Even though the magnetocaloric material has a fine porous structure, with a high thermal conductivity and a large heat transfer area, all the experimental results and numerical calculations indicate (see e.g. [15, 37, 146]) that such solutions will not lead to compact devices. The compactness has a strong correlation with the cost of a device. Furthermore, low porous structures with a very high heat transfer surface lead to pump losses and heat generation due to the fluid friction. Therefore, we can expect that future applications will not necessarily apply a conventional AMR, or the AMR will by integrated together with some other mechanisms to increase the heat transfer rate, for example thermal diodes. Because of their importance we dedicated a whole chapter to thermal diodes in this book.

References

1. Gschneidner KA Jr, Pecharsky VK, Tsokol AO (2005) Recent development in magnetocaloric materials. Rep Prog Phys 68:1479–1539
2. Brück E, Tegus O, Thanh DTC et al (2006) Magnetocaloric refrigeration near room temperature (invited). J Magn Magn Mater 310:2793–2799
3. Schmidt FW, Willmott AJ (1981) Thermal energy storage and regeneration. Hemisphere Publication Corp, Washington
4. Brown G (1976) Magnetic heat pumping near room temperature. J Alloy Compd 47:3673–3680
5. Yu B, Liu M, Egolf PW et al (2010) A review of magnetic refrigerator and heat pump prototypes built before the year 2010. Int J Refrig 33:1029–1060
6. Steyert WA (1978) Stirling-cycle rotating magnetic refrigerators and heat engines for use near room temperature. J Appl Phys 49:1216–1226
7. Barclay JA, Steyert WA (1982) Active magnetic regenerator. US Patent No 4.332.135
8. Yu B, Gao Q, Zhang B et al (2003) Review on research of room temperature magnetic refrigeration. Int J Refrig 26:622–636
9. Plaznik U, Tušek J, Kitanovski A et al (2012) Numerical and experimental analyses of different magnetic thermodynamic cycles with an active magnetic regenerator. Appl Therm Eng 59(1–2):52–59
10. Kitanovski A, Plaznik U, Tušek J et al (2014) New thermodynamic cycles for magnetic refrigeration. Int J Refrig 37:28–35
11. Hall JL, Reid CE, Spearing IG et al (1996) Thermodynamic consideration for the design of active magnetic regenerative refrigerators. Adv Cryog Eng 41:1653–1659
12. Kitanovski A, Tušek J, Poredoš A (2014) Magnetic cooling: from fundamentals to high efficiency refrigeration. In: Sandeman K, Gutfleisch O (ed) Magnetic thermodynamic cycles, Wiley (in press)

13. Egolf PW, Kitanovski A, Vuarnoz A et al (2008) Magnetic groundsource heat pump–optimized prototype. End year report, Swiss Federal Office of Energy, p 12
14. Vuarnoz D, Kitanovski A, Gonin C et al (2010) Modeling of a two-stage magnetic refrigerator with wavy-structure gadolinium heat exchangers. Int J Refrig 33(4):745–752
15. Kitanovski A, Egolf PW, Poredoš A (2012) Rotary magnetic chillers with permanent magnets. Int J Refrig 35(4):1055–1066
16. Tušek J, Kitanovski A, Prebil I et al (2011) Dynamic operation of an active magnetic regenerator (AMR): numerical optimization of a packed-bed AMR. Int J Refrig 34:1507–1517
17. Tušek J, Kitanovski A, Poredoš A (2013) Geometrical optimization of packed-bed and parallel-plate active magnetic regenerators. Int J Refrig 36:1456–1464
18. Bjørk R, Engelbrecht K (2011) The influence of the magnetic field on the performance of an active magnetic regenerator (AMR). Int J Refrig 34:192–203
19. Tušek J (2012) Thermomagnetic properties of active magnetic regenerator (In Slovene). PhD thesis, University of Ljubljana, Faculty of Mechanical Engineering, Slovenija
20. Tsukagoshi T, Matsumoto K, Hashimoto T et al (1997) Optimum structure of multilayer regenerator with magnetic materials. Cryogenics 37:11–14
21. Richard MA, Rowe A, Chahine R (2004) Magnetic refrigeration: single and multimaterial active magnetic regenerator experiments. J Appl Phys 95(4):2146–2150
22. Rowe A, Tura A (2006) Experimental investigation of a three-material layered active magnetic regenerator. Int J Refrig 29:1286–1293
23. Zimm C, Boeder A, Chell J et al (2006) Design and performance of a permanent magnet rotary refrigerator. Int J Refrig 29(8):1302–1306
24. Jacobs S (2009) Modeling and optimal design of a multiplayer active magnetic refrigeration system. Paper presented at the 3rd international conference of IIR on magnetic refrigeration at room temperature, Des Moines, Iowa, USA, 11–15 May 2009, pp 267–273
25. Nielsen KK, Bahl CRH, Engelbrecht K et al (2010) Numerical modeling of graded active magnetic regenerators. Paper presented at the 4th international conference of IIR on magnetic refrigeration at room temperature, Baotou, Inner Mongolia, China, 23–28 Aug 2010, pp 445–453
26. Russek S, Auringer J, Boeder A et al (2010) The performance of a rotary magnet magnetic refrigeration with layered beds. Paper presented at the 4th international conference of IIR on magnetic refrigeration at room temperature, Baotou, Inner Mongolia, China, 23–28 Aug 2010, pp 339–349
27. Aprea C, Greco A, Maiorino A (2011) A numerical analysis of an active magnetic regenerative refrigerant system with a multi-layer regenerator. Energy Convers Manage 52 (1):97–107
28. Engelbrecht K, Nielsen KK, Bahl CRH (2011) Experimental results for a magnetic refrigerator using three different types of magnetocaloric material regenerators. Int J Refrig 34(4):1132–1140
29. Aprea C, Greco A, Maiorino A (2013) The use of the first and of the second order phase magnetic transition alloys for an AMR refrigerator at room temperature: a numerical analysis of the energy performances. Energy Convers Manage 70(1):40–55
30. Legait U, Guillou F, Kedous-Lebouc A et al (2014) An experimental comparison of four magnetocaloric regenerators using three different materials. Int J Refrig 37:147–155
31. Tušek J, Kitanovski A, Tomc U (2014) Experimental comparison of multi-layered La–Fe–Co–Si and single-layered Gd active magnetic regenerators for use in a room-temperature magnetic refrigerator. Int J Refrig 37:117–126
32. Bjørk R, Bahl CRH, Katter M (2010) Magnetocaloric properties of $LaFe_{13-x-y}Co_xSi_y$ and commercial grade Gd. J Magn Magn Mater 322:3882–3888
33. Nielsen KK, Tušek J, Engelbrecht K et al (2011) Review on numerical modeling of active magnetic regenerators for room temperature applications. Int J Refrig 34:603–616
34. Yan Z, Chen Y (1992) The effect of field-dependent heat capacity on the characteristics of the ferromagnetic Ericsson refrigeration cycle. J Appl Phys 72(1):1–5

35. Xia ZR, Ye XM, Lin GX et al (2006) Optimization of the performance characteristics in an irreversible magnetic Ericsson refrigeration cycle. Phys B 381:246–255
36. He JZ, Wu X, Deng XF (2007) Performance characteristics of an irreversible magnetic Brayton refrigeration cycle. Int J Refrig 31(1):138–144
37. Kitanovski A, Egolf PW (2009) Application of magnetic refrigeration and its assessment. J Magn Magn Mater 321(7):777–781
38. DeGregoria A (1991) Modeling the active magnetic regenerator. Adv Cryog Eng 37:867–873
39. Smailli A, Chahine R (1998) Thermodynamic investigations of optimum active magnetic regenerators. Cryogenics 38:247–252
40. Shir F, Della Torre E, Bennett LH et al (2004) Modeling of magnetization and demagnetization in magnetic regenerative refrigeration. IEEE Trans Magn 40(4):2098–2100
41. Engelbrecht K (2008) Numerical Model of an Active Magnetic Regenerator Refrigerator with Experimental Validation. PhD thesis, University of Wisconsin, Madison, USA
42. Šarlah A, Poredoš A (2010) Dimensionless numerical model for simulation of active magnetic regenerator refrigerator. Int J Refrig 33(6):1061–1067
43. Tagliafico G, Scarpa F, Canepa F (2010) A dynamic 1-d model for a reciprocating active magnetic regenerator: influence of the main working parameters. Int J Refrig 33(2):286–293
44. Risser M, Vasile C, Engel T et al (2010) Numerical simulation of magnetocaloric system behavior for an industrial application. Int J Refrig 33:973–981
45. Vuarnoz D, Kawanami T (2012) Numerical analysis of a reciprocating active magnetic regenerators made of gadolinium wires. Appl Therm Eng 37:388–395
46. Burdyny T, Rowe A (2013) Simplified modeling of active magnetic regenerators. Int J Refrig 36(3):932–940
47. Nikkola P, Mahmed C, Balli M et al (2014) 1D model of an active magnetic regenerator. Int J Refrig 37:43–50
48. Chiba Y, Smaili A, Mahmed C et al (2014) Thermal investigations of an experimental active magnetic regenerative refrigerator operating near room temperature. Int J Refrig 37:36–42
49. Petersen TF, Pryds N, Smith A et al (2008) Two-dimensional mathematical model of a reciprocating room temperature active magnetic regenerator. Int J Refrig 31:432–443
50. Nielsen KK, Bahl CRH, Smith A et al (2009) Detailed numerical modeling of a linear parallel-plate active magnetic regenerator. Int J Refrig 32:1478–1486
51. Liu M, Yu B (2011) Numerical investigations on internal temperature distribution and refrigeration performance of reciprocating active magnetic regenerator of room temperature magnetic refrigeration. Int J Refrig 34(3):617–627
52. Oliveira PA, Trevizioli PV, Barbosa JR Jr et al (2012) A 2D hybrid model of the fluid flow and heat transfer in a reciprocating active magnetic regenerator. Int J Refrig 35:98–114
53. Li J, Numazawa T, Nakagome H et al (2011) Numerical modeling on a reciprocating active magnetic regenerator refrigeration in room temperature. Cryogenics 51:347–352
54. Šarlah A, Kitanovski A, Poredoš A et al (2006) Static and rotating active magnetic regenerators with porous heat exchangers for magnetic cooling. Int J Refrig 29:1332–1339
55. Anzelius A (1926) Über Erwärmung vermittels durchströmender Medien. Z Angew Math Mech 6:291–294
56. Nusselt W (1928) Der Beharrungszustand in Winderhitzen. Z des Ver Deut Ing 7:1052
57. Hausen H (1929) Über Die Theorie Von Warmeaustauches in Regeneratoren. Z Angew Math Und Mech 9:173
58. Coppage JE, London AL (1953) The periodic-flow regenerator-summary of design theory. Trans ASME 75:779–787
59. Hill A, Willmott AJ (1987) A robust method for regenerative heat exchanger calculations. Int J Heat Mass Transf 30:241–249
60. Foumney EA, Heggs PJ (1991) Heat exchange engineering. In: Design of heat exchangers, vol 1. Ellis Horwood, New York
61. Dragutinovic GD, Baclic BS (1998) Operation of counterflow regenerators. WIT Press, Computational Mechanics Publications, London

62. Saasamoinen JJ (1999) Heat transfer in crossflow regenerators. Int J Heat Mass Transf 42:3205–3216
63. Kuppan (2000) Heat exchanger design handbook. Marcel Dekker, New York
64. Shah RK, Sekulič DP (2003) Fundamentals of heat exchanger design. Wiley, New York
65. Kaviany M (1995) Principles of heat transfer in porous media, 2nd edn. Springer, Berlin
66. Kunii D, Smith JM (1960) Heat transfer characteristics of porous rocks. AIChE J 6:71–78
67. Vortmeyer D, Adam W (1984) Steady-state measurements and analytical correlations of axial effective thermal coductivities in packed beds at low gas flow rates. Int J Heat Mass Transf 27:1465–1472
68. Hadley GR (1986) Thermal conductivity of packed metal powders. Int J Heat Mass Transf 29:909–919
69. Hsu CT, Cheng P, Wong KW (1994) Modified Zehner-Schlunder models for stagnant thermal conductivity of porous media. Int J Heat Mass Transf 37:2751–2759
70. Vafai K (2005) Handbook of porous media, 2nd edn. Taylor & Francis, New York
71. Wakao N, Keguei S (1982) Heat and mass transfer in packed beds. Routledge, London
72. Yuan ZG, Somerton WH, Udell KS (1991) Thermal dispersion in ticked-walled tubes as model of porous media. Int J Heat Mass Transf 34:2715–2726
73. Hsiao KT, Advani SG (1999) Modified effective thermal conductivity due to heat dispersion in fibrous porous media. Int J Heat Mass Transf 42:1237–1254
74. Incropera FP, DeWitt DP (1996) Fundamentals of heat and mass transfer. Wiley, New York
75. Engelbrecht K, Nellis GF, Klein SA (2006) The effect of internal temperature gradients on regenerator matrix preformance. J Heat Transf 128:1060–1069
76. Šarlah A, Tušek J, Poredoš A (2012) Comparison of thermo-hydraulic properties of heat regenerators applicable to active magnetic refrigerators. Stroj Vest–J Mech Eng 58(1):16–22
77. Morrish AH (1965) The Physical principles of magnetism. Wiley, New York
78. Ashcroft NW, Mermin D (1976) Solid state physics. Saunders College, New York
79. Risser M, Vasile C, Keith B et al (2012) Construction of consistent magnetocaloric materials data for modelling magnetic refrigerators. Int J Refrig 35:459–467
80. Tagliafico G, Scarpa F, Tagliafico LA (2013) A dimensionless description of active magnetic regenerators to compare their performance and to simplify their optimization. Int J Refrig 36:941–949
81. Šarlah A (2008) Thermohydraulic properties of heat regenerators in magnetic refrigerators. PhD thesis, University of Ljubljana, Faculty of Mechanical Engineering, Slovenia
82. Petersen TF, Engelbrecht K, Bahl CRH et al (2008) Comparison between a 1D and a 2D numerical model of an active magnetic regenerative refrigerator. J Phys D Appl Phys 41:105002
83. Nielsen KK, Bahl CRH, Smith A et al (2010) A comprehensive parameter study of an active magnetic regenerator using a 2D numerical model. Int J Refrig 33:753–764
84. Nickolay M, Martin H (2002) Improved approximation for the Nusselt number for hydrodynamically developed laminar flow between parallel plates. Int J Heat Mass Transf 45 (15):3263–3266
85. Bejan A, Kraus AD (2003) Heat transfer handbook. Wiley, Hoboken
86. Kandlikar SG, Garimella S, Li D et al (2006) Heat transfer and fluid flow in minichannels and microchannels. Elsevier, Oxford
87. Bavie're R, Favre-Marinet M, LePerson S (2006) Bias effects on heat transfer measurements in micro-channel flows. Int J Heat Mass Transf 49:3325–3337
88. Keys WM, London AJ (1984) Compact heat exchangers, 3rd edn. McGraw-Hill, New York
89. Rohsenow WM, Hartnett JP, Cho YI (1998) Handbook of heat transfer. McGraw-Hill, New York
90. Dikeos J, Rowe (2013) Validation of an active magnetic regenerator test apparatus model. Int J Refrig 36:921–931
91. Whitaker S (1972) Forced convection heat transfer correlations for flow in pipes, past flat plate, single cylinders, single spheres, and for flow in packed beds and tube bundles. AIChE J 18:361–371

92. Engelbrecht K, Tušek J, Nielsen KK et al (2013) Improved modelling of a parallel plate active magnetic regenerator. J Phys D Appl Phys 46:255002
93. Dai W, Shen BG, Li DX et al (2000) Application of high-energy Nd-Fe-B magnet in the magnetic refrigeration. J Magn Magn Mater 218:25–30
94. Peksoy O, Rowe A (2005) Demagnetizing effects in active magnetic regenerators. J Magn Magn Mater 288:424–432
95. Smith A, Nielsen KK, Christensen DV et al (2010) The demagnetizing field of a nonuniform rectangular prism. J Appl Phys 107:103910
96. Christensen D, Nielsen KK, Bahl CRH et al (2011) Demagnetizing effects in stacked rectangular prisms. J Phys D Appl Phys 44(21):215004
97. Bjørk R, Bahl CRH (2013) Demagnetization factor for a powder of randomly packed spherical particles. Appl Phys Lett 103:102403
98. Bouchard J, Nesreddine H, Galanis N (2009) Model of a porous regenerator used for magnetic refrigeration at room temperature. Int J Heat Mass Transf 52(5–6):1223–1229
99. Nielsen KK, Smith A, Bahl CRH et al (2012) The influence of demagnetizing effects on the performance of active magnetic regenerators. J Appl Phys 112(9):094905
100. Lipsø KW, Nielsen KK, Christensen DV et al (2011) Measuring the effect of demagnetization in stacks of gadolinium plates using the magnetocaloric effect. J Magn Magn Mater 323:3027–3032
101. Tušek J, Kitanovski A, Zupan S et al (2013) A comprehensive experimental analysis of gadolinium active magnetic regenerators. Appl Therm Eng 53:57–66
102. Chang ZC, Chen PH (1998) Flow channeling effect on a regenerator's thermal performance. Cryogenics 38(2):191–196
103. Nemec D, Levec J (2005) Flow through packed bed reactors: 1. single-phase flow. Chem Eng Sci 60(24):6947–6957
104. du Toit C (2008) Radial variation in porosity in annular packed beds. Nucl Eng Des 238 (11):3073–3079
105. Hsu CH (2005) Dynamic modeling of convective heat transfer in porous media. Taylor & Francis, New York
106. Heggs PJ, Burns D (1988) Single-blow experimental prediction of heat transfer coefficients: a comparison of four commonly used techniques. Exp Therm Fluid Sci 1:243–251
107. Nielsen KK, Engelbrecht K, Bahl CRH (2013) The influence of flow maldistribution on the performance of inhomogeneous parallel plate heat exchangers. Int J Heat Mass Transf 60:432–439
108. Jensen JB, Bahl CRH, Engelbrecht K et al (2011) Analysis of single blow effectiveness in non-uniform parallel plate regenerators. Int J Heat Mass Transf 54:4746–4751
109. Nielsen KK, Engelbrecht K, Christensen DV et al (2012) Degradation of the performance of microchannel heat exchangers due to flow maldistribution. Appl Therm Eng 40:236–247
110. Lozano JA, Engelbrecht K, Bahl CRH et al (2013) Performance analysis of a rotary active magnetic refrigerator. Appl Energ 111:669–680
111. Nielsen KK, Nellis GF, Klein SA (2013) Numerical modeling of the impact of regenerator housing on the determination of Nusselt numbers. Int J Heat Mass Transf 65:552–560
112. Engelbrecht K, Nielsen KK, Bahl CRH et al (2013) Material properties and modeling characteristics for $MnFeP_{1-x}As_x$ materials for application in magnetic refrigeration. J Appl Phys 113(17):173510
113. von Moos L, Nielsen KK, Engelbrecht K et al (2014) Experimental investigation of the effect of thermal hysteresis in first order material MnFe(P, As) applied in an AMR device. Int J Refrig 37(1):303–306
114. Basso V, Sasso CP, Bertotti G et al (2006) Effect of material hysteresis in magnetic refrigeration cycles. Int J Refrig 29:1358–1365
115. von Moos L, Bahl CRH, Nielsen KK et al (2014) Preisach approach to modeling partial phase transitions in the first order magnetocaloric material MnFe(P, As). Phys B 435:144–147

116. Romero Gomez J, Ferreiro Garcia R, Carbia Carril J (2013) A review of room temperature linear reciprocating magnetic refrigerators. Renew Sustain Energy Rev 21:1–12
117. Risser M, Vasile C, Müller C et al (2013) Improvement and application of a numerical model for optimizing the design of magnetic refrigerators. Int J Refrig 36:950–957
118. Li P, Gong M, Wu J (2008) Geometric optimization of an active magnetic regenerative refrigerator via second-law analysis. J Appl Phys 104:103536
119. Petersen TF (2007) Numerical modelling and analysis of a room temperature magnetic refrigeration system. PhD thesis, Technical University of Denmark, Departmant of eneregy conversion and storage, Denmark
120. Bjørk R, Bahl CRH, Smith A (2010) An optimized magnet for magnetic refrigeration. J Magn Magn Mater 322(21):3324–3328
121. Bjørk R, Bahl CRH, Smith A (2011) Improving magnet designs with high and low field regions. IEEE Trans Magn 47(6):1687–1692
122. Zhang ZY, Long Y, Ye RC et al (2005) Corrosion resistance of magnetic refrigerant gadolinium in water. Paper presented at the 1st international conference of IIR on magnetic refrigeration at room temperature, Montreaux, Switzerland, 2010, pp 85–92
123. Zhang M, Long Y, Ye RC (2011) Corrosion behavior of magnetic refrigeration material La–Fe–Co–Si in distilled water. J Alloy Compd 509:3627–3631
124. Zamni L, Maudni El Alami S, Balli M et al (2012) Corrosion tests of gadolinium in various heat conducting fluids. Paper presented at the 5th international conference of IIR on magnetic refrigeration at room temperature, Grenoble, France, 17–20 Sept 2012, pp 137–144
125. Tang Y, Chen Y, Zhang E (2012) The corrosion behaviour of $LaFe_{11.6}Si_{1.4}$ alloy in inorganic and organic water solution. Paper presented at the 5th international conference of IIR on magnetic refrigeration at room temperature, Grenoble, France, 17–20 Sept 2012, pp 187–192
126. Wu J, Liu C, Hou P et al (2014) Fluid choice and test standardization for magnetic regenerators operating at near room temperature. Int J Refrig 37:135–146
127. Silva DJ, Bordalo BD, Pereira AM et al (2014) The effect of coolants on the performance of magnetic micro-refrigerators. J Nanosci Nanotechnol 14:4337–4340
128. Tuchinsky L (2008) Novel manufacturing process for metal and ceramic microhoneycombs. Adv Eng Mater 10(3):793–802
129. Bahl CRH, Petersen TF, Pryds N et al (2008) A versatile magnetic refrigeration test device. Rev Sci Instrum 79:093906
130. Trevizoli PV, Barbosa JR, Ferreira RTS (2011) Experimental evaluation of a Gd-based linear reciprocating active magnetic regenerator test apparatus. Int J Refrig 34:1518–1526
131. Tura A, Nielsen KK, Rowe A (2012) Experimental and modeling results of a parallel plate-based active magnetic regenerator. Int J Refrig 35:1518–1527
132. Engelbrecht K, Nielsen KK, Pryds N (2011) An experimental study of passive regenerator geometries. Int J Refrig 34:1817–1822
133. Jian L (2014) Optimizing and fabricating magnetocaloric materials. Chin Phys B 23 (4):047503
134. Katter M, Zellmann V, Reppel GW et al (2009) Magnetocaloric and mechanical properties of reactively sintered $La(Fe, Co, Si)_{13}$. Paper presented at the 3rd international conference of IIR on magnetic refrigeration at room temperature, Des Moines, Iowa, USA, 11–15 May 2009, pp 83–88
135. Katter M, Zellmann V, Barcza A (2010) Sintering behaviour and thermally induced decomposition and recombination (TDR) process of $LaFe_{13\text{-}x\text{-}y}Co_xSi_y$ alloys. Paper presented at the 4th international conference of IIR on magnetic refrigeration at room temperature, Baotou, Inner Mongolia, China, 23–28 Aug 2010, pp 49–62
136. Moore JD, Klemn D, Lindackers D et al (2013) Selective laser melting of La(Fe Co, Si)13 geometries for magnetic refrigeration. J Appl Phys 114:043907
137. Tura A, Schopher S, Rowe A et al (2010) Heat transfer and viscous losses in micro-channel passive regenerators. Part I–Apparatus and experimental results. Paper presented at the 4th international conference of IIR on magnetic refrigeration at room temperature, Baotou, Inner Mongolia, China, 23–28 Aug 2010, pp 285–293

138. Sui Y, Teo CJ, Lee PS (2012) Direct numerical simulation of fluid flow and heat transfer in periodic wavy channels with rectangular cross-sections. Int J Heat Mass Transf 55 (1–3):73–88
139. Manglik RM, Zhang JH, Muley A (2005) Low Reynolds number forced convection in three-dimensional wavy-plate-fin compact channels: fin density effects. Int J Heat Mass Transf 48 (8):1439
140. Bahl CRH, Velázquez D, Nielsen KK et al (2012) High performance magnetocaloric perovskites for magnetic refrigeration. Appl Phys Lett 100:121905
141. Bahl CRH, Engelbrecht K, Bjørk R et al (2011) Design concepts for a continuously rotating active magnetic regenerator. Int J Refrig 34:1792–1796
142. Jabbari M, Spangenberg J, Hattel J (2013) Modeling of the interface behavior in tape casting of functionally graded ceramics for magnetic refrigeration parts. Int J Refrig 36:2403–2409
143. Pryds N, Clemens F, Menon M et al (2011) A monolithic perovskite structure for use as a magnetic regenerator. J Am Ceram Soc 94(8):2549–2555
144. Pulko B, Tušek J, Moore JD et al (2014) Epoxy-bonded La-Fe-Co-Si magnetocaloric plates. J Magn Magn Mater 375:65–73
145. Skokov KP, Karpenkov DY, Kuz'min MD et al (2014) Heat exchangers made of polymer-bonded La(Fe, Si)13. J Appl Phys 115:17A941
146. Tušek J, Pulko B, Tomc U et al (2013) Development of the magnetocaloric regenerator: final report of industrial project. Ljubljana, Faculty of Mechanical Engineering, Laboratory for refrigeration and district energy, p 77

Chapter 5
Magnetocaloric Fluids

Magnetocaloric fluids represent a class of magnetic fluids that can generally be subdivided into two main categories, i.e. ferrofluids (FF) (or magnetic nanofluids) and magnetorheological (MR) fluids. The general characteristic of all these fluids is that they change their physical properties when exposed to an external magnetic field. Ferrofluids are suspensions of magnetic nanoparticles dispersed in certain base liquids. Magnetorheological fluids are particle suspensions of micron-sized particles consisting of a magnetic material dispersed in a carrier liquid. The size difference between the particles in ferrofluids and the particles in magnetorheological fluids results in the most important differences concerning the properties and the behaviour of these fluids.

There was not much experimental work done in the domain of magnetocaloric fluids. On the other hand, a variety of studies and applications can be found in the more general field of ferrofluids and magnetorheological fluids. Therefore, we will take advantage of these studies and briefly describe models that can be applied for magnetocaloric fluids as well. Since both magnetocaloric ferrofluids and magnetocaloric magnetorheological fluids are viscous, Newtonian and/or non-Newtonian fluids, an important feature that concerns their flow properties relates to the rheology. Therefore, we begin the chapter with the rheology of magnetic fluids and the various rheological, constitutive models. Then we introduce the equations that relate to the thermo-magneto–fluid dynamics of such fluids. Since this concerns complex investigations and the aim of this book is to provide information that may serve for a basic understanding, we also refer to the literature that may give the most comprehensive overview of the phenomenon of ferrofluid-dynamics. In addition, a review of the existing studies and applications of magnetocaloric fluids has been performed. At the end of the chapter, a discussion and guidelines are given on the potential design of devices that concern magnetocaloric energy conversion.

Note that this chapter may provide important information also for the application of magnetic fluids that may also be successfully applied in magnetocaloric energy conversion as the heat transfer fluids. Moreover, a smart design of magnetocaloric devices with the application of magnetic fluids can bring new solutions for valve systems, sealing elements, thermal diode mechanisms or even the parts of fluid propulsion systems (pumps).

A. Kitanovski et al., *Magnetocaloric Energy Conversion*,
Green Energy and Technology, DOI 10.1007/978-3-319-08741-2_5

5.1 Rheology of Suspensions

Fluids can be divided according to their rheological behaviour into Newtonian and non-Newtonian fluids. Newtonian fluids follow Newton's law of viscosity, which is expressed in one-dimensional form as

$$\tau = \eta_L \frac{\mathrm{d}v}{\mathrm{d}y} \tag{5.1}$$

It is clear from Eq. (5.1) that the relation between the velocity gradient (shear rate) and the shear stress in a Newtonian fluid is linear. In order to account for the solids content in a carrier or base fluid, Einstein [1, 2] derived an equation for the "effective" viscosity of a homogeneous Newtonian suspension, comprising spherical particles with a small volume concentration:

$$\eta_{\mathrm{eff}} = \eta_L(1 + 2.5\ \phi_V) \tag{5.2}$$

In this equation, ϕ_V is the volume fraction of the solid phase in a solid-liquid suspension. Note that Eq. (5.2) does not take the sizes or positions of the particles into account, and the theory neglects the effects of particle interaction. The shear stress in this case is expressed by replacing the liquid-phase viscosity η_L with the effective viscosity η_{eff}.

$$\tau = \eta_{\mathrm{eff}} \frac{\mathrm{d}v}{\mathrm{d}y} \tag{5.3}$$

After Einstein [1, 2], several models for effective viscosity were proposed. They comprise additional terms to describe the interaction between particles, account for high concentration of solid particles, their size and their shape. Figure 5.1 shows the difference in dynamic viscosity between the carrier fluid and effective viscosity of a Newtonian suspension. Because of the complexity of a theoretical description, these models are based on empirical evaluation.

The rheological behaviour of a homogeneous, non-Newtonian fluid in the case of highly turbulent flow or in the case of small concentrations of a solid phase

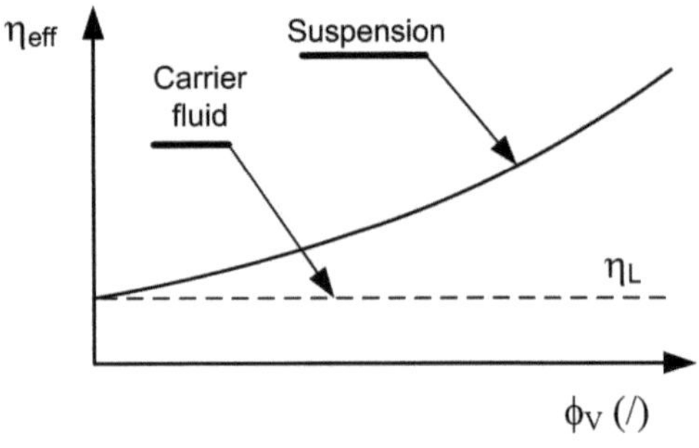

Fig. 5.1 An example of the variation of the effective viscosity with the concentration of solids

(where the carrier fluid is a Newtonian fluid) is similar to that of a Newtonian fluid, so the effective viscosity in such cases may be used as an approximation to describe the relation between the shear stress and the shear rate.

In Table 5.1 (which comprises Eqs. 5.4–5.13) we show some of the expressions that can be applied for the Newtonian suspension's viscosity as a function of the solid's volume fraction ϕ_V.

For higher solids concentrations, with the effect of the shape and particle size accounted for, the suspensions usually show non-Newtonian behaviour. Figure 5.2 shows an example of the shear stress depending on the velocity gradient for different types of non-Newtonian fluids.

Different rheological models for laminar flow of homogeneous non-Newtonian suspensions are shown in Table 5.2 (which comprises Eqs. 5.14–5.18).

Since higher concentrations of a solid phase in a non-Newtonian suspension, as well as the flow rate, strongly influence the relation between the shear stress and the shear rate, the relation in Eq. (5.1) is not linear any more. When the rheological behaviour (or model) of a particular suspension is not known, it is useful to determine the apparent viscosity as

$$\tau = \eta_{\mathrm{app}} \cdot \frac{\mathrm{d}v}{\mathrm{d}y} = \eta_{\mathrm{app}} \cdot \dot{\gamma} \tag{5.19}$$

where η_{app} is a function of the shear rate $\dot{\gamma}$. The relation in Eq. (5.19) can be simply generalized for homogeneous non-Newtonian fluids. Figure 5.3 shows the apparent viscosity of non-Newtonian suspensions depending on the shear rate. It is clear that the apparent viscosity is approaching the viscosity of a carrier fluid at high shear rates and low concentrations of solid particles. When the shear rate is lowered, apparent viscosity rapidly increases. In order to determine the apparent viscosity, experiments are required.

A method used to determine the shear rate from experimental data was given by Mooney [3] and Rabinowitch [4]. Note that in this case the shear rate is defined at the wall of a pipe, since the experimental data gives values of the wall shear stress. The apparent viscosity is, therefore, determined from data that show the dependence of the wall shear stress on the wall shear rate. The so-called Mooney-Rabinowitch equation can be derived from the following expression, where the plot of the wall shear stress τ_w versus the shear rate at the wall may help in the selection of an appropriate rheological model:

$$\left(\frac{\mathrm{d}v}{\mathrm{d}y}\right)_w = \frac{8 \cdot v}{d} \cdot \left[\frac{3}{4} + \frac{1}{4} \cdot \frac{\mathrm{d}\left(\ln \frac{8v}{d}\right)}{\mathrm{d}\left(\ln \frac{\Delta p \cdot d}{4 \cdot L}\right)}\right] \tag{5.20}$$

If the suspension is assumed to be homogeneous, the rheological parameters are considered constant all over the pipe cross-section. For the non-homogeneous suspension, one has to consider the flow patterns [5]. However, also for such a flow one may define the effective and apparent viscosity, as demonstrated by Kitanovski and Poredoš [6].

Table 5.1 Some expressions for Newtonian suspension's viscosity as a function of a solids volume fraction (see also Darby [175], Hanks [176], Steffe [177], Bica et al. [46])

Author	Equation	Comment	Equation number
Guth and Simha [178]	$\eta_{\text{eff}} = \eta_L\left(1 + 2.5\,\phi_V + 14.1\,\phi_V^2 + \cdots\right)$	Interaction between particles	(5.4)
Vand (after Darby) [179]	$\eta_{\text{eff}} = \eta_L \cdot \exp\left(\frac{2.5\cdot\phi_V}{1-0.609\cdot\phi_V}\right)$	No interparticle forces	(5.5)
	$\eta_{\text{eff}} = \eta_L \cdot \exp\left(\frac{2.5\cdot\phi_V+2.7\cdot\phi_V^2}{1-0.609\cdot\phi_V}\right)$	Includes doublet collisions, but not triplet	(5.6)
Mooney [180]	$\eta_{\text{eff}} = \eta_L \cdot \exp\left(\frac{2.5\cdot\phi_V}{1-K\cdot\phi_V}\right)$ $0.75 < K < 1.5$	K depends upon the system and is defined experimentally	(5.7)
Frankel and Acrivos [181]	$\eta_{\text{eff}} = \eta_L \cdot \frac{9}{8}\left[\frac{(\phi_V/\phi_{\max})^{\frac{1}{3}}}{1-(\phi_V/\phi_{\max})^{\frac{1}{3}}}\right]$	Concentrated suspensions only	(5.8)
Jeffrey and Acrivos [182]	$\eta_{\text{eff}} = \eta_L(1 + A\cdot\phi_V)$ $2.5 < A < 10$	Ellipsoid particles	(5.9)
Thomas [183]	$\eta_{\text{eff}} = \eta_L\left(1 + 2.5\,\phi_V + 10.05\,\phi_V^2 + 0.00273e^{16.6\,\phi_V}\right)$ d_p from 0.099 to 435 μm $\phi_V < 0.625$	Also includes interaction between the solid particles	(5.10)
Batchelor [184]	$\eta_{\text{eff}} = \eta_L\left(1 + 2.5\phi_V + 6.2\,\phi_V^2\right)$	Comprises particle interactions. Brownian motion and inertia of higher importance	(5.11)
Bicerano [185]	$\eta_{\text{eff}} = \eta_L\left(1 + \langle\eta\rangle\,\phi_V + k_H\,\phi_V^2\right)$	$<\eta>$ is the intrinsic viscosity, k_H is the dimensionless Huggins coefficient	(5.12)
Brinkman [186]	$\eta_{\text{eff}} = \eta_L \frac{1}{(1-\phi_V)^{2.5}}$	Extended Einstein's equation for higher concentrations	(5.13)

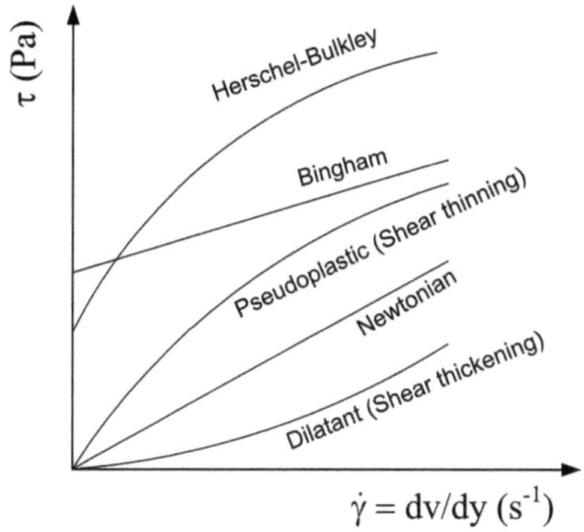

Fig. 5.2 A rheogram showing different non-Newtonian fluids

Table 5.2 Different rheological models for laminar suspension flow (see also Darby [175], Hanks [176], Steffe [177])

Rheological model	Equation	Equation number
Bingham	$\tau = \tau_0 + \eta_B \cdot \left(\frac{\mathrm{d}v}{\mathrm{d}y}\right)$	(5.14)
Power law (Ostwald-de-Waele)	$\tau = K_1 \cdot \left(\frac{\mathrm{d}v}{\mathrm{d}y}\right)^{n_1}$	(5.15)
Casson	$\tau^{0.5} = (\tau_0)^{0.5} + \left(\eta_C \frac{\mathrm{d}v}{\mathrm{d}y}\right)^{0.5}$	(5.16)
Herschel–Bulkley	$\tau = (\tau_0) + K_1 \cdot \left(\frac{\mathrm{d}v}{\mathrm{d}y}\right)^{n}$	(5.17)
Papanastasiou [187]	$\tau = \tau_0 \left(1 - e^{-m\frac{\mathrm{d}v}{\mathrm{d}y}}\right) + \eta_P \frac{\mathrm{d}v}{\mathrm{d}y}$	(5.18)

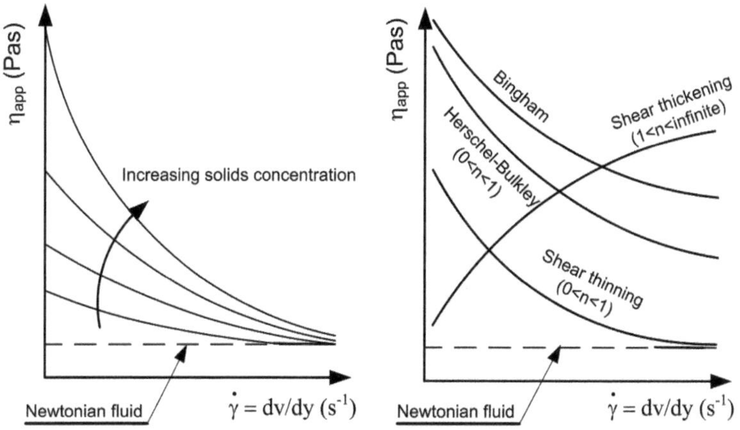

Fig. 5.3 An apparent viscosity of non-Newtonian suspensions depending on a shear rate

5.2 Rheology of Magnetic Fluids

Ferrofluids are magnetically controllable nanofluids and have been synthesized since the 1960s [7–9]. They can be used in a variety of applications [10, 11]. A typical ferrofluid consists of magnetic particles (about 5–10 % in terms of volume fraction) having size of about 3–15 nm and monodispersed in a base liquid. The basic type of ferrofluid is the colloidal ferrofluid. This is a suspension of finely dispersed particles in a base medium, including suspensions in which particles tend to settle. Despite the fact that a small concentration gradient can be established after longer exposure to a magnetic or gravitational field, the particles will stay suspended [7]. Particles will keep suspended because of Brownian motion. However, they can agglomerate. To prevent agglomeration due to van der Waals forces the particles can be coated with an adsorbed layer of surfactant with a thickness of approximately 2 nm [7]. Macroscopically looking, these fluids manifest themselves as magnetizable single-phase liquid media.

The applied magnetic field may cause the formation of particle chains in the direction of the magnetic field. An increased magnetic field will increase interaction between the particles. Also, particle chains will become longer. The longer the chain of particles, the larger the resistance of the fluid to flow and, consequently, the larger will be the viscosity. Therefore, with the induced magnetic field, this will certainly affect the viscosity of the ferrofluid, but will in most cases induce very small yield stress. In most cases these fluids do not exhibit any magnetorheological behaviour.

Magnetorheological (MR) fluids also represent magnetically controllable fluids. However, in contrast to ferrofluids the MR fluids are suspensions of micrometre range ferro- or ferrimagnetic multi-domain particles in a liquid matrix [12, 13]. MR fluids were introduced in the early 1940s; however, the majority of research activities relate to the past two decades. Today, these fluids represent an important part of many applications. One of them is in damping devices [14]. A typical magnetorheological (MR) fluid consists of solid particles, typically in the range from 1 to 20 μm, with volume fraction of about $\phi_V = 0.4 - 0.5$ [15]. Because of the micron-sized particles the Brownian motion in MR fluids is negligible. With respect to MR fluids the coupling constant represents an important parameter, which relates the level of tendency towards the formation of chains and agglomerates. In MR fluids, the coupling constant is high, which means that the fluid tends to create particle chains or forms agglomerates. The result is a yield stress that depends not only on the common rheological parameters, but also on the applied magnetic field. As a consequence, the chaining of solid particles, which is established along the suspension (depends on the magnetic field direction), transforms the liquid form of MR fluid into a solid (the apparent viscosity can increase up to about 1,000 times). If the shear stress (induced by the pressure difference on the MR fluid) overcomes the yield stress, the MR fluid will start to flow. Note that the particle aggregation processes in MR fluids are reversible. Therefore, the result of the removed magnetic

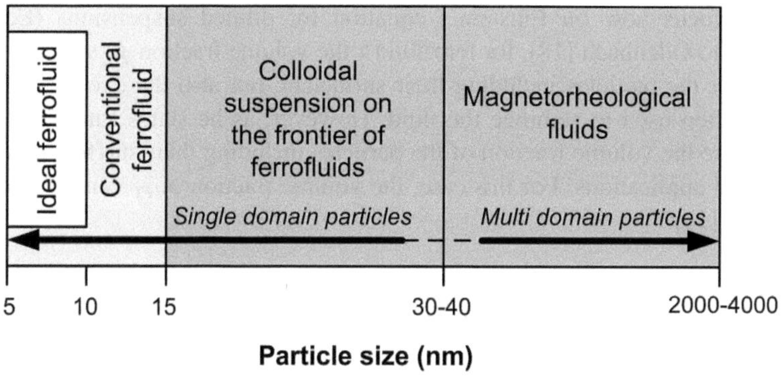

Fig. 5.4 The classification of magnetic fluids according to Lopez et al. [16]

field will be the reduced influence of the magnetically induced yield stress, by changing the solid MR fluid into its liquid suspension form.

Note that there is no clear definition of the transition from a ferrofluid to an magnetorheological fluid. Lopez et al. [16] presented an example for such a transition, as shown in Fig. 5.4. They have defined the ideal ferrofluid as a colloidal suspension of nanoparticles of ferro- or ferrimagnetic materials dispersed in a liquid carrier which do not settle, despite the potential long-term exposure to the applied (gravitational or magnetic) field. As denoted by Lopez et al. [16], in ideal ferrofluids, dispersed particles should be smaller than approximately 10 nm and coated with a molecular layer of surfactant to prevent the van der Waals attractive forces. Under these conditions, Brownian motion dominates over the forces of interaction. The real ferrofluids are polydispersed. The dipole–dipole energy of the interaction between them overcomes Brownian motion and causes clustering under the applied magnetic field. The latter increases the viscosity and can also lead to yield stress.

Genc and Derin [17] also defined ideal ferrofluids as colloidal suspensions of ultrafine (5–10 nm) single-domain magnetic nanoparticles and for real ferrofluids, the size would be in the range from 5 to 15 nm. For a particle size between 15 and 40 nm they considered the magnetic fluid to be a ferrofluid. Above this limit, the fluid would exhibit the characteristics of an MR fluid.

5.2.1 *Rheology of Ferrofluids*

Since the first theories on the rheology of ferrofluids have neglected the interaction between solid particles, according to Odenbach [18], these theories can only be used for quantitative predictions of the behaviour of highly diluted fluids. This is especially because experimental results have shown that a strong influence on viscosity can be induced by a magnetic field, not only in magnetorheological fluids, but also in ferrofluids with sufficient particle–particle interaction.

Let us focus now on Einstein's equation for diluted suspensions (Eq. 5.2). According to Odenbach [18], for ferrofluids, the volume fraction ϕ_V should account not only for the particles including their surfactant, but also the excess surfactant, which is often used to stabilize the fluid. However, as he states further, it is sufficient to use the volume fraction of the particles including their surfactant layer for all practical applications. For this case, the volume fraction ϕ_{VFF} can be calculated as follows [18]:

$$\phi_{\text{VFF}} = \phi_V \left(\frac{d + 2s}{d} \right)^3 \tag{5.21}$$

where d denotes the diameter of the magnetic particles, and s represents the thickness of the surfactant layer. As reported by Odenbach [18], only few efforts have been made to determine the real value of s in given samples of ferrofluids with higher precision. The error resulting from a lack of knowledge about the thickness of the surfactant layer can therefore be larger than that from neglecting the excess surfactant. Since for higher concentrations one also has to consider the interparticle relation, Einstein's equation (Eq. 5.2) will not be an appropriate choice any more for volume fractions of solid particles above 10 %.

For the application of ferrofluids, Rosensweig [7] proposed a modified Batchelor's equation (Eq. 5.11). For this he assumed that the suspension's viscosity should diverge for a certain critical volume fraction $\phi_{\max}$ of suspended material:

$$\eta_{\text{eff}} = \eta_L \left(1 - 2.5\phi_{\text{VFF}} + (2.5\,\phi_{\max} - 1) \left(\frac{\phi_{\text{VFF}}}{\phi_{\max}} \right)^2 \right) \tag{5.22}$$

where $\phi_{\max}$ represents the maximum volume (packing) fraction, which Rosensweig [7] estimated to be 0.74. Note that this is a rather high value, especially if the packing is random or if clustering of the magnetic particles appears. On the other hand, suspended particles with different sizes can lead to high packing fraction. Table 5.3 shows some examples of packing fractions for monodispersed particles.

When the ferrofluid is under the influence of a shear flow, the magnetic particles will tend to rotate. This is due to the torque produced by the viscous forces. Under

Table 5.3 The maximum packing fraction of various arrangements of monodispersed particles [188]

Arrangement	Maximum packing fraction $\phi_{\max}$
Simple cube	0.52
Maximum thermodynamically stable configuration	0.548
Hexagonally packed sheets just touching	0.605
Random close packing	0.637
Face-centred cubic/hexagonal close packed	0.68
Body-centred cubic/hexagonal close packed	0.74

the applied field, the magnetic moments of the particles will align together with the field direction. If the field direction and the vorticity of the flow are collinear, the influence on the motion of the particle and therefore on the flow of the fluid does not appear [18].

However, if the vorticity and field direction are perpendicular, there will be a magnetic torque that will act in opposition to the torque produced by the viscous forces. The free rotation of the particle in the flow will be damped and this will increase the fluid flow resistance and will result in increased viscosity. This effect is associated with the so-called rotational viscosity, which is related to highly diluted suspensions with very small particle interaction.

According to Shliomis [19], this can be defined as follows:

$$\eta_{\mathrm{rot}} = \frac{3}{2} \phi_{\mathrm{VFF}}\, \eta_L \left(\frac{\alpha - \tanh \alpha}{\alpha + \tanh \alpha} \right) \cdot \overline{\sin^2 \beta} \tag{5.23}$$

where α is the ratio between the magnetic and thermal energy of the particles

$$\alpha = \frac{\mu_0\, m\, H}{k\, T} \tag{5.24}$$

where μ_0 is the vacuum permeability, m is the magnetic moment of a particle, k represents Boltzmann's constant and T is the temperature. The coefficient β represents the angle between the vorticity and the field direction, and its term in Eq. (5.23) is shown as the spatial average [7].

More information can be obtained in Rosensweig [7], Shliomis [19], McTague [20], Rosensweig et al. [21], Ambacher et al. [22], and Patel et al. [23].

Note again that rotational viscosity is a property of highly diluted suspensions. There the particle–particle interactions are not so intensive (changes of viscosity induced by the magnetic field are in the order of a few percent as a maximum [20]). However, in concentrated ferrofluids with higher volume fractions of solids (e.g. a volume fraction of 10 %), the interactions between the particles have an important role and cause the so-called magneto-viscous effect.

Note that the model of Shliomis [19] does not take into account the interaction between the particles; therefore, it can be used only for calculation of the rotational viscosity with the absence of the magneto-viscous effect.

An example of models that take into account the magneto-viscous effects of magnetic fluids can be found in Zubarev et al. [24, 25], who assumed that the formation of the chains of magnetic particles dominate in the magneto-viscous effects in magnetic fluids. The results of their study also show good agreement with the experimental data from Odenbach [18]. Zubarev and Iskakova [26] also developed a theoretical model for prediction of the magneto-viscous effect for the drop-like aggregates in ferrofluids.

The magneto-viscous effect (MVE) can be defined according to Odenbach [18] as

$$\mathrm{MVE} = \frac{\eta_H - \eta_{\mathrm{eff}}}{\eta_{\mathrm{eff}}} \tag{5.25}$$

where η_H represents the viscosity of the ferrofluid under an applied magnetic field H and η_{eff} represents the viscosity of the ferrofluid in the absence of the magnetic field.

5.2.1.1 Rheologic Models Applied for Ferrofluids

The following examples provide brief information about the different research activities in which the rheology of ferrofluids have been investigated.

Title	Synthesis, rheological properties and magneto-viscous effect of Fe_2O_3/paraffin ferrofluids (Hezaveh et al. [27])
Materials used for investigation	Fe_2O_3 magnetic nanoparticle powder with average size of <50 nm, pure liquid paraffin as base fluid, oleic acid (OA) used as surfactant of nanoparticles Fe_2O_3 nanoparticles in different weight fractions (5, 10, 15, 20, 25, and 30 %) were mixed with oleic acid by 15 % weight relative to the solid powder
Constitutive model	Bingham model (Eq. 5.14): better predictions in lower concentrations (Bingham plastic region) rather than higher concentrations Casson model (Eq. 5.16): in comparison to the Bingham model, the Casson model shows poor predictability for the system
Viscosity	The authors compared the measured viscosity of Einstein's equation (Eq. 5.2), Brinkman's equation (Eq. 5.13) and Roscoe's model [28]. It was shown that these models do not offer a good prediction of the viscosity of the suspension. The viscosity of ferrofluids has been investigated for different weight fractions at a constant shear rate of 5 s^{-1} and under different magnetic fields. The results show that at a constant shear rate the viscosity increases with increased magnetic field. However, the authors have noted that this increment was followed by a decrease in viscosity after a certain peak value and pointed out that at high mass concentrations (30 %) this phenomenon was more sensible. The authors have explained this as the result of the phase separation in the ferrofluid, e.g. due to particle aggregation

As can be seen from these examples, different constitutive models have been applied for characterization of the rheology of different ferrofluids. Most of these models regard **Bingham** (see for instance López et al. [16], Hezaveh et al. [27], Hong et al. [29], Hosseini et al. [30], Rodríguez-Arco et al. [31], Shah et al. [32]). The **Casson** model has been applied by Hezaveh et al. [27] and Hong et al. [29], however, it did not provide the best fitting. The **Herschel–Bulkley** model was applied by Hong et al. [29]. In the Bingham model (Eq. 5.14), when the shear stress τ is smaller than the yield stress τ_B, there will be no fluid motion. The shear rate in the Bingham model is directly proportional to the difference between the shear stress and yield stress. With an increased shear rate ferrofluids can show Newtonian and shear thinning behaviour. The Casson model (Eq. 5.16) considers the yield stress as well as the shear thinning behaviour. The same is also true for the Herschel–Bulkley model (Eq. 5.17). Note that the yield stress is associated with the particle size, the volume fraction and the magnetic field. The higher these are, the higher will be the yield stress. This also depends on the potential agglomeration of the particles.

Title	Magnetic nanofluid properties and some applications (Vékás [33])
Materials used for investigation	Magnetic fluids based on Fe_3O_4 nanoparticles
Viscosity	Vand's equation (Eq. 5.6), and the equation given by Krieger and Dougherty [34]
Title	Field-induced rotational viscosity of a ferrofluid: effect of capillary size and magnetic field direction (Andhariya et al. [35])
Materials used for investigation	Nanomagnetic particles of iron ferrite in water and kerosene-based ferrofluids. The particles were coated with a bi-layer of surfactant. For the kerosene-based ferrofluid the magnetite particles were coated with a single layer of surfactant and dispersed in kerosene. The volume fraction of particles in the water-based magnetic fluid was 0.64 %, while in the kerosene-based magnetic fluid the particles had a volume fraction of 2.24 %. The average particle diameter was 10.4 nm. The work is concentrated into field-induced rotational viscosity when a ferrofluid flows through a capillary placed in a magnetic field
Viscosity	The authors considered the rotational viscosity by Shliomis (Eq. 5.23)
Title	Design method for automatic energy transport devices based on the thermomagnetic effect of magnetic fluids (Lian et al. [36])
Materials used for investigation	A kerosene-based Mn–Zn ferrite magnetic fluid with a saturation magnetization of 6.5×10^4 Am^{-1}. The average diameter of the suspended magnetic particles was about 6.8 nm and the volume fraction of the ferrite magnetic nanoparticles was 0.045
Viscosity	The authors applied Einstein's equation (Eq. 5.2), which they extended with a linear form to account for the influence of the magnetic field on the viscosity. Using a combination of two equations they proposed a new equation, which takes into account the influence of the magnetic field
Title	Influence of large size magnetic particles on the magneto-viscous properties of a ferrofluid (Shah et al. [32])
Materials used for investigation	Three ferrofluids were investigated. The first ferrofluid contained magnetite particles with an average size of 10 nm dispersed in transformer oil. The second fluid, denoted by authors as the magnetorheological fluid, consisted of large magnetite particles, having a 30-nm particle size dispersed in transformer oil. The third fluid was labelled as the nano- magnetorheological fluid and represented a mixture of the first and the second fluid in different weight proportions. The particles were coated with oleic acid surfactant and dispersed in transformer oil. The magnetic field was increased from 0 to 1 T

(continued)

(continued)

Title	Influence of large size magnetic particles on the magneto-viscous properties of a ferrofluid (Shah et al. [32])
Constitutive model	The authors applied the Bingham model (Eq. 5.14). The rheological properties of the first two suspensions were measured for varying shear and field values. The yield stress varied from 2.2 to 5.5 Pa for a magnetic field from 0 to 1 T. The second fluid with 30-nm particles exhibited magneto-viscous behaviour with an increased magnetic field. The yield stress obtained was almost 15 times higher than that of the first fluid
Viscosity	For the ferrofluid: the magneto-viscous effect was measured for a fixed shear rate of 10 s^{-1} and a varying magnetic field. A clear dependence of the magneto-viscosity on the field is observed. The increase in the MVE with magnetic field in the small range of shear rate was calculated with the help of relations given by Martsenyuk et al. [37] and Shliomis [19]
	For MRF: the authors expected strong dipolar interactions between the particles. The suspension showed a magneto-viscous effect, an increase in the viscosity and the occurrence of yield stress
Title	Rheological properties of water-based Fe_3O_4 ferrofluids (Hong et al. [29])
Materials used for investigation	Water-based ferrofluids were prepared from Fe_3O_4 ferromagnetic nanoparticles, and oleate sodium and Polyethylene Glycol (PEG) surfactants. For bilayer-stabilized ferrofluids, oleate sodium was used to modify the surface of a ferromagnetic nanoparticle as the first layer. Then PEG was utilized to cover the second layer. The average size of 98 % of nanoparticles was about 23 nm. The viscosity and yield stress were analysed for the ferrofluid with a high solid mass fraction (>25 %)
Constitutive model	Bingham (Eq. 5.14), Casson (Eq. 5.16) and Herschel–Bulkley model (Eq. 5.17) were applied to describe the rheological properties of high-concentration ferrofluids without the application of a magnetic field. Einstein's equation (Eq. 5.2), the best fitting with the experimental results was obtained with the Herschel–Bulkley model
	The authors investigated the magnetic field with intensities of 22, 49 and 73 mT. It was found that the Herschel-Bulkley (Eq. 5.17) model gives the best fitting results with respect to the experiments. In this model the authors separated the yield stress caused in the absence of a magnetic field and the yield stress associated with the magnetic field. They also proposed a correlation for the prediction of the magnetic-field-dependent yield stress
Viscosity	The authors applied Einstein's equation (Eq. 5.2)
Title	Rheological properties of a γ-Fe_2O_3 paraffin-based ferrofluid (Hosseini et al. [30])
Materials used for investigation	γ-Fe_2O_3 magnetic nanoparticle powder, (mean particle size of <50 nm), coated with oleic acid (OA) and dispersed in a suitable amount of paraffin. The concentration was adjusted to a 30 % mass fraction of particles. The average particle size of the aggregates was 398 nm, which varied in the range 210–686 nm. Referring to this, it

(continued)

(continued)

Title	Rheological properties of a γ-Fe_2O_3 paraffin-based ferrofluid (Hosseini et al. [30])
	was estimated that the aggregates contain about 27–280 magnetic nanoparticles. All the particles and aggregates were assumed to be spherical and the surfactant layer diameter was considered to be 2 nm
Constant model	Authors propose the Bingham model (Eq. 5.14)
Viscosity	The viscosity of the ferrofluid decreases with increasing shear rate, (i.e. shear-thinning behaviour). The authors conclude that when the applied shear rate increases, the nanoparticles begin to arrange their orientation in the shearing direction. The increasing shear rate destroys the initial bonds existing between the nanoparticles, which results in a decrease in the viscosity. The magneto-viscous effect increases versus an increasing magnetic field for every shear rate. The magnetic field also caused aggregation and the formation of drop-like structures, which could lead to an increase in the viscosity in accordance with the size of the structures
Title	Colloids on the Frontier of Ferrofluids (Lopez et al. [16])
Materials used for investigation	Two new kinds of ferrofluids: one was composed of CoNi nanospheres with a diameter of 24 nm, while the other was composed of CoNi nanofibres 56 nm long and 6.6 nm wide. The spherical and fibre-like CoNi particles were considered to be single-domain particles. The ferrofluids were prepared by dispersing proper amounts of the synthesized powders in a mineral oil. L-α-Phosphatidylcholine was used as the surfactant to avoid irreversible particle aggregation. The particle volume fraction was 5 % in all cases
Constitutive model	The authors applied the Bingham model (Eq. 5.14). The trends in the static yield stress were different for both ferrofluids. In the case of the nanosphere ferrofluid, the yield stress resulted in a strong increase with an increase in the magnetic field. The static yield stress is the shear stress required to induce the flow of a material. However, the magnetic field dependence of the static yield stress of the nanofibre ferrofluid was rather weak at low and medium field ($H < 100\ \mathrm{kAm}^{-1}$) and became strong at higher field values. At high shear rates the relationship between the shear stress and the shear rate was linear. The dynamic yield stress, which is needed to continuously break the aggregates that reform in the presence of the magnetostatic forces, was higher than the static yield stress
Viscosity	For viscosity of the nanosphere ferrofluid, Batchelor's equation (Eq. 5.11) was applied. The authors, however, denoted the difference between the experimental viscosity and the calculations using Batchelor's equation. They assumed that the reason for this is due to the existence of magnetostatic interactions between the particles, as well as due to the deviation from a spherical shape. For viscosity of the nanofibre ferrofluid, the predictions for the suspensions of spheroidal particles were used with the help of expressions from Hinch and Leal [38] and Larson [39]
Title	Stability and magnetorheological behaviour of magnetic fluids based on ionic liquids (Rodrigez-Arco et al. [31])
Materials used for investigation	Magnetic fluids consisting of magnetite nanoparticles dispersed in a quaternary onium cation-based ionic liquid. Citric, humic and oleic acid-additives were used to stabilize the fluids. The results of these

(continued)

(continued)

Title	Stability and magnetorheological behaviour of magnetic fluids based on ionic liquids (Rodrigez-Arco et al. [31])
	tests showed that a true ferrofluid was only obtained when the nanoparticles were coated with a layer of surfactant compatible with the ionic liquid. The mean diameter of the nanoparticles for different samples was about 10 nm, while the volume fractions were between 5 and 10 vol%, depending on the type of material analysed
Constitutive model	The authors applied the Bingham model (Eq. 5.14)
Viscosity	The authors used Einstein's (Eq. 5.2) and Batchelor's (Eq. 5.11) equation

5.2.2 Rheology of Magnetorheological Fluids

In contrast to ferrofluids, in magnetorheological fluids, even a small magnetic field will have very strong influence on the interactions between the particles and on the viscosity. Magnetorheological fluids are suspensions that contain micron-sized magnetic particles. The carrier liquids are usually oils; however, other applications, such as for instance liquid metals should not be neglected. The volume fraction of solid particles in these fluids varies between 10 and 50 %. Note again that the primary use of magnetorheological fluids is not the manipulation of their flow, but magnetically controllable viscosity and yield stress. We will come back to this issue when we speak about the applications in magnetocaloric energy conversion. The primary applications for magnetorheologic fluids regard the so-called direct shear mode (application in brakes and clutches) and the valve mode (application in dampers). However, due to the fast response and the precise controllability, magnetorheological fluids are becoming attractive for many different applications (see, e.g. Olabi and Grunwald [40]).

In order to stabilize a dispersion of particles, small amounts of dispersants are added to the carrier liquid. However, in contrast to ferrofluids, surfactant techniques, which avoid the agglomeration of particles, are not usually applied. Other methods relate to polymer core–shell magnetic particles or composite magnetic particles, applications of nanoparticles, nanotubes or nanowires (see, e.g. Lim et al. [41], Park et al. [42], Cho et al. [43], Fang et al. [44]). Because the particles are large, they can no longer be considered as single-domain particles, as is the case with ferrofluids. Namely, nanosized particles, which can be considered as magnetic single domains, are in a permanent state of magnetization. Therefore, they can always possess a magnetic dipole, even in the absence of an applied field. On the other hand, micron-sized particles represent a magnetic polydomain [45]. Therefore, the soft ferromagnetic micron-sized particle will not retain a remanent magnetization after the removal of the magnetic field. Such particles will have a zero overall magnetic moment and, consequently, the magnetic interactions between the particles will be small. When the magnetic field is applied, however, this will result in large magnetic moments and strong magnetic interaction between the particles, which will tend to create chain-like structures. In magnetorheologic fluids the so-called coupling constant represents the characteristics upon which the particles tend to form chains or agglomerates under the applied magnetic field. In the absence of the magnetic field, the coupling constant will be very small. This is especially so because the particles have no permanent magnetic moment without an applied magnetic field [18]. Therefore, only the remanence will keep the magnetic moment; however, in most applications this will be very small. As a result, in the absence of a magnetic field, an magnetorheological fluid will behave as an ordinary Newtonian or non-Newtonian, solid-liquid suspension or slurry. The latter is a strong function of the particles' volume fraction as well as the properties of the carrier fluid. The viscosity can be determined by viscosity of the carrier liquid and the volume fraction of the suspended material, i.e. the effective viscosity [18]. However, the magnetic field will induce an interparticle interaction and will strongly affect the rheological behaviour.

5.2.2.1 Rheologic Models Applied for Magnetorheological Fluids

The flow of a magnetorheologic fluid, especially in the presence of a magnetic field, will lead to non-Newtonian behaviour. This is particularly the case when the formation of particle aggregates occurs. In most cases of magnetorheological fluids, the **Bingham** constitutive model was applied (see, e.g. Bica et al. [46], Carlson and Jolly [14], Odenbach [18], Olabi and Grunwald [40], Park et al. [47], Engin et al. [48], Jiang et al. [49], Iglesias [50], Omidbeygi and Hashemabadi [51], Serano et al. [52]). However, in certain cases also the **Herschel–Bulkey** model was applied to fit post-yield shear thinning and shear thickening (see, e.g. Bica et al. [46], Burguera [53], Wang and Gordaninejad [54], Resiga et al. [55], Yamanaka et al. [56], Mrlik et al. [57]). Also, the **Casson** model was applied for magnetorheological fluids (Bica et al. [46], Sidpara et al. [58], Gabriel and Laun [59], Kim et al. [60]). An interesting rheological model is that given by Papanastasiou (Eq. 5.18). This model was also applied for MR fluids (see, for e.g. Farjoud et al. [61], Resiga et al. [62]). An advantage of using such a model could be the fact that it can cover both domains; those with a yield stress and those without (i.e. a constitutive equation for a magnetic-field- and particle-size-dependent model). This also provides the possibility to define the rheological transition from ferrofluids to magnetorheologic fluids.

In 1995–1996 Ginder et al. [63, 64] proposed a relation for the yield stress in a magnetorheologic fluid. For very low applied magnetic fields, where the relationship between the magnetization M and the magnetic field intensity H is considered to be linear, the relationship between the yield stress and applied field was given by the following proportion:

$$\tau_0 \propto \phi_V \, \mu_0 \, H^2 \tag{5.26}$$

where ϕ_V is the particle volume fraction, H is the applied field and μ_0 is the permeability of free space (vacuum). For magnetic flux densities that are above the linear region but lower than the saturation, Ginder et al. [63, 64] proposed the following relation:

$$\tau_0 = \sqrt{6}\,\phi_V\,\mu_0\,H^{\frac{3}{2}}\sqrt{M_S} \qquad (5.27)$$

where the term $\mu_0 M_S$ represents the saturation magnetization. For the saturation magnetic field, Ginder et al. [63, 64] proposed another relation for the yield stress:

$$\tau_0(H_S) = \frac{4}{5^{\frac{5}{2}}}\,\phi_V\,\mu_0\,M_S^2\,\xi(3) \qquad (5.28)$$

for which $\xi(3)$ is the Riemann zeta function and was defined to be 1.202 as a constant. The verification of the model was provided by Genc and Phulé [65].

Note also that the yield stress of a bidisperse suspension will represent higher values than that of a more sedimentation-stable monodisperse suspension (see, e.g. Charles and See [66], Kittipoomwong et al. [67], Wereley et al. [68]).

For more information on the rheology of magnetorheologic fluids one is also referred to the chapter by Bossis et al. [12]. There, the authors present a chain model for the yield stress in magneto-rheology that was very similar to Eq. (5.27):

$$\tau_0 = 2.31\,\phi_V\,\mu_0\,H^{\frac{3}{2}}\sqrt{M_S} \qquad (5.29)$$

As Bossis et al. [12] noted, the dependence between the yield stress and the volume fraction was experimentally confirmed for not very concentrated suspensions with $\phi_V < 0.2 - 0.3$ by Ginder et el. [69], See et al. [70], and Brady et al. [71]. Furthermore, with regard to the exponent of the magnetic field intensity H, different authors have obtained similar values to 3/2 (e.g. Ginder et al. [69], Orihara et al. [72] and Bonnecaze and Brady [73]). As denoted by Bossis [12], this exponent will decrease with an increase in the magnetic field.

Some researchers (see, e.g. Fang et al. [74], Hato et al. [75], Hong et al. [76]) have also applied a correlation basically developed by Cho et al. [77] and Choi et al. [78] for electro-rheological fluids, but modified for use in MR fluid.

The examples in the following pages provide brief information about some research activities in which the rheology of magnetorheological fluids is investigated.

Title	A new generation of magneto-rheological fluid dampers (Gordanlnejad et al. [79])
Materials used for investigation	Novel magnetorheological-polymer gels (MPRG) were investigated, with 81 % mass fraction of carbonyl iron particles added to the synthesized crosslinked polyimide gel with addition of the additives
Constitutive model	Authors applied the Bingham model (Eq. 5.14). Chains of particles in the form of columnar structures, parallel to the applied field hindered the flow of the fluid and increased the apparent viscosity of the suspension. The pressure needed to yield these chain-like structures increased with applied magnetic field resulting in a field-

(continued)

(continued)

Title	A new generation of magneto-rheological fluid dampers (Gordanlnejad et al. [79])
	dependent yield stress. In the case when the yield stress is to high (pre-yield region), MRPG behaves as a viscoelastic solid, and therefore, the shear below the yield stress was given by the complex material modulus and the shear strain, which was also field dependent. The dynamic yield stress increased significantly with an increase in magnetic flux density. Authors applied the Ginder's model (Eqs. 5.26–5.28) for the yield stress
Viscosity	Apparent viscosity was derived by the ratio between shear stress and shear rate, and decreased with increasing shear rate, which show the shear-thinning behaviour and it appears due to the presence of micron-sized particles. At higher shear rates the apparent viscosity decreases and then reaches a constant value
Title	Magnetorheology of submicron diameter iron microwires dispersed in silicone oil (Bell et al. [80])
Materials used for investigation	Magnetorheological fluid containing iron microwires with 260 nm diameter and two distinct length distributions of 5.4 ± 5.2 μm and 7.6 ± 5.1 μm suspended in silicone oil (0.45 Pa s). The micro-wire-based suspensions compared to spherical suspensions lead to higher yield stresses at low magnetic fields. The sedimentation effect can be also substantially reduced compared to conventional MR fluids. However, maximum volume fraction of microwires was much less than the desired 30–40 vol% achievable with spherical particles. Authors conclude that suspensions comprised of microwires are more useful in applications where sedimentation would be extremely detrimental and low yield stresses are sufficient
Constitutive model	Authors applied the Bingham model (Eq. 5.14). For microwire-based suspensions, the trend in the yield stress data was found to be proportional to the square root of the applied field
Title	Viscosity behaviour of magnetic suspensions in fluid-assisted finishing (Cheng et al. [81])
Materials used for investigation	Magnetic suspensions of micrometre-sized carbonyl iron (CI) particles with and without abrasive cerium oxide (CeO_2) particles are studied for their ensuing polishing effectiveness. Volumetric component ratios of the magnetorheological fluid in the study were 33.84 % CI particles, 57.34 % silicone oil, 2.82 % stabilizing agent, and 6 % CeO_2 as abrasive particle
Constitutive model	Authors applied the Bingham model (Eq. 5.14)
Viscosity	The relation between the magnetic intensity H and the viscosity of the fluid was taken from Shulman et al. [82]
Title	Synthesis and magnetorheology of suspensions of submicron-sized cobalt particles with tunable particle size (Lopez et al. [83])
Materials used for investigation	Synthesized cobalt powders with diameters in the range 50 nm–1 μm were used for the preparation of MR suspensions (solid concentration 5 vol%). These suspensions were prepared by dispersing

(continued)

(continued)

Title	Synthesis and magnetorheology of suspensions of submicron-sized cobalt particles with tunable particle size (Lopez et al. [83])
	appropriate amounts of cobalt powders in silicone oil. Aluminium stearate was used as dispersant
Constitutive model	Authors applied the Bingham model (Eq. 5.14)
Title	Average particle magnetization as an experimental scaling parameter for the yield stress of dilute magnetorheological fluids (Vereda et al. [84])
Materials used for investigation	Authors have applied a large number of different particles suspended in the carrier fluid. The properties of the different types of particles used for the preparation of MR fluids are shown below
	Solid iron particles; spherical; size (μm) 0.76 ± 0.40; M_S (kAm^{-1}) = 1,600
	Porous iron spheres; spherical; size (μm) 0.7 ± 0.2; M_S (kAm^{-1}) = 766 (de Vicente et al. [85])
	Porous iron plates; plate-like; diameter (μm) 2.1 ± 0.5; thickness (μm): 0.25 ± 0.05; M_S (kAm^{-1}) = 766 (de Vicente et al. [85])
	Porous iron rods; rod-like; diameter (μm): 0.45 ± 0.08; length (μm): 4.7 ± 2.2; M_S (kAm^{-1}) = 707 (de Vicente et al. [85])
	Solid magnetite spheres; spherical; size (μm) 0.68 ± 0.15; M_S (kAm^{-1}) = 475 (de Vicente et al. [86], Vereda et al. [87])
	Solid magnetite rods; rod-like; diameter (μm) 0.56 ± 0.12; length (μm): 0.7 ± 0.4; M_S (kAm^{-1}) = 475 (de Vicente et al. [86], Vereda et al. [87])
	The MR fluids were prepared by dispersing particles silicone oil with viscosity of in 20 m Pa s. Particle volume fraction was in all cases low, ranging from 0.5 to 2.1 %, mainly because the authors wanted to work with dilute dispersions for which the yield stress is expected to depend linearly on the volume fraction of particles
Constitutive model	The experimental parameter that was chosen for the definition of the yield stress was an average volumetric particle magnetization as the function of magnetic field, which was calculated from magnetization curves taken either from powder samples or from the suspensions. In the case of suspensions the following equation was applied for the average particle magnetization $\overline{M_p(H)} = \frac{M_{\text{suspension}}(H)}{\phi_V}$ (5.30)
	The static yield stress of a dilute suspensions was defined by the following relation $\tau_{0\,\text{static}} = \phi_V \cdot 0.00219 \cdot \overline{M_p}^2$ (5.31)
	The static yield stress is defined as the minimum stress required to start the flow. Authors have also defined the dynamic yield stress, which is the stress needed to continuously break the aggregates that reform by the influence of the field once the flow has started (de Vicente et al. [88])

(continued)

(continued)

Title	Average particle magnetization as an experimental scaling parameter for the yield stress of dilute magnetorheological fluids (Vereda et al. [84])
	The dynamic yield stress was estimated by extrapolating the fitting of a Bingham equation (Eq. 5.14) to zero shear rate (determined by a fit to data points of non-negligible shear rate). By this the static yield stress depends only on the interparticle magnetic interactions (with the exception of the rod-based MR fluids), whereas the dynamic yield stress depends on hydrodynamic interactions between particles or aggregates. The dynamic yield stress was defined by the following relation: $\tau_{0\,\mathrm{dynamic}} = \phi_V \cdot 0.001824 \cdot \overline{M_p}^{2.1}$ (5.32)
Title	Improved thermooxidation and sedimentation stability of covalently-coated carbonyl iron particles with cholesteryl groups and their influence on magnetorheology (Mrlik et al. [89])
Materials used for investigation	Bare carbonyl iron (CI) microparticles and carbonyl iron (CI) microparticles coated with a low density substance, cholesteryl chloroformate (CI-chol) were suspended in silicone oil with 40, 60 and 80 % of mass particle concentrations. The size of most of the bare CI particles was in the range of 0.5–2 μm, and the surfaces of these particles were roughened with small pieces of milled iron
Constitutive model	Both suspensions exhibited pseudoplastic behaviour with a certain level of yield stress. The increase of the magnetic field resulted in stiffer internal chain-like structures and consequently increased shear stress. Authors have applied Herschel–Bulkley model (Eq. 5.17). Yield stress was defined by the Ginder's equation (see Eqs. 5.26–5.28). There, the exponent of the magnetic field intensity was defined to be 1.60, 1.67 and 1.75 for suspensions consisting of mass concentration of 40, 60 and 80 % of modified CI particles (CI-chol), respectively
Title	A low sedimentation magnetorheological fluid based on platelike iron particles, and verification using a damper test (Shah et al. [90])
Materials used for investigation	Iron micro-sized particles of different sizes having a plate-like structure. The average particles size was about 2 μm (small size) and 19 μm (large size). Both the particles and heavy paraffin oil were mixed using a mechanical stirrer. Less than 4 % of commercial grease was used to prevent sedimentation. The total particle volume fraction of the bidisperse magnetorheological fluid was 16 %
Constitutive model	The applied magnetic field strength varied from 0 to 228 kAm^{-1}. The yield stress was obtained using the Bingham model (Eq. 5.14) by linear extrapolation of the flow curve to zero shear. Shear stress increased faster at a low shear rate and then changed very slowly at high shear rate, where the viscosity became field independent
Viscosity	The viscosity versus shear rate was measured for the magnetorheological fluid in the zero field at various magnetic field strengths in the range of shear rate of 0.01–800 s^{-1}. A strong shear thinning effect was observed as viscosity decreased with increased shear rate

5.3 Ferrohydrodynamics and Heat Transfer in Magnetic Fluids

Most of the work that concerns the motion of a magnetic fluid under a static, rotating or varying magnetic field was performed by Rosensweig [7], the "father" of ferrohydrodynamics. Today, a number of studies have been performed and we address some of them in this subsection. We start this subsection by providing the basic governing equations for the homogeneous flow of Newtonian or non-Newtonian magnetic fluids. These are based on the available literature (Bird et al. [91], Neuringer and Rosensweig [92], Rosensweig [7, 93–95]). Furthermore, since we deal in such equations with the homogeneous distribution of solid particles in the base (nanofluids) or the carrier fluid (particle suspensions), we also present a few models that relate to the definition of the effective thermal conductivity.

The equation of continuity can be written as

$$\frac{\partial \rho}{\partial t} + (\nabla \cdot \rho \vec{v}) = 0 \tag{5.33}$$

where v represents the velocity of the magnetic fluid and ρ represents the density of the magnetic fluid, which may be calculated with help of the solids volume fraction ϕ_V or the solids mass fraction ϕ_m as

$$\rho = \phi_V \cdot \rho_{\text{solid}} + (1 - \phi_V) \cdot \rho_{\text{liquid}} = \frac{1}{\frac{\phi_m}{\rho_{\text{solid}}} + \frac{(1-\phi_m)}{\rho_{\text{liquid}}}} \tag{5.34}$$

The equation of motion for a (non)Newtonian magnetic fluid can be written as [91]

$$\rho \frac{D\vec{v}}{Dt} = -\nabla p + \left[\nabla \cdot \bar{\bar{\tau}}\right] + \rho \vec{g} \tag{5.35}$$

where $\bar{\bar{\tau}}$ represents the total stress tensor and may be divided into the viscous part and the magnetic part:

$$\bar{\bar{\tau}} = \bar{\bar{\tau}}_{\text{viscous}} + \bar{\bar{\tau}}_{\text{magnetic}} \tag{5.36}$$

In order to include the non-Newtonian constitutive model, the viscous stress tensor may be applied for an incompressible fluid with the generalized Newtonian model [91]:

$$\bar{\bar{\tau}}_{\text{viscous}} = \eta \left(\nabla \vec{v} + \nabla \vec{v}^T\right) = \eta \, \dot{\bar{\bar{\gamma}}} \tag{5.37}$$

where $\nabla \vec{v}$ is the velocity gradient tensor with components $\left(\frac{\partial}{\partial x_i}\right) v_j$ and $\nabla \vec{v}^T$ is the "transpose" velocity gradient tensor with the components $\left(\frac{\partial}{\partial x_j}\right) v_i$, and where the term $\overline{\overline{\dot{\gamma}}}$ represents the rate of strain tensor (rate of deformation tensor). Therefore, in the generalized Newtonian model one can simply replace the constant viscosity by the non-Newtonian viscosity as the function of shear rate, which can be written as the function of the magnitude of the rate of strain tensor [91]:

$$\dot{\gamma} = \sqrt{\frac{1}{2}\left(\overline{\overline{\dot{\gamma}}} : \overline{\overline{\dot{\gamma}}}\right)} \tag{5.38}$$

Note that in magnetic fluids, both the viscosity and the potential yield stress may depend not only on the conventional viscous forces, but also on the magnetic field intensity. The magnetic component of the stress can be expressed according to Rosensweig [7] as

$$\overline{\overline{\tau}}_{\text{magnetic}} = -\mu_0 \frac{H^2}{2} \cdot \overline{\overline{u}} + \mu_0 \vec{H}\vec{H} \tag{5.39}$$

where $\overline{\overline{u}}$ represents the unit tensor and $\vec{H}$ represents the vector of the magnetic field. Based on Eq. (5.39), Rosensweig [7] expressed the magnetic force density on the body as

$$\nabla \cdot \overline{\overline{\tau}}_{\text{magnetic}} = \mu_0 \vec{M} \cdot \nabla \vec{H} \tag{5.40}$$

For which in the case that the magnetization M is collinear with H, the force density is defined by magnitudes of M and H as [7, 92]

$$\nabla \cdot \overline{\overline{\tau}}_{\text{magnetic}} = \mu_0 M \cdot \nabla H \tag{5.41}$$

Based on the equations above, we now write the equation of motion for an incompressible magnetic fluid in the following form:

$$\frac{\partial}{\partial t}(\rho \vec{v}) = -\nabla \cdot \rho \vec{v}\vec{v} - \nabla p + \left[\nabla \cdot \overline{\overline{\tau}}_{\text{viscous}}\right] + \mu_0 M \nabla H + \rho \vec{g} \tag{5.42}$$

The energy equation in the form of the equation of change of internal energy is expressed according to Bird et al. [91] for an incompressible magnetic fluid as

$$\underbrace{\frac{Du}{Dt}}_{\substack{\text{total rate of increase of}\\ \text{internal energy per unit}\\ \text{volume}}} = \underbrace{-\nabla \cdot \vec{q}}_{\substack{\text{rate of internal}\\ \text{energy addition by}\\ \text{heat conduction per}\\ \text{unit volume}}} + \underbrace{\left[\overline{\overline{\tau}}_{\text{viscous}} : \nabla \vec{v}\right]}_{\substack{\text{irreversible rate}\\ \text{of internal energy}\\ \text{increase per unit}\\ \text{of volume by viscous}\\ \text{dissipation}}} + \left\{\text{rate of magnetic work}\right\} \tag{5.43}$$

According to the work of Neuringer and Rosensweig [92], the rate of magnetic work done on the system can be expressed by the substantial derivative as

$$\begin{bmatrix}\text{rate of magnetic work} \\ \text{done on fluid per unit} \\ \text{volume}\end{bmatrix} = \mu_0 H \frac{DM}{Dt} \tag{5.44}$$

Then the energy equation Eq. (5.43) can be defined as

$$\frac{Du}{Dt} = -\nabla \cdot \vec{q} + \left[\overline{\overline{\tau}}_{\text{viscous}} : \nabla \vec{v}\right] + \mu_0 H \frac{DM}{Dt} \tag{5.45}$$

By expressing the internal energy u as a function of enthalpy h and the magnetic energy (HM) (see the Chap. 1 on thermodynamics):

$$u = h + \mu_0 MH \tag{5.46}$$

and substituting Eq. (5.46) into Eq. (5.45), we can obtain the following relation for the energy equation:

$$\frac{Dh}{Dt} + \mu_0 \frac{D}{Dt}(MH) = -\nabla \cdot \vec{q} + \left[\overline{\overline{\tau}}_{\text{viscous}} : \nabla \vec{v}\right] + \mu_0 H \frac{DM}{Dt} \tag{5.47}$$

From the relation:

$$d(MH) = MdH + HdM \tag{5.48}$$

Equation (5.47) can be rearranged into:

$$\frac{Dh}{Dt} + \mu_0 M \frac{DH}{Dt} = -\nabla \cdot \vec{q} + \left[\overline{\overline{\tau}}_{\text{viscous}} : \nabla \vec{v}\right] \tag{5.49}$$

The total derivative of the enthalpy equals (see the Chap. 1 on thermodynamics):

$$dh(s, H) = \left(\frac{\partial h}{\partial s}\right)_H ds + \left(\frac{\partial h}{\partial H}\right)_s dH = T\,ds - \mu_0 MdH \tag{5.50}$$

Since the total derivate of the entropy equals (see the Chap. 1 on thermodynamics):

$$ds(T, H) = \left(\frac{\partial s}{\partial T}\right)_H dT + \left(\frac{\partial s}{\partial H}\right)_T dH = \frac{c_H}{T} dT + \left(\frac{\partial s}{\partial H}\right)_T dH \tag{5.51}$$

where Eq. (5.51) can be rewritten using the Maxwell relation:

$$\left(\frac{\partial s}{\partial H}\right)_T = \mu_0 \left(\frac{\partial M}{\partial T}\right)_H \tag{5.52}$$

Then the total derivative of the enthalpy can be expressed as

$$dh(s,H) = c_H dT + \mu_0 \left[T \left(\frac{\partial M}{\partial T} \right)_H - M \right] dH \tag{5.53}$$

The specific heat c_H in this case represents the specific heat in a constant magnetic field for the magnetic fluid. This means that is takes into account both, the specific heat of the magnetic material and the specific heat of the base or carrier liquid. The magnetocaloric material in a static magnetic field exhibits a phase change in the form of latent heat. In such a case, it is more convenient to keep the energy equation in a form with the enthalpy (i.e. Eq. 5.49).

The enthalpy of the magnetic fluid (at a constant magnetic field) may be defined as follows:

$$h = \frac{m_{\text{solid}} h_{\text{solid}} + m_{\text{liquid}} h_{\text{iiquid}}}{m} = \phi_m \cdot h_{\text{solid}} + (1 - \phi_m) \cdot h_{\text{liquid}} \tag{5.54}$$

If the liquid phase does not undergo a phase change, then Eq. (5.54) may be written as

$$h = \frac{m_{\text{solid}} h_{\text{solid}} + m_{\text{liquid}} c_{p\,\text{liquid}} \vartheta_{\text{liquid}}}{m} = \phi_m \cdot h_{\text{solid}} + (1 - \phi_m) \cdot c_{p\,\text{liquid}} \vartheta_{\text{liquid}} \tag{5.55}$$

In this case, a special continuous-properties model can be applied, given by Egolf and Mantz [96]. The contributions of the magnetic field change can be added to such a model.

In the subsequent text we apply equations that consider the magnetic material to possess ordinary specific heat properties. For this particular case, the specific heat c_H of the magnetic fluid can be calculated using the simple relation (see also Xuan and Roetzel [97]):

$$c_H = \frac{m_{\text{solid}} c_{H\,\text{solid}} + m_{\text{liquid}} c_{p\,\text{liquid}}}{m} = \phi_m \cdot c_{H\,\text{solid}} + (1 - \phi_m) \cdot c_{p\,\text{liquid}} \tag{5.56}$$

With the help of Eqs. (5.49–5.53) the energy equation can be now written as

$$c_H \frac{DT}{Dt} + \mu_0 \left[T \left(\frac{\partial M}{\partial T} \right)_H - M \right] \frac{DH}{Dt} + \mu_0 M \frac{DH}{Dt} = -\nabla \cdot \vec{q} + \left[\overline{\overline{\tau}}_{\text{viscous}} : \nabla \vec{v} \right] \tag{5.57}$$

By rearranging Eq. (5.57) the energy equation can now be expressed as:

$$c_H \frac{\partial T}{\partial t} + c_H(\vec{v} \cdot \nabla T) + \mu_0 \left[T \left(\frac{\partial M}{\partial T} \right)_H \right] \frac{\partial H}{\partial t} + \mu_0 \left[T \left(\frac{\partial M}{\partial T} \right)_H \right] (\vec{v} \cdot \nabla H) \\ = -\nabla \cdot \vec{q} + \left[\overline{\overline{\tau}}_{\text{viscous}} : \nabla \vec{v} \right] \tag{5.58}$$

Since in a magnetic fluid, one may experience the anisotropy and the field-dependent effective thermal conductivity (see some examples in Fang et al. [98, 99], Reinecke et al. [100], Parekh and Lee [101], and Nkurikiyimfura et al. [102, 103]), we define the heat flux vector to be [91]:

$$\vec{q} = -[\kappa \cdot \nabla T] \tag{5.59}$$

where κ represents the second-order tensor, i.e. the thermal conductivity tensor.

If the magnetic material is isotropic, then Eq. (5.59) takes a conventional form:

$$\vec{q} = -\lambda \nabla T \tag{5.60}$$

For the isotropic effective thermal conductivity, the energy equation Eq. (5.58) takes the following form:

$$c_H \frac{\partial T}{\partial t} + c_H \vec{v} \cdot \nabla T + \mu_0 \left[T \left(\frac{\partial M}{\partial T} \right)_H \right] \frac{\partial H}{\partial t} + \mu_0 \left[T \left(\frac{\partial M}{\partial T} \right)_H \right] \vec{v} \cdot \nabla H \\ = \nabla \cdot (\lambda \nabla T) + \left[\overline{\overline{\tau}}_{\text{viscous}} : \nabla \vec{v} \right] \tag{5.61}$$

5.3.1 A Short Note on the Effective Thermal Conductivity

The thermal conductivity is a well-defined physical property for monophase fluids. In the case of a suspension of solid particles and the carrier or base fluid, the effective thermal conductivity becomes a model quantity. The basic approaches apply thermal resistivity networks in analogy to the calculation methods for electric circuits [104].

Since Maxwell [105], who first investigated the effective thermal conductivity, many different models have been developed, as shown in Table 5.4.

The effective thermal conductivity for the magnetic fluid can be defined by use of different models, which have been developed for suspensions, solid composites as well as nanofluids. In most of the correlations, the effective thermal conductivity is expressed as a function of the volume fraction of the solid phase, and by taking into account the thermal conductivities of the solid phase and the liquid phase, respectively:

$$\lambda = f(\lambda_{\text{liquid}}, \lambda_{\text{solid}}, \phi_V) \tag{5.62}$$

Table 5.4 Examples of models for effective thermal conductivity of suspensions

Author	Model	Comment
Maxwell [105]	$\lambda = \lambda_{\text{liquid}} \left(1 + \frac{3 \cdot \left(\frac{\lambda_{\text{liquid}}}{\lambda_{\text{solid}}} - 1 \right) \phi_V}{\left(\frac{\lambda_{\text{liquid}}}{\lambda_{\text{solid}}} + 2 \right) - \left(\frac{\lambda_{\text{liquid}}}{\lambda_{\text{solid}}} - 1 \right) \phi_V} \right)$ (5.63)	Very dilute suspensions of spheres, does not account for interaction between particles
Rayleigh [189]	$\lambda = \lambda_{\text{liquid}} \left(\frac{[(2+\alpha)/(1-\alpha)] - 2\phi_V - 0.525[(3-3\alpha)/(4+3\alpha] \cdot \phi_V^{\frac{10}{3}}}{[(2+\alpha)/(1-\alpha)] + \phi_V - 0.525[(3-3\alpha)/(4+3\alpha] \cdot \phi_V^{\frac{10}{3}}} \right)$ $\alpha = \frac{\lambda_{\text{solid}}}{\lambda_{\text{liquid}}}$ (5.64)	Extended Maxwell's model to higher solid fractions
Jeffrey [190]	$\lambda = \lambda_{\text{liquid}} \left(1 + 3 \cdot \phi_V \cdot \beta + 3 \cdot \phi_V^2 \cdot \beta^2 \cdot \gamma \right)$ $\gamma = 1 + \frac{\beta}{4} + \frac{3\beta}{16} \left(\frac{\alpha + 2}{2\alpha + 3} \right)$ $\beta = \frac{\alpha - 1}{\alpha + 2} \quad \alpha = \frac{\lambda_{\text{solid}}}{\lambda_{\text{liquid}}}$ (5.65)	Spherical particles $0 \leq \alpha \leq \infty$

A review of different correlations useful for suspensions, including nanofluids and magnetic fluids, can be found in Nkurikiyimfura et al. [103], Kaviany [106], Kandula [107], Kleinstreuer and Feng [108], Wang and Mujumdar [109] and Singh [110].

Besides the particle shapes and packing formations, researchers also evaluated the influence on the effective conductivity of other physical properties and parameters. This is especially so when considering nanofluids. For instance, Das et al. [111] investigated the influence of temperature in the case of nanofluids. They observed a strong increase in the effective thermal conductivity with the increased temperature, which could be a consequence of the kinetics of molecules and solid nanoparticles. This was also confirmed by Abareshi et al. [112].

However, as denoted by Kleinstreuer and Feng [108], a large number of researchers did not find the strong dependence of the effective thermal conductivity on the temperature. Therefore, the dependence of the effective thermal conductivity on temperature stays an open field for future research efforts. Kleinstreuer and Feng [108] also reviewed the measuring techniques for effective thermal conductivity in nanofluids.

Some researchers have also investigated the influence on the effective thermal conductivity by *pH* value, the type of the base fluid, the nanoparticle shape, the degree of nanoparticle dispersion/interaction, and various additives (see, e.g., Wang and Choi [113], Li and Peterson [114], Zhu et al. [115], Jang and Choi [116], Timoefeva et al. [117, 118], Murshed et al. [119]).

Based on the existing research, we can conclude that the thermal conductivity of magnetic fluids (nanofluids or magnetorheological fluids) will depend not only on the basic parameters, defined in Eq. (5.62), but will also take into account the following dependencies:

- **Effect of Brownian-motion** (ferrofluids only) (see also Lee et al. [120], Wang et al. [113], Keblinski et al. [121], Jang and Choi [122], Kleinstreuer and Li [123], Prasher [124], Li [125], Kumar et al. [126], Koo and Kleinstreuer [127], Bao [128], and Feng and Kleinstreuer [129]),
- **Effect of magnetic field** as well as the related anisotropy of the effective thermal conductivity (see, e.g. Blums [130], Xuan and Wang [131], Kirchler and Odenbach [132], Philip et al. [133, 134], Wright et al. [135], Wensel et al. [136], Shima et al. [137], Gavili et al. [138], Nkurikiyimfura et al. [139] (note stronger effects can be expected in magnetorheological fluids),
- **Effect of surfactants,**
- **Effect of particle size and shape** including potential agglomeration or clustering effects (see, e.g. Odenbach [18], Keblinski et al. [140], Keblinski [141], Eapen et al. [142], Prasher et al. [143], Tillman and Hill [144], Wang et al. [145], Tillman and Hill [146], Wang and Fan [147], Bishop et al. [148], Mendelev and Ivanov [149]),
- **Effect of temperature,** which can also affect the magnetic properties of magnetic fluids in the presence of a magnetic field (not to be misunderstood as the magnetic convection effect).

5.4 Review of Research on Magnetocaloric Fluids

The earliest work performed in the field of magnetocaloric energy conversion for the application of magnetocaloric fluids can be found in Resler and Rosensweig [150, 151], who published their articles in 1964 and 1967, respectively. In 1969, another study was conducted by Van der Voort [152]. In 1985, Rosensweig [7] published a book on ferrohydrodynamics, which in a large part dealt also with the magnetocaloric energy conversion using ferrofluids. This work can actually be considered as the basis for the study of magnetic fluids, including those that may be applied in magnetocaloric energy conversion.

From 1993, an idea to perform heating and cooling based on magnetocaloric fluids can be found in the patent of Burnett [153]. In 1996, Shao [154] published work on the preparation of magnetocaloric nanofluids for room-temperature magnetic refrigeration. Later in 2003, also Shir et al. [155] reported on the research and production of magnetic nanocomposites that could consist of a solid phase in a nanofluid.

Since then, a number of publications can be found concerning magnetocaloric fluids; however, there is no evidence that these magnetocaloric fluids have been applied in refrigeration, heat pumping or power generation.

On the other hand, this domain is still developing, and magnetocaloric fluids are the subject of research efforts for different applications. For instance, today's well-known hyperthermia cancer treatment (still in the phase of trials), which applies the hysteresis effect of particles in ferrofluids induced by alternating magnetic field (see, e.g. Jordan et al. [156], Hiergeist et al. [157], Rosensweig [158]), was upgraded by the idea of Prof. Tishin [159] (see also Tishin et al. [160]), who proposed the use of the magnetocaloric effect to enhance the heating of cancer cells by applying a variable magnetic field. In this case, the magnetocaloric particle would act as the nano-heat pump, providing the required heat to destroy the cancer cells. Later, the idea of drug release based on the magnetocaloric effect was investigated by Li et al. [161].

The majority of articles that are related to the use of magnetocaloric fluids are related to propulsion (pump) systems, and the majority of them regard the (self) heat-driven thermal management systems.

5.4.1 *Magnetocaloric Fluid Propulsion*

We first present the basic principle of magnetocaloric pumping, which was described by Rosensweig [7], by developing the ferrohydrodynamic Bernoulli equation. Figure 5.5 shows an element of the magnetic fluid with the mass $\rho \cdot a \cdot ds$ in the gradient of the magnetic field. The fluid moves with a velocity v along the distance s, and with the height h over the reference ground level. Rosensweig [7] proposed the following generalized Bernoulli equation:

$$\frac{dp}{ds} + \rho v \frac{dv}{ds} + \rho g \frac{dh}{ds} - \mu_0 M \frac{dH}{ds} = 0 \tag{5.66}$$

If Eq. (5.66) is now integrated from the section, e.g. denoted by 1 to the section denoted by 2, then the following form will be obtained:

$$\int_1^2 \frac{dp}{\rho} + \frac{v_2^2 - v_1^2}{2} + g(h_2 - h_1) - \mu_0 \int_1^2 \frac{M}{\rho} dH = 0 \tag{5.67}$$

Since ferrofluid can be considered as an incompressible fluid with a constant density, Eq. (5.67) takes the following form, with dimensions per unit of volume:

$$p_1 + \rho \frac{v_1^2}{2} + \rho g h_1 - \mu_0 \int_0^{H_1} M\, dH = p_2 + \rho \frac{v_2^2}{2} + \rho g h_2 - \mu_0 \int_0^{H_2} M\, dH \tag{5.68}$$

Let us consider now the example shown in Fig. 5.5b. The ferrofluid is placed in a tube, which is cooled at one end (cold part with temperature of $T_1 = T_2$) and heated at the other end (hot part with temperature of $T_3 = T_4$). In the middle of the tube, the constant magnetic field region is given by the magnetic field source, by the direction of the magnetic field along the tube axis (direction from cold to hot part). Because of the higher magnetization in the cold part, the ferrofluid will be attracted to the magnetic field. In the hot part, since the ferrofluid is heated, its magnetization will be reduced. If one now applies the ferrohydrodynamic Bernoulli equation (Eq. 5.68), by neglecting gravitational forces and keeping in mind that the kinetic energy for the constant cross-section of the tube will remain constant, the following expression can be made between the stations 1 and 2, which are placed in the free space:

$$p_1 = p_2 - \mu_0 \left(\overline{M} H\right)_2 \tag{5.69}$$

Analogously to Eq. (5.69), the pressure difference can be shown for stations 3 and 4:

$$p_3 - \mu_0 \left(\overline{M} H\right)_3 = p_4 \tag{5.70}$$

where:

$$\overline{M} = \frac{1}{H} \int_0^H M\, dH \tag{5.71}$$

As denoted by Rosensweig [7], the part of the tube inside the magnetic field is not applicable to ferrohydrodynamic Bernoulli equation, since the assumption of the isothermal flow field is inherent and its derivation does not hold there. However, by neglecting the acceleration of the fluid, friction and gravity Rosensweig [7] shows the following relationship inside the solenoid:

$$0 = -\nabla p^{*} + \mu_0 M \, \nabla H \tag{5.72}$$

where Rosensweig [7] denotes the pressure $p*$ to be the composite pressure. Since the magnetic field is considered to be uniform, it follows that $p_2* = p_3*$. The pressure differences between stations 4 and 1 in Fig. 5.5b can now be defined according to Rosensweig [7] as

$$\Delta p = p_4 - p_1 = \mu_0 H \left[\overline{M}(T_1) - \overline{M}(T_4)\right] = \mu_0 H \, \Delta\overline{M} \tag{5.73}$$

The pressure difference in Eq. (5.73) can actually be considered as the basis for the propulsion of the magnetocaloric fluid.

The propulsion of the magnetocaloric fluid may not be (with regard to the present knowledge in this domain) interesting for power generation (see the Chap. 8), but results in the development of the heat-driven pumping system, especially for the thermal management of electronic devices, seem to be some good directions for near-future market applications.

A comprehensive research project was performed by Love et al. [162] in 2003, where they developed and experimentally investigated a magnetocaloric pump (see also Love et al. [163]). Soon after in 2004, Yamaguchi et al. [164] performed a study of the characteristics of a thermomagnetic motor based on magnetic fluids.

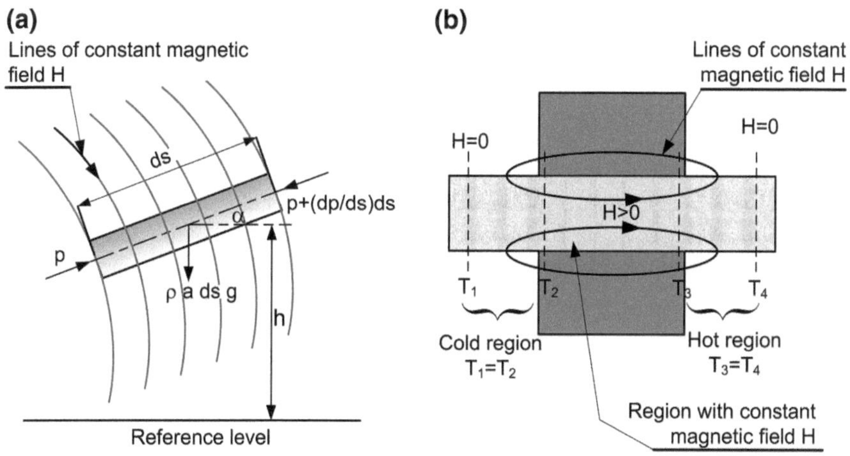

Fig. 5.5 **a** An element of magnetic fluid in the gradient of a magnetic field, **b** the magnetocaloric pump principle (see also Rosensweig [7])

In 2009, Fumoto et al. [165] investigated the use of a magnetocaloric fluid for thermal management. The same year, experimental research on the application of propulsion and thermal management with the ferrofluid was performed by Lian and Xuan [166] and reported later also in 2011 [167], where they simulated chip cooling with such an application.

In 2010, Pal et al. [168] did experimental research on a ferrofluid-based pump. For this purpose, a table-top version of thermomagnetic pump was constructed and its performance was experimentally evaluated for two different types of ferrofluids. The authors emphasized that thermomagnetic pumps are basically low-head high-heat-flux discharge devices (see also Pal et al. [169]).

In 2010, Kitanovski and Egolf [170] described and showed an idea for applying a magnetocaloric fluid for cooling the concentrated photovoltaic system.

In 2011, Xia et al. [171] reported on a theoretical study of a micro-pump based on the magnetocaloric effect of a magnetic fluid. A number of different concepts for applying ferrohydrodynamic micro-pump systems were presented by Nguyen [172] in 2012, including magnetocaloric fluids.

In 2014, Petit et al. [173] reported on an experimental investigation of a ferrofluid (based on MnZn) pump for thermal management. They pointed out that the results from the scientific literature generally conclude that the experimental static pressure is lower than the theoretical one. A conclusion has been made according to the author's observations, that the magnetostatic pressure is limited due to magneto-convective motion of the ferrofluid, which causes a decrease in the temperature gradient in the ferrofluid. A further reduction in the pressure was assumed to be due to the agglomeration of magnetic particles.

5.4.2 Refrigeration and Heat Pumping by the Application of a Magnetocaloric Fluid

Following earlier research described at the beginning of this chapter, in the past decade there were a small number of research activities performed in the field of magnetic refrigeration and heat pumping concerning the magnetocaloric fluids.

The first such study was published by Rosensweig in 2006 [15], who investigated the room-temperature refrigeration with magnetocaloric suspensions by the application of a permanent magnetic field source (Fig. 5.6). As the author pointed out, the advantage of applying this kind of system (which includes regenerative heat transfer) over systems with solid magnetocaloric materials avoids wear, drag and leakage problems from mechanical sliding seals.

In the theoretical study [15], the regenerative refrigeration Brayton cycle was investigated for the system providing 100 W of cooling and operating between the heat-source temperature of 273 K and the heat-sink temperature of 298 K. A modified Halbach permanent-magnet array has been proposed as the magnetic field source with a high magnetic flux density of 3 T. The working medium evaluated in the study of

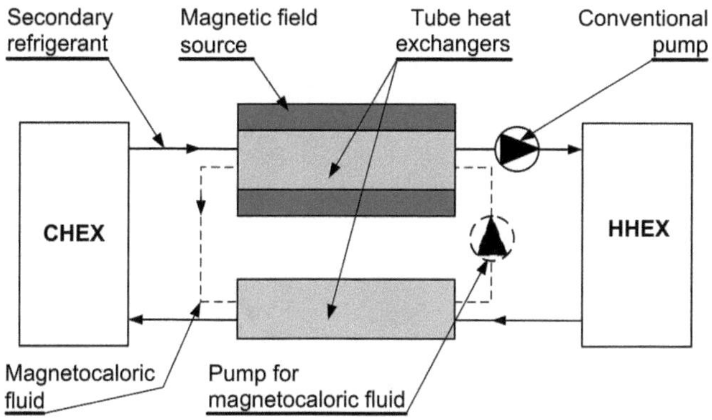

Fig. 5.6 A regenerative magnetocaloric-fluid-based refrigerator concept proposed by Rosensweig [7, 15]

Rosensweig was a suspension of 5.5-nm gadolinium particles suspended in NaK. As pointed out by Rosensweig [15], highly concentrated suspensions of magnetic particles are necessary to achieve a reasonable power density for the device. However, the disadvantages of this application are certainly the rheological influences of yield stress and viscosity that increase the energy dissipation and demand, therefore, an increase in the pump power. The results of the study [15] reveal the exergy efficiency (denoted as Carnot efficiency) of such a system, depending on different parameters, could be between 24 and 50 %.

In 2009 a group from the University of Applied Sciences of Western Switzerland [174] patented the concept of performing refrigeration or heat pumping with magnetocaloric fluids.

In 2010, from the same group, Kitanovski and Egolf [170] described a new system for magnetic refrigeration and heat pumping using magnetocaloric fluids (Fig. 5.7). This regenerative system would apply different magnetocaloric materials in several magnetic fluids. Each fluid would correspond to a certain temperature range of operation, thus providing the "layering" of magnetocaloric materials in the direction of the temperature gradient. Figure 5.7 shows the conceptual principle of operation for such a device.

According to Fig. 5.7, a static quadruple permanent magnet assembly provides the magnetic field (note also the different number of poles and also other types of magnetic field sources could be applied). Different magnetocaloric fluids with magnetocaloric particles having different Curie temperatures flow in the azimuth direction. The flow of magnetocaloric fluids is provided by a pump system, which may also be based on magnetohydrodynamic propulsion (note also that this kind of application might also be used for the magnetization of the magnetocaloric particles). The system is fully static, except for the fluid flow of magnetocaloric fluids, as well as the secondary refrigerant (secondary fluid). The fluid flow of the secondary fluid and the magnetocaloric fluids is performed through an internal cross-flow heat exchanger.

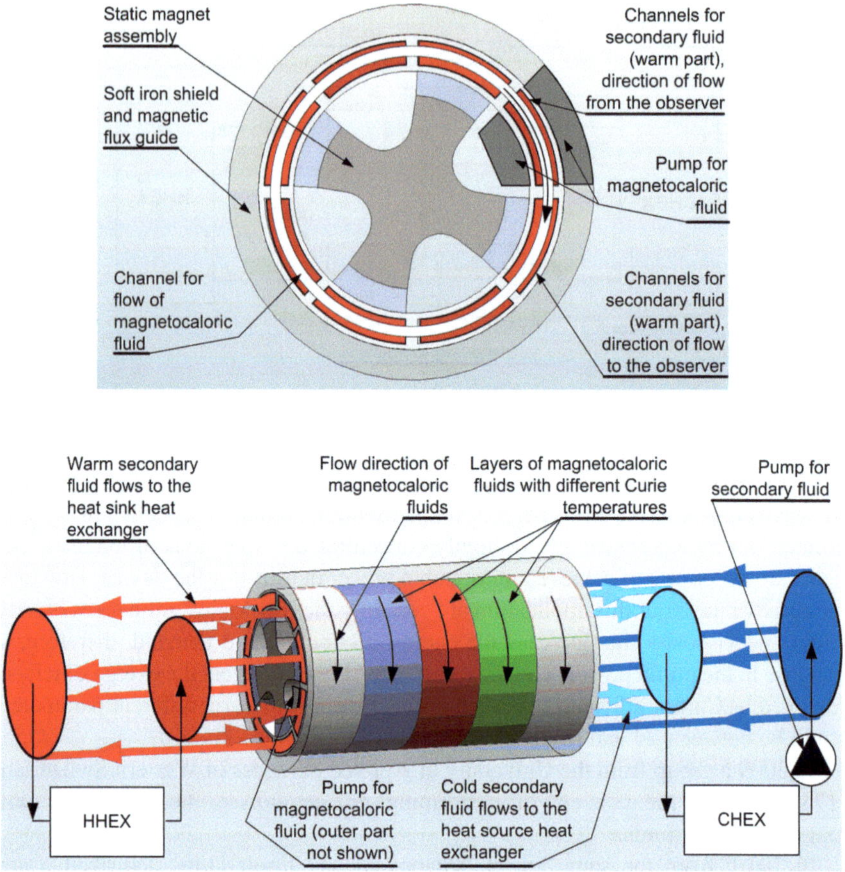

Fig. 5.7 The concept of a magnetic refrigerator or a heat pump based on magnetocaloric fluids

5.5 A Note on the Design of Magnetocaloric Refrigeration or Heat-Pump Devices Based on Magnetocaloric Fluids

The great advantage of magnetocaloric energy conversion based on magnetocaloric fluids is in the following facts:

- No moving parts;
- Simplified construction of magnetic field sources;
- No need for dynamic seals or valves;
- Potential application also in small or micro-sized devices.

The disadvantages of such systems relate to the rheology of magnetocaloric fluids and the related pumping power consumption and the overall magnetocaloric effect, which is reduced by the thermal mass of the base or the carrier fluid.

Based on previous subsections, we can now draw certain conclusions as well as guidelines for the future development of magnetocaloric devices, based on magnetocaloric fluids. As pointed out by Rosensweig (Sect. 5.4.2), the application of a magnetorheologic suspension is doubtful because of the drawbacks, which concern high losses due to pumping. On the other hand, a high volume fraction of the solid magnetocaloric material in the carrier fluid can provide good power density for a device. Therefore, there is still an open question as to whether there exists a solution that could make the application of magnetocaloric magnetorheologic suspensions feasible.

An important issue which needs to be addressed is also the thermal capacity of the carrier or base liquid compared to that of the magnetocaloric material. Also, the thermal conductivity of the carrier fluid and the magnetocaloric material plays an important role.

Another important issue regards the pumping system, which should operate with high efficiency. This also concerns different magnetic field sources, which may be applied in applications with magnetocaloric fluids.

Because of the above-mentioned problems, as well as other important design issues, the text below discusses a particular problem or application.

5.5.1 Applications of Magnetorheologic Fluids (Including Magnetocaloric Suspensions)

Below we describe potential applications not only for magnetocaloric suspensions, but also for other types of magnetic suspensions; applications that may be successfully applied as part of the magnetocaloric energy-conversion system.

5.5.1.1 Magnetorheologic Fluids as Seals or Valves

This domain is already well applied in different applications. Since the magnetic field source in magnetocaloric devices is always present it makes sense to take this advantage and apply magnetorheologic fluids for seals (especially in rotating devices) or valve system, which would provide switching of the fluid flow direction. Both the magnetorheological valve and the sealing concern applications of magnetocaloric energy conversion, which is based on active magnetic regeneration (AMR). Note that in devices with thermal diode mechanisms (see Chap. 6), valves or dynamic sealing are not actually required.

5.5.1.2 Magnetorheologic Fluids as Thermal Diode Mechanisms

A whole chapter in this book has been dedicated to special heat-transfer mechanisms—thermal diodes. Therefore, the reader should read this first to become familiar with thermal diode mechanisms, so as to understand this particular application of magnetorheological fluids. The magnetorheological fluid can be successfully applied as a thermal switch or thermal valve to provide oscillating heat flux to or from the magnetocaloric material. The activation or deactivation of such a mechanism will be based on the oscillation (alternation) of the magnetic field intensity. In order to provide an efficient heat transfer, the magnetorheologic fluid should possess high effective thermal conductivity and, on the other hand, small thermal mass. The latter is required to enable rapid transfer of heat from or to the magnetorheological thermal diode. Note that the application of the magnetorheological fluid in this case does not necessarily concern the macroscopic, but rather a microscopic system.

5.5.1.3 Magnetorheologic Fluids as Actuators for Pump Systems

An advantage of applying a magnetorheologic fluid as pump actuators for pump systems in magnetocaloric devices is that they may be integrated into the system, as well as because of the fact that the variation of the magnetic field in magnetocaloric devices, based on the AMR principle, are strongly related to the fluid flow through the porous structure of the AMR. The magnetorheologic fluid can therefore serve as the piston in such systems, or the actuator of the peristaltic pump, membrane pump or other mechanisms. Note that the movement of fluids (e.g. water) in the AMR relates to rather small oscillations. In terms of distances that the water has to pass in a single process, this corresponds to about 10–40 mm, e.g. for a higher frequency of the operation of a device and depending on the effectiveness of the AMR.

5.5.1.4 Magnetorheologic Magnetocaloric Fluids as Refrigerants

In this case one should take care of the thermal mass of the carrier fluid. The specific heat of magnetocaloric materials is small, compared to, e.g. water or oils. Therefore, the carrier fluid should possess high thermal conductivity and small thermal mass. The volume fraction of the magnetocaloric material should be high, but one should also note that this strongly influences the rheology of such a fluid. Also, the magnetic field in such devices should be rather high, certainly above 1–1.5 T. If the magnetic field is too low, the magnetocaloric effect will not provide sufficient heat to affect the carrier fluid; therefore, the overall temperature change due to the magnetocaloric effect will be rather small or not sufficient.

If some surfactants are applied to provide better dispersion of the magnetocaloric particles, one should take care of their thermal conductivity, since it may represent the thermal resistance for heat diffusion from the magnetocaloric particle to the carrier fluid. This holds true also for magnetocaloric particles, where the smaller size will have a strong influence on the desired rapid diffusion of the magnetocaloric effect towards the carrier fluid and vice versa.

The best carrier fluids for the magnetorheologic magnetocaloric refrigerant are actually liquid metals. Since mercury is toxic, one could think of other fluids, based on NaK or gallium-indium alloys. For the first, one should take care about the hazard of applying NaK and its potential contact with water. For the second, one should keep in mind the high costs of such a fluid.

The pumping system for these kinds of fluids should be carefully selected, and one should not neglect the possibility of applying the magnetohydrodynamic principle.

5.5.2 Applications of Ferrofluids (Including Magnetocaloric Ferrofluids)

In this particular case, the applications will be restricted to a small volume fraction of the solid particles. Therefore, the effect that can be provided by magnetorheologic fluids for seals, valve or pump systems, would not actually be very efficient with the application of ferrofluids. Despite what we may think about the application of ferrofluids for heat-transfer fluids, our investigations unfortunately show that the effect of the higher heat-transfer coefficient will be neglected by the higher friction losses. Therefore, the overall efficiency of the magnetocaloric device will not be better than that of applying a cheaper and natural fluid, for example, water.

Because of the above facts, we may think about a few basic applications where magnetocaloric fluids can be applied.

5.5.2.1 Magnetocaloric Ferrofluids as Refrigerants

In this case the facts that we have pointed out for magnetorheologic refrigerants have to be considered. However, the rheological properties of the ferrofluid offer better possibilities for application of different pumping systems. The friction during the flow of such a fluid will be smaller compared to magnetorheologic fluids. Because of the small volume fraction of the solid magnetocaloric material in the ferrofluid, in most cases when the ferrofluid is applied as refrigerant, the base fluid should be liquid metal (when speaking about near-room-temperature applications). It is therefore not a coincidence that Rosensweig [15] considered the NaK fluid to be a base fluid for the magnetocaloric ferrofluid. In this case, a permanent-magnetic field sources can also be applied, but they should provide very high magnetic flux density of about 2.5–3 T. So high magnetic flux density will certainly increase the efficiency. However, the cost and the weight restriction of permanent magnets are too large. Oils or other water-based fluids represent too large thermal mass, which will be reflected in the small overall magnetocaloric effect of the magnetocaloric ferrofluid. In order to compensate for the small volume fraction and still apply ordinary base fluids, higher magnetic fields should be applied (e.g. above 3 T), which directs one to the application of superconducting magnets.

5.5.2.2 Magnetocaloric Ferrofluids in Thermal Management

As could be seen in the previous subsection, thermal management can be successfully provided by application of magnetocaloric fluids. Since the heat source in this case is anyway available, such a system will be self-operating, similar to the case of heat pipes. Considering the findings of researchers we may emphasize that the application of magnetocaloric fluids for thermal management regards rather small devices (e.g. applications in electronics) with a high power density.

5.5.2.3 Magnetocaloric Ferrofluids in Medicine

Magnetocaloric fluids may be successfully applied in cancer treatment, where they would act as heat pumps (see Sect. 5.4). Compared to other types of magnetic induction treatment, this type of treatment certainly represents an advantage because of the magnetocaloric effect.

References

1. Einstein A (1906) Eine neue Bestimmung der Moleküldimensionen. Ann Physik 19 (2):289–306
2. Einstein A (1911) Berichtigung. Ann Phys 34:591–592
3. Mooney M (1931) Explicit formulas for slip and fluidity. J Rheol 2:210–215
4. Rabinowitsch B (1929) Über die Elastizität von Solen. Z Phys Chem A 145:1–7
5. Kitanovski A, Vuarnoz D, Ata-Caesar D et al (2005) The fluid dynamics of ice slurry. Int J Refrig 28(1):37–50
6. Kitanovski A, Poredoš A (2002) Concentration distribution and viscosity of ice-slurry in heterogeneous flow. Int J Refrig 25:827–835
7. Rosensweig RE (1985) Ferrohydrodynamics. Cambridge University Press, New York
8. Blums E, Cebers A, Mayorov MM (1997) Magnetic fluids. Walter de Gruyters, Berlin
9. Odenbach S (2006) Ferrofluids. In: Buschow KHJ (ed) Handbook of magnetic materials, vol 16. Elsevier, Amsterdam, pp 127–208
10. Anton I, De Sabata L, Vekas J (1990) Application orientated researches on magnetic fluids. Magn Magn Mater 85:219–226
11. Vékás L (2008) Ferrofluids and magnetorheological fluids. Adv Sci Tech 54:127–136
12. Bossis G, Volkova O, Lacis S et al (2002) Magnetorheology: fluids, structures and rheology. In: Odenbach S (ed) Ferrofluids: magnetically controllable fluids and their applications, vol 594. Lecture notes in physics. Springer, Heidelberg, pp 202–230
13. Shulman ZP, Kordonsky VI, Zaltsgendler EA et al (1986) Structure, physical properties and dynamics of magnetorheological suspensions. Int J Multiph Flow 12(6):935–955

14. Carlson JD, Jolly MR (2000) MR fluid, foam and elastomer devices. Mechatronics 10 (4):555–569
15. Rosensweig RE (2006) Refrigeration aspects of magnetic particle suspensions. Int J Refrig 29:1250–1258
16. López-López MT, Gómez-Ramírez A, Rodríguez-Arco L et al (2012) Colloids on the frontier of ferrofluids. Rheological properties. Langmuir 28:6232–6245
17. Genc S, Derin B (2014) Synthesis and rheology of ferrofluids: a review. Curr Opin Chem Eng 3:118–124
18. Odenbach S (2002) Magnetoviscous effect in ferrofluids M 71. Springer, Berlin
19. Shliomis MI (1972) Effective viscosity of magnetic suspensions. Soviet Phys JETP 34 (6):1291–1294
20. McTague JP (1969) Magnetoviscosity of magnetic colloids. J Chem Phys 51(1):133–136
21. Rosensweig RE, Kaiser R, Miskolczy G (1969) Viscosity of magnetic fluid in a magnetic field. J Colloid Interface Sci 29(4):680–686
22. Ambacher O, Odenbach S, Stierstadt K (1992) Rotational viscosity in ferrofluids. Z Phys B Condens Matter 86:29–32
23. Patel R, Upadhyay RV, Mehta RV (2003) Viscosity measurements of a ferrofluid: comparison with various hydrodynamic equations. J Colloid Interface Sci 263:661–664
24. Zubarev A, Odenbach S, Fleisher J (2002) Rheological properties of dense ferrofluids. Effect of chain-like aggregates. J Magn Magn Mater 252:241–243
25. Zubarev A (2002) In: Odenbach S (ed) Ferrofluids, magnetically controllable fluids and their aplications. Lecture notes in physics. Springer, Berlin, p 143
26. Zubarev AY, Iskakova LY (2004) To the theory of rheological properties of ferrofluids: influence of drop-like aggregates. Phys A 343:65–80
27. Hezaveh H, Fazlali A, Noshadi I (2012) Synthesis, rheological properties and magnetoviscos effect of Fe_2O_3/paraffin ferrofluids. J Taiwan Inst Chem E 43:159–164
28. Roscoe R (1952) The viscosity of suspensions of rigid spheres. J Appl Phys 3(8):267–269
29. Hong RY, Ren ZQ, Han YP et al (2007) Rheological properties of water-based Fe_3O_4 ferrofluids. Chem Eng Sci 62:5912–5924
30. Hosseini SM, Fazlali A, Ghasemi E et al (2010) Rheological properties of a y-Fe_2O_3 paraffin-based ferrofluid. J Magn Magn Mater 322:3792–3796
31. Rodríguez-Arco L, López-López MT, Durán JDG et al (2011) Stability and magnetorheological behaviour of magnetic fluids based on ionic liquids. J Phys Condens Matter 23:15
32. Shah K, Upadhyay RV, Aswal VK (2012) Influence of large size magnetic particles on the magneto-viscous properties of ferrofluid. Smart Mater Struct 21:12
33. Vékás L (2004) Magnetic nanofluids properties and some applications. Rom J Phys 49:707–721
34. Krieger IM, Dougherty TJ (1959) A mechanism for non-Newtonian flow in suspensions of rigid spheres. T Soc Rheol 3:137–152
35. Andhariya N, Chudasama B, Patel R et al (2008) Field induced rotational viscosity of ferrofluid: effect of capillary size and magnetic field direction. J Colloid Interface Sci 323:153–157
36. Lian W, Xuan Y, Li Q (2009) Design method of automatic energy transport devices based on the thermomagnetic effect of magnetic fluids. Int J Heat Mass Transf 52:5451–5458
37. Martsenyuk MA, Raikher YL, Shliomis MI (1974) On the kinetics of magnetization of suspensions of ferromagnetic particles. Sov Phys JETP 38:413
38. Hinch EJ, Leal LG (1972) The effect of Brownian motion on the rheological properties of a suspension of non-spherical particles. J Fluid Mech 52:683–712
39. Larson RG (1999) The structure and rheology of complex fluids. Oxford University Press, New York
40. Olabi AG, Grunwald A (2007) Design and application of magneto-rheological fluid, technical report. Mater Des 28:2658–2664

41. Lim ST, Cho MS, Jang IB et al (2004) Magnetorheological characterization of carbonyl iron based suspension stabilized by fumed silica. J Magn Magn Mater 282:170
42. Park JH, Kwon MH, Park OO (2001) Rheological properties and stability of magnetorheological fluids using viscoelastic medium and nanoadditives. Korean J Chem Eng 18:580–585
43. Cho MS, Lim ST, Jang IB et al (2004) Encapsulation of spherical iron-particle with PMMA and its magnetorheological particles. IEEE Trans Magn 40(4):3036–3038
44. Fang FF, Choi HJ, Choi WS (2010) Two-layer coating with polymer and carbon nanotube on magnetic carbonyl iron particle and its magnetorheology. Colloid Polym Sci 288:359–363
45. Charles SW (2002) The preparation of magnetic fluids. In: Odenbach S (ed) Ferrofluids: magnetically controllable fluids and their applications, vol 594. Lecture notes in physics. Springer, Berlin, pp 3–18
46. Bica I, Liu YD, Choi HJ (2013) Physical characteristics of magnetorheological suspensions and their applications. J Ind Eng Chem 19:394–406
47. Park BJ, Park CW, Yang SW et al (2009) Core-shell typed polymer coated-carbonyl iron suspensions and their magnetorheology. J Phys Conf Ser 149:012078
48. Engin T, Evrensel C, Gordanineja F (2005) Numerical simulation of laminar flow of water-based magneto-rheological fluids in microtubes with wall roughness effect. Int Commun Heat Mass 32:1016–1025
49. Jiang W, Zhang Y, Xuan S et al (2011) Dimorphic magnetorheological fluid with improved rheological properties. J Magn Magn Mater 323:3246–3250
50. Iglesias GR, López-López MT, Durán JDG et al (2012) Dynamic characterization of extremely bidisperse magnetorheological fluids. J Colloid Interface Sci 377:153–159
51. Omidbeygi F, Hashemabadi SH (2012) Experimental study and CFD simulation of rotational eccentric cylinder in a magnetorheological fluid. J Magn Magn Mater 324:2062–2069
52. Serrano AR, Donado F, Rubio-Rosas E (2013) Preparation and characterization of magnetite/dextran nanocomposite used as a precursor of magnetic fluid. J Magn Magn Mater 335:149–158
53. Burguera EF, Love BJ, Sahul R et al (2008) A physical basis for stability in bimodal dispersions including micrometer-sized particles and nanoparticles using both linear and non-linear models to describe yield. J Intel Mat Syst Str 19:1361–1367
54. Wang X, Gordaninejad F (1999) Flow analysis of field-controllable, electro-and magneto-rheological fluids using Herschel-Bulkley model. J Intel Mat Syst Str 10:601–608
55. Resiga DS, Vékás L, Susan-Resiga R et al (2007) A rheological model for magneto-rheological fluids. In: Proceedings of the 3rd German–Romanian workshop on turbomachinery hydrodynamics, Timisoara, May 10–12, pp 141–148
56. Yamanaka S, Abe H, Naito M et al (2012) Colloidal dispersibility of fatty acid-capped iron nanoparticles and its effect on static and dynamic magnetorheological response. Colloid Surf A 415:239–246
57. Mrlik M, Sedlacik M, Pavlinek V et al (2013) Magnetorheology of carbonyl iron particles coated with polypyrrole ribbons: the steady shear study. J Phys Conf Ser 412:012016
58. Sidpara A, Das M, Jain VK (2009) Rheological characterization of magnetorheological finishing fluid. Mater Manuf Process 24:1467–1478
59. Gabriel C, Laun HM (2009) Combined slit and plate–plate magnetorheometry of a magnetorheological fluid (MRF) and parameterization using the Casson model. Rheol Acta 48:755–768
60. Kim IG, Song KH, Park BO et al (2011) Nano-sized Fe soft-magnetic particle and its magnetorheology. Colloid Polym Sci 289:79–83
61. Farjoud A, Ahmadian M, Mahmoodi N et al (2011) Nonlinear modeling and testing of magneto-rheological fluids in low shear rate squeezing flows. Smart Mater Struct 20:085013
62. Resiga DS (2009) A rheological model for magneto-rheological fluids. J Intel Mat Syst Str 20:1001–1010

63. Ginder JM, Davis LC, Elie LD (1995) Rheology of magnetorheological fluids: models and measurements. In: 5th International conference on ER fluids and MR suspensions, Singapore, pp 504–514
64. Ginder JM, Davis LC, Elie LD (1996) Rheology of magnetorheological fluids: models and measurements. Int J Mod Phys B 10(23–24):3293–3303
65. Genc S, Phulé PP (2002) Rheological properties of magnetorheological fluids. Smart Mater Struct 11:140–146
66. Charles E, See H (2009) Microstructural investigations of the yielding behaviour of bidisperse magnetorheological fluids. Rheol Acta 48:19–32
67. Kittipoomwong D, Klingenberg DJ, Ulincy JC (2005) Dynamic yield stress enhancement in bidispersion magnetorheological fluids. J Rheol 49:1521
68. Wereley NM, Chaudhuri A, Yoo JH et al (2006) Bidisperse magnetorheological fluids using Fe particle at nanometer and micron scale. J Intell Mater Syst Struct 17:393–401
69. Ginder JM, Nichols ME, Elie LD et al (1999) Magnetorheological elastomers: properties and applications. In: Wutttig M (ed) Proceedings of SPIE. Smart structures and materials, vol 3675. Smart Materials Technologies, Newport Beach, 12 July 1999
70. See H (2001) Mechanisms of magneto-and electro-rheology: recent progress and unresolved issues. Appl Rheol 11(2):70–82
71. Brady JF, Bossis G (1988) Stokesian dynamics. Ann Rev Fluid Mech 20:111–157
72. Orihara H, Doi M, Ishibashi Y (1999) Two types of mechanism of electrorheological effect in polymer blends. Int J Mod Phys B 13:1949–1955
73. Bonnecaze RT, Brady JF (1992) Yield stresses in electrorheological fluids. J Rheol 36:73–115
74. Fang FF, Choi HJ, Jhon MS (2009) Magnetorheology of soft magnetic carbonyl iron suspension with single-walled carbon nanotube additive and its yield stress scaling function. Colloid Surf A 351:46–51
75. Hato MJ, Choi HJ, Sim HH et al (2011) Magnetic carbonyl iron suspension with organoclay additive and its magnetorheological properties. Colloid Surf A 377:103–109
76. Hong CH, Liu YD, Choi HJ (2013) Carbonyl iron suspension with halloysite additive and its magnetorheology. Appl Clay Sci 80:366–371
77. Cho MS, Choi HJ, Jhon MS (2005) Shear stress analysis of a semiconducting polymer based electrorheological fluid system. Polymer 46:11484–11488
78. Choi HJMS, Cho JW, Kim CAMS et al (2001) A yield stress scaling function for electrorheological fluids. Appl Phys Lett 78:3806
79. Gordanlnejad F, Fuchs A, Dogrour U et al (2004) A new generation of magneto-rheological fluid dampers. Final progress report
80. Bell RC, Karli JO, Vavreck AN et al (2008) Magnetorheology of submicron diameter iron microwires dispersed in silicone oil. Smart Mater Struct 17:015028
81. Cheng H, Yeung Y, Tong H (2008) Viscosity behavior of magnetic suspensions in fluid-assisted finishing. Prog Nat Sci 18:91–96
82. Shulman ZP, Kordonskii VI, Zaltsgendler EA (1984) Structure and magnetic and rheological characteristics of a ferrosuspension. Magnetohydrodynamics 20(3):223–229
83. López-López MT, Kuzhir P, Meunier AG et al (2010) Synthesis and magnetorheology of suspensions of submicron-sized cobalt particles with tunable particle size. J Phys Condens Matter 22:324106
84. Vereda F, Vicente J, Segovia-Gutierrez JP et al (2011) Average particle magnetization as an experimental scaling parameter for the yield stress of dilute magnetorheological fluids. J Phys D Appl Phys 44:425002
85. Vicente J, Vereda F, Segovia-Gutierrez JP et al (2010) Effect of particle shape in magnetorheology. J Rheol 54:1337–1343
86. Vicente J, Segovia-Gutierrez JP, Andablo-Reyes E et al (2009) Dynamic rheology of sphere- and rod-based magnetorheological fluids. J Chem Phys 131:10

87. Vereda J, Vicente J, Hidalgo-Alvarez R (2007) Influence of a magnetic field in the formation of magnetite particles via two precipitation methods. Langmuir 23:3581–3589
88. Vicente J, Klingenberg DJ, Hidalgo-Alvarez R (2011) Soft Matter 7:3701
89. Mrlík M, Ilčíková M, Pavlínek V et al (2013) Improved thermooxidation and sedimentation stability of covalently-coated carbonyl iron particles with cholesteryl groups and their influence on magnetorheology. J Colloid Interface Sci 396:146–151
90. Shah K, Phu DX, Seong MS et al (2014) A low sedimentation magnetorheological fluid based on platelike iron particles, and verification using a damper test. Smart Mater Struct 23:027001
91. Bird RB, Stewart WE, Lightfoot EN (2006) Transport phenomena, 2nd edn. Wiley, New York
92. Neuringer JL, Rosensweig RE (1964) Ferrohydrodynamics. Phys Fluids 7(12):1927
93. Rosensweig RE (2004) Continuum equations for magnetic and dielectric fluids with internal rotations. J Chem Phys 121(3):1228–1242
94. Rosensweig RE (2002) Basic equations for magnetic fluids with internal rotations. In: Odenbach (ed) Ferrofluids. Lecture notes in physics, vol 594. Springer, Berlin, pp 61–84
95. Rosensweig RE (1987) Magnetic fluids. Ann Rev Fluid Mech 19:437–453
96. Egolf PW, Manz H (1994) Theory and modelling of phase change materials with and without mushy regions. Int J Heat Mass Transf 37(18):2917–2924
97. Xuan Y, Roetzel W (2000) Conceptions for heat transfer correlation of nanofluids. Int J Heat Mass Transf 43:3701–3708
98. Fang X, Xuan Q, Li Q (2009) Anisotropic thermal conductivity of magnetic fluids. Prog Nat Sci 19(2):205–211
99. Fang X, Xuan Y, Li Q (2009) Anisotropic thermal transport in magnetic fluids. In: ASME 2009 second international conference on micro/nanoscale heat and mass transfer, vol 1. Shanghai, pp 18–21
100. Reinecke BN, Shan JW, Suabedissen KK et al (2008) On the anisotropic thermal conductivity of magnetorheological suspensions. J Appl Phys 104(2):023507
101. Parekh K, Lee HS (2010) Magnetic field induced enhancement in thermal conductivity of magnetite nanofluid. J Appl Phys 107(9):09A310
102. Nkurikiyimfura I, Wanga Y, Pan Z (2013) Effect of chain-like magnetite nanoparticle aggregates on thermal conductivity of magnetic nanofluid in magnetic field. Exp Therm Fluid Sci 44:607–612
103. Nkurikiyimfura I, Wanga Y, Pan P (2013) Heat transfer enhancement by magnetic nanofluids: a review. Renew Sust Energ Rev 21:548–561
104. Egolf PW, Ata-Caesar D, Kitanovski A et al (2005) Thermal energy transport and storage with multi-functional fluids: the effective thermal conductivity. In: The sixth KSME-JSME thermal and fluids engineering conference, Jeju City, Korea
105. Maxwell JC (1873) A treatise on electricity and magnetism. Clarendon Press, Oxford
106. Kaviany M (1995) Principles of heat transfer in porous media. Springer, New York
107. Kandula M (2011) On the effective thermal conductivity of porous packed beds with uniform spherical particles. J Porous Media 14(10):919–926
108. Kleinstreuer C, Feng Y (2011) Experimental and theoretical studies of nanofluid: thermal conductivity enhancement: a review. Nanoscale Res Lett 6(1):1–13
109. Wang XQ, Mujumdar AS (2008) A review on nanofluids-part I: theoretical and numerical Investigations. Braz J Chem Eng 25(4):613–630
110. Singh AK (2008) Thermal conductivity of nanofluids. Defence Sci J 58:600–607
111. Das SK, Putra N, Theisen P et al (2003) Temperature dependence of thermal conductivity enhancement for nanofluids. J Heat Transf 125(4):567–574
112. Abareshi M, Goharshiadi EK, Zebarjad SM et al (2010) Fabrication, characterization and measurement of thermal conductivity of Fe_3O_4 nanofluids. J Magn Magn Mater 322 (24):3895–3901

113. Wang XW, Xu XF, Choi SUS (1999) Thermal conductivity of nanoparticle-fluid mixture. J Thermophys Heat Transf 13(4):474–480
114. Li CH, Peterson GP (2006) Experimental investigation of temperature and volume fraction variations on the effective thermal conductivity of nanoparticle suspensions (nanofluids). J Appl Phys 99(8):084314
115. Zhu D, Li X, Wang N et al (2009) Dispersion behavior and thermal conductivity characteristics of Al_2O_3-H_2O nanofluids. Curr Appl Phys 9:131–139
116. Jang SP, Choi SUS (2004) Role of Brownian motion in the enhanced thermal conductivity of nanofluids. Appl Phys Lett 84:4316–4318
117. Timofeeva EV, Gavrilov AN, McCloskey JM et al (2007) Thermal conductivity and particle agglomeration in alumina nanofluids: experiment and theory. Phys Rev E 76:061203
118. Timofeeva EV, Smith DS, Yu W et al (2010) Particle size and interfacial effects on thermo-physical and heat transfer characteristics of water-based alpha-SiC nanofluids. Nanotechnology 21:215703
119. Murshed SMS, Leong KC, Yang C (2005) Enhanced thermal conductivity of TiO_2-water based nanofluids. Int J Therm Sci 44:367–373
120. Lee S, Choi SUS, Li S et al (1999) Measuring thermal conductivity of fluids containing oxide nanoparticles. J Heat Transf 121:280–289
121. Keblinski P, Phillpot SR, Choi SUS et al (2002) Mechanisms of heat flow in suspensions of nanos-sized particles (nanofluids). Int J Heat Mass Transf 45:855–863
122. Jang SP, Choi SUS (2007) Effects of various parameters on nanofluid thermal conductivity. J Heat Transf 129:617–623
123. Kleinstreuer C, Li J (2008) Discussion: effects of various parameters on nanofluid thermal conductivity. J Heat Transf 130:025501
124. Prasher (2006) Brownian-motion-based convective-conductive model for the effective thermal conductivity of nanofluids. J Heat Transf 128:588–595
125. Li J (2008) Computational analysis of nanofluid flow in microchannels with applications to micro-heat sinks and bio-MEMS. PhD thesis, North Carolina State University
126. Kumar DH, Patel HE, Kumar VRR et al (2004) Model for heat conduction in nanofluids. Phys Rev Lett 93:144301
127. Koo J, Kleinstreuer C (2004) A new thermal conductivity model for nanofluids. J Nanopart Res 6:577–588
128. Bao Y (2008) Thermal conductivity equations based on Brownian motion in suspensions of nanoparticles (nanofluids). J Heat Transf 130:042408
129. Feng Y, Kleinstreuer C (2010) Nanofluid convective heat transfer in a parallel disk system. Int J Heat Mass Transf 53:4619–4628
130. Blums E (2002) Heat and mass transfer phenomena. In: Odenbach S (ed) Ferrofluids: magnetically controllable fluids and their applications, vol 594. Lecture notes in physics. Springer, Berlin
131. Li Q, Xuan Y, Wang J (2005) Experimental investigations on transport properties of magnetic fluids. Exp Therm Fluid Sci 30:109–116
132. Krichler M, Odenbach S (2013) Thermal conductivity measurements on ferrofluids with special reference to measuring arrangement. J Magn Magn Mater 326:85–90
133. Philip J, Shima P, Raj B (2008) Evidence for enhanced thermal conduction through percolating structures in nanofluids. Nanotechnology 19:305706
134. Philip J, Shima P, Raj B (2007) Enhancement of thermal conductivity in magnetite based nanofluid due to chainlike structures. Appl Phys Lett 91:203108
135. Wright B, Thomas D, Hong H et al (2007) Magnetic field enhanced thermal conductivity in heat transfer nanofluids containing Ni coated single wall carbon nanotubes. Appl Phys Lett 91:173116
136. Wensel J, Wright B, Thomas D et al (2008) Enhanced thermal conductivity by aggregation in heat transfer nanofluids containing metal oxide nanoparticles and carbon nanotubes. Appl Phys Lett 92:023110

137. Shima PD, Philip J, Raj B (2009) Magnetically controllable nanofluid with tunable thermal conductivity and viscosity. Appl Phys Lett 95:133112
138. Gavili A, Zabihi F, Isfahani TD et al (2012) The thermal conductivity of water base ferrofluids under magnetic field. Exp Therm Fluid Sci 41:94–98
139. Nkurikiyimfura I, Wang Y, Pan Z et al (2011) Thermal conductivity enhancement of magnetic nanofluid in magnetic field. In: Proceedings of the international conference on materials for renewable energy and environment (ICMREE), Shanghai, 20–22 May
140. Keblinski P, Prasher R, Eapen J (2008) Thermal conductance of nanofluids: is the controversy over? J Nanopart Res 10:1089–1097
141. Keblinski P (2009) Thermal conductivity of nanofluids. In: Volz S (ed) Thermal nanosystems and nanomaterials. Springer, Berlin, pp 213–221
142. Eapen J, Rusconi R, Piazza R et al (2010) The classical nature of thermal conduction in nanofluids. J Heat Transf 132:102402
143. Prasher R, Evans W, Meakin P et al (2006) Effect of aggregation on thermal conduction in colloidal nanofluids. Appl Phys Lett 89:143119
144. Tillman P, Hill JMA (2006) New model for thermal conductivity in nanofluids. In: Proceedings of the International conference on nanoscience and nanotechnology, Brisbane, 3–7 July
145. Wang BX, Sheng WY, Peng XFA (2009) Novel statistical clustering model for predicting thermal conductivity of nanofluid. Int J Thermophys 30(6):1992–1998
146. Tillman P, Hill JM (2010) Modeling the thermal conductivity of nanofluids. In: Bai YL, Zheng QS, Wei YG (eds) Proceedings of the IUTAM symposium on mechanical behavior and micro-mechanics of nanostructured materials, Beijing, 27–30 June. Springer, Heidelberg, pp 111–118
147. Wang L, Fan J (2011) Toward nanofluids of ultra-high thermal conductivity. Nanoscale Res Lett 6:153
148. Bishop KJM, Wilmer CE, Soh S et al (2009) Nanoscale forces and their uses in self-assembly. Small 5(14):1600–1630
149. Mendelev VS, Ivanov AO (2004) Ferrofluid aggregation in chains under the influence of a magnetic field. Phys Rev E 70:051502
150. Resler EL Jr, Rosensweig RE (1964) Magnetocaloric power. AIAA J 2(8):1418–1422
151. Resler EL Jr, Rosensweig RE (1967) Regenerative thermomagnetic power. J Eng P 89:399–406
152. Van der Voort E (1969) Ideal magnetocaloric conversion. Appl Sci Res 20:98–114
153. Burnett JE (1993) Magnetic heating and cooling systems. US patent 5.231.834
154. Shao YZ, Lai JKL, Shek CH (1996) Preparation of nanocomposite working substances for room-temperature magnetic refrigeration. J Magn Magn Mater 163:103–108
155. Shir F, Yarik L, Bennett LH et al (2003) Room temperature active regenerative magnetic refrigeration: magnetic nanocomposites. J Appl Phys 93(10):8295–8297
156. Jordan A, Wust P, Fahling H et al (1993) Inductive heating of ferrimagnetic particles and magnetic fluids: physical evaluation of their potential for hyperthermia. Int J Hyperth 9 (1):51–68
157. Hiergeist R, Andra W, Buske N et al (1999) Application of magnetite ferrofluids for hyperthermia. J Magn Magn Mater 201:420–422
158. Rosensweig RE (2002) Heating magnetic fluid with alternating magnetic field. J Magn Magn Mater 252:370–374
159. Tishin AM (2006) Method for carrying out a magnetic therapy of malignant neoplasms. WO 2006135270:A1
160. Tishin AM, Zatsepina EV, Egolf PW et al (2009) Magnetocaloric effect applied for a cancer tumour defeat: an improved hyperthermia method. In: Proceedings of third IIF-IIR international conference on magnetic refrigeration at room temperature, Des Moines, 11–15 May

161. Li J, Qu Y, Ren J et al (2012) Magnetocaloric effect in magnetothermally-responsive nanocarriers for hyperthermia-triggered drug release. Nanotechnology 23(50):505706
162. Love LJ, Jansen JF, McKnight TE et al (2003) Magnetocaloric pump for lab-on-a chip technology: phase I report ORNL/TM-2003/245. Report, U.S. Department of Energy
163. Love LJ, Jansen JF, McKinght TE et al (2004) A magnetocaloric pump for microfluidic applications. IEEE T Nanobiosci 3(2):101–110
164. Yamaguchi H, Sumiji A, Shuchi S et al (2004) Characteristics of thermo-magnetic driven motor using magnetic fluid. J Magn Magn Mater 272:2362–2364
165. Fumoto K, Ikegawa M, Kawanami T et al (2009) Heat transfer characteristics of a thermosensitive magnetic fluid in microchannel. J Therm Sci Tech-JPN 4(3):332–339
166. Lian W, Xuan Y, Li Q (2009) Design method of automatic energy transport devices based on the thermomagnetic effect of magnetic fluids. Int J Heat Mass Transf 52:5451–5458
167. Xuan Y, Lian W (2011) Electronic cooling using an automatic energy transport device based on thermomagnetic effect. Appl Therm Eng 31:1487–1494
168. Pal S, Chakraborty S, Datta A et al (2010) Experimental investigation of a ferrofluid based thermomagnetic pump. In: Proceedings of the 37th national and 4th international conference on fluid mechanics and fluid power, IIT Madras, Chennai, 16–18 Dec
169. Pal S, Datta A, Sen S et al (2011) Characterization of a ferrofluid-based thermomagnetic pump for microfluidic applications. J Magn Magn Mater 323:2701–2709
170. Kitanovski A, Egolf PW (2010) Innovative ideas for future research on magnetocaloric technologies. Int J Refrig 33:449–464
171. Xia C, Liu T, Qin X et al (2011) Research on the micropump based on the magnetocaloric effect of magnetic fluid. In: Second international conference on mechanic automation and control engineering, Inner Mongolia, 15–17 July
172. Nguyen NT (2012) Micro magnetofluidics: interactions between magnetism and fluid flow on the microscale. Microfluid Nanofluid 12(1–4):1–16
173. Petit M, Avenas Y, Kedous-Lebouc A et al (2014) Experimental study of a static system based on a magneto-thermal coupling in ferrofluids. Int J Refrig 37:201–208
174. Kitanovski A, Egolf PW, Sari O (2009) Method and device for the generation of cold and heat by magneto-calorific effect. US patent 7481063 B2
175. Darby R (1986) Hydrodynamics of slurries and suspensions. In: Cheremisinoff NP (ed) Encyclopedia of fluid mechanics: slurry flow technology. Gulf Publishing Company, New York, pp 49–92
176. Hanks RW (1986) Principles of slurry pipeline hydraulics. In: Cheremisinoff NP (ed) Encyclopedia of fluid dynamics: slurry flow technology. Gulf Publishing Company, Houston, pp 213–276
177. Steffe JF (1992) Rheological methods in food process engineering, 2nd edn. Freeman press, Michigan
178. Guth E, Simha R (1936) Untersuchungen über die Viskosität von Suspensionen und Lösungen. 3. Über die viskosität von kugelsuspensionen. Kolloid Z 74(3):266–275
179. Vand V (1948) Viscosity of solutions and suspensions. 1. Theory. J Phys Colloid Chem 52:277–299
180. Mooney M (1951) The viscosity of a concentrated suspension of spherical particles. J Colloid Sci 6:162–170
181. Frankel NA, Acrivos A (1967) On the viscosity of a concentrated suspension of solid spheres. Chem Eng Sci 22:847–853
182. Jeffrey DJ, Acrivos A (1976) The rheological properties of suspensions of rigid particles. AIChE J 22:417–432
183. Thomas DG (1965) Transport characteristics of suspension: VIII. A note on the viscosity of Newtonian suspensions of uniform spherical particles. J Colloid Sci 20:267–277
184. Batchelor GK (1977) The effect of Brownian motion on the bulk stress in a suspension of spherical particles. J Fluid Mech 83(1):97–117

185. Bicerano J, Douglas JF, Brune DA (1999) Model for the viscosity of particle dispersions. JMS Rev Macromol Chem Phys C 39(4):561–642
186. Brinkman HC (1952) The viscosity of concentrated suspensions and solutions. J Chem Phys 20:571
187. Papanastasiou TC (1987) Flow of materials with yield. J Rheol 31:385–404
188. Barnes HA, Hutton JE, Walters FRSK (1989) An introduction to rheology. Elsevier, Amsterdam
189. Rayleigh Lord (1892) On the influence of obstacles in rectangular order upon the properties of a medium. Philos Mag 56:481–502
190. Jeffrey DJ (1973) Conduction through a random suspension of spheres. P Roy Soc A Math Phys A 335:355–367

Chapter 6
Special Heat Transfer Mechanisms: Active and Passive Thermal Diodes

Many types of systems and devices require specific thermal management. For example, they can operate in such a manner that a rapid redirection of the heat flux or its intensity is required. This relates to different kinds of electronic devices, sensors, actuators, energy conversion devices, as well as some emerging technologies. The latter include ferroic energy conversion technologies based on magnetocaloric, electrocaloric, elastocaloric and other effects related to solid-state physics. With certain new design ideas it is perhaps even possible to reconsider existing systems and devices, which apply heat regenerators or require periodic heat-flux variations (i.e. Stirling devices).

This chapter presents mechanisms and devices that can be applied as solid-state and (micro)fluidic thermal diodes. The thermal diode (heat "semiconductor", thermal switch, heat valve, thermal rectifier) is a physical phenomenon, mechanism or device in which it is possible to manipulate and control the direction of the heat flux and sometimes also the intensity of the heat flux. In this chapter the characteristics of such thermal diodes are described and presented with respect to different potential applications.

A careful reader of this book will realize that the application of thermal diode mechanisms can be crucial for the future development of ferroic solid-state technologies. Conventional heat transfer mechanisms, such as heat convection, are too limited by the heat transfer surface and the properties of the working fluid. In order to boost the power density of the magnetocaloric device by one order of magnitude or more, a different research direction is required that may take us out of the scope of the familiar and conventional active magnetic regeneration, which is actually based on heat convection. One solution is the introduction of thermal diode mechanisms. These can also lead to the design and construction of a new and advanced generation of magnetocaloric energy conversion or other ferroic (e.g., electrocaloric) devices.

A. Kitanovski et al., *Magnetocaloric Energy Conversion*,
Green Energy and Technology, DOI 10.1007/978-3-319-08741-2_6

6.1 Introduction

The operation of a thermal diode mechanism can be managed by the application of an external energy source (active thermal diodes) or simply by influencing the characteristics of materials or fluids (passive thermal diodes) without the need for an additional power source. The thermal diode can have the function of a thermal switch or valve, i.e. to block or enable heat to flow. One of the most basic mechanisms that can be applied as a thermal diode is a thermosyphon [1]. Another possibility is the heat pipe [2, 3], which can be found in numerous different systems, from space applications, electronic devices, to applications in buildings. Despite the fact that a heat pipe can provide efficient heat transfer, it cannot operate as a fast switching mechanism or thermal diode for both the heat-flux direction and the heat-flux intensity (i.e. changes that need to be implemented in a certain very short time interval between a millisecond and approximately a second).

Figure 6.1 shows an example of the application of a thermal diode mechanism in the case of an application in magnetocaloric, barocaloric, electrocaloric or elastocaloric energy conversion. In the case of Fig. 6.1 a material is sandwiched between two thermal diodes. When the material is being magnetized (magnetocaloric), polarized (electrocaloric), pressurized (barocaloric) or stretched (elastocaloric), it heats up. For a thermodynamic cycle we need to remove this generated heat. In this particular case the thermal diode's operation is set to ON in order to guide the generated heat to the heat sink. Consequently, the material's temperature will decrease. Another thermal diode, which is connected with the heat source at the same time, needs to be set OFF (or it should represent a closed heat valve). When the material is being demagnetized (magnetocaloric), depolarized (electrocaloric), depressurized (barocaloric) or released (elastocaloric), it cools down. For a thermodynamic cycle we need to bring the heat from the heat source. In a continuous process these steps will form a thermodynamic cycle.

This is only one example where a thermal diode mechanism operates as a thermal switch or heat valve. However, thermal diodes may also possess other mechanisms or a combination of these mechanisms. Figure 6.2 shows an example

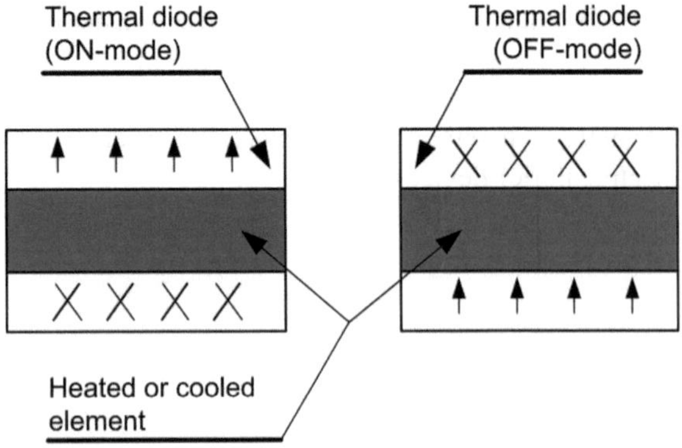

Fig. 6.1 Thermal diode operation in magnetocaloric, electrocaloric, elastocaloric or barocaloric energy conversion

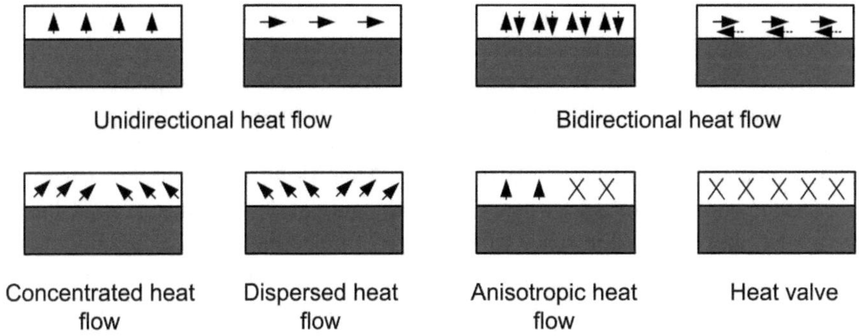

Fig. 6.2 Examples of the operating characteristics of a thermal diode

of different thermal diode characteristics. A particular thermal diode can possess several such mechanisms. Unidirectional flow relates to thermal diodes, which allow a heat flux in only one direction, i.e. laterally or vertically with respect to the heated or cooled element. Another principle relates to thermal diodes that allow a bidirectional heat flux. In certain cases some mechanisms may provide a concentrated and some a dispersed heat flux (a similar situation relates to the application of so-called nanostructured metamaterials that apply transformation thermodynamics). The anisotropic heat flux may relate to different structures, material properties, as well as mechanisms. The heat valve is a typical thermal diode mechanism that acts as the thermal switch (insulator–conductor).

Important Note: The alternation of the magnetic field in magnetic refrigeration can also serve as the driving source for different kinds of thermal diode mechanisms. This fact must not be neglected in the design of future magnetocaloric devices.

6.2 Active Solid-State Thermal Diodes

This kind of thermal diode requires an external energy source for its operation, which can be provided by a magnetic field, electric current, applied voltage or electric field, pressure, mechanical stress, etc.

6.2.1 Thermoelectrics

Thermoelectrics can be applied as an effective thermal switch or heat-valve mechanism in magnetic refrigeration [4–6]. Thermoelectric modules are used for electricity generation or for cooling and heat pumping. They work on two different principles. If the purpose is electricity generation, then the phenomenon called the

Seebeck effect [8] is taken into account. Here the heat is delivered to start the motion of the electrons, thus generating electricity. This kind of device is called a thermoelectric generator. However, if the purpose is to cool or to pump heat, then the electricity has to be delivered to the device, which starts the transport of heat from one side of the device to the other. This kind of cooling device is called a Peltier element or Peltier module (Fig. 6.3). Thermoelectric electricity generators are used in various processes where a sufficient amount of waste heat is available. In particular, they are used in space technology, for powering satellites and unmanned space probes. The power plant consists of the radioactive isotopes Uranium 235 or Plutonium 238 and thermoelectric generators, which use the heat released from the radioactive core decay [9, 10]. The field of refrigeration and cooling using Peltier elements is even wider. From the military and aerospace industries through everyday consumer applications (food storage, etc.) to industrial and laboratory needs [11]. In some special industrial and laboratory processes there is a need for fast and precise temperature control and management, especially on extremely small surfaces. This has not been possible until recently, when thin-film Peltier elements were designed [12, 13]. One of the leading research institutions in this domain is the German Fraunhofer Institute for Physical Measurement Techniques [14–16]. The problem of thermoelectrics is primarily its low efficiency, with maximum values reaching approximately 15 % [17–19]. However, at the moment this is the most accessible commercial technology for thermal diodes. A comprehensive review of thermoelectrics and their properties can be found in the book written by Nolas [20].

A simple schematic of a Peltier module is shown in Fig. 6.3. The electrons move in the opposite direction to the electric current density, denoted by *J*. In the warm part of the *p*-type semiconductor (at the heat sink), which is doped with excess holes, the electrons tend to fulfil holes (the holes are also carriers, but they migrate

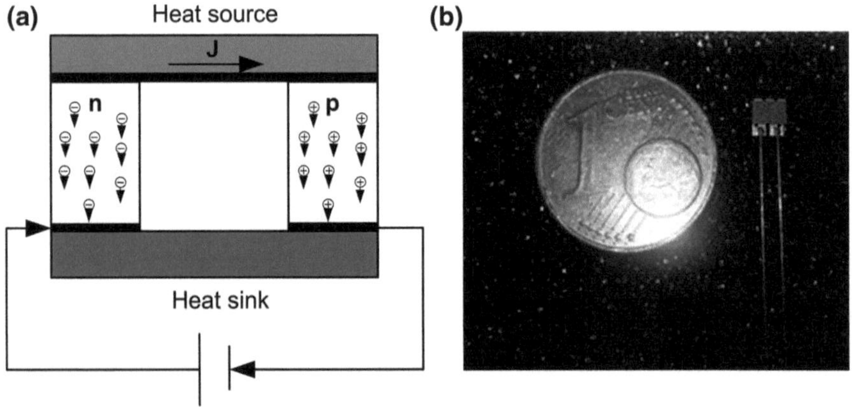

Fig. 6.3 **a** Example of a Peltier element consisting of an *n*-type of semiconductor, doped with electrons, and a *p*-type semiconductor, doped with holes, both being charge carriers, **b** Photograph of a small, thin-film Peltier module in our laboratory

in the direction of the electric current density). When the electrons pass the junction of a metal/*p*-type semiconductor, the energy of the electrons drops and the heat is released. This is because they charge the metal with negative charges, which then act by repelling of the electron flow. When the electrons enter a, e.g. copper, conductor on the cold side of a *p*-type semiconductor (*p*-*n* metal junction) they are excited to a higher energy level. They absorb heat from the metal, which consequently cools down. Then, when the electrons leave the *n*-type semiconductor and enter the copper, their energy level drops (the heat is released).

In the following description we will try to avoid the common use of Peltier modules as coolers or heat pumps. Their broad application in thermal management (i.e. electronic devices) and also in most other cases represents only a cooling device, and not really a real thermal switch or a heat valve. So, how do we distinguish between the two cases?

It is well known that the efficiency of a Peltier module strongly depends on the temperature range in which it operates. This also holds true for the exergy efficiency. In thermal management, the Peltier modules usually provide the necessary cooling power with a certain larger temperature span between the heat source and the heat sink.

However, a larger temperature difference will influence the COP of the Peltier module, as can also be seen in Fig. 6.4. So in order to operate with the highest possible efficiency of the magnetocaloric device, the Peltier module needs to operate at small currents as well as small temperature differences between the heat source and the heat sink. In this particular case, the cooling power of the module is rather low. It also demonstrates the operation where a thermal diode plays a role. A Peltier thermal diode does not necessarily require a high cooling power (as in thermal management), which also relates to high currents. In particular, when applied in combination with other solid-state refrigeration technologies (magnetocaloric, electrocaloric, barocaloric, elastocaloric, etc.), the large temperature span of the Peltier thermal diode drastically influences the overall efficiency of the cooling device. So in such a case it should represent a solid-state heat transporter rather than a heat pump. Comprehensive information about the necessity for such operating characteristics can be found in references [4–6].

A Peltier thermal diode can provide fast switching (depending on whether we have bulk ceramics or thin films) of the heat flux as well as operating as a heat valve, with certain losses, which in the case of the closed heat valve relates to the thermal mass and the thermal conductivity of the Peltier module.

6.2.1.1 Why Must the Efficiency of a Thermal Diode with Peltier Modules Be High?

With regard to the calculation of the exergy efficiency from the data in Fig. 6.4 we may wrongly estimate that the efficiency of a thermal diode, based on Peltier modules, will be very low. Note again that it is not the Peltier element that creates the temperature difference between the heat source and the heat sink of the magnetic

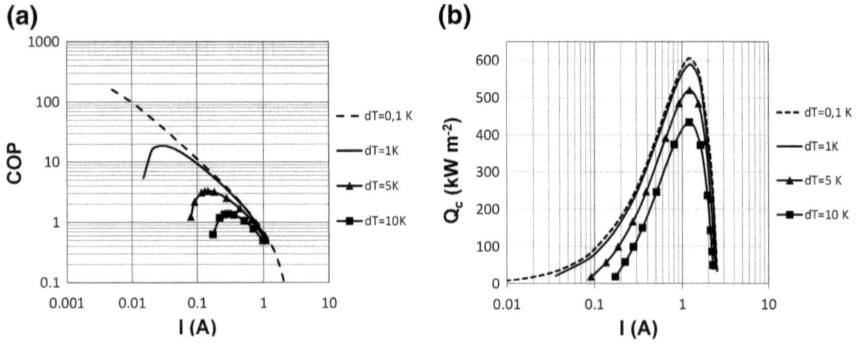

Fig. 6.4 **a** The coefficient of performance (COP), **b** The specific cooling power of thin-film Peltier modules (calculation based on information in [7])

refrigerator. This is provided by the magnetocaloric effect. The thermal diode mechanism is therefore applied only for the heat transport.

In our particular case we assume the operation of a magnetic refrigerator that is based on AMR with thermal diodes (Fig. 6.5). In this particular case, thin-film Peltier modules serve for the heat transport to and from the magnetocaloric material. For the purpose of analysis, we fix the temperature difference on the hot and cold sides of the Peltier module to be only 0.1 K. During the operation a temperature profile is built up along the magnetocaloric material with the thermal diodes. The coldest part (during the demagnetization) of the magnetocaloric material in our particular case is estimated to be 273.1 K, and the hottest part (during the demagnetization) is estimated to be 300.1 K. We assume now an isofield cooling process, during which the magnetocaloric material is in a demagnetized state and the thermal diodes, which are connected to the heat source (via the fluid flow), are active. From these, the evaluated thermal diode mechanisms (activated Peltier modules) are represented by A, B and C. The $T-x$ diagram in Fig. 6.5 also presents an approximate temperature profile established over selected thermal diodes in the steady-state condition.

We estimate that the temperatures of the cold parts of the selected Peltier thermal diodes A, B, C correspond to 273, 288 and 303 K, respectively. Each of these thermal diodes has a temperature span between the heat source and the heat sink equal to 0.1 K. Therefore, the temperatures of the hot parts of the selected Peltier thermal diodes A, B, C correspond to 273.1, 288.1 and 300.1 K, respectively.

In Fig. 6.5 (right), the exergy flow diagram is shown. The element A in this case represents the Peltier module. The exergy of cooling energy from the cold magnetocaloric material flows into the element A. If this is not considered to be Peltier element, but an element with the infinite thermal conductivity, the exergy flux which enters element A equals the exergy flux which exits element A (i.e. $\dot{Q}_H = \dot{Q}_C$ and $\dot{E}_H = \dot{E}_C$). However, since the Peltier module in this particular case is considered for transport of heat, the example in Fig. 6.5 (right) becomes more complex. In order to provide the operation of the Peltier module, this requires certain input of

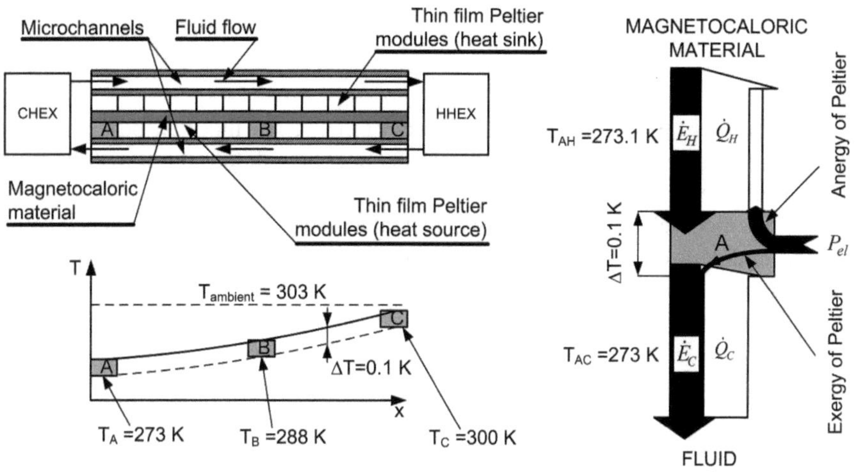

Fig. 6.5 Example of a magnetocaloric material embodied within Peltier thermal diodes. CHEX and HHEX represent the heat-source and heat-sink exchangers, respectively

electric power P_{el} (i.e. pure exergy). Because of the low exergy efficiency of the Peltier module, most of this exergy will be destroyed and will represent an additional anergy flow to the cold magnetocaloric material. Only a small portion of exergy from P_{el} will be added to the exergy flux $\dot{E}_H$ and therefore the exergy flux from the Peltier element to the working fluid will be larger than the exergy flux from the cold magnetocaloric material to the Peltier element (i.e. $\dot{E}_C > \dot{E}_H$).

Now, according to the diagram in Fig. 6.4 and the data obtained from the commercial programme [7], each of the thin-film Peltier modules A, B and C will provide 4 $\mathrm{kWm^{-2}}$ of cooling. Note that these are not optimal conditions, but just an example to demonstrate the difference between the Peltier module and the thermal diode efficiency.

Let us consider the temperature of the ambient (not to be misunderstood as the heat-sink temperature of the Peltier) is 303 K. The COP of the Peltier module under these conditions is COP = 160.8 (note that the COP depends on the temperature difference between the heat source and the heat sink of a device, which in our case is 0.1 K for the Peltier modules).

The specific exergy flux which exits selected Peltier modules equals:

$$\dot{e}_{\mathrm{CA}} = \left|\left(1 - \frac{T_{\mathrm{amb}}}{T_C}\right)\right| \cdot \dot{q}_C = \left|\left(1 - \frac{303}{273}\right)\right| \cdot 4 = 0.439\,\mathrm{kW/m^2} \tag{6.1}$$

$$\dot{e}_{\mathrm{CB}} = \left|\left(1 - \frac{T_{\mathrm{amb}}}{T_C}\right)\right| \cdot \dot{q}_C = \left|\left(1 - \frac{303}{288}\right)\right| \cdot 4 = 0.208\ \mathrm{kW/m^2} \tag{6.2}$$

$$\dot{e}_{CC} = \left|\left(1 - \frac{T_{amb}}{T_C}\right)\right| \cdot \dot{q}_C = \left|\left(1 - \frac{303}{300}\right)\right| \cdot 4 = 0.04\,\text{kW/m}^2 \tag{6.3}$$

We consider this exergy flux to be a sum of the exergy flux transported through the Peltier module from cold magnetocaloric material with the addition of the small amount of exergy flux provided by the electric power input to the Peltier module (by Peltier module). The latter can be calculated using the exergy efficiency of the Peltier module itself and the electric power input. With the specific cooling power of 4 kWm^{-2}, the input specific electric power for each of the selected Peltier modules A, B and C equals:

$$COP_{Peltier} = \frac{\dot{q}_C}{\dot{w}} \Rightarrow \dot{w} = \frac{P_{el}}{A} = \frac{\dot{q}_C}{COP_{Peltier}} = \frac{4}{160.8} = 0.025\,\text{kW/m}^2 \tag{6.4}$$

The exergy efficiency of the Peltier module alone represents the ratio between the COP of the Peltier and the COP of the Carnot cycle for the same conditions. Therefore, it follows for the selected modules A, B and C that:

$$\xi_A = \left(\frac{COP_{Peltier}}{COP_C}\right)_A = \frac{160.6}{\left(\frac{273}{273.1-273}\right)} = 0.059 \tag{6.5}$$

$$\xi_B = \left(\frac{COP_{Peltier}}{COP_C}\right)_B = \frac{160.6}{\left(\frac{288}{288.1-288}\right)} = 0.056 \tag{6.6}$$

$$\xi_C = \left(\frac{COP_{Peltier}}{COP_C}\right)_C = \frac{160.6}{\left(\frac{300}{300.1-300}\right)} = 0.053 \tag{6.7}$$

According to the results in Eqs. (6.5)–(6.7) we could estimate that the application of the Peltier thermal diode leads to an inefficient solution.

However, we should again note that it is not the Peltier modules A, B and C that provide the temperature difference between their cold parts and the ambient. The Peltier modules operate at a very small temperature difference. All the remaining temperature difference for each of the selected thermal diodes A, B and C is created by the magnetocaloric material and the corresponding magnetocaloric effect. Now, the exergy flux that exits the cold magnetocaloric material and enters the Peltier module can be defined as:

$$\dot{e}_{HA} = \dot{e}_{CA} - \xi_A \cdot \dot{w} = 0.439 - 0.059 \cdot 0.025 = 0.437\ \text{kW/m}^2 \tag{6.8}$$

$$\dot{e}_{HB} = \dot{e}_{CB} - \xi_B \cdot \dot{w} = 0.208 - 0.056 \cdot 0.025 = 0.207\ \text{kW/m}^2 \tag{6.9}$$

$$\dot{e}_{HC} = \dot{e}_{CC} - \xi_C \cdot \dot{w} = 0.04 - 0.053 \cdot 0.025 = 0.039\ \text{kW/m}^2 \tag{6.10}$$

In order to calculate the exergy efficiency of each of the Peltier mechanisms as thermal diodes we refer to the right-hand side of Fig. 6.5. The exergy efficiency of the thermal diode mechanism, which is based on a Peltier module, can be defined as the ratio between the output specific exergy flux and the input specific exergy flux. In our particular case this means that the output exergy represents the exergy flux on the cold side of the Peltier and the input exergy flux represents the sum of the exergy flux from the cold magnetocaloric material, plus the input specific electric power:

$$\xi_{\mathrm{TDi}} = \frac{\dot{e}_{Ci}}{\dot{w} + \dot{e}_{\mathrm{Hi}}} \tag{6.11}$$

Now, for each of the selected thermal diodes A, B and C, we calculate the exergy efficiency as:

$$\xi_{\mathrm{TDA}} = \frac{\dot{e}_{\mathrm{CA}}}{\dot{w} + \dot{e}_{\mathrm{HA}}} = \frac{0.439}{0.025 + 0.437} = 0.95 \tag{6.12}$$

$$\xi_{\mathrm{TDB}} = \frac{\dot{e}_{\mathrm{CB}}}{\dot{w} + \dot{e}_{\mathrm{HB}}} = \frac{0.208}{0.025 + 0.207} = 0.9 \tag{6.13}$$

$$\xi_{\mathrm{TDC}} = \frac{\dot{e}_{\mathrm{CC}}}{\dot{w} + \dot{e}_{\mathrm{HC}}} = \frac{0.04}{0.025 + 0.039} = 0.63 \tag{6.14}$$

We can show from the results in Eqs. (6.12)–(6.14) that the exergy efficiency of the thermal diode mechanism in this particular case is very high, i.e. 95 % for thermal diode A and 90 % for thermal diode B. However, much lower, 63 % for thermal diode C, which operates in the vicinity of the ambient temperature. In the case that the warm side of the thermal diode corresponds to the ambient temperature, the exergy efficiency of the thermal diode equals the exergy efficiency of the Peltier element. This is due to the fact that all the temperature difference between the ambient temperature and the cold part of the Peltier element is established by the Peltier element itself and without the exergy flux from the magnetocaloric effect.

This example above is similar, as we would compare two heat pumps: air/water and ground source water/water. In this particular case, the first heat pump uses the heat source as the pure *anergy*, whereas in the second case, the heat source already represents a certain *exergy*. The same is true in the case of Peltier thermal diodes with an embodied magnetocaloric material. The magnetocaloric material represents the exergy potential, whereas the Peltier module can somehow be treated in this case as the heat transporter.

With the above results we prove that Peltier modules, despite their low exergy efficiency, can be efficiently applied as the thermal diodes; however, under particular conditions, where the temperatures of both sides of the Peltier have to be kept at a small difference (with regard to adiabatic temperature change) and the thermal diode should still enable a sufficient transport of heat flux.

If we consider a realistic temperature profile (between 273 and 300 K) along the demagnetized (zero field) magnetocaloric material in the magnetic refrigerator, it is possible, based on the approach described above, to define the local exergy efficiency of a particular Peltier thermal diode mechanism. Furthermore, we define the integrated average exergy efficiency of all the Peltier thermal diode mechanisms along such a magnetocaloric material (the exergy was defined at a temperature of the ambient equal to 303 K). The results for this example are shown in Fig. 6.6.

As can be seen in Fig. 6.6, different temperature profiles, which also correspond to different mass flows of the working fluid, do not drastically influence the average exergy efficiency of all the thermal diode mechanisms embodied in the magnetocaloric device (100 thermal diodes were estimated in this particular case). The average values of the exergy efficiency of the thermal diodes are high (above 85 %), which again points to the high efficiency of such a mechanism, despite the fact that the exergy efficiency of the Peltier module is very low.

The value 85 % of exergy efficiency denotes that 15 % of exergy will be destroyed due to application of the Peltier thermal diodes on the cold side of the magnetocaloric material (positioned below the magnetocaloric material in Fig. 6.5). Since thermal diodes are placed also on the warm side of the magnetocaloric material (positioned above the magnetocaloric material in Fig. 6.5), one can expect that Peltier thermal diode mechanism in the magnetocaloric device will reduce its exergy efficiency for about 15–30 % (depending on the characteristics of Peltier modules and their efficiency, which in our case was taken to be rather low). Despite reduction in the exergy efficiency due to the application of thermal diodes, the same mechanism avoids exergy losses in other parts of device: valves, seals, dead volume, bidirectional pumping, etc.

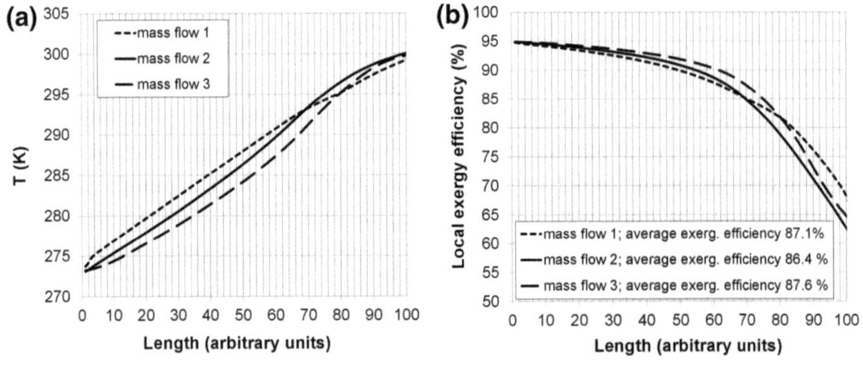

Fig. 6.6 **a** Temperature profiles along the magnetocaloric material in the demagnetized state, **b** Corresponding local and average exergy efficiency of a thermal diode mechanism on the "cold" side of a magnetic refrigerator

6.2.2 Thermionics

The concept of thermionic energy conversion is not really a new scientific area. The basic principle of operation is the flow of an electric current through the vacuum gap between the hot cathode and the cold anode, due to the voltage potential. When the cathode is exposed to the heat source it starts to emit electrons, which are absorbed by the anode on the other side of the vacuum gap. The electrons are, in this case, heat carriers [21]. The problem of thermionics is in its high-temperature operating range (above 230 °C [22]), which is at present thought to be useless for devices operating at or near room temperature. This is a great weakness, especially because thermionic devices show higher efficiencies than thermoelectrics (above 30 %) [19, 23]. Therefore, researchers have spent the last 15 years involved in extensive studies and designed a new principle of thermionics operating at or near room temperature. This new technology is called Hetero-Structured Integrated Thermionics (HIT) [24–26]. Furthermore, researchers have also developed thin-film thermionic cooling devices (especially HIT devices). These micro devices work at/near room temperature with high cooling-power densities (600 Wm^{-2}, temperature span 4.1 K) [26–29] and with fast response times (40 μs) [30], which is used in high-speed micro on-spot cooling.

6.2.3 Spincaloritronics

Spintronics [31] is an emerging technology in the field of microelectronics and nano-electronics and is based on the combination of electron spin and its charge. Mainly, the focus of the research is in the areas of computer, military and space technologies [32, 33]. At the moment, spintronics technology is not directly applicable as thermal diodes; however, it contains some segments that could be combined with some other mechanisms of thermal diodes, discussed in this chapter. One of the directions, in which the research of spintronics is heading and has some potential for synergy with thermal diodes, is the manipulation of the so-called magnetoresistance effect. This phenomenon occurs in some metals when they are exposed to a magnetic field. The electrical resistance decreases significantly, thus increasing the electrical conductivity. This phenomenon has become even more interesting since the discovery of the "giant" magnetoresistance effect (GMR), which was independently discovered by Albert Fert [34] and Peter Grünberg [35]. The 2007 Nobel Prize in physics was awarded to them for the discovery of GMR. The GMR exhibits an even larger change in electrical resistance in the presence of a magnetic field. The problem of spintronics is in materials, which for GMR is useful only at extremely low temperatures (below −150 °C), so it has no applicable value. Again, researchers are struggling with the development and discovery of new materials that are useful at/near room temperature [36–42].

Spincaloritronics [43–45] is a branch of spintronics [46]. Whereas the latter refers to coupled electron spin and charge transport in condensed-matter structures and devices, spincaloritronics deals with the interaction of spins with heat currents, with the spin-Seebeck effect and the spin-Peltier effect being the two leading physical phenomena. Spincaloritronics has only recently emerged and is thus still a fundamental field of research. However, it shows a great deal of potential and rapid developments on the path to useful applications [43]. The spin-Seebeck effect [47–50] is a combination of spintronics and the Seebeck effect, where electrical voltage is generated in the material due to the temperature gradient across the material. This means that one side of the material is colder than the other. The first to demonstrate the spin-Seebeck effect was Uchida in 2008 [47]. The phenomenon of spin-Seebeck is therefore useful for generating electricity. This physical phenomenon is similar to thermoelectric generators. Of course, the spin-Seebeck effect can be inversely applied by delivering an electric current in order to obtain cooling and this is called the spin-Peltier effect [51–54].

Figure 6.7 is a schematic view of the measurement system for the spin-Peltier effect [55, 56].

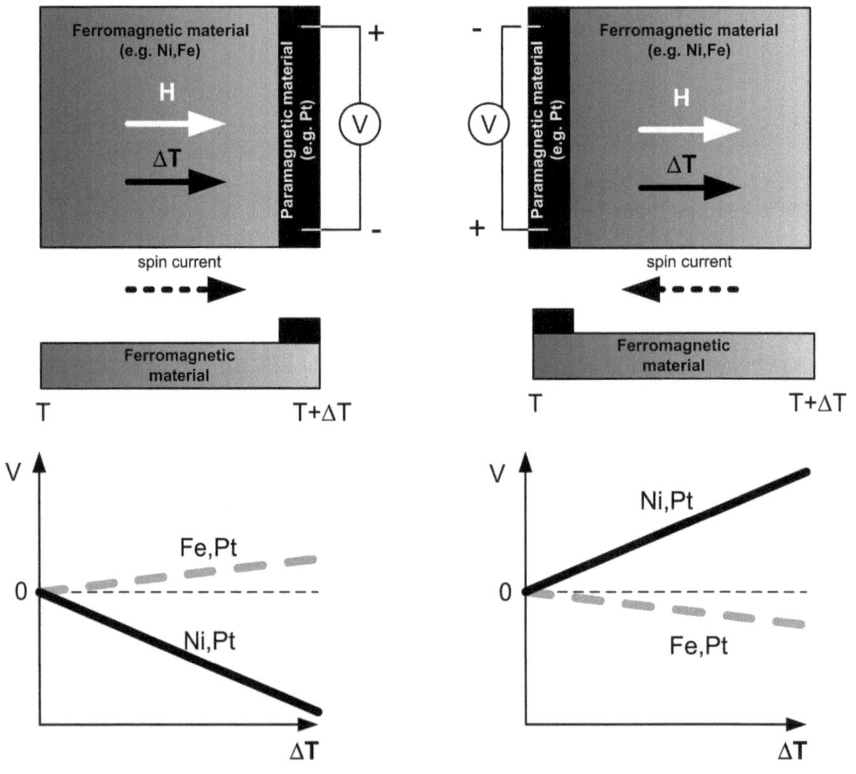

Fig. 6.7 An example of a measurement of the spin-Seebeck effect in a Ni–Pt or Fe–Pt pair (see also 55, 56)

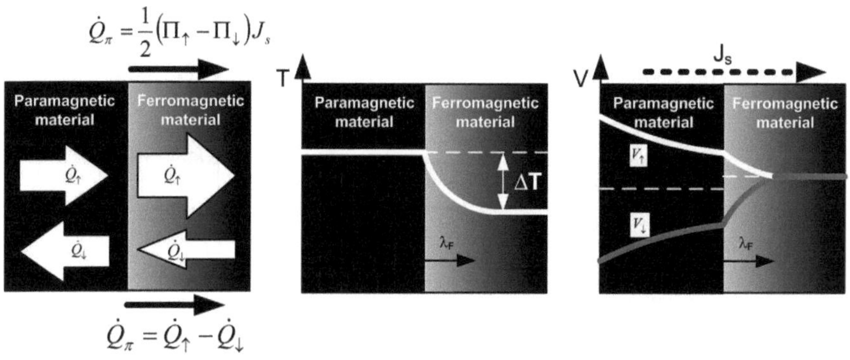

Fig. 6.8 An example of the spin-Peltier effect in a paramagnetic/ferromagnetic pair (see also 45, 51, 52)

The system in the Fig. 6.7 consists of a thin-film ferromagnetic material (e.g. Ni, Fe) and a thin-film Pt material. In this particular case the magnetic field H is applied along the ferromagnetic sample, as well as the temperature gradient. In a ferromagnetic metal under a temperature gradient, the thermally induced spin voltage $\mu\uparrow - \mu\downarrow$ induces a spin current in the direction of the interface surface ferromagnetic material (Ni or Fe in this case)—paramagnetic material Pt. This spin current in the paramagnetic material (i.e. Pt) generates the electromotive force or electric voltage.

In the case of the spin-Peltier effect (Fig. 6.8) the spin current is driven through the interface between the paramagnetic metal and the ferromagnetic metal [45, 51, 52].

In a paramagnetic metal, the Peltier heat current for both spin species is equal [45, 51, 52]. In two spin channels, the flow direction of two heat currents ($\dot{Q}_\uparrow$ and $\dot{Q}_\downarrow$) is opposite, so the total heat current equals zero. However, in a ferromagnetic material, the heat currents are different. A consequence of this is the net heat current from the interface into the ferromagnetic material or the net heat current from the ferromagnetic material towards the interface. In Fig. 6.8 the spin current $Js = J_\uparrow - J_\downarrow$ is driven from the paramagnetic material across the interface, into the ferromagnetic material. In this particular case there is no charge current ($J_\uparrow + J_\downarrow = 0$) and a net heat flow can be observed in the ferromagnetic material. This leads to a temperature gradient between the paramagnetic and the ferromagnetic material. Note that the net heat flow rapidly drops due to the spin relaxation length λ_F. The spin relaxation length (or spin diffusion length) denotes a kind of electron memory for the spin direction and it significantly exceeds the mean free path.

6.2.4 Active and Passive Mechanical Contact-Based Thermal Diodes

These types of thermal diodes relate to the different properties of materials and mechanisms, for example, the piezoelectric, shape-memory effect and others. The

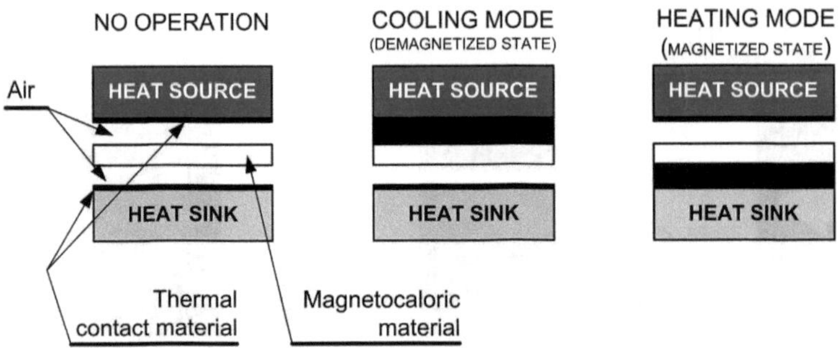

Fig. 6.9 Example of a thermal switch based on a mechanical contact

mechanical thermal switch serves for the thermal contact between the heat source and the heat sink. The mechanisms behind these kinds of switches are based on thermal actuation (the temperature-dependent elongation of materials), as well as other types of actuation (electric, magnetic, chemical, etc.). Figure 6.9 shows an example of a thermal switch based on a mechanical contact.

According to Fig. 6.9 there is a gap between the heat source and the heat sink. In our particular case we will consider that air is used to fill this gap. In the cooling mode, when the magnetocaloric material is in the demagnetized state, the thermal contact material (thermal switch) is active, which means that it is elongated and in such a way that it is in contact with the heat source. The heat is transferred through the thermal switch by thermal diffusion (conduction). In the heating mode, the lower thermal switch is active and the upper thermal switch is inactive. Now the heat flux can be transferred to the heat sink.

The thermally rectifying materials, which are based on thermally activated elongation, must exhibit high thermal conductivity. Most of these mechanisms can be referred to as passive and thermally actuated. This group of materials includes elastomers (solid, liquid), bimetals, shape-memory alloys, etc.

Shape-memory metallic alloys exhibit thermomechanical actuation. One of the most representative alloys is the nickel–titanium alloy; however, there are a number of other potential materials, e.g. Co–Ni–Al, Co·Ni–Ga, Fe–Mn–Si, Cu–Al–Ni, Cu–Zn and Fe–Pt. In the group of non-metallic materials we should not neglect the shape-memory polymers [57–62], which can be activated by many different mechanisms, i.e. thermal, electrical, radiative (light), magnetic or chemical [62]. The selection of shape-memory materials generally depends on the mode of actuation, the operating temperature and the desired behaviour.

Other types of materials can be referred to as bimetals. This kind of material is made by bonding different metals in a bimetallic strip, e.g. steel and copper or steel and brass, whereas each metal possesses a different coefficient of thermal expansion. The temperature gradient will lead to the mechanical displacement of such material.

The major drawback of all contact thermal switch mechanisms, besides the thermal contact resistance between the thermal switch and the magnetocaloric material, is the thermal contact resistance between the surface of the thermal switch and the wall of the, e.g. microchannel heat exchanger (heat source or heat sink). This can significantly influence the heat-transfer irreversible losses as well as the response time. One should therefore, before focusing on a particular mechanism, evaluate the feasibility of its application.

Therefore, in order to be applied in magnetic refrigeration, the contact thermal switch mechanism should operate over a very small distance, e.g. of the order of a few μm. This will be required to minimize the irreversible heat losses, and at the same time still provide high contact resistance ratio between the ON and OFF operations. According to the analysed data below, and considering an occupancy volume of 20 % for the thermal switch material and 80 % for the air in the OFF mode, the ratio between the contact resistance in the OFF and ON operations varied from very small values at 1 mm thickness up to maximum of 2 % at a thickness of 10 microns.

In the case below (Fig. 6.10) we show a simple example of how the thermal resistance and distance between the contact thermal switch and the wall of the microchannel heat exchanger influences the temperature difference required for the heat transfer. For this purpose we will estimate the macroscopic (bulk) mechanism of the thermal diffusion.

We consider the magnetocaloric material to be embodied within the contact thermal diodes (the thermal contact resistance between the diodes and the magnetocaloric material is neglected). For this purpose we investigate gadolinium as the magnetocaloric material exposed to a magnetic field change of about 1 T (an adiabatic temperature change 4 K). For this case the maximum specific cooling capacity is defined to be 1,200 Jkg^{-1}. This has been calculated as the product of the isothermal entropy change of 4.1 $Jkg^{-1}K^{-1}$ and a temperature of 292 K.

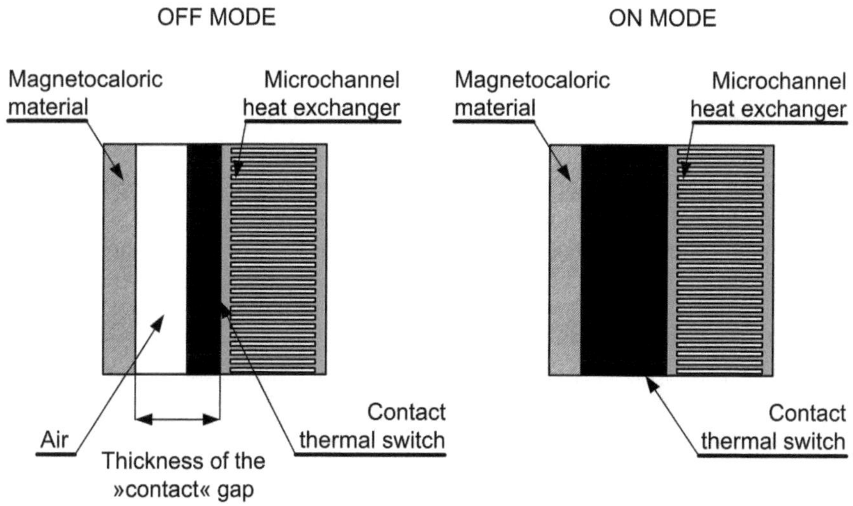

Fig. 6.10 Example of a contact thermal switch (figure shows only the "hot" part of the device)

The magnetocaloric material represents a small plate with dimensions of 120 mm × 40 mm × 0.1 mm (density 7,900 kgm^{-3}), resulting in a maximum cooling capacity of 4.55 J. Since the adiabatic temperature change of 4 K is estimated, this can also be considered as the maximum possible temperature difference between the gadolinium and the micro heat exchangers. The conditions under which the irreversible heat-transfer losses will lead to temperature differences between the magnetocaloric material and the micro heat exchanger larger than 4 K actually define the operation of a device for which no cooling will be performed any more. Note that for efficient applications, the temperature difference should actually be substantially lower.

The resistance of the thermal diode mechanism in the ON operation has been defined as:

$$R = \frac{\Delta T \cdot A}{\dot{Q}} = \frac{\Delta T}{\dot{q}} = R_{\mathrm{TD}} + R_{\mathrm{contact}} \tag{6.15}$$

where ΔT represents the temperature difference across the thermal diode, A represents the heat-transfer surface (the cross-section of the thermal diode) and $\dot{Q}$ represents the heat flux. The resistance due to heat diffusion through the thermal diode material has been defined as:

$$R_{\mathrm{TD}} = \frac{\delta}{\lambda} \tag{6.16}$$

where δ represents the thickness of the thermal diode and λ represents its thermal conductivity. The R_{contact} represents the thermal contact resistance between the thermal switch and the wall of the heat exchanger. The selected values (see Table 6.1) for the thermal contact resistance have been chosen on the basis of the available literature [63–65]. Note that these values strongly depend on the material, its surface roughness, the contact pressure as well as the fluid, which fills the contact voids.

Figure 6.11 shows the temperature difference between the magnetocaloric material and the wall of the microchannel heat exchanger for the case when we consider the thermal contact resistance between the thermal diode and the wall of the microchannel heat exchanger to be 0.01 m^2KkW^{-1}. Note that this kind of thermal contact resistance corresponds to applications where the pressure is applied

Table 6.1 The parameters applied in the analysis

Frequency (Hz)	Corresponding cooling power $\dot{Q}$(W)	Corresponding specific cooling power $\dot{q}$ (kWm^{-2})	Thermal switch conductivity λ ($Wm^{-1}K^{-1}$)	Total thickness of thermal switch when operating δ (μm)	R_{contact} (m^2KkW^{-1})
20	91	18.96	1, 10, 100	1, 10, 100, 1000	0.01, 0.1, 1
50	228	47.4	1, 10, 100	1, 10, 100, 1000	0.01, 0.1, 1

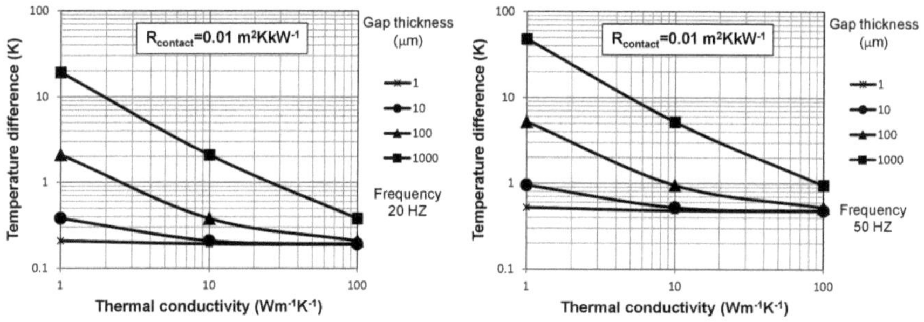

Fig. 6.11 The temperature difference between the magnetocaloric material and the wall of the microchannel heat exchanger for the thermal contact resistance between the thermal diode and the wall of the microchannel heat exchanger $R_{contact} = 0.01\ m^2KkW^{-1}$

on the contact (e.g. 3 MPa for the smooth surface of aluminium material) [66]. As can be seen from Fig. 6.11, in order to achieve a temperature difference smaller than the considered adiabatic temperature change of 4 K, the gap thickness should be as small as possible.

A strong influence, especially for a gap thickness larger than 100 μm, comes from the thermal conductivity of the contact thermal switch. Since we consider the thermal diode mechanism to be applied for high-frequency operation of the magnetic refrigerator, a frequency of 20 Hz can be taken as a minimum value. For such a frequency, using the case considered in Fig. 6.11, the thermal switch will provide an efficient operation over a wide range of thermal conductivities as well as gap thicknesses. However, the increase in the frequency of the operation to 50 Hz will drastically reduce the applicability to gap thicknesses, which correspond to dimensions that are smaller than 100 microns.

The contact thermal resistance of $R_{contact} = 0.05\ m^2KkW^{-1}$ will, in most cases, still require a contact pressure; however, smaller than in the case of $R_{contact} = 0.01\ m^2KkW^{-1}$. In this particular case, at a frequency of the operation equal to 20 Hz, the contact thermal switch will also operate efficiently for a broad range of parameters; however, an increase in the frequency to 50 Hz will limit its application, mostly to a gap thickness of about 10 microns or smaller, especially for thermal conductivities of the thermal switch that are lower than 10 $Wm^{-1}K^{-1}$ (Fig. 6.12).

A further increase in the thermal contact resistance to $R_{contact} = 0.1\ m^2KkW^{-1}$ will further decrease the required contact pressure. However, such a contact resistance will enable the efficient application of a thermal switch only for frequencies of 20 Hz and gap thicknesses below 100 microns. For a frequency of operation equal to 50 Hz, the contact thermal switch application is feasible only at thicknesses up to 100 microns for high thermal conductivities (i.e. between 10 and 100 $Wm^{-1}K^{-1}$) for any thermal conductivities below 10 $Wm^{-1}K^{-1}$, the thickness should be up to a few tens of microns (Fig. 6.13).

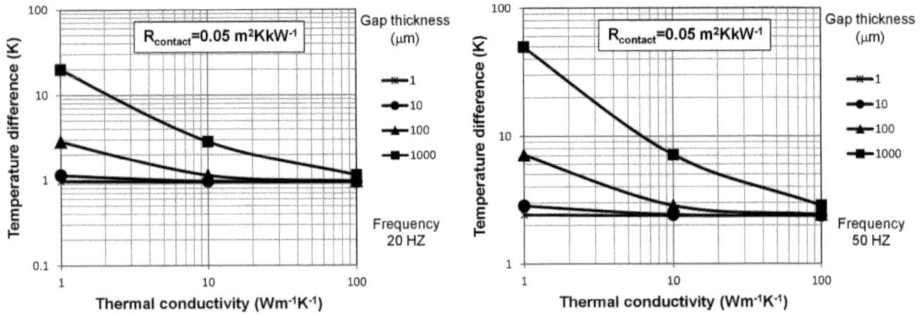

Fig. 6.12 The temperature difference between the magnetocaloric material and the wall of the microchannel heat exchanger for the thermal contact resistance between the thermal diode and the wall of the microchannel heat exchanger $R_{\text{contact}} = 0.05\ \text{m}^2\text{KkW}^{-1}$

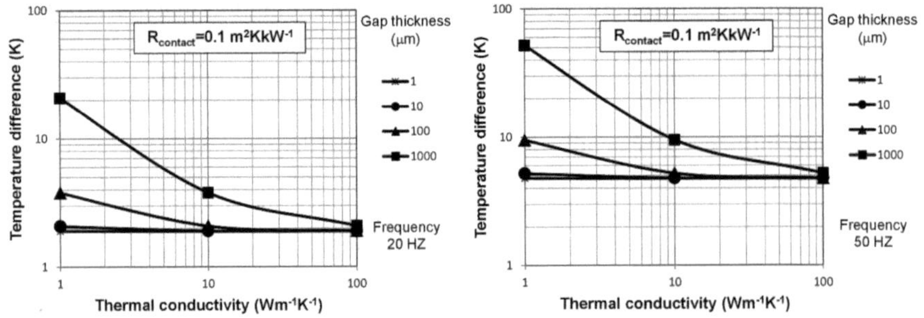

Fig. 6.13 The temperature difference between the magnetocaloric material and the wall of the microchannel heat exchanger for the thermal contact resistance between the thermal diode and the wall of the microchannel heat exchanger $R_{\text{contact}} = 0.1\ \text{m}^2\text{KkW}^{-1}$

Based on the results of this simple example we can show that the operation of the contact thermal switch should be limited to rather small distances (e.g. less than 100 microns) between the thermal switch in the off mode and the wall of the microchannel heat exchanger. In most cases we should try to employ thermal conductivities of the thermal switch that are higher than 10 $\text{Wm}^{-1}\text{K}^{-1}$. Furthermore, almost all applications of contact thermal switches will require pressure contacts. These facts can significantly limit the applicability of contact thermal switches, especially because we have evaluated a rather simplified case by neglecting, e.g. the thermal mass of the system and the contact resistances between the thermal switch and the magnetocaloric material. Furthermore, we have neglected the heat flux in the opposite direction, especially because the rectification factors (the ratio between the thermal resistance in the ON and OFF operations) were obtained as rather high, i.e. around 8 for a gap thickness of 1 micron, around 80 for a gap thickness of 10 microns, and between 260 and 800 for a gap thickness of 100 microns.

6.3 Passive Solid-State Thermal Rectificators

In most publications that relate to passive thermal diode mechanisms the authors refer to thermal rectificators. It was as long ago as 1935 when Starr [67] published a paper on thermal and electrical rectification using a copper/cuprous oxide combination and where a rectification efficiency of 39 % was achieved. There are a number of different mechanisms for thermal rectification, such as the anisotropic thermal conduction of some materials, a change in the material's thermal conductivity due to different influential mechanisms, memory-shape materials [68], the temperature dependence of thermal conductivity at interfaces [69], asymmetrical nanostructured geometry, nanostructured interfaces, quantum thermal systems, elastomer liquid crystals, etc. (see Ref. [70]). A vast number of different research activities are being carried out in the field of carbon-nanostructured thermal rectification [71–73].

Thermal rectification is a phenomenon in which the thermal transport along a specific axis is dependent on the temperature gradient or the heat flux. The thermally rectifying "materials", which apply anisotropy, must exhibit a high thermal conductivity in one direction and a low thermal conductivity in the other direction. Similarly, some materials must exhibit a high thermal conductivity in, e.g. the z plane, while representing heat insulators in the plane x-y. This kind of property can be, for instance, provided by the application of graphene and it may even be found on the market.

A review of the different mechanisms and models for thermal rectification can be found in Roberts and Walker [74] and Walker [75].

Most publications that refer to the anisotropic behaviour of a material do not relate to bulk materials. Of these, we should refer to some findings from the article of Kobayashi et al. [76], who have experimentally demonstrated thermal rectification in a device comprising two perovskite cobalt oxides, i.e. $LaCoO_3$ and $La_{0.7}Sr_{0.3}CoO_3$, with different thermal conductivities (see Fig. 6.14 for a better presentation of the problem). The authors reported on a rectifying coefficient of 1.43 for a temperature span of 60 K (the ratio between the two fluxes in different directions—see Fig. 6.14).

The dimensions of the two bars resulted in a total length of 12.4 mm. In Fig. 6.14 the heat flux is denoted by the heat flux density $\dot{q}_r$ (Wm^{-2}). The materials A and B are bonded at the centre. The characteristic of the material A is that it exhibits high thermal conductivity at a low temperature T_C, and low thermal conductivity at a high temperature T_H. The material B exhibits the inverse characteristics.

When materials A and B are in contact with the thermal baths T_C (heat sink) and T_H (heat source), where T_H is attached to material B, then the total thermal resistance of the bar should be high. The heat flux $\dot{q}_{r1}$ flows through material B to material A. In contrast, when the conditions for material A are reversed, the total thermal resistance becomes higher resulting in a heat flux $\dot{q}_{r2}$ that is smaller than $\dot{q}_{r1}$. The rectifying coefficient is defined by the ratio of $|\dot{q}_{r1}|$ to $|\dot{q}_{r2}|$.

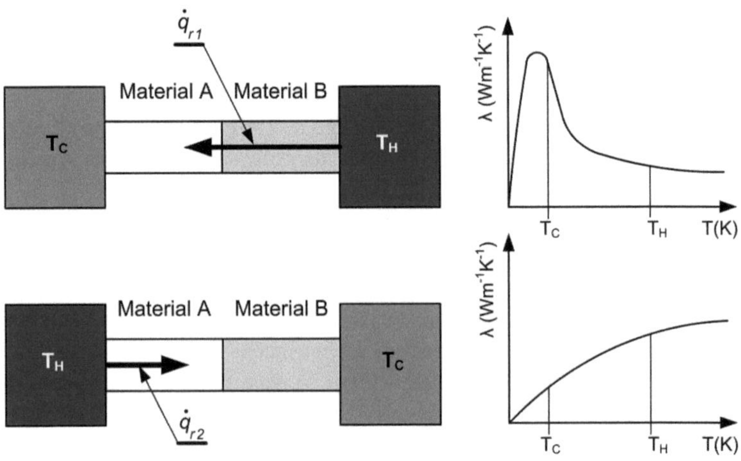

Fig. 6.14 Schematic figure of two bonded materials with different, temperature-dependent thermal conductivities and the variation of the thermal conductivity (above, for a material such as $LaCoO_3$ and below for a material such as $La_{0.7}Sr_{0.3}CoO_3$), which is required for the thermal rectification

Kurabayashi [77] published an article on the anisotropic thermal properties of solid polymers. He also reviewed the existing theories and experimental methods for analysing the polymer orientation. The basis for the anisotropic thermal conductivity of oriented polymers is in the difference between the thermal energy transport mechanisms in the parallel and perpendicular directions with respect to molecules. We should therefore not neglect the possibility of applying polymers. Moreover, the recent developments in polymer science show [78] that such materials may be serious candidates for future thermal conductors.

Most of the phenomena described in publications on thermal rectification by anisotropic thermal conductivity relate to the nanoscale, and most of the work performed has been based on theoretical investigations. Because of the large number of publications on anisotropic thermal conductivity we refer below only to some of them.

Li et al. [79] proposed the coupling of two nonlinear one-dimensional lattices. They theoretically analysed mechanisms that enable heat flux through a system in one direction and where the system acts as an insulator when the temperature gradient is reversed. They also discussed the potential experimental realization of such nanoscale systems.

In 2006, Chang et al. [80] demonstrated nanoscale thermal rectification with the use of high thermal conductivity carbon and boron nitride nanotubes. These were deposited externally and non-homogeneously using heavy molecules [80] of $C_9H_{16}Pt$. The resulting nanoscale system yielded asymmetric thermal conductance with higher heat flux in the direction of decreasing mass density. The authors predicted that solitons may be responsible for the observed phenomenon. They also reported on a rectification of 2 % at room temperature; however, higher values of 7 % were also measured.

Wang et al. reported [81] on the important role of interface collisions on thermal rectification. Schmotz et al. [82] demonstrated a thermal diode mechanism based on a standard silicon processing technology using the rectification of phonon transport. They developed a thermal diode that consists of an array of differently shaped holes milled into a thin silicon membrane. The authors reported on a rectification ratio for the heat flux of 1.7 at a temperature of 150 K. Based on their experiences the authors suggested that the thermal diode should serve as a building block in full analogy to electrical circuits, such as memory, gates and transistors. The heat flux in this particular device is carried mainly by phonons, while leaving the electronic degree of freedom completely unaffected [82].

To generalize the effect we refer to the work of Roberts and Walker [74], who divided the thermal rectification mechanisms into:

- bulk mechanisms,
- molecular—nanoscale mechanisms.

6.3.1 Bulk Mechanisms

6.3.1.1 Metal/Insulator Coupling

The heat flux is transported mainly by electrons in the metal and phonons in the insulator. The two materials transport heat with different carriers. Therefore, the transfer between the electrons and phonons occurs [74]. In the vicinity of the interface in the metal, the electrons scatter with the phonons. These are then transmitted into the insulator or reflected at the interface into the metal. The electron-phonon scattering and the phonon transmission lead to an effective contact resistance [74].

6.3.1.2 Thermal Strain/Warping at Interfaces

The interface is composed of materials with different temperature-dependent properties. If the temperature on each side of the interface varies, this will result in different effective contact areas.

6.3.1.3 Thermal Potential Barrier at Interfaces

Electronic effects at interfaces can be observed in both the metal-oxide interfaces and the metal–metal interfaces. According to Rogers [83] the differences in the work function produced a thermally rectifying effect. The work function is the energy required to remove an electron from the highest filled level of a solid. In the case of a junction of two metals with different work functions, the electrons will leave the metal with a lower work function and travel to the metal with a higher work function.

6.3.1.4 Temperature Dependence of the Thermal Conductivity at Interfaces

In this particular case the rectification is based on the difference in the temperature dependence of the thermal conductivity between two materials (see Fig. 6.14)

6.3.2 Molecular-Nanoscale Mechanisms

- *Non-uniform mass loading*
  Non-uniform deposition of mass on nanotubes. See above in the text by Chang et al. [80]. The higher thermal conductivity is observed in the direction of high-mass to low-mass.
- *Asymmetric nanostructured geometry*
  Most applications relate to carbon, boron nitride nanotubes, asymmetric graphene sheets, carbon nanocones or nanohorns [72–74, 83–86].
- *Nanostructured interfaces*
  Thermal transport at solid–solid interfaces and related effects.
- *Anharmonic/non-linear lattices in 1D chains*
  A thermal rectifier using a nonlinear 1D lattice and connecting it to two thermal reservoirs at different temperatures [74, 87].
- *Quantum (active) thermal systems*

Generally, passive solid-state thermal rectificators do not represent some extraordinary features for the thermal diode mechanism. The effect of the so-called rectification is low and is limited to large temperature gradients, and the effect itself is not in terms of an order or larger. However, we should not forget that this is an emerging and rapidly developing domain, which can perhaps bring nanostructured-built-up-to-bulk materials with very high rectificitation effects in the future.

6.4 Micro Fluidic Thermal Diodes

In this subsection we present microfluidic mechanisms that can be applied as thermal diodes in magnetic refrigeration. Microfluidics is a new interdisciplinary domain involving many different scientific fields. In our particular case we will refer to microfluidic systems, which incorporate electrical engineering, mechanical engineering and physics. These systems further refer to the domains of electrohydrodynamics, ferrohydrodynamics, magnetohydrodynamics, and a special domain of electro- and magneto-rheology.

For each of these particular domains, we have focused mostly on solutions that have the potential to be applied in magnetic refrigeration. Since all these mechanisms represent fluidic contact thermal switches, we will focus on those which apply liquids, since they represent a better potential for high heat fluxes compared to gases.

6.4.1 Electrohydrodynamics

6.4.1.1 Electrowetting

Electrowetting is a process in which the surface tension of the fluid can be manipulated by an externally applied electric field. This kind of process can influence the shape of a particular drop, as well as activate the drop's movement. The latter we denote as electrocapillary flow.

Electrowetting as the Manipulation of the Static Thermal Diode Mechanism

The electrowetting effect was first described by Lippmann [88]. He applied his findings in the development of a capillary electrometer and several other applications. It was not until 1993 that Berge [89] suggested that an insulating layer between the electrodes and the fluid should be applied to prevent the electrolysis. The electrowetting process can be divided into four main mechanisms [90]:

- The electrocapillary effect (we will consider this as electrowetting electrocapillary flow),
- electrowetting on a dielectric (EWOD),
- electrowetting on insulator-coated electrodes (EICEs),
- (spontaneous) electrowetting on line electrodes (ELEs).

In Fig. 6.15 we show the EICE and EWOD mechanisms. In the EICE (Fig. 6.15a) a coating insulating film with a thickness from a few μm to about 200 μm is deposited on the surface of the electrode. The electrostatic potential is applied between the electrode and the drop, which is on the insulator. In Fig. 6.15b the EWOD electrowetting is provided by the dielectric film that is coated on the top and the bottom of the electrodes. The result is the alternation of the static changes in the liquid–solid contact angle. With an increase in the voltage, the contact angle is reduced.

An example of an electrowetting EWOD application in magnetic refrigeration is shown in Fig. 6.16. In this particular case the Brayton-type magnetic refrigeration cycle is considered. The electrodes, which are separated (not shown in the figure), serve for the alternation of the voltage and thus control the manipulation of the surface tension of electrolyte liquid drops. The microchannels serve as the expanded surface for the heating and cooling of the working fluid, which then flows

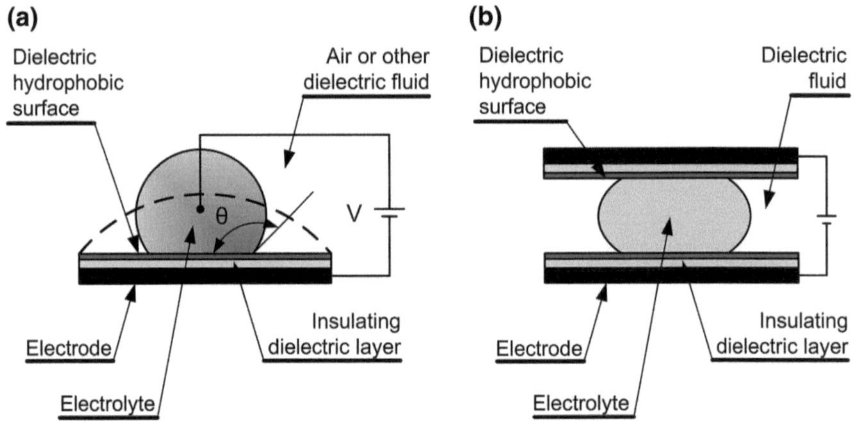

Fig. 6.15 Examples of electrowetting manipulation of the surface tension of the fluid, **a** electrowetting on insulator-coated electrodes (EICEs), **b** electrowetting on a dielectric (EWOD)

to the heat-sink and heat-source heat exchangers, respectively. During the magnetization cycle, the thermal diode mechanism is in the OFF-position on both sides of the magnetocaloric material. The magnetocaloric material heats during the magnetization. During the process of the isofield cooling, the upper series of thermal diodes is switched ON, thus providing the contact between the heat sink and the magnetocaloric material. The thermal diodes are then switched OFF and the process of demagnetization is performed. Therefore, the magnetocaloric material cools down. The lower series of thermal diodes is switched ON, providing a contact for the heat flux from the heat source.

Electrowetting as the Manipulation the Electrocapillary Flow of the Thermal Diode Mechanism

In the electrocapillary-induced flow (Fig. 6.17), the surface tension is based on the electric potential that acts across an interface. Here, the velocity of the fluid can reach up to 10 cms^{-1}, or even higher in some applications. This is very high if we take into account the microscale in which the thermal diode acts. The electrocapillary flow is many times referred to as one of the basic mechanisms for digital microfluidics [91, 92].

In Fig. 6.17a, planar parallel line electrodes (ELE) provide the spontaneous electrowetting flow of the liquid film [93]. A film with a thickness of several microns flows due to the elongated drop when the electric potential is applied over the electrodes. The film velocity is higher than the macroscopic spreading of the drop itself [94]. The reason for this is the bulk electric pressure gradient in the

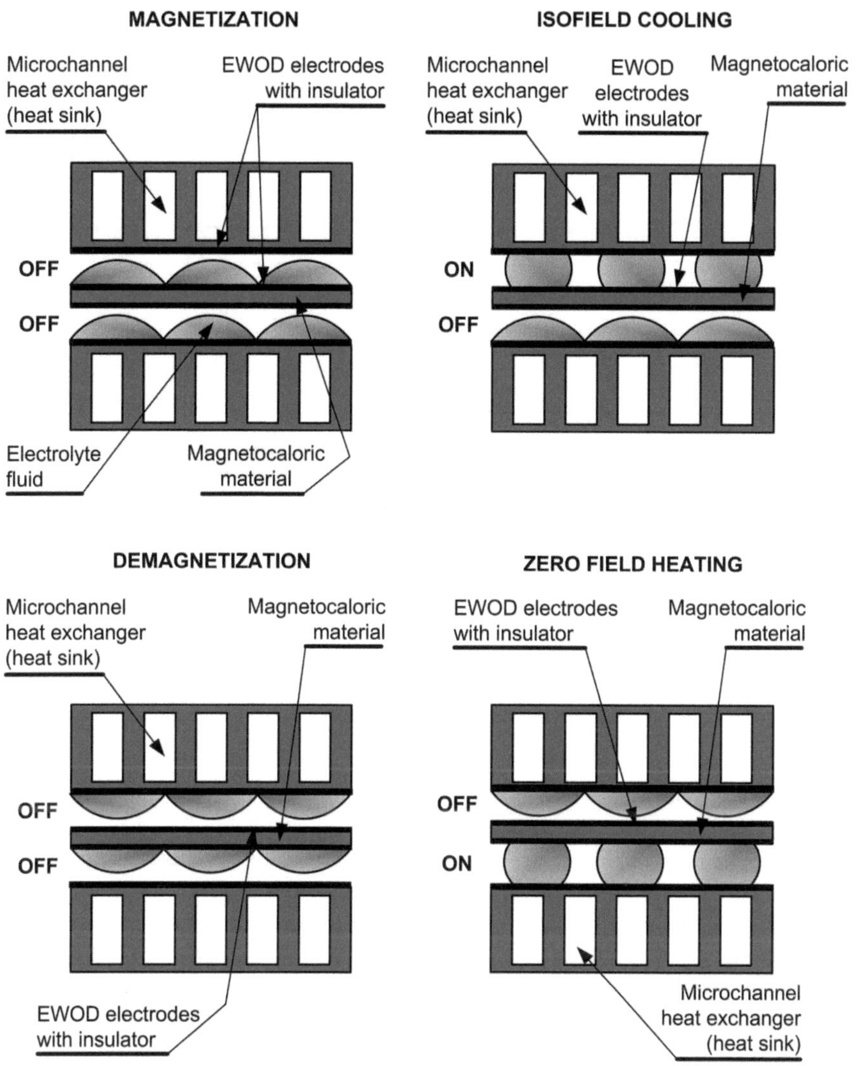

Fig. 6.16 An example of the application of the EWOD principle in magnetic refrigeration

contact region. A negative capillary pressure is induced, which leads to the fluid flow and pushes a thin electrowetting film in front of the macroscopic drop. Figure 6.17b shows an example of an application in which the voltage across the two ends induces the motion of the droplet. The motion is actuated by the pressure difference and the asymmetric change of the surface tension of two menisci. According to Yeo and Chang [93], the application of such a pair (liquid metal and electrode) may be problematic in microchannels, especially because of the

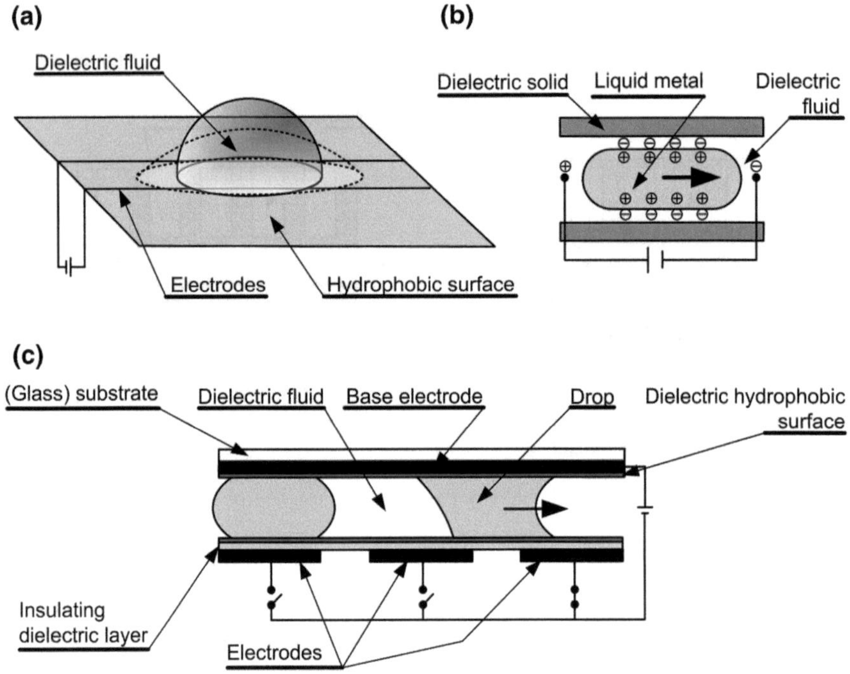

Fig. 6.17 Examples of the microfluidic electrocapillary flow

relatively high electrical conductance of the fluid system, which can lead to high currents and interactions between the neighbouring channels.

The application of the electrowetting principle in capillary flow seems to be a much better solution for magnetic refrigeration. Figure 6.17c shows the application of the EWOD principle for the capillary flow of the droplet. In this particular case the electrode's array patterns can be created by photolithography on, e.g. the glass substrate [92, 94]. A constant ground potential is applied to the droplet of the polarizable and/or conductive liquid, which is embodied between the bottom plate of electrodes and a top plate. On the bottom, the electrodes are covered by the insulating layer in order to prevent the electrical current from passing through the droplet. Both surfaces on the top and bottom are covered by the hydrophobic coating. Usually, the filler oil (e.g. silicone oil) surrounds the droplet; however, this is not necessary in all applications. There are two aims of using the filler liquid. One is to support the droplet's motion mechanism and the other is to prevent the evaporation of the droplet [92]. The oil–water surface tension gradient induces the flow of the droplet.

An example of applying the EWOD principle or the manipulation of a drop's movement in magnetic refrigeration is shown in Fig. 6.18. In this particular case, two drops are applied with their movement, which depends on a certain thermodynamic process in the magnetic refrigeration (e.g. the Brayton refrigeration

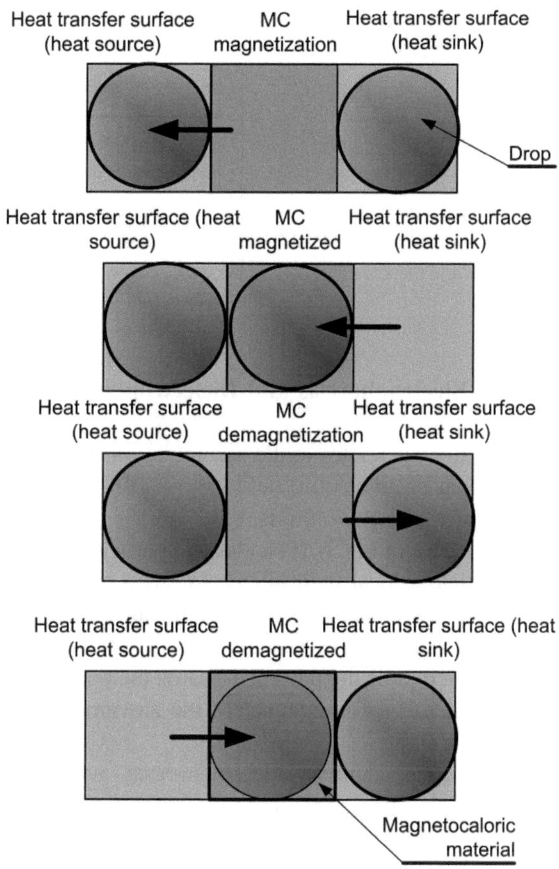

Fig. 6.18 A simple example of the application of the EWOD principle in the capillary flow of drops, used as the thermal diode mechanism in magnetic refrigeration

process in this particular case). When the magnetocaloric material is being magnetized, both drops have a position on the surface, which corresponds to the heat sink and the heat source, respectively. The magnetocaloric material heats up, so the heat flux needs to be transferred to the drop on the right-hand side. Therefore, in the process of isofield cooling, the drop on the right-hand side moves on the surface of the magnetocaloric material, absorbs the heat and moves back to the heat sink surface. In the demagnetization process both drops are at the position on the surface of the heat sink and the heat source. The magnetocaloric material cools, and the drop on the left-hand side moves on the surface of the magnetocaloric material in order to be cooled. Later, it moves back to the initial position in order to accept heat from the heat source.

This example is very primitive and does not concern some valuable solution; however, its purpose it to indicate to the reader a way of applying the micro-fluidic thermal diode mechanism in magnetic refrigeration.

More information on electrohydrodynamic and potential mechanisms can be found in work of Zhao and Yang [95] and Squires and Quake [96].

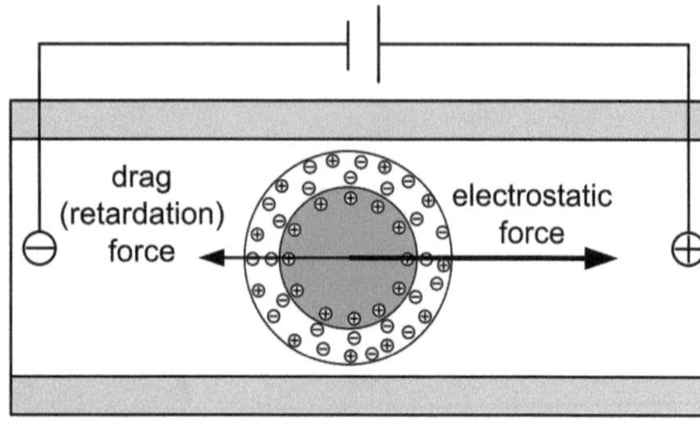

Fig. 6.19 The electrophoretic flow of a charged particle

6.4.1.2 Electrophoresis and Dielectrophoresis

Electrophoresis relates to the manipulation of charged particles or molecules by the application of an electric field [90, 97, 98]. The fluids in this case can be various electrolytes, dielectric fluids, etc.

In the case of Fig. 6.19, a charged particle is immersed in the electrolyte. Around this particle a local field is created. As a result, a double electric layer (EDL, see also Electroosmosis) is created. The influence of the applied electric field on a particular particle will lead to two forces acting on the particle: the electrostatic force, which pulls the particle towards the positive electrode; and the friction (drag) force, which is a consequence of the movement of the particle through the viscous liquid [90, 97].

In practice, capillary electrophoresis, which is mostly used as a separation technique, is often applied for the separation of ionized molecules.

Dielectrophoresis is similar to electrophoresis; however, in this case the polarizable (neutral) particles are immersed in an electrolyte. When a non-uniform electric field is applied, the dielectrophoretic force is established. Similar to the situation with electrophoresis, a viscous drag force will also act on the particle [90, 99, 100]. Figure 6.20 shows three different examples of an electric field acting on a polarizable particle.

In the case of Fig. 6.20a, a uniform electric field is established over the particle. Therefore, the net force on such a particle will lead to no movement. In the case that the fluid in which the neutral particle is immersed has a higher polarization than the particle itself, the net force will act on the particles in such a way that they will migrate towards the region with the lowest electric field (Fig. 6.20b). Figure 6.20c shows an example of positive dielectrophoresis, in which a neutral particle is immersed in a less polarizable fluid (than the polarization of the particle) or vacuum. In this case, the particles will migrate towards the region where the electric field is the highest, i.e. towards the electrodes.

Electrophoresis and dielectrophoresis, as principles, do not have a great potential in applications as a thermal diode mechanism in magnetic refrigeration. More interesting, perhaps, is the domain of electroosmosis and related fluid flow, which is presented in the following subsection.

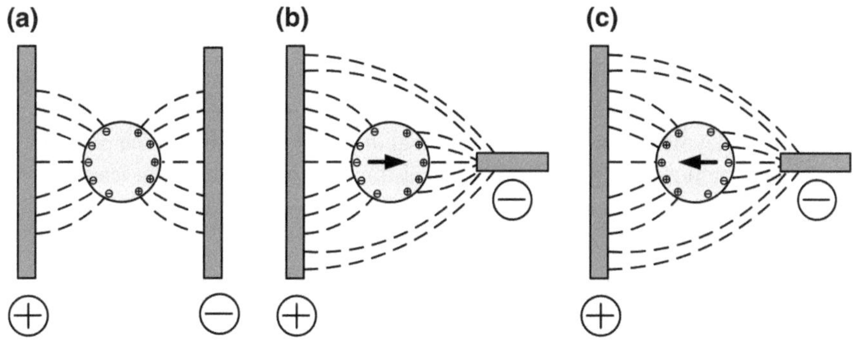

Fig. 6.20 The dielectrophoretic flow of a neutral particle, **a** no net force is established over the particle due to the uniform electric field, **b** negative dielectrophoresis, **c** positive dielectrophoresis

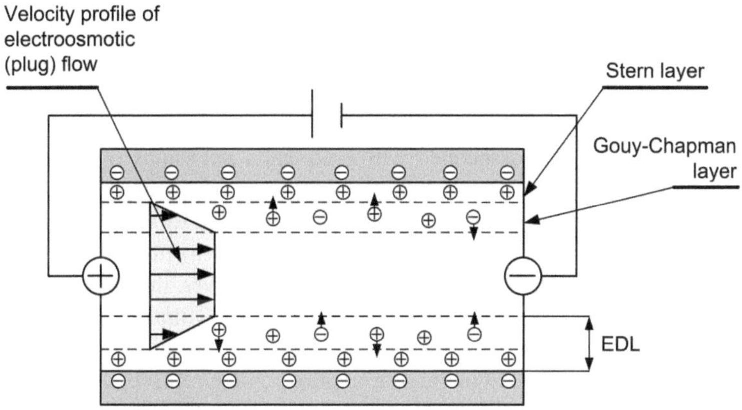

Fig. 6.21 An example of electroosmotic flow

6.4.1.3 Electroosmosis

Electroosmotic flow (Fig. 6.21) relates to a bulk ionized fluid motion under the influence of an externally applied electric field that is parallel to the surfaces in the presence of an established double layer (EDL), which represents a thin region of net charge density near the two-phase interface.

The EDL layer is formed by a static Stern layer and a Gouy-Chapman layer, respectively, and is developed between the electrolyte liquid and the charged surface. Negative ions from the liquid are attracted to the surface. A thin static charge layer is formed (Stern layer) on the surface due to the electrostatic force. The second, thicker and mobile (can be moved by an electric field) layer is called the Gouy-Chapman layer. The thickness of the EDL region can vary from 10 nm to a few 100 nm [90]. The EDL layer is moved by the electric field as a result of the

Coulomb forces. However, because of the viscous forces, the rest of the fluid also flows with it.

The electroosmotic principle can be used in many different applications in microfluidics, e.g. pumping, valving, mixing, splitting, etc. However, this kind of mechanism may not serve well as the thermal diode mechanism, but the potential to manipulate the flow and to enhance the heat-transfer coefficient might lead to some future solutions for thermal diode mechanisms.

6.4.2 Ferrohydrodynamics

Ferrohydrodynamics (FHD) relates to the fluid motion induced by a magnetic field. This domain should not be confused with magnetohydrodynamics (MHD) (see also the next subsection on MHD). In FHD there is no need to apply an electrical current to the fluid. The driving force that induces the fluid flow is related to the material magnetization in an alternating magnetic field. More information about the physics behind ferrohydronamics can be found in Chap. 5 on magnetocaloric fluids. We also refer to the book of Rosensweig [101], which describes important, pioneering work carried out since the 1960s in this area.

Ferrohydrodynamics is associated with collodial suspensions of nanofluids that are formed by magnetic nanosized particles, suspended in the base liquid (e.g. organic solvents, water). These suspensions are also referred to as ferrofluids. The particles do not settle (stratify) due to the presence of Brownian motion. They are also coated with a surfactant to prevent clumping. However, the volume fraction of nanoparticles in the base liquid is rather small (high volume fractions lead to clustering), which prevents ferrofluids from being used as successful liquid magnetocaloric refrigerants for magnetic refrigeration, as the replacement of the active magnetic regenerator (of course this relates to the present state of development of magnetocaloric materials and the associated magnetocaloric effect).

Nanofluids are used as heat-transfer fluids in many applications [102–104]. This also holds true for ferrofluids [105–107]. For more information on ferrofluids, the reader is referred to the chapter on magnetocaloric fluids, as well as books by Rosensweig [101] and Odenbach [108, 109]. In this subsection we do not focus on the behaviour of ferrofluids or their applications to enhance heat transfer, but instead deal with mechanisms that can be used as ferrofluid thermal switches or ferrofluid thermal diodes.

A review of the principles of micro magnetofluids can be found in Nguyen [110]. Among the various applications of ferrofluids in magnetic refrigeration, an interesting solution that represents an enhanced heat transfer as well as the application of a thermal switch by the ferrofluid flow was given by Kitanovski and Egolf [4]. This example is shown in Fig. 6.22. In the case of Fig. 6.22, four active magnetic regenerators (AMRs) are placed inside the cylinder and are filled with the ferrofluid. Two cross-flow micro heat exchangers connected to the heat sink (not shown in the figure) comprise the water fluid channels and voids that contain the

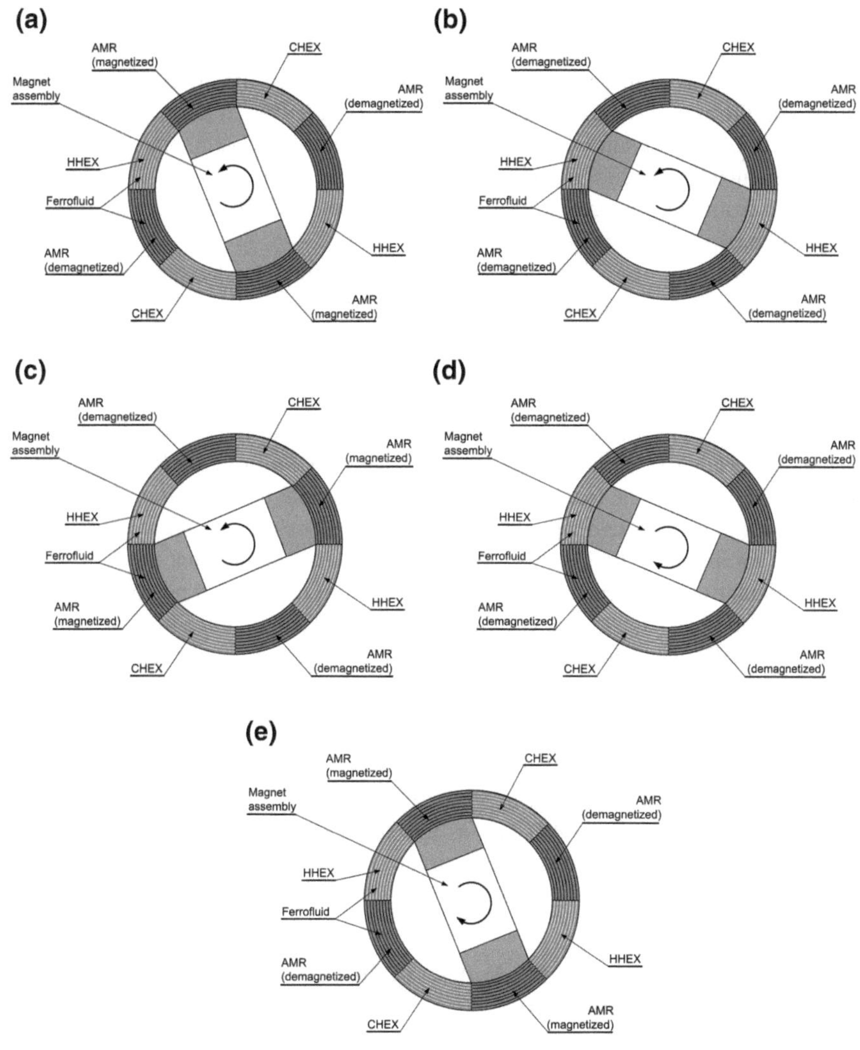

Fig. 6.22 Example of the application of a ferrofluid in magnetic refrigeration

ferrofluid. We denote this heat exchanger as HHEX (hot heat exchanger). Another pair of cross-flow micro heat exchangers is connected to the heat source (not shown in the figure). They also comprise water fluid channels and voids, which contain the ferrofluid. We denote this heat exchanger as CHEX (cold heat exchanger). A magnet assembly rotates as the pendulum in each direction. The maximum angle (azimuth path) that the magnet assembly performs in one direction is 90°. Since the ferrofluid is incompressible, as well as responding to the magnetic field (magnetic fluid), each movement of the magnet assembly will pull as well as push the ferrofluid through the voids in the AMRs, as well as the heat exchangers. In the case of

Fig. 6.22a, two AMRs are magnetized. Therefore, the magnetocaloric material heats, and the heat is partially transferred to the ferrofluid by heat conduction.

The magnet assembly rotates in the counter-clockwise direction in order to pull the ferrofluid from the magnetized AMR to the HHEX (Fig. 6.22b). There, the heat is rejected to the heat sink via the water. Since the ferrofluid is incompressible, such a movement also pushes the rest of the ferrofluid in a cylinder. Therefore, the ferrofluid from the demagnetized AMRs flows to the CHEX, where it absorbs heat from the heat source (via water).

The rotation of the magnet assembly is continued in the counter-clockwise direction until the second pair of AMRs is being magnetized (Fig. 6.22c). The second pair of AMRs heats up, and the ferrofluid, which absorbed the heat in the CHEX, moves towards the demagnetized AMRs. Now, the rotation of the magnet assembly starts in the clockwise direction (Fig. 6.22d). The ferrofluid from the void of the magnetized AMRs is now pulled to the HHEX to reject the heat. Simultaneously, the other thermodynamic processes are also performed in the cylinder. At the end (Fig. 6.22e), the magnet assembly is at its initial position. It is clear that the described processes above will lead to the simultaneous production of cold, without the need for any switching valve mechanism. Despite the fact that the ferrofluid in this particular case represents the heat-transfer fluid, it may also be considered as a thermal diode mechanism that exploits the ferrofluid flow induced by the magnetic field (especially when the consider system in the Fig. 6.22 is small).

Based on the possibility of using ferrofluids as thermal diode mechanisms, we can distinguish between the following solutions:

- Ferrofluid thermal diode that applies the anisotropy of thermal conductivity,
- ferrofluid thermal diode that applies the magnetically induced fluid flow,
- ferrofluid thermal diode that acts as the thermal contact switch and
- ferrofluid and magnetowetting principle.

6.4.2.1 Ferrofluid Thermal Diode that Applies the Anisotropy of Thermal Conductivity

This principle is similar to solid-state thermal rectificators. However, in this particular case the magnetic field induces thermal rectification due to the magnetic-field-dependent thermal conductivity of the ferrofluid. Ferrofluids can therefore be used as thermal diode rectificators, where the anisotropy of the thermal conductivity is manipulated with the magnetic field strength as well as with the direction of the magnetic field [111–114]. The rectification factor is rather low (e.g. up to 300 %) to be efficiently applied in magnetic refrigeration. In magnetic refrigeration, the minimum rectification factor (the ratio between the thermal resistance in the OFF and ON operation of the thermal diode) should be at least 10 (1,000 %), or preferably even higher.

6.4.2.2 Ferrofluid Thermal Diode that Applies the Magnetically Induced Fluid Flow

This kind of principle can be driven by a combination of temperature and magnetic field gradients. Since this principle can also be applied for magnetocaloric fluids, it is described in detail in the chapter on magnetocaloric fluids. Another principle is to manipulate the ferrofluid flow by a magnetic field. One of such solutions is presented in Fig. 6.22.

6.4.2.3 Ferrofluid Thermal Diode that Acts as the Thermal Contact Switch

Manipulation of the ferrofluid drop flow with a magnetic field can lead to migration of the drop to the desired heated or cooled surface. This mechanism is similar to magnetowetting, where the surface tension of the ferrofluid drop is manipulated by the magnetic field.

6.4.2.4 Ferrofluid and Magnetowetting Principle

The magnetowetting principle is similar to the electrowetting principle; however, here the manipulation of the surface tension of the ferrofluid drop is, in most cases, based on the application of a magnetic field [110, 115–119]. This can lead to an efficient application of the thermal contact ferrofluid switch. According to Nguyen [110], there are different principles for magnetowetting: magnetowetting in a uniform magnetic field; magnetowetting in a nonuniform magnetic field; magnetowetting with a diamagnetic droplet and the magnetically controllable surface of magnetic nanotubes; the sliding motion of a ferrofluid drop; and the rolling motion of a ferrofluid covered by hydrophobic particles. Figure 6.23 shows examples of electrowetting with a uniform and nonuniform magnetic field. In this case the dashed lines show the initial state of the ferrofluid drop. When applying the uniform magnetic field over the drop, its surface tension will change from the angle θ' to the angle θ. If the drop was placed between two surfaces, such a manipulation could lead to the migration of the drop from the bottom to the upper surface and vice versa. The non-uniform field, when applied to the ferrofluid drop, will lead to better wetting ($\theta' > \theta$). Also in this case, the drop can act as a thermal contact switch, or when placed between two surfaces, both being manipulated with the magnetic field, such a drop would transport heat from one plate to another.

The magnetowetting principle can also be associated with the Rosensweig instability [101, 108–110], which presents a multiple-peaked structure (like a hedgehog) of the ferrofluid when subjected to a magnetic field.

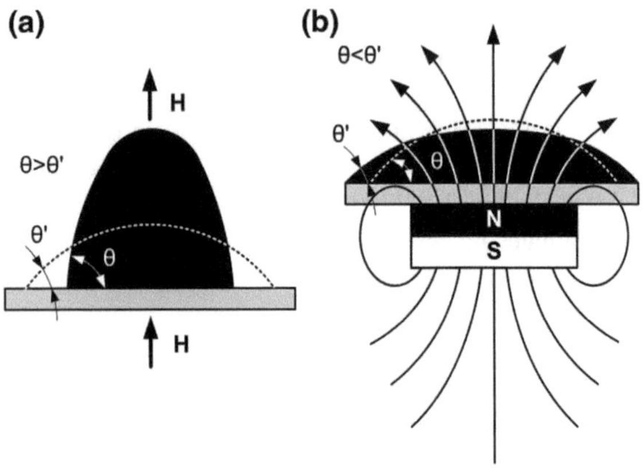

Fig. 6.23 **a** Manipulation of the ferrofluid drop by a uniform magnetic field, **b** Manipulation of the ferrofluid drop by a nonuniform magnetic field

6.4.3 Magnetohydrodynamics

Magnetohydrodynamics (MHD) represents the interaction of an electric and a magnetic field with electrically conducting fluids (e.g. plasma, electrolytes, liquid metals, etc.). This results in a Lorentz force on the fluid, and as a consequence, the fluid flow (Fig. 6.24). With respect to magnetic refrigeration, the most interesting applications represent the use of liquid metals, such as Hg, NaK and liquid metal alloys based on Ga and In. The MHD principle can be also applied for the liquid metal bi-directional flow, which is common in active magnetic regenerators (AMRs). Therefore, it could also serve for the principle shown in Fig. 6.22.

In this case we place the electrodes over the two regions of the AMRs (Fig. 6.25). Instead of a ferrofluid, a liquid metal will be applied as the working fluid. In the case of Fig. 6.25a, one pair of AMRs is being magnetized. Since the magnetic field is applied to this region, the electrodes are switched ON and therefore propel the liquid metal from the magnetized AMR to the HHEX. Since the

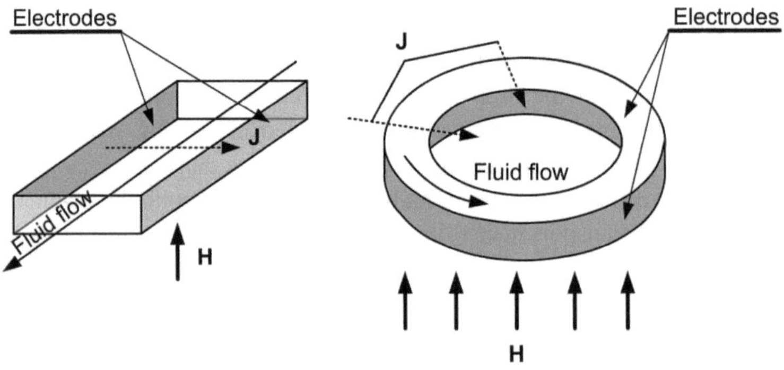

Fig. 6.24 Example of the MHD fluid flow induced by an electric and magnetic field

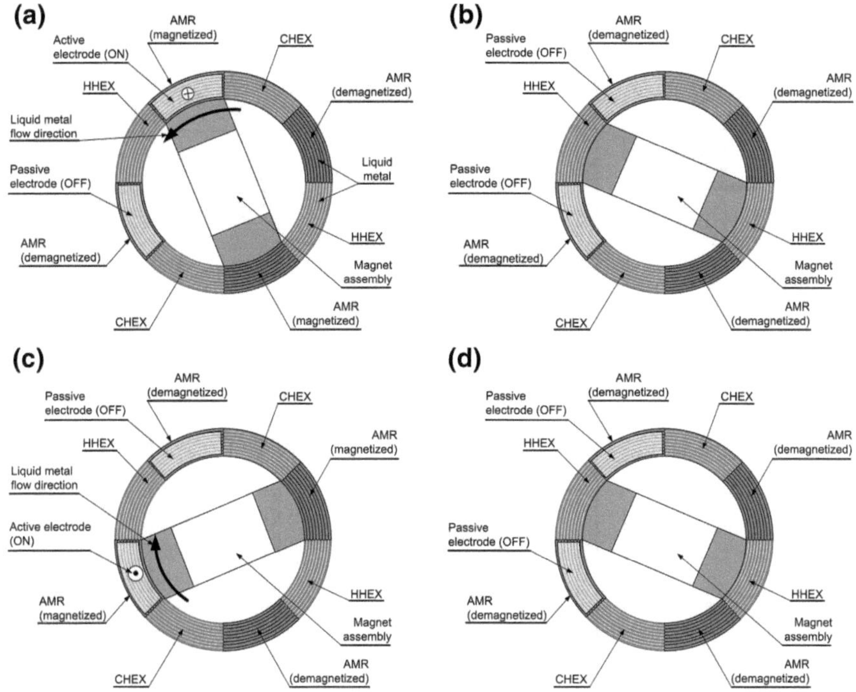

Fig. 6.25 An example of applying the MHD principle in magnetic refrigeration

liquid metal is incompressible, it will therefore move in an azimuth direction in all the elements of the device. When the magnet assembly passes to the HHEX, the liquid metal is at rest, so no fluid flows take place (Fig. 6.25b). When the second pair of AMRs is being magnetized (Fig. 6.25c), the liquid metal fluid flow is reversed by activating the second electrodes. The fluid now flows in the counter direction. Note that this could also be done with a single MHD system and not two, as shown in Fig. 6.25.

The MHD can also be applied as the thermal contact diode, which uses the liquid flow propelled by the Lorenz force to or from the heat sink or heat source (Fig. 6.26). Note that the magnetic field source denoted by N–S can be provided by any kind of magnet. Of course, in this case one should find a solution for how to avoid the magnetization of the magnetocaloric material when this is in the process related to its demagnetized state. Therefore, Fig. 6.26 should serve only as an illustration and not as a solution.

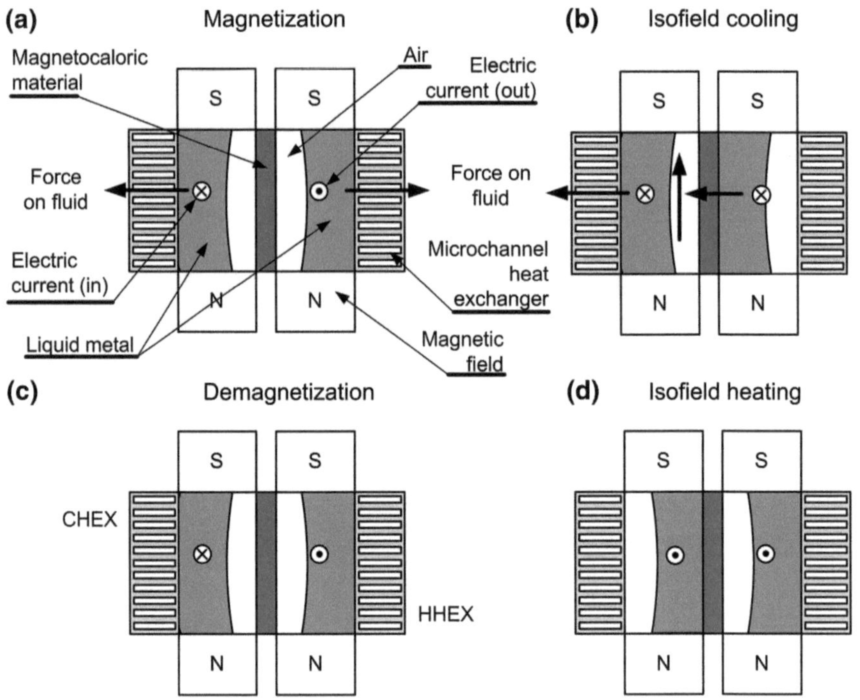

Fig. 6.26 An example of applying the MHD principle for the contact thermal diode mechanism

6.4.4 *Magnetorheology and Electrorheology*

Despite the fact that magnetorheology deals with magnetic fluids as well, one should distinguish between ferrofluids and magnetorheological fluids. The latter are suspensions of larger particles (see Chap. 5), and where the particles are not suspended any more by Brownian motion. Also, the volume fraction of solid magnetic particles in the carrier fluid (mineral or silicone oils, kerosene, water and water solutions, etc.) can be substantially higher (e.g. 20–50 %) than is the case for ferrofluids. Magnetorheologic fluids can be successfully applied in dampers, brakes, valves, clutches, etc. [120–124]. They are becoming an indispensable part of the future automotive industry [120], aerospace industry, medicine, etc. [125]. A special research direction is also focused on magnetisable gels (ferrogels) and polymers [126–128], which can be designed with nano- or micro-sized particles.

Magnethorheologic (MR) fluids exhibit a substantial change in their rheological properties in an alternating magnetic field. Under an applied magnetic field, the micron-sized magnetic particles tend to form chain-like structures in the direction of the magnetic field (Fig. 6.27). This drastically influences the axial velocity field (along the channel). The magnetorheologic suspensions behave as non-Newtonian fluids in the absence of a magnetic field. Such fluids are mostly dependent on the

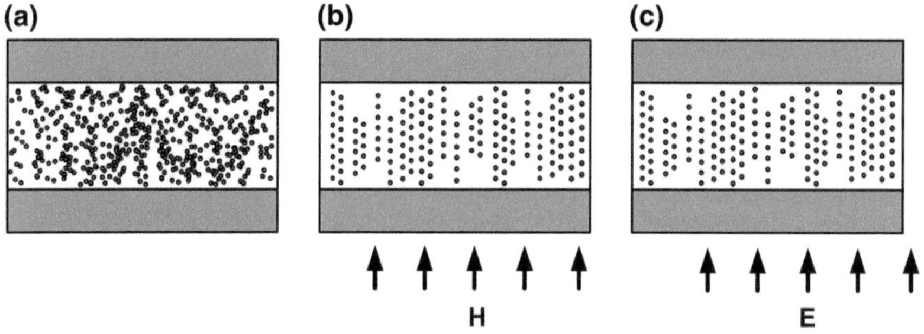

Fig. 6.27 **a** The magnetorheologic or electrorheologic fluid in the absence of electric and magnetic fields, **b** The magnetorheologic fluid in the presence of an applied magnetic field, **c** The electrorheologic fluid in the presence of applied electric field

volume fraction of the solid particles, the particle shape and size, the ratio of the density of the two phases and the viscosity of the carrier fluid. In most cases, the shear stress also depends on the yield stress, which defines the conditions under which the fluid will start to flow. However, with the magnetic field being present, the yield stress will drastically increase with the magnetic field strength, transforming the liquid magnetorheologic fluid into a solid structure [129]. This transformation can occur in a time period of milliseconds [130, 131].

The magnetic particles can also be coated to prevent or reduce the effect of sedimentation. Fang et al. [132] made analyses on two different MR suspensions, among which the first MR comprised carbonyl iron particles, and in the second, the carbonyl iron particles were coated with multiwall carbon nanotubes (MWCNT) in order to decrease the sedimentation effect. The addition of nanoparticles to the MR can substantially influence the magnetorheology and other properties of MR fluids.

Electrorheologic (ER) fluids [133–136] exhibit similar behaviour to MRs. These suspensions consist of a non-aqueous carrier fluid with suspended micron-sized dielectric particles. As with the MR fluids, the ER fluids react strongly with the applied electric field, which leads to a substantial increase in the apparent viscosity, yield stress and shear stress up to several orders of magnitude and the transition from a liquid to a solid-like substance [134, 135]. Also in this case, a chain-like structure of particles is formed in the direction of the electric field (Fig. 6.27). The application field of electrorheologic fluids is similar as that for magnetorheologics [133]. There are also international conferences on Electrorheological Fluids and Magnetorheological Suspensions, which deal with this important domain [137].

Since the aim of this subsection is to present the available technologies for thermal diodes, magnetorheology and electrorheology can be potentially applied for the following principles of thermal diodes:

- MR or ER thermal diode that applies a magnetically induced fluid flow,
- MR or ER thermal diode that acts as a thermal contact switch and
- MR or ER thermal diode that applies the anisotropy of thermal conductivity.

6.4.4.1 MR or ER Thermal Diode that Applies a Magnetically Induced Fluid Flow

The application of the MR or ER fluid flow, due to very large pressure gradients, is not a mechanism that could be used for thermal diodes. Therefore, especially under an electric field or magnetic field gradients, this kind of fluid should not be considered as a heat transfer fluid.

6.4.4.2 MR or ER Thermal Diode that Acts as a Thermal Contact Switch

The time response in MR or ER fluids is in the time period of milliseconds [130, 131, 138, 139]. This is sufficient for such a fluid to be efficiently applied as a thermal contact switch. However, one should combine the phenomena of the liquid to solid state of MR or ER fluids into smart solutions, especially because some thermodynamic processes will require the simultaneous demagnetization and the operation of the contact thermal switch.

6.4.4.3 MR or ER Thermal Diode that Applies the Anisotropy of Thermal Conductivity

The anisotropy of the thermal conductivity is more obvious in MR and ER suspensions with a smaller volume fraction [140]. The rectification factor is rather low for these fluids to be efficiently applied as thermal rectificators with an anisotropic, electric field [141], or magnetic field [140, 142–144] dependent thermal conductivity.

6.5 Review of the Research on Thermal Diodes in Magnetic Refrigeration

The idea on application of a thermal diode mechanism in magnetic refrigeration at room temperature was first introduced by Kitanovski and Egolf in 2010 [4]. They realized that the application of an AMR with convective heat transfer and all the corresponding losses will strongly restrict the realization of magnetic refrigeration technology in cost-and energy-efficient market applications. As the thermal diode mechanisms, Kitanovski and Egolf proposed two domains: the solid state and the microfluidic thermal diodes [4].

As the reader of this book will realize, magnetocaloric energy conversion faces some major obstacles. Namely, the best prototypes can efficiently work at a frequency of operation of less than 5 Hz (the number of thermodynamic cycles per unit of time). This low frequency means higher efficiency, but low power density of the device. Because of the expensive permanent magnet materials, as well as expensive magnetocaloric materials, an efficient magnetic refrigerator operating with the basic principle of AMR will be very expensive [145] (see also the chap. 9). Most of the losses related to devices that apply the basic AMR principle are related to the low manufacturability of magnetocaloric materials, high heat-transfer irreversible losses, viscous losses, mechanical friction losses, losses related to valve systems (dead volume and internal leakage) [4–6].

Most of these difficulties can be overcome by the introduction of thermal diode mechanisms [4]. After the publication by Kitanovski and Egolf in 2010 [4] others also began with research on thermal diodes in magnetic refrigeration at or near room temperature. Almost all these investigations were directed towards solid-state thermal diodes. Since there were not many investigations, we describe the different approaches of the various authors.

Silva et al. [146, 147] theoretically investigated the use of solid-state thermal rectificators (with controllable thermal conductivity) to be used in magnetic refrigeration. They proved that such a principle can be applied to magnetic refrigeration as the cascade system of thermal diodes. They showed the great potential of the proposed technology with high cooling-power densities and very high operating frequencies (above 100 Hz) [146].

Olsen et al. [148], together with Tasaki et al. [149–151], evaluated the principle of the solid-state thermal diode mechanism in which they also evaluated a kind of fictive type of solid rectificator or solid contact thermal switch. The work was focused on the use of such mechanisms for a magnetic cooling system in vehicles. In the concept of a solid-state magnetocaloric refrigerator the device consisted of several layers of magnetocaloric material and thermal diodes (which formed together a cascade system). Based on the concept a number of numerical simulations were performed. The results of Olsen et al. [148] revealed that the proposed concept could operate with a high volumetric cooling power even for a temperature span of 60 K. When they compared an ideal gadolinium regenerator with plates of 50 microns thickness (which is not yet achieved in magnetic refrigeration, and therefore the authors referred to such an AMR as a “dream pipe”) and the concept of cascaded embodied gadolinium between thermal diodes, the last showed about 3.5-times higher cooling power density [148].

Most of the work relating to applications of thermal diode mechanisms in magnetic refrigeration has been done in the field of solid-state Peltier thermal diodes. For instance, Egolf et al. [152, 153] presented a detailed study of the operation of thin-film thermoelectric modules with Ni-nanowires and concluded that they could be a potential solution to be used as thermal diodes for magnetic refrigeration; however, they pointed out certain aspects that have to be taken into consideration, i.e. the thermal and electrical resistance, durability and low-cost production.

In a recent study by Egolf et al. [154], which was focused on thermal switches built with Ni-nanowire Peltier modules, the performance of these modules was evaluated on a theoretical and an experimental basis. An approximate estimate of the performance of a magnetic refrigerator of the types built so far was compared to that of a magnetic refrigerator with nanowire thermal switches. The authors compared the operation of the AMR-based magnetic refrigerator with a refrigerator based on a thermal diode mechanism. The results reveal that for the same cooling power of 50 W and the same exergy efficiency of 52 % (for the given operating parameters and a magnetic field change of 2 T), the AMR-based refrigerator would require five times more magnetocaloric material than a refrigerator based on thermal diodes. Consequently, the mass of permanent magnets would be substantially smaller.

In the study of Tomc et al. [5, 6, 155], thin-film Peltier thermal diodes were considered to be applied with a thin plate of gadolinium. The value for the magnetic field change was taken to be 1 T. The results of their study reveal that the application of thin Peltier thermal diodes in combination with a magnetocaloric material could lead to high exergy efficiency for the whole device, e.g. up to 65 % for an operating frequency of about 20 Hz (corresponding to a specific cooling power of about 2 kW per kg of gadolinium). This would strongly depend on the thickness of the magnetocaloric material, the operating frequency (the higher the frequency is, the lower efficiency will be), as well as the temperature span between the heat source and the heat sink.

The only work that relates to the application of microfluidic thermal diodes in magnetic refrigeration at room temperature was reported in a diploma thesis at the University of Ljubljana [156]. The results of this particular work showed that fluidic thermal switches can be successfully applied for very high operating frequencies (above 100 Hz).

6.6 Potential Configurations of Thermal Diodes in Magnetic Refrigeration

As noted before, the reason that today's state-of-the-art magnetocaloric technology is not competitive with conventional refrigeration technologies is based on one of the most important characteristics of a magnetocaloric device, i.e. the power density. Not having a sufficiently high power density directly affects the cost of a device. However, to decrease the costs of the device (which is predominantly a consequence of its large mass and the high costs of the magnetic and magnetocaloric materials) the power density should be increased. And the best way to achieve this is to increase the operating frequency of the device. However, an increase in the operating frequency usually leads to low efficiency due to the poor heat-transfer characteristics, which the classical active magnetic regenerator has to face. In the preceding chapters on AMR research and development a number of suggestions were given about how to improve the more efficient AMR operation at higher frequencies, i.e. from improving the AMR geometry, layering different magnetocaloric materials, using different working fluids, to using different design concepts for the magnetic device (reciprocating or rotary) as well as to apply different thermodynamic cycles. However, all these potential improvements still revolve around one more fundamental issue, which is the conventional concept of the active magnetic regenerator itself. This concept remains the focus of research and it most definitely has some more room for improvement. However, it is becoming increasingly clear that this course of magnetic refrigeration research will most probably not give us the desired results (high operating frequency, large temperature span, high cooling power and simultaneously high efficiency). The problem is that even very advanced active magnetic regenerator has its limitations with regard to efficient heat transfer at higher operating frequencies, which is clear from the presented state-of-the-art devices in previous years (Chap. 7) as well as some theoretical studies [157].

Hence, it is important to try some other potential research directions, especially in terms of new magnetocaloric device concepts as a whole: different approaches to AMR research or even completely new concept ideas should be welcomed and encouraged.

In the previous chapters we provided comprehensive information about the different mechanisms that can be applied as thermal diodes. The following subsections give certain theoretical aspects and suggestions about how to approach the design configuration of such systems. Note again that the research of magnetocaloric energy conversion with the application of thermal diodes is still a very fresh and new field of research and therefore still needs a lot more work and attention.

The aim of this subsection is to describe and show how thermal diodes should be implemented regarding magnetocaloric material and what operation is to be expected. Different design concepts are presented, each with an explanation of the operation.

Note that these are strictly theoretical suggestions, the sole purpose of which is to give the reader an idea behind the technology and especially to try to encourage her/his creative thinking towards different approaches to magnetocaloric energy conversion.

6.6.1 Single-Stage Magnetocaloric Device with Thermal Diodes

A single-stage device incorporating thermal diodes is presented in Fig. 6.28. It is a thin layer of magnetocaloric (MC) material stacked between two thermal diodes 1 and 2. Each thermal diode faces one heat exchanger, cold and hot, respectively. Similar concepts were already presented in some works in the field of electrocaloric energy conversion [158–160]. The notion of "single stage" means that only the

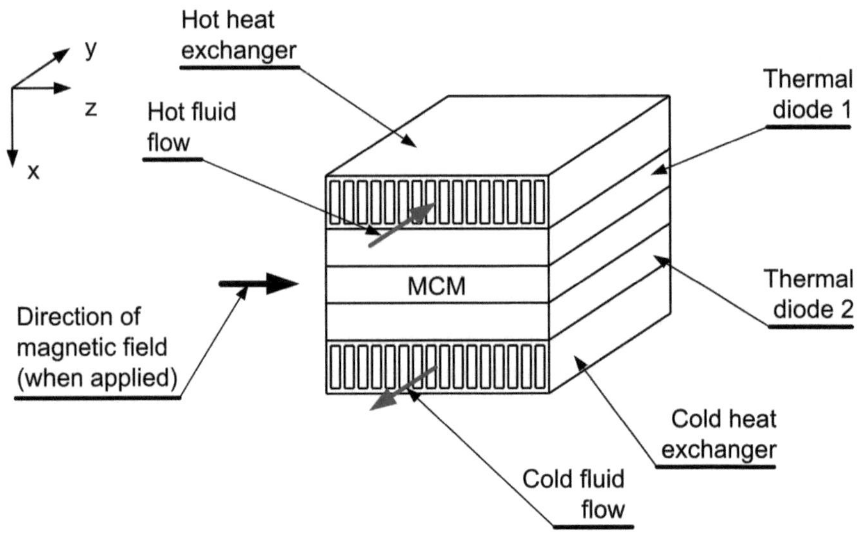

Fig. 6.28 Design concept of a single-stage magnetocaloric device with thermal diodes

magnetocaloric effect of the MC material is what produces the temperature span between the cold and hot heat exchanger. Therefore, there is no heat regeneration in the MC material and such a device, depending on the intensity of the magnetic field change, will provide a rather small temperature span between the heat source and the heat sink.

Figure 6.29 shows the operation for such a single-stage device. It is clear that the operating temperature span along the length (Fig. 6.28, *y*-direction) is mostly produced by the MC material, implying the single-stage operation. There is a certain increase/decrease in the temperature span on the hot/cold side of fluid flows, respectively. However, this is only due to the fluid absorbing/releasing the heat to the MC material and should not be regarded as a process of heat regeneration. Furthermore, two different working fluid loops are applied, one on the hot side and one on the cold side, which also disregards the option of active magnetic regeneration. It is clear that the majority of the temperature span between the hot and cold sides is done by the MC material with its MC effect. The thermal diodes should not contribute to the temperature span of the device, or at least this contribution should be as small as possible, since this decreases the irreversible losses or the required amount of work for the operation of the thermal diodes. Therefore, the major part of the overall input work should be magnetic work done on the MC material.

The thermodynamic cycle of the single-stage operation in Fig. 6.29 is as follows. When the magnetic field is applied to the MC material it heats up due to the MC effect. At this moment, thermal diode 1 becomes operational and starts to transfer heat from the MC material to the hot heat exchanger. Meanwhile, thermal diode 2 should not operate and should act as a thermal insulator, thus preventing the heat from being transferred from the MC material to the cold heat exchanger. When the

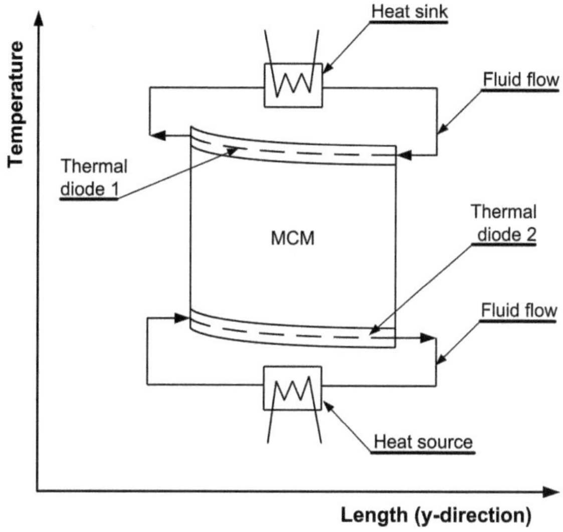

Fig. 6.29 Temperature distribution of a single-stage magnetocaloric device with thermal diodes along the length (y-direction in Fig. 6.28) of the device

process of demagnetization occurs, the temperature of the MC material decreases (due to the MC effect).

At this point thermal diode 2 should start to operate, transporting the heat from the cold heat exchanger to the MC material. During this process thermal diode 1 should not operate and should act as a thermal insulator, preventing heat from the hot heat exchanger from being transferred back to the MC material. All four phases of the magnetization, thermal diode 1 ON, demagnetization, thermal diode 2 ON represent a single thermodynamic cycle for such a device. The fluid flow through the heat exchangers may be continuous or periodic. In the last case, this is difficult to perform without very large losses. Furthermore, the periodic fluid flow should be well synchronized with the operation of the device.

6.6.1.1 Active and Passive Thermal Diodes in a Single-Stage Operation

According to the previous subsections, these two types of thermal diodes should operate in a different manner regarding their temperature profile of operation.

Figures 6.30 and 6.31 show the steady-state operation of the magnetocaloric device with passive and active thermal diodes, respectively. Both figures show all four phases of a single-stage thermodynamic cycle, regarding the temperature profile transversal to the fluid flow (see Fig. 6.28, x-direction).

Figure 6.30 presents the operation of passive thermal diodes coupled with magnetocaloric material. In Fig. 6.30a the magnetization phase occurs, which leads to a temperature increase in the MC material. Both thermal diodes (TD1 and TD2) are at that point non-operational and just a weak temperature response of the thermal diodes can be seen due to the MC material temperature increase. Both thermal diodes have, in this example, low effective thermal conductivities.

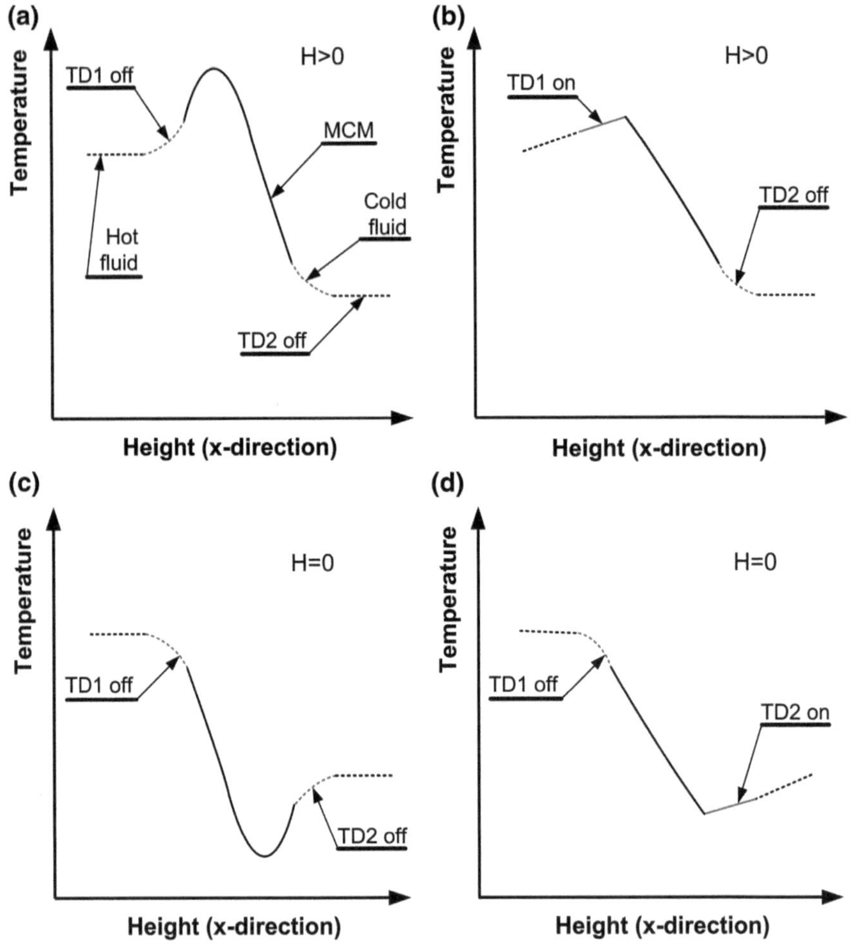

Fig. 6.30 Four phases of operation in the perpendicular direction to the fluid flow of the magnetocaloric device with passive thermal diodes

In Fig. 6.30b the passive thermal diode 1 becomes operational. Since it is passive, its thermal properties are influenced in a manner that they result in a higher effective thermal conductivity. The thermal diode 1 (Fig. 6.30b) has a higher thermal conductivity than the non-operational thermal diode 2. This, as an example in the figure, will lead to linear temperature distribution of thermal diode 1. Meanwhile, thermal diode 2, which is at this moment inactive, responds as a non-perfect thermal insulator between the MC material and the cold fluid. Therefore, most of the heat flux will be transported from the MC material to the hot fluid. Figure 6.30c presents the process of the demagnetization. The temperature of the MC material decreases due to the MC effect. Both thermal diodes are, in this phase, again inactive, with low thermal conductivities (act as non-perfect insulators), and thus slowly respond to the temperature changes in the MC material. In the next process, shown in Fig. 6.30d, the passive thermal diode 2 becomes operational. A substantial increase in its effective thermal conductivity makes it more susceptible to the temperature levels of the cold fluid and MC material. A fast heat transport via thermal conduction occurs, which is, as an example, presented as a linear temperature profile of thermal diode 2. Meanwhile, thermal diode 1 is non-operational with low effective thermal conductivity (acts as non-perfect heat insulator), which leads to a slower temperature response to the surrounding temperature changes. In this case most of the heat flux is transported from the cold fluid to the MC material.

Figure 6.31 presents the operation of active thermal diodes coupled with a magnetocaloric material. All four phases of the operation are practically the same as in Fig. 6.30.

However, the difference is in the operating temperature profile of the active thermal diodes. The notion "active" in this case means that the active thermal diodes should operate as a sort of heat pump. When they are non-operational, they should exhibit only a low heat transfer due to their low thermal conductivity. Whereas in operating mode, when their characteristics are influenced directly by some energy input (e.g. electric energy), they operate as heat pumps. This reflects in rapid heat transport in the direction of a temperature gradient, from the cold side to the hot side of the thermal diode (similar to the heat pump). This can be seen from Figs. 6.31b, d.

Figure 6.31b shows the operation of an active thermal diode 1, while the active thermal diode 2 is non-operational. Furthermore, the operation of thermal diode 1 is shown with the opposite temperature profile (with respect to analogue passive thermal diodes shown in Fig. 6.30) since they operate in a similar way to heat pumps; therefore, the heat in a thermal diode is transported from a lower temperature (MC material side) to a higher temperature (hot fluid side). Meanwhile, the non-operational active thermal diode 2 responds to the temperature changes only by its low thermal conductivity. Figure 6.31d shows similar operation to that in Fig. 6.31b. However, now the active thermal diode 2 is operational. Thus it works as a heat pump, and rapidly transports heat from the lower temperature (cold fluid side) to the higher temperature (MC material side).

At this point it has to be emphasized that even though the active thermal diodes can operate as heat pumps (i.e. Peltier), thus having the ability to cool/heat, they should not be used to increase the temperature span in a magnetocaloric device. Their sole purpose is to work as fast heat-transport mechanisms between the fluid and the MC material. The temperature span of the device should be, in any case, performed by the MC material. The active thermal diode should just help to transport the heat faster from/to the MC material at the lowest possible temperature differences. In this case the efficiency of the active thermal diodes can be kept high, which will not substantially affect the total efficiency of a device.

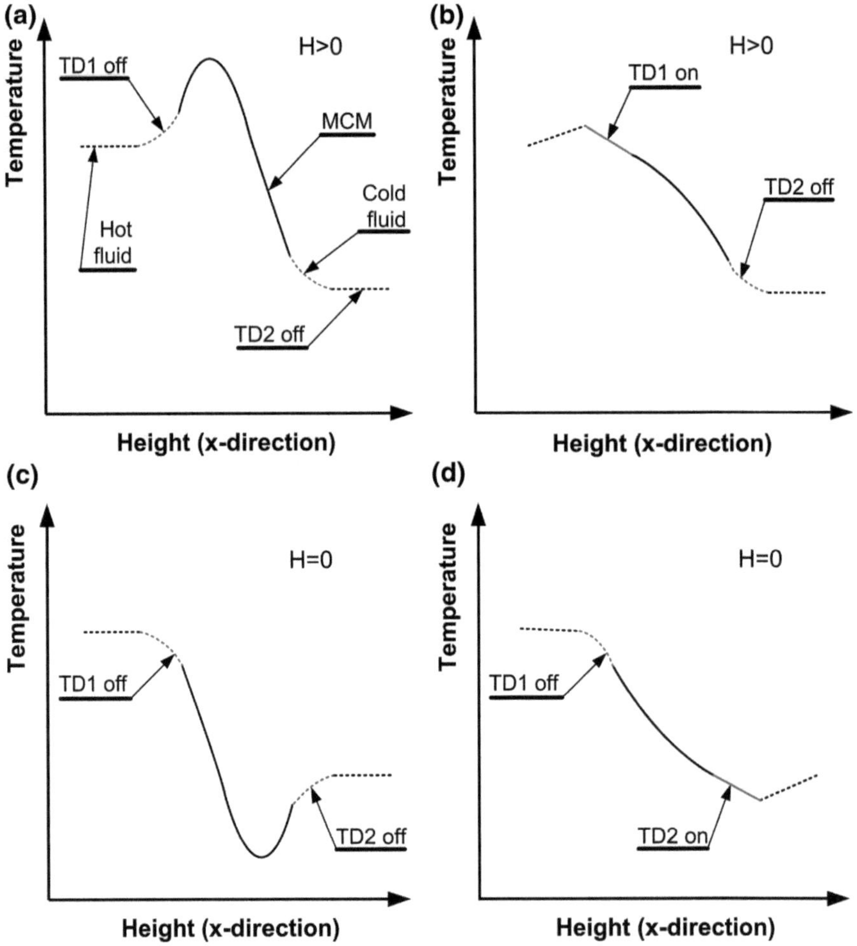

Fig. 6.31 Four phases of operation in the perpendicular direction to the fluid flow of the magnetocaloric device with active thermal diodes

6.6.2 Cascade Magnetocaloric Device with Thermal Diodes

A cascade magnetocaloric device with thermal diodes comprises a number of single-stage devices stacked on each other, as shown in Fig. 6.32. The purpose of such a design is to increase the operating temperature span of the whole device. From Fig. 6.33 it is clear that each MC material operates in a single-stage mode at different temperature levels. However, the heat from the MC material at lower temperature is transported via a thermal diode to the MC material at higher temperature. In this manner the heat is transported from the cold heat exchanger to the hot heat exchanger.

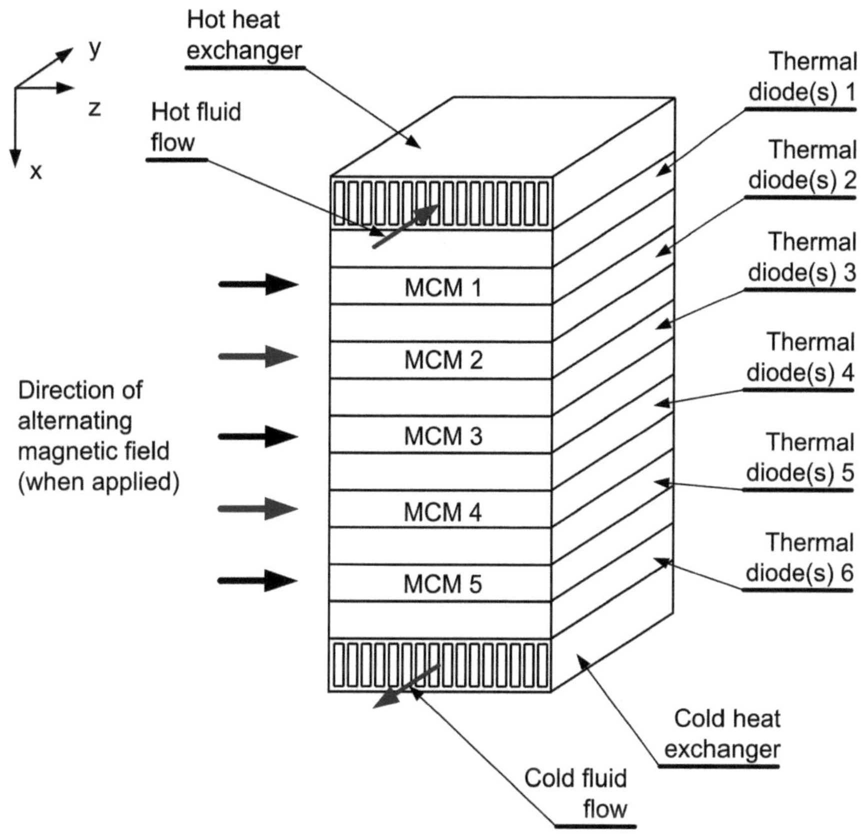

Fig. 6.32 Design concept of a cascade magnetocaloric device with thermal diodes: *black arrows* magnetization, *grey arrows* demagnetization

As seen from Fig. 6.32, five MC materials are applied, each embodied within thermal diodes on the top and bottom. Each MC material shares one thermal diode with the neighbouring MC material. Only two MC materials at the hot and cold ends have one thermal diode facing the hot or cold heat exchanger, respectively. The magnetic field is applied to MC materials in a particular way: every second MC material is magnetized or demagnetized. In Fig. 6.32 the black arrows represent the applied magnetic field, whereas the grey arrows represent the demagnetization phase. Therefore, in the given example in Fig. 6.32, the MCMs denoted by numbers 1, 3 and 5 are magnetized, while the MCMs denoted by numbers 2 and 4 are demagnetized. After this phase, the magnetic field should switch in a manner that it magnetizes the MCMs denoted by numbers 2 and 4, while the MCMs denoted by numbers 1, 3 and 5 are demagnetized. The operation of such a cascade system is thoroughly shown and explained in Fig. 6.34.

The alternating of the magnetic field in the cascade system is of great importance. Only in this manner can heat be transported via thermal diodes between two

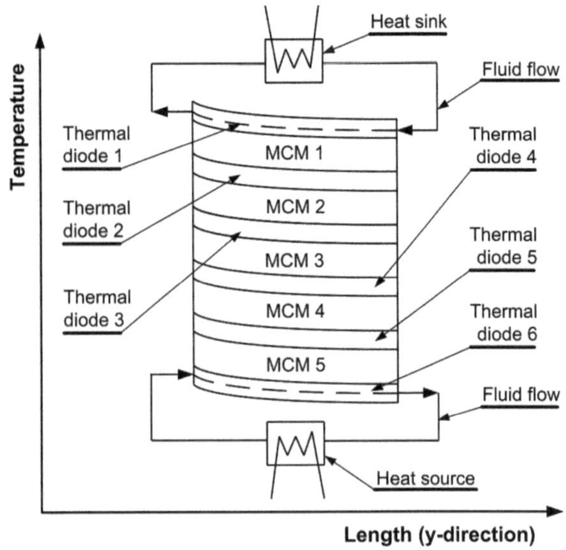

Fig. 6.33 Temperature distribution of a cascade magnetocaloric device with thermal diodes along the length (*y*-direction in Fig. 6.32) of the device

neighbouring MC materials, as shown in Fig. 6.34. Figure 6.34 presents all four phases of the cascade device's operation in the steady state. It shows the operation for a given example from Fig. 6.32. In Fig. 6.34a the processes of alternating magnetization and demagnetization can be observed. The MCMs denoted by numbers 1, 3 and 5 are magnetized, thus their temperature increases due to the MC effect. Meanwhile, the MCMs denoted by numbers 2 and 4 are demagnetized; therefore, their temperatures decrease due to the MC effect. All six thermal diodes are, at this point, non-operational. Figure 6.34b presents the second phase where the thermal diodes denoted by numbers 1, 3 and 5 become operational (in this example all the thermal diodes are considered to be passive thermal diodes).

Thermal diode 5 transports heat from the magnetized MCM 5 to the demagnetized MCM 4. Thermal diode 3 transports the heat from MCM 3 to MCM 2. Thermal diode 1, on the other hand, transports heat from MCM 1 to the hot fluid. Thermal diodes 2, 4 and 6 are non-operational during this process, preventing the heat from being transported from the magnetized MCMs or the cold fluid back to the neighbouring demagnetized MCMs, which are at a lower temperature.

In the third process presented in Fig. 6.34c the processes of magnetization and demagnetization are altered. The MCMs denoted by numbers 2 and 4 are magnetized, while the MCMs denoted by numbers 1, 3 and 5 are demagnetized, which results in a temperature increase or decrease, respectively. All the thermal diodes are non-operational in this phase. In the last and fourth process (Fig. 6.34d), the thermal diodes denoted by numbers 2 and 4 start to transport heat from the magnetized MCMs 2 and 4, to the demagnetized MCMs 1 and 3. Thermal diode 6 transports heat from the cold fluid to the demagnetized MCM 6. All four phases of operation represent one operating thermodynamic cycle of the device.

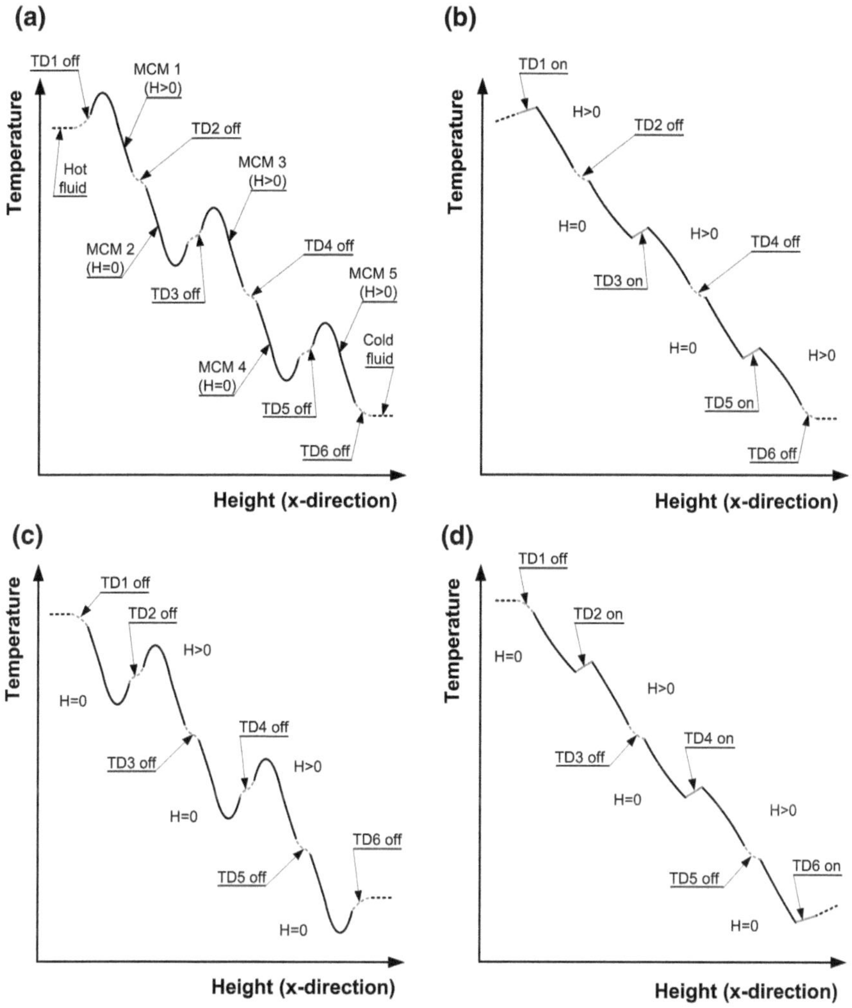

Fig. 6.34 Four phases of operation along the direction perpendicular to the fluid flow (*x*-direction in the Fig. 6.32) of the cascade magnetocaloric device with passive thermal diodes

The example of a cascade magnetocaloric device with thermal diodes presented in this subsection is just one of the possibilities for a cascade design. Here, five MC materials with six passive thermal diodes stacked in a cascade were presented. However, the design allows any other number of MC materials to be used in a cascade as well as the use of active thermal diodes. The choice of the number of MC materials depends on the desired operating temperature span, as well as the irreversible losses that occur at each stage. Furthermore, since each MC material operates at a certain temperature within the operating temperature span it would be wise to consider MC materials with different Curie temperatures.

6.6.3 Active Magnetic Regeneration with Thermal Diodes

In the previous subsections single-stage and cascade operations with thermal diodes were presented. However, when considering different designs of magnetocaloric devices with thermal diodes, we can also apply the active magnetic regeneration process to such a device. In this case the operation and the design of the active magnetic regenerator with thermal diodes are different from the classic AMR. Figure 6.35 shows a basic design concept for such a device.

A long and thin magnetocaloric plate is stacked between two sets of thermal diodes on each side. The thermal diodes denoted by number 1 face a hot heat exchanger, while thermal diodes denoted by number 2 face a cold heat exchanger. From Fig. 6.35 this looks similar to the basic single-stage device, only with a longer MC material.

However, from Fig. 6.36 it is clear that only one heat-transfer fluid is used in the AMR device with thermal diodes, suggesting the counter-current flow as in the classic AMR. The difference between the classic AMR and the AMR with thermal diodes is in the asymmetrical counter-current flow of the AMR with thermal diodes. This means that in the case of the AMR with thermal diodes, the fluid flows from the heat source to the heat sink on the side of the thermal diodes 1, as well as it flows from the heat sink to the heat source on the side of thermal diodes 2. The classic AMR requires periodic flow, whereas in the case of AMR the fluid flow may be continuous, without the need for fluid-switching valves.

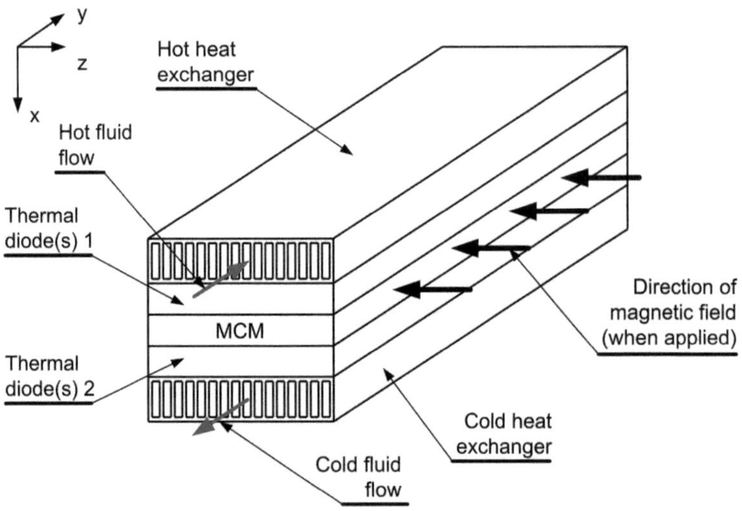

Fig. 6.35 Design concept of an active magnetic regenerator with thermal diodes

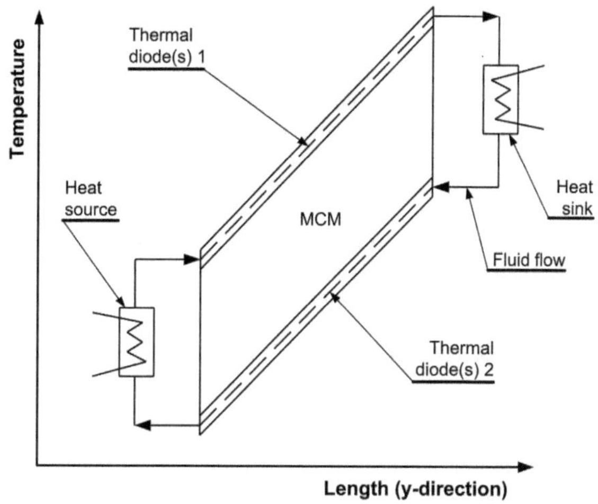

Fig. 6.36 Temperature distribution of the active magnetic regenerator with thermal diodes along the length (*y*-direction) of the device

The operation of an AMR with thermal diodes also consists of four processes, which are actually the same as in a single-stage device. However, in the case of the AMR with thermal diodes, the counter-current flow of the same working fluid also allows the process of regeneration along the length (Fig. 6.35, *y*-direction) of the magnetocaloric material.

References

1. Kirkpatrick JP (1973) Variable conductance heat pipes–from the laboratory to space. Paper presented at the 1st International heat pipe conference, Stuttgart, 15–17 Oct 1973
2. Chaudhrya HN, Hughes BR, Ghanib SA (2012) A review of heat pipe systems for heat recovery and renewable energy applications. Renew. Sust. Energ. Rev 16(4):2249–2259
3. Reay D, McGlen R, Kew P (2013) Heat pipes: theory, design and applications. Butterworth-Heinemann, Oxford
4. Kitanovski A, Egolf PW (2010) Innovative ideas for future research on magnetocaloric technologies. Int J Refrig 33(3):449–464
5. Tomc U, Tušek J, Kitanovski A et al (2013) A new magnetocaloric refrigeration principle with solid-state thermoelectric thermal diodes. Appl Therm Eng 58(1–2):1–10
6. Tomc U, Tušek J, Kitanovski A et al (2014) A numerical comparison of a parallel-plate AMR and a magnetocaloric device with embodied micro thermoelectric thermal diodes. Int J Refrig 37:185–193
7. http://micropelt.com
8. Rowe DM (ed) (2006) Thermoelectrics handbook: macro to nano. CRC Press, Boca Raton
9. Velidi G, Guven U (2011) Usage of nuclear power as a powerful source for space stations and for space development missions. Paper presented at the 62nd International astronautical congress, Cape Town, South Africa
10. Fleurial JP, Caillat T, Nesmith BJ et al (2011) Thermoelectrics: from space power systems to terrestrial waste heat recovery applications. In: NASA 2011 Thermoelectrics applications workshop

11. Riffat SB, Ma X (2003) Thermoelectrics: a review of present and potential applications. Appl Therm Eng 23(8):913–935
12. Böttner H, Nurnus J, Gavrikov A et al (2004) New thermoelectric components using microsystem technologies. J Microelectromech S 13(3):414–420
13. Tritt TM (2002) Thermoelectric materials: principles, structure, properties, and applications. Encyclopedia of Materials: Science and Technology, pp 1–11
14. Böttner H (2005) Micropelt miniaturized thermoelectric devices: small size, high cooling power densities, short response time. In: ICT 2005. 24th International conference on thermoelectrics, IEEE, pp 1–8
15. Böttner H, Nurnus J, Braun M et al (2004) Micopelt: state of the art. Road map and applications. Micropelt GmbH
16. Böttner H, Nurnus J, Schubert A et al (2007) New high density micro structured thermogenerators for stand alone sensor systems. Micropelt GmbH
17. Jeffrey Snyder G, Ursell TS (2003) Thermoelectric efficiency and compatibility. Phys Rev Lett 91:148301
18. Wee D (2011) Analysis of thermoelectric energy conversion efficiency with linear and nonlinear temperature dependence in material properties. Energy Convers Manage 52:3383–3390
19. Bean JA (2010) Thermionic refrigeration. Lectures on advanced semiconductor devices. Department of Electrical Engineering, University of Notre Dame, Notre Dame
20. Nolas GS, Sharp J, Goldsmid HJ (2001) Thermoelectrics: basic principles and new materials developments, 1st edn. Springer, Berlin
21. Hatsopoulos GN, Kaye J (1958) Measured thermal electron engine. J Appl Phys 29:1124–1125
22. Yeom J, Shannon MA (2008) Micro-coolers. Elsevier, New York
23. Lough BC, Lee SP, Lewis RA et al (2001) Numerical calculation of thermionic cooling efficiency in a double barrier semiconductor heterstructure. Physica E 11:287–291
24. Shakouri A, Bowers JE (1997) Heterostructure integrated thermionic coolers. Appl Phys Lett 71:1234–1236
25. Mahan GD, Woods LM (1998) Multilayer thermionic refrigeration. Phys Rev Lett 80:4016–4019
26. Shakouri A, Zhang Y (2005) On-chip solid-state cooling for integrated circuits using thin-film microrefrigerators. IEEE Trans Comp Packag Technol 28:65–69
27. Zeng G, Fan X, LaBounty C et al (2004) Cooling power density of SiGe/Si superlattice micro refrigerators. Mater Res Soc Symp Proc 793:S2.2.1–2.2.7
28. Zhang Y, Shakouri A, Zeng G (2004) High-power-density spot cooling using bulk thermoelectrics. Appl Phys Lett 85:2977–2979
29. Hishinuma Y, Geballe TH, Moyzhes BY et al (2002) Vacuum thermionic refrigeration with a semiconductor heterojunction structure. Appl Phys Lett 81(22):4242–4244
30. Zhang Y, Christofferson J, Shakouri A et al (2006) On-chip high speed localized cooling using superlattice microrefrigerators. IEEE Trans Comp Packag Technol 29:395–401
31. Xu YB, Thompson SM (ed) (2006) Spintronic materials and technology. Taylor & Francis, Boca Raton, p 423
32. Pearton SJ, Abernathy CR, Norton DP et al (2003) Advances in wide bandgap materials for semiconductors. Mater Sci Eng R 40:137–168
33. Wolf SA, Awschalom DD, Buhrman RA et al (2001) Spintronics: a spin-based electronics vision for the future. Science 294:1488–1495
34. Baibich MN, Broto JM, Fert A et al (1988) Giant magnetoresistance of (001)Fe/(001)Cr magnetic superlattices. Phys Rev Lett 61(21):2472–2475
35. Binasch G, Grünberg P, Saurenbach F et al (1989) Enhanced magnetoresistance in layered magnetic structures with antiferromagnetic interlayer exchange. Phys Rev B 39 (7):4828–4830
36. Medvedkin GA, Ishibashi T, Nishi T et al (2000) Jpn J Appl Phys 39:L949
37. Medvedkin GA, Hirose K, Ishibashi T et al (2002) J Cryst Growth 236:609

38. Choi S, Cha GB, Hong SC et al (2002) Solid-State Commun 122:165
39. Ueda K, Tahata H, Kawai T (2001) Appl Phys Lett 79:988
40. Matsumoto Y, Murakami M, Shono T et al (2001) Science 291:854
41. Holtzberg F, von Molnar S, Coey JMD (1980) Handbook on semiconductors, Moss T (ed). North-Holland, Amsterdam
42. Dietl T, Ohno H, Matsukura F et al (2000) Science 287:1019
43. Bauer GEW, Saitoh E, van Wees BJ (2012) Spin caloritronics. Nat Mater 11:391–399
44. Bauer GEW, MacDonald AH, Maekawa S (2010) Spin caloritronics. Solid State Commun 150:459–460
45. Slachter A (2011) Thermal spin transport and spin transfer torque in ferromagnetic/non-magnetic nanoscale devices. Zernike Institute PhD thesis series 2011–23, Physics of Nanodevices of the Zernike Institute for Advanced Materials at the University of Groningen, the Netherlands, p 152
46. Žutić I, Fabian J, Das Sarma S (2004) Spintronics: fundamentals and applications. Rev Mod Phys 76:323–410
47. Uchida K, Takahashi S, Harii K et al (2008) Observation of the spin Seebeck effect. Nature 455:778–781
48. Uchida K, Xiao J, Adachi H et al (2010) Spin seebeck insulator. Nat Mater 9:894–897
49. Adachi H, Uchida K, Saitoh E et al (2013) Theory of the spin Seebeck effect. Rep Prog Phys 76:036501
50. Jaworski CM, Myers RC, Johnston-Halperin E et al (2012) Giant spin Seebeck effect in a non-magneticmaterial. Nature 210:487
51. Flipse J, Bakker FL, Slachter A et al (2012) Cooling and heating with electron spins: observation of the spin Peltier effect. Nat Nanotechnol 7:166–168
52. Flipse J, Bakker FL, Slachter A et al (2012) Direct observation of the spin-dependent Peltier effect. Nat Nanotechnol 7:166–168
53. Gravier L, Serrano-Guisan S, Reuse F et al (2006) Spin-dependent Peltier effect of perpendicular currents in multilayered nanowires. Phys Rev B 73:052410
54. Gravier L, Serrano-Guisan S, Ansermet JP (2005) Spin-dependent Peltier effect in Co/Cu multilayer nanowires. J Appl Phys 97:10C501
55. Ota T, Uchida K, Kitamura Y et al (2010) Electric detection of the spin-Seebeck effect in Ni and Fe thin films at room temperature. J Phys: Conf Ser 200:062020
56. Uchida K, Ota T, Harii K et al (2010) Spin-seebeck effects in $Ni_{81}Fe_{19}$/Pt films. Solid State Commun 150:524–528
57. Kheirikhah MM, Rabiee S (2011) A review of shape memory alloy actuators in robotics. Lect Notes Comput Sci 6556:206–217
58. Nespoli A, Besseghini S, Pittaccio S et al (2010) The high potential of shape memory alloys in developing miniature mechanical devices: a review on shape memory alloy mini-actuators. Sens Actuators, A 158(1):149–160
59. Jani JM, Leary M, Subic A et al (2014) A review of shape memory alloy research, applications and opportunities. Mater Des 56:1078–1113
60. Huang WM, Ding Z, Wang CC et al (2010) Shape memory materials. Material Today 13 (7–8):54–61
61. Liu C, Qinb H, Mather PT (2007) Review of progress in shape-memory polymers. J Mater Chem 17:1543–1558
62. Liu Y, Du H, Liu L et al (2014) Shape memory polymers and their composites in aerospace applications: a review. Smart Mater Struct 23:023001
63. Fletcher LS (1972) A review of thermal control materials for metallic junctions. J Spacecraft Rockets 9(12):849–850
64. Fletcher LS (1988) Recent developments in contact conductance heal transfer. ASME J Heat Transf 110(4B):1059–1070
65. Miki BB, Rohsenow WM (1966) Thermal contact resistance, MIT Report No. DSR 74542-41

66. Fletcher LS (1991) Conduction in solids—Imperfect metal-to-metal contacts: thermal contact resistance, Section 502.5, heat transfer and fluid mechanics data books. Genium Publishing Company, Schenectady, New York
67. Starr C (1936) The copper oxide rectifier. Physics 7(1) (J Appl Phys):15–19
68. Vaidyanathan R (2007) Development of cryogenic shape memory actuator materials for switches, seals and valves. Report NASA/CR—2009-215441
69. Wang J, Zheng Z (2010) Heat conduction and reversed thermal diode: the interface effect. Phys Rev E Stat Nonlin Soft Matter Phys 81(1)
70. Roberts NA, Walker DG (2011) A review of thermal rectification observations and models in solid materials. Int J Therm Sci 50:648–662
71. Chang CW, Okawa D, Majumdar A et al (2006) Solid-state thermal rectifier. Science 314
72. Yang N, Zhang G, Li B (2008) Carbon nanocone: a promising thermal rectifier. Appl Phys Lett 93
73. Yang N Carbon nano-cone: an excellent thermal rectifier, presentation. National University of Singapore, Department of Physics and Centre for Computational Science and Engineering
74. Roberts NA, Walker DG (2011) A review of thermal rectification observations and models in solid materials. Int J Therm Sci 50(5):648–662
75. Walker G (2006) Thermal rectification mechanisms including noncontinuum effects, 18th National & 7th ISHMT-ASME Heat and Mass Transfer Conference, IIT Guwahati, India, 4–6 Jan 2006, p 8
76. Kobayashi W, Teraoka Y, Terasaki Y (2009) An oxide thermal rectifier. Appl Phys Lett 95:171905
77. Kurabayashi K (2001) Anisotropic thermal properties of solid polymers. Int J Thermophys 22(1):277–288
78. Shen S, Henry A, Tong J et al (2010) Polyethylene nanofibres with very high thermal conductivities. Nat Nanotechnol 5:251–255
79. Li B, Wang L, Casati G (2004) Thermal diode: rectification of heat flux. Phys Rev Lett 93 (18):184301
80. Chang CW, Okawa D, Majumdar A (2006) Solid-state thermal rectifier. Science 314:1121
81. Wang J, Zheng Z (2010) Heat conduction and reversed thermal diode: The interface effect. Phys Rev E 81:011114
82. Schmotz M, Maier J, Scheer E (2011) A thermal diode using phonon rectification. New J Phys 13:113027
83. Rogers GFC (1961) Heat transfer at the interface of dissimilar metals. Int J Heat Mass Transf 2:150–154
84. Roberts N, Walker D (2008) Monte Carlo study of thermal transport of direction and frequency dependent boundaries in high Kn systems. In: Proceedings of ITherm 2008, Walt Disney World Resort Orlando, FL USA
85. Wu G, Li B (2008) Thermal rectifiers from deformed carbon nanohorns. J Phys: Condens Matter 20:175211
86. Hu J, Ruan X, Chen Y (2009) Thermal conductivity and thermal rectification in graphene nanoribbons: a molecular dynamics study. Nano Lett
87. Terraneo M, Peyrard M, Casati G (2002) Controlling the energy flow in non-linear lattices: a model for a thermal rectifier. Phys Rev Lett 88(9):4302.1–4302.4
88. Lippmann MG (1875) Relation entre les ph´enom`enes ´electriques et capillaires. Ann Chim Phys 5:494
89. Berge B (1993) Electrocapillarit´e et mouillage de films isolants par l'eau. CR Acad Sci III 317:157
90. Li D (2008) Encyclopedia of microfluidics and nanofluidics, Dongqing Li (ed) Springer Science Business Media, p 600
91. Berthier J (2008) Microdrops and digital microfluidics. William Andrew Inc., p 434
92. Paik PY, Chakrabarty K, Pamula VK (2007) Adaptive cooling of integrated circuits using digital microfluidics, Artech House Inc., p 203

93. Yeo L, Chang H-C (2006) Electrowetting films on parallel line electrodes. Phys Rev E 73:011605
94. Pollack MG, Shenderov AD, Fair RB (2002) Electrowetting-based actuation of liquid droplets for microfluidic applications. Lab Chip 2:96–101
95. Zhao C, Yang C (2013) Electrokinetics of non-Newtonian fluids: a review. Adv Colloid Interface Sci 201–202:94–108
96. Squires TM, Quake SR (2005) Microfluidics: fluid physics at the nanoliter scale. Rev Mod Phys 77(3):977–1026
97. Nguyen NT, Wereley ST (2006) Fundamentals and applications of microfluidics, 2nd edn. Artech House Inc., Norwood
98. Bruus H (2008) Theoretical Microfluidics. Department of Micro and Nanotechnology Technical University of Denmark, Oxford, New York
99. Ramos A (ed) (2011) Electrokinetics and electrohydrodynamics in microsystems, Springer, New York, p 297
100. Tabeling P (2005) Introduction to microfluidics, translated by Suelin Chen. Massachusetts Institute of Technology, Cambridge, Oxford
101. Rosensweig RE (1985) Ferrohydrodynamics. Dover Publications, New York, p 344
102. Philip J, Shima PD (2012) Thermal properties of nanofluids. Adv Colloid Interface Sci 183–184:30–45
103. Sundar LS, Singh MK, Sousa ACM (2013) Investigation of thermal conductivity and viscosity of Fe_3O_4 nanofluid for heat transfer applications. Int Commun Heat Mass 44:7–14
104. Esmaeilnejad A, Aminfar H, Neistanak MS (2014) Numerical investigation of forced convection heat transfer through microchannels with non-Newtonian nanofluids. Int J Therm Sci 75:76–86
105. Azizian R, Doroodchi E, McKrell T et al (2014) Effect of magnetic field on laminar convective heat transfer of magnetite nanofluids. Int J Heat Mass Transf 68:94–109
106. Nkurikiyimfura I, Wang Y, Pan Z (2013) Heat transfer enhancement by magnetic nanofluids—A review. Renew Sust Energy Rev 21:548–561
107. Motozawa M, Sekine T, Sawada T et al (2013) Variation of forced convective heat transfer in rectangular duct flow of a magnetic fluid under magnetic field. In: 13th International conference on electrorheological fluids and magnetorheological suspensions (ERMR2012). J Phys: Conf Ser 412:012025
108. Odenbach S (2002) Magnetoviscous effects in ferrofluids. Springer, Berlin, p 151
109. Odenbach S (ed) (2009) Colloidal magnetic fluids. Basics, development and application of ferrofluids. Springer, Berlin, p 430
110. Nguyen NT (2012) Micro magnetofluidics-interactions between magnetism and fluid flow on the microscale. Microfluid Nanofluid 12(1–4):1–16
111. Krichler M, Odenbach S (2013) Thermal conductivity measurements on ferrofluids with special reference to measuring arrangement. J Magn Magn Mater 326:85–90
112. Fu HL, Gao L (2011) Theory for anisotropic thermal conductivity of magnetic nanofluids. Phys Lett A 375:3588–3592
113. Gavili A, Zabihi F, Isfahani TD et al (2012) The thermal conductivity of water base ferrofluids under magnetic field. Exp Therm Fluid Sci 41:94–98
114. Fang X, Xuan Y, Li Q (2009) Anisotropic thermal conductivity of magnetic fluids. Prog Natl Sci 19:205–211
115. Berim GO, Ruckenstein E (2011) Nanodrop of an ising magnetic fluid on a solid surface. Langmuir 27(14):8753–8760
116. Zhou Q, Ristenpart WD, Stroeve P (2011) Magnetically induced decrease in droplet contact angle on nanostructured surfaces. Langmuir 27:11747–11751
117. Lee CP, Yang ST, Wei ZH (2012) Field dependent shape variation of magnetic fluid droplets on magnetic dosts. Department of Power Mechanical Engineering, National Tsing Hua University, Hsinchu
118. Knieling H, Rehberg I, Richter R (2010) The growth of localized states on the surface of magnetic fluids. University of Bayreuth, Bayreuth

119. Nguyen NT, Zhu G, Chua YC et al (2010) Magnetowetting and sliding motion of a sessile ferrofluid droplet in the presence of a permanent magnet. Am Chem Soc Langmuir 26 (15):12553–12559
120. Choi SB, Han YM (2012) Magnetorheological fluid technology: applications in vehicle systems. CRC Press, Boca Raton, p 322
121. Kavlicoglu NC, Kavlicoglul BM, Liu Y et al (2007) Response time and performance of a high-torque magneto-rheological fluid limited slip differential clutch. Smart Mater Struct 16:149–159
122. Salloom MY (2013) Intelligent magneto-rheological fluid directional control valve. Int J Innov Manag Technol 4(4):406–409
123. Nguyen QH, Choi SB, Wereley NM (2008) Optimal design of magnetorheological valves via a finite element method considering control energy and a time constant. Smart Mater Struct 17:025024
124. Karakoc K, Park EJ, Suleman A (2008) Design considerations for an automotive magnetorheological brake. Mechatronics 18:434–447
125. Wereley NM (ed) (2013) Magnetorheology: advances and applications. Royal Society of Chemistry: RSC Smart Materials, London, p 396
126. Ahn S, Kasi RM, Kim SC et al (2008) Stimuli-responsive polymer gels. Soft Matter 4:1151–1157
127. Filipcsei G, Zrínyi M (2010) Magnetodeformation effects and the swelling of ferrogels in a uniform magnetic field. J Phys: Condens Matter 22:276001
128. Zubarev A-Y (2012) On the theory of the magnetic deformation of ferrogels. Soft Matter 8:3174–3179
129. Yamanaka S, Abe H, Naito M et al (2012) Colodial dispersibility of fatty acid-capped iron nanoparticles and its effects on static and dynamic magnetorheological response. College of Environmental Technology, Muroran Institute of Technology, Muroran
130. Weiss KD, Duclos TG, Carlson JD et al (1993) High strength magneto-and electro-rheological fluids. Soc Autom Eng:932451
131. Carlson JD, Weiss KD (1994) A growing attraction to magnetic fluids. Mach Des 66 (15):61–64
132. Fang F, Liu YD, Choi HJ (2012) Carbon nanotube coated magnetic carbonyl iron microspheres prepared by solvent casting method and their magneto-responsive characteristic. Collodials and Surfaces A. Physicochem Eng Aspects
133. Hao T (2002) Electrorheological suspensions. Adv Colloid Interface Sci 97:1–35
134. Winslow WM (1949) Induced fibration of suspensions. J Appl Phys 20:1137–1140
135. Halsey TC (1992) Electrorheological fluids. Science 258:761–766
136. Block H, Kelly JP (1988) Electro-rheology. J Phys D Appl Phys 21:1661
137. Unal HI (ed) (2013) 13th International conference on electrorheological fluids and magnetorheological suspensions (ERMR2012). J Phys: Conf Ser 412:011001
138. Nava R, Ponce MA, Rejon L et al (1997) Response time and viscosity of electrorheological fluids. Smart Mater Struct 6:67–75
139. Martin JE, Odinek J, Halsey TC et al (1998) Structure and dynamics of electrorheological fluids. Phys Rev E 57(1):756–775
140. Yildirim G, Genc S (2013) Experimental study on heat transfer of the magnetorheological fluids. Smart Mater Struct 22:085001
141. Wang YC, Sun XX, Tang XR et al (2010) Investigation of thermal conductivity of alumina/silicone oil electrorheological fluids. Adv Mater Res 129–131:421–425
142. Reinecke BN, Shan JW, Suabedissen KK et al (2008) On the anisotropic thermal conductivity of magnetorheological suspensions. J Appl Phys 104(2):023507
143. Kao J, Sung WP, Chen R (2012) Research on the electrical and thermal conductivity of magnetorheological fluid. Adv Mater Res 581–582:786–789
144. Bica I (2006) Thermal conductivity of a magnetoreheological suspension based on mineral oil and iron micro-particles. J Ind Eng Chem 12(4):620–625

145. Kitanovski A, Egolf PW, Poredoš A (2012) Rotary magnetic chillers with permanent magnets. Int J Refrig 35:1055–1066
146. Silva DJ, Bordalo BD, Pereira AM et al (2012) Solid state magnetic refrigerator. Appl Energy 93:570–574
147. Silva DJ, Ventura J, Araújo JP et al (2014) Maximizing the temperature span of a solid state active magnetic regenerative refrigerator. Appl Energy 113:1149–1154
148. Olsen UL, Bahl CRH, Engelbrecht K et al (2014) Modeling of in-vehicle magnetic refrigeration. Int J Refrig 37:194–200
149. Tasaki Y, Takahashi H, Murakami R et al (2013) Study on in-vehicle magnetic refrigerator (verification of heat transport characteristics for a new systems). Japan Society of Mechanical Engineers 2013 Annual meeting–Kaisaiyokoku, No. 13-1, paper no. S081014, p 5
150. Tasaki Y, Takahashi H, Yasuda Y et al (2012) Heat transport mechanism study of the new system–the study of automotive magnetic air conditioning cooling and heating, 2012 Japan Society of Refrigerating and Air Conditioning Engineers Annual Conference program, C142-p 4
151. Tasaki Y, Takahashi H, Yasuda Y et al (2012) A study on the fundamental heat transport potential of an in-vehicle magnetic refrigerators,in: Fifth IIF-IIR international conference on magnetic refrigeration at room temperature, Thermag V, Grenoble, France, 2012, pp 445–452
152. Egolf PW, Gravier L, Francfort T et al (2012) High-frequency magnetocaloric modules with heat gate operating with the Peltier effect. Fifth IIF-IIR International conference on magnetic refrigeration at room temperature, Thermag V, Grenoble, France 2012:61–72
153. Egolf PW, Vuarnoz D, Gravier L et al (2012) The thermodynamics of devices with solid layered-bed magnetocaloric refrigerants. Int J Refrig 35(6):1506–1517
154. Egolf PW, Gravier L, Francfort T et al (2014) High-frequency magnetocaloric modules with heat gates operating with the Peltier effect. Int J Refrig 37(1):176–184
155. Tomc U, Tušek J, Kitanovski A et al (2012) Thermoelectric-magnetocaloric energy conversion. Fifth IIF-IIR international conference on magnetic refrigeration at room temperature, Thermag V, Grenoble, France 2012:469–476
156. Kraščzek B (2014) Analysis of operating active magnetic regenerator when using liquid thermal switches (in Slovene). University of Ljubljana, Diploma Thesis, Faculty of Mechanical Engineering, p 87
157. Kitanovski A, Egolf PW (2009) Application of magnetic refrigerationand its assessment. J Magn Magn Mater 321:777–781
158. Epstein RI, Malloy KJ (2009) Electrocaloric devices based on thin-film heat switches. J Appl Phys 106
159. Hehlen MP, Mueller AH, Weisse-Bernstein NR et al (2013) Electrocaloric refrigerator using electrohydrodynamic flows in dielectric fluids. Laser Refrigeration of Solids VI, Proc of SPIE, vol 8638
160. Jia Y, Ju YS (2012) A solid-state refrigerator based on the electrocaloric effect. Appl Phys Lett 100

Chapter 7
Overview of Existing Magnetocaloric Prototype Devices

Since Brown [1] built the first magnetic refrigerator prototype working near room temperature in 1976 there has been a large number of different prototypes [2] designed and built over the past 40 years. However, despite there being 4 decades of scientific input into the field of magnetocaloric energy conversion near room temperature, the technology is only now at the brink of commercialization. With every new prototype that is built, we see slow but measureable improvements toward the desired goal. Figure 7.1 shows the number of prototypes built until 2014. It is clear that the number of prototypes increases with each new year, pointing to the fact that the magnetocaloric energy conversion research community is expanding research activities together with the S-curve of technology development.

The reasons why magnetocaloric technology near room temperature is only now slowly becoming market-ready lie in the fact that there are a number of different obstacles regarding the device's design, which need to be overcome before finalizing the idea of commercializing magnetocaloric technology. Addressing all the design issues is, of course, a difficult task. With each prototype built researchers try to solve different design issues, ranging from the heat-transfer problems of the AMRs and the working fluids, the design issues of the AMRs and magnet assemblies, to the problems related to peripheral elements, such as pump and valves systems. In this manner, each prototype built up until now has its own special feature. Each of them is trying to address one or more of the engineering barriers, but almost never a system as a whole.

However, there is one major characteristic that can be distinguished for all the existing prototypes. There are two groups of devices, i.e. the ones operating in a reciprocating (linear motion) manner and the others operating in a rotary manner. The reciprocating and rotary operations of the magnetocaloric device are more or less related to the way how the AMR is exposed to the alternating magnetic field, as is described in Chap. 4 on Active Magnetic Regeneration (see also Chap. 8). One possibility is to move the AMR or the magnet in a reciprocal direction back and forth, and the other is to rotate the AMR or the magnet. As you will see in the following sections, each of the two methods has its pros and cons, depending on the functionality of the device. Is it an experimental testing device? Or is it a real

A. Kitanovski et al., *Magnetocaloric Energy Conversion*,
Green Energy and Technology, DOI 10.1007/978-3-319-08741-2_7

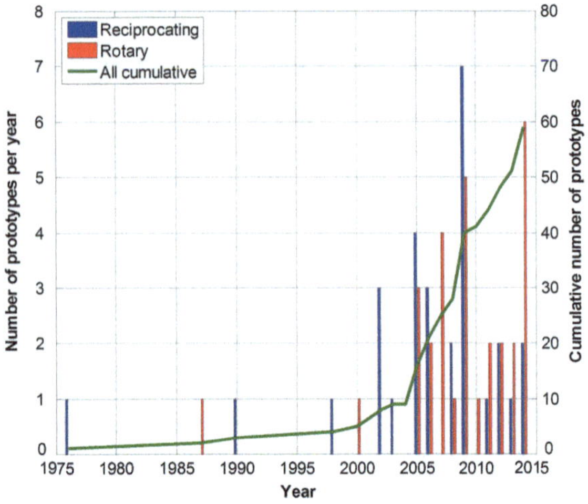

Fig. 7.1 Number of prototypes presented each year since 1976

magnetic refrigerator/heat pump? In any case, the bottom line is clear. The best solution for a real magnetocaloric application is yet to come.

The following sections are intended to describe all the existing prototypes built so far. They are divided into two sections on reciprocating and rotary designs. The main characteristics of each prototype are described regarding its design and operation. Usually, the authors present the operating characteristics in the form of no-load temperature spans and zero-temperature-span specific cooling powers. The no-load temperature span is the maximum temperature span a device can obtain. The notion "no-load" means a zero cooling power. On the other hand, zero-temperature-span cooling power is the maximum cooling power the device can produce, which is for the zero temperature span of the device. Each section also contains tables with pictures and the general characteristics of certain prototypes, along with information regarding the institutions and corresponding researchers. The information in the text was summarized after published papers, whereas the tables also include some information from personal correspondence with the authors. We therefore have to thank again the magnetocaloric research community for taking the time to help us with this book. This chapter is followed by Chap. 8 on Design Issues and Future Perspectives for Magnetocaloric Energy Conversion.

7.1 Reciprocating Prototypes

7.1.1 USA Prototypes

The first reciprocating magnetocaloric prototype operating near room temperature, which was also the first prototype ever built, was built by NASA and presented by Brown in 1976 [1]. Brown applied regeneration to increase the temperature span in the form of magnetic Stirling-like thermodynamic cycle. The regenerative magnetocaloric materials used were 1-mm-thick Gd plates. The plates were immersed in the heat-transfer fluid, which was a mixture of 80 % water and 20 % ethyl alcohol. The magnetic regenerative material was moved reciprocally from the hot end to the cold end of the heat-transfer fluid and vice versa. A magnetic field of 7 T was applied using a superconducting magnet. The highest no-load temperature span achieved was 47 K. Brown proposed a cascade system to further increase the temperature span of his magnetocaloric device.

The second reciprocating magnetocaloric device was designed at David Taylor Research Center by Green et al. in 1990 [3]. The two magnetocaloric materials used in the AMR were Gd ($T_c = 293$ K) and Tb ($T_c = 235$ K). Rolled ribbons of 1/3 of Tb, 1/3 of Gd–Tb alloy and 1/3 of Gd were layered along the length of the AMR (the direction in which the temperature gradient occurs due to the fluid flow). The AMR was 140 mm long, 40 mm in diameter and contained 500 g of magnetocaloric material. A superconducting magnet was used to induce 7 T of magnetic field. The heat-transfer fluid was nitrogen gas. The device could operate at a frequency of 0.0143 Hz and achieved a maximum no-load temperature span of 24 K. At 5 K of temperature span, the device produced 10 W kg^{-1} of specific cooling power (cooling power per mass of magnetocaloric material), whereas at 20.5 K the specific cooling power was 4 W kg^{-1}.

The third reciprocating prototype was built by Astronautics Technology Center and Ames Laboratory from Iowa State University (Table 7.1) and presented by Zimm et al. in 1998 [4]. The prototype consisted of two AMRs moving reciprocally in and out from the Nb–Ti superconducting magnet. The induced magnetic field was 5 T. Each AMR contained 1.5 kg of Gd spherical particles with a diameter varying between 0.15 and 0.3 mm. The heat-transfer fluid was water. Such a device could produce a specific cooling power of 200 W kg^{-1} for a temperature span of 9 K and at a magnetic field change of 5 T. At 1.5 T and 7 K, the specific cooling power was 70 W kg^{-1}.

In 2002 researchers from Los Alamos National Laboratory presented a reciprocating prototype in an article by Blumenfeld et al. [5]. The prototype consisted of a high-temperature superconducting magnet and a spherical Gd packed-bed AMR. In this way the magnet and the AMR were static, while the heat-transfer fluid was oscillatory, flowing through the AMR bed with the help of a piston. The heat-transfer fluid used was water. The magnet's magnetic field strength was 1.7 T. The cylindrical AMR was 160 mm long, with a diameter of 25 mm and the packed-bed spheres had a diameter of 0.2 mm. The total mass of the AMR was 64.6 g. The operating characteristics were measured at two operating frequencies, i.e., 0.017 and 0.034 Hz. At 0.017 Hz the no-load temperature span achieved was 23 K. When a specific cooling load of 12.4 W kg^{-1} was applied, the temperature span in the steady state was 18 K. At 0.034 Hz, a no-load temperature span was not reported. However, when applying 46.4 W kg^{-1} of specific cooling load, a temperature span of 13.5 K was achieved.

In 2012 researchers from Oak Ridge National Laboratory also presented their first reciprocating magnetic prototype in Shassere et al. [6]. The testing device consisted of electromagnet assembly that could induce magnetic field of 0.75 T. Two AMRs were constructed from 1-mm-thick Gd sheets with spacings of 0.25 and 0.5 mm. This accounted for total AMR masses of 575 and 475 g, respectively.

Table 7.1 Prototype built by Astronautics Technology Center and Ames Laboratory

Name and address of institute	Astronautics Corporation of America, Astronautics Technology Center, 718 Walsh Rd, Madison WI 53714
Name of contact person/email	Dr. Steven L. Russek/s.russek@astronautics.com
Year of production	1997
Type	Reciprocating
Maximum frequency	0.16 Hz
Maximum cooling power	600 W at $\Delta T = 9$ K
Maximum temperature span	23 K at 100 W
Type of AMR	Packed bed
MC material(s)	Gd spheres
Type of magnets	Superconducting, 5 T

The heat-transfer fluid was a mixture of water and a commercial ethylene glycol. The counter current flow was managed by reversible flow direction gear pump. The best performance showed the AMR with spacing of 0.5 mm. It could produce 8.8 W kg^{-1} of specific cooling power at the temperature span of 5.6 K and operating frequency of 0.17 Hz. However, only some initial testing results were showed, thus this was not the maximum performance. Furthermore, the AMR testing device had one interesting feature. It was designed in a manner that thermal imaging could be applied to measure the transient temperature profile along the AMR between the hot and the cold side during the operation of the device.

7.1.2 Canadian Prototypes

Rowe and Barclay [7], researchers from the University of Victoria, Canada, presented their first prototype in 2002. The general characteristics are presented in Table 7.2. The device consisted of two packed-bed AMRs moving alternately in and out of a Nb−Ti superconducting magnet, which induced a magnetic field of 2 T. The device operated at frequencies of 0.2–1.2 Hz. The heat-transfer fluid was helium. The magnetocaloric material used in the AMRs was gadolinium (Gd). Three different masses of magnetocaloric material were tested. These masses were 125, 250 and 375 g, respectively. The achieved maximum no-load temperature spans were 13, 18 and 20 K, respectively. The device was also designed for experiments in cryogenics, e.g., temperatures of about 20 K.

Table 7.2 Reciprocating prototype from the University of Victoria

Name and address of institute	Mechanical Engineering, University of Victoria
Name of contact person/email	Andrew Rowe/arowe@uvic.ca
Year of production	2001
Type	Reciprocating [Compressed gas (He, N_2, CO_2) 0–9 atm.]
Maximum frequency	1 Hz
Maximum cooling power	40 W
Maximum temperature span	85 K
Type of AMR	Spheres, plates, particulate
	(1–3 layers)
MC material(s)	Gd, Gd_xTb_{1-x}, Gd_xEr_{1-x}, Tb
Type of magnets	Superconducting, 0–5 T

In the 2004 Richard et al. [8] report on further experimental results. The experimental set up was generally the same as in the previous case; however, different compositions of AMRs were tested. First, two AMRs were packed with 180 g of 0.12 × 0.7 × 1 mm Gd flakes. The maximum no-load temperature span achieved was 16 K. Cooling loads were also applied in this experiment. The device produced 5.6 W kg^{-1} of specific cooling power at a temperature span of 14 K and at an operating frequency of 0.8 Hz. In the second experimental test, layered bed AMRs were evaluated. Each AMR was filled with 45 g of Gd and 40 g of $Gd_{0.74}Tb_{0.26}$ flakes (T_c = 278 K). In this manner, the operating characteristics of the device were enhanced. 20 K of no-load temperature span was achieved. The device produced 11.8 W kg^{-1} of specific cooling power for the 14 K temperature span at the operating frequency of 0.8 Hz.

In 2006, researchers from University of Victoria presented new experimental results on the described experimental magnetocaloric device [9]. The investigation was focused on the two AMRs layered with three different magnetocaloric materials (Gd, $Gd_{0.74}Tb_{0.26}$ and $Gd_{0.85}Er_{0.15}$, respectively). The mass of both AMRs together was 270 g. The experiment was carried out at two different magnetic fields (1.5 and 2 T). The maximum no-load temperature span at the lower magnetic field was 44 K at the operating frequency of 1 Hz. At the higher magnetic field, a 50 K temperature span was achieved at the operating frequency of 0.8 Hz. These results can definitely be considered as a milestone in magnetic refrigeration research, showing the potential of applying layered AMRs.

The most recent results of the Canadian reciprocating prototype were published in 2011 by Arnold et al. [10]. In this case, each of the AMRs was constructed from two magnetocaloric materials, i.e. 33 g of Gd and 35 g of $Gd_{0.85}Er_{0.15}$ crushed particles. The characteristics of the AMRs were tested in two magnetic fields of 2 and 5 T, respectively. The frequency of the operation was 0.85 Hz. A lower field maximum no-load temperature span of 33 K has been achieved. A maximum specific cooling power of 110.3 W kg^{-1} was obtained for a zero temperature span. In the high magnetic field, the maximum no-load temperature span was 59 K. The specific cooling power in this case for a zero temperature span was 274.1 W kg^{-1}. The device produced specific cooling powers of 95.6 and 257.4 W kg^{-1} at temperature spans of 39 and 7 K, respectively.

7.1.3 Japanese Prototypes

The first reciprocating prototype from Japan was presented in 2002 by the Tokyo Institute of Technology in collaboration with Chubu Electric Power Co. Inc. [11]. The two moving AMRs consisted of Gd spheres with diameter of 0.3 mm, and with the total mass of spheres being 2.2 kg. A magnetic field of 4 T was induced by a superconducting magnet. The heat-transfer fluid was water. At frequency of 0.167 Hz, the device produced a specific cooling power of 45.5 W kg^{-1} at a temperature span of 23 K. In a lower magnetic field of 1 T, a specific cooling power of 18.2 W kg^{-1} was obtained for a temperature span of 11 K.

In 2005 researchers from Hokkaido University presented their first prototype in the article by Kawanami et al. [12]. It was a Gd packed-bed AMR with a permanent-magnet assembly that produced a magnetic field of 0.9 T. The heat-transfer fluid was air. In 2007 Kawanami [13] presented an improved device. The Gd particles of 0.6 mm diameter were filled in a packed bed with a total packing mass of 62 g. A magnetic field of 1 T was provided by a permanent magnet. The highest possible operating frequency was approximately 0.35 Hz. The heat-transfer fluid used was distilled water. The maximum temperature span achieved was approximately 12 K.

Later, in 2008, a joint paper was published by researchers from Hokkaido University, Kobe University and Kushiro National College of Technology by Nakamura et al. [14]. Their device consisted of 33.4 g of Gd spheres packed in an AMR, using a permanent-magnet assembly with a magnetic field of 2 T. They applied water and air as the heat-transfer fluids. In the case of air as the heat-transfer fluid, the maximum no-load temperature span was 7 K, whereas with water, the temperature span was increased to 21 K.

A year later, in 2009, another prototype was built in a joint project involving the above-mentioned institutions by Hirano et al. [15]. They presented a reciprocating device where a static permanent-magnet assembly was magnetizing and demagnetizing a moving AMR bed with a magnetic field of 2.3 T. They tested AMRs with La–Fe–Si spheres having diameters of 1.6, 1 and 2 mm. In additional tests they also experimentally evaluated an AMR consisting of Gd spheres with a diameter of 0.6 mm. The total mass of each AMR was 12, 13, 15 and 33 g, respectively. The applied heat-transfer fluid flowing through all four AMRs was air. The highest no-load temperature span was achieved with the gadolinium-based AMR and was equal to 4.2 K. The highest no-load temperature span of the La–Fe–Si AMRs was achieved with the smallest diameter of spheres. This temperature span was 3.7 K. In the case of spheres with a diameter of 1.6 mm, a 3.1 K of temperature span was obtained, while in the case of the largest spheres of 2 mm, the temperature span was 2.7 K.

Table 7.3 presents the reciprocating magnetocaloric device designed by Toshiba Corporation in 2006.

7.1.4 Chinese Prototypes

In the past decade, China became one of the most active countries regarding the number of research papers as well as the number of prototypes for magnetic refrigeration devices. The first recorded Chinese reciprocating prototypes were presented in 2005 at the first International Conference on Magnetic Refrigeration at Room Temperature in Montreaux, Switzerland [16, 17].

Table 7.3 A reciprocating prototype designed by Toshiba Corporation

Name and address of institute	Corporate Research and Development Center, Toshiba Corporation 1, Komukai-Toshiba-cho, Saiwai, Kawasaki, 212-8582, Japan
Name of contact person/email	Akiko Takahashi Saito/aki.saito@toshiba.co.jp
Year of production	2006
Type	Linear
Maximum frequency	1 Hz
Maximum cooling power	N/A
Maximum temperature span	46 K
Type of AMR	Packed bed with spherical particles, (also tested with parallel Gd plates)
MC material(s)	Gd, Gd alloys, La–Fe–Si based
Type of magnets	Halbach-type permanent magnet (Nd–Fe–B)

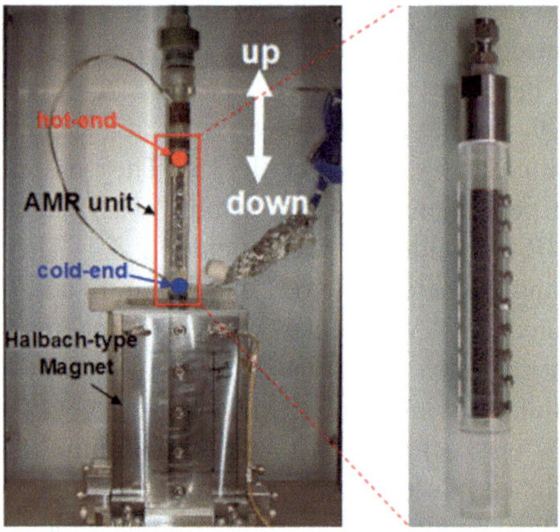

Researchers from Xi'an Jiaotong University presented experimental results on the operation of the reciprocating magnetic refrigeration prototype in Yu et al. [16] and Gao et al. [18]. The testing device consisted of a moving AMR and a water-cooled electromagnet, which induced a magnetic field of 2.18 T. The heat-transfer fluid was water. In this experiment three different AMRs were tested. The two of those were packed-bed AMRs, filled with gadolinium particles with average diameters of 0.3 and 0.56 mm. The masses of the AMRs were 930 and 1,109 g, respectively. The third AMR consisted of 1,213 g of Gd–Si–Ge particles with average diameters ranging between 0.3 and 0.75 mm. Using the AMR with Gd particles of 0.3 mm diameter, a specific cooling power of 20.2 W kg^{-1} at 3 K of temperature span was obtained. In the case of the AMR filled with Gd particles with a diameter of 0.56 mm, the specific cooling power was 16.1 W kg^{-1} at a temperature span of 3 K. The lowest performance was obtained with the Gd–Si–Ge-based AMR, where the specific cooling power was 8.5 W kg^{-1} for a temperature span of 3 K.

The researchers from Nanjing University presented experimental results from their first AMR testing device in 2005 in Lu et al. [17]. In the device, two AMRs were applied. These were moving in and out of the 1.4 T magnetic field provided by a Halbach permanent-magnet assembly. The heat-transfer fluid was water. The experiments were conducted using three different magnetocaloric materials: Gd, Gd–Si–Ge and Gd–Si–Ge–Ga, respectively. The AMRs consisted of spherical particles with a diameter of 0.2 mm (the masses were not reported). The performance of the two AMRs (Gd–Si–Ge and Gd–Si–Ge–Ga) was similar to the gadolinium-based AMR. In all three cases, a maximum no-load temperature span of 23 K was obtained. For a cooling power of 40 W the temperature span decreased to 5 K.

In 2006, the Technical Institute of Physics and Chemistry from Beijing presented an experimental device in the article of Yao et al. [19]. This experimental device had two AMR beds (each consisting of 583.7 g of gadolinium particles), and two permanent-magnet assemblies with 1.5 T of magnetic field. The movement of the AMRs was alternating, i.e. when one AMR was magnetized, the other AMR was demagnetized. Helium was used as the heat-transfer fluid. The maximum no-load temperature span achieved was approximately 42 K. The device produced 43.9 W kg^{-1} of specific cooling power at 18.2 K of temperature span, at a frequency of 1 Hz.

The Batou Research Institute of Rare Earths presented the results of a magnetocaloric testing device in the article by Huang et al. [20]. They designed a device with a 1.5-T Nd–Fe–B Halbach permanent-magnet assembly. This was moved back and forth over two, three-layered AMR beds at an operating frequency of 0.178 Hz. The particle size of the magnetocaloric material in layers, i.e. Gd, $LaFe_{10.82}Co_{0.78}Si_{1.1}B_{0.3}$ and $LaFe_{11.12}Co_{0.12}Si_{1.1}$, varied between 0.2 and 0.5 mm. The three materials were layered along each AMR with respect to their Curie temperatures. The total mass of the two beds together was 950 g. This device was capable of achieving a maximum temperature span of 18 K. When specific cooling powers of 13.3 and 21 W kg^{-1} were applied, the temperature spans decreased to 10 and 5 K, respectively. The first reciprocating magnetocaloric prototype from the Baotou Research Institute of Rare Earths is presented in Table 7.4.

The fifth Chinese reciprocating prototype was developed by the South China University of Technology in 2009 and was presented by Zheng et al. [21]. The design of the device was similar to that of Huang et al. [19]. In this case, a single 1.5-T Nd–Fe–B permanent-magnet assembly was moving between two AMRs packed with Gd grains. This device was designed to perform an AMR Ericsson-like thermodynamic cycle. Other operating characteristics and the performance of this device were not reported.

In 2013, the Batou Research Institute of Rare Earths presented their second reciprocating magnetic refrigeration testing device in the article of Cheng et al. [22]. This was definitely an improved version of the 2006 prototype [20]. The general design remained the same, with one 1.5-T Nd–Fe–B permanent-magnet assembly moving over two AMR beds. However, the operating frequency was increased to 0.9 Hz. They tested three AMRs with three different magnetocaloric materials [Gd, $LaFe_{11.0}Co_{0.9}Si_{1.1}B_{0.25}$ (T_c = 291 K) and $LaFe_{11.08}Co_{0.82}Si_{1.1}B_{0.25}$ (T_c = 279 K)]. The AMRs consisted of packed beds, with the particle diameters between 0.42 and

Table 7.4 First reciprocating device from the Batou Research Institute of Rare Earths

Name and address of institute	Baotou Research Institute of Rare Earths. Address: No. 36, Huanghe Street, RE High-Tech Zone, Baotou, Inner Mongolia, P.R. China, 014030
Name of contact person/email	Jiaohong Huang/jiaohongh@sina.cn
Year of production	2002–2004
Type	Reciprocating
Maximum frequency	0.16 Hz
Maximum cooling power	35 W
Maximum temperature span	17.8 K
Type of AMR	2 × 350 g packed bed
MC material(s)	Gd irregular particles, 0.5–1 mm
Type of magnets	Reciprocating permanent magnet Nd–Fe–B 1.5 T

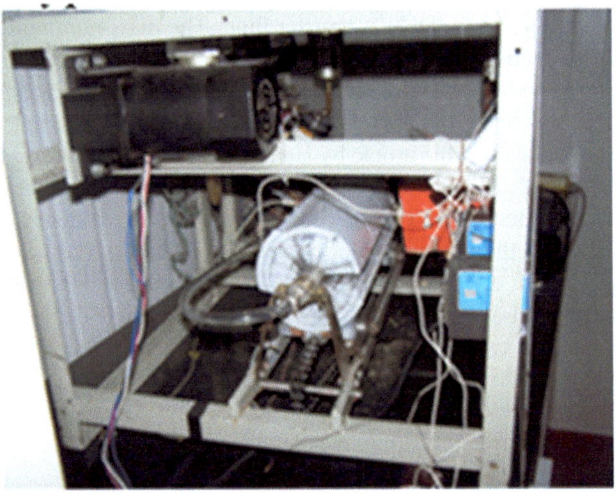

0.85 mm. The two AMRs were applying a single material, one with Gd and the other based on the La–Fe–Co–Si–B material (T_c = 291 K). The third was a two-layer AMR based on La–Fe–Co–Si–B material. The masses of the two La–Fe–Co–Si–B-based AMRs were 580 g, each. The mass of the Gd AMR was 785 g. In the case of the AMR consisting of Gd, the 1 % Na(OH) solution in water was used as a heat-transfer fluid, whereas in the AMR consisting of La–Fe–Co–Si–B the heat-transfer fluid was a mixture of Na_2MoO_4, Na_3PO_3, $NaCr_2O_7$ and Na_2SiO_3 fluids to prevent corrosion. All three AMRs were compared in relation to the maximum no-load temperature span achieved. The highest temperature span of 15.3 K was obtained by the two-layered La–Fe–Co–Si–B-based AMR, whereas in the AMR with the single-material La–Fe–Co–Si–B, the maximum temperature achieved was 12.7 K. The performance of the Gd-based AMR was poorer, with only 8.1 K of maximum temperature span.

7.1.5 French Prototypes

The first French magnetocaloric prototype was presented in 2003 by Grenoble Electrical Engineering Laboratory in the article of Clot et al. [23]. They designed a reciprocating device with a 0.8-T permanent-magnet Halbach assembly. The AMR was made of 1-mm Gd plates with a spacing of 0.15 mm. The total mass of the AMR was 223 g. The heat-transfer fluid was water. The operating frequency of the device was approximately 0.42 Hz. The maximum no-load temperature span obtained with this device was 7 K. The specific cooling power was 39.5 W kg^{-1} at 4 K of temperature span.

Based on experience from the 2003 device [23] researchers from the Grenoble Electrical Engineering Laboratory built a second experimental device (Table 7.5). This was presented in 2009 by Dupuis et al. [24]. The Nd–Fe–B Halbach magnet assembly provided a magnetic field of 0.8 T by moving over the static AMR. The static AMR consisted of 1-mm Gd plates. The device allowed operating frequencies up to 1 Hz. The maximum no-load temperature span achieved was approximately 7.8 K.

Another experimental study was conducted on the described testing device. In 2014, Legait et al. [25] presented the results of an extensive study on six different magnetocaloric materials. Three single-material AMRs were made of Gd, $Pr_{0.65}Sr_{0.35}MnO_3$ and La–Fe–Co–Si (T_c = 293 K), respectively. The fourth AMR was layered with four different La–Fe–Co–Si materials (Curie temperatures of 283, 288, 293 and 298 K). The total mass of the Gd-based AMR was 190 g, for the $Pr_{0.65}Sr_{0.35}MnO_3$-based AMR it was 91 g, and the mass of both AMRs based on La–Fe–Co–Si was 150 g. All the plates were 1 mm thick, with Gd having a spacing of 0.3 mm, and in the AMRs with the other materials, the spacing for the fluid flow was 1 mm. The highest no-load temperature span obtained by the Gd AMR was 11.5 K, and with the layered La–Fe–Co–Si AMR, a 10.5 K temperature span was obtained. In the AMRs with a single material, the temperature span was the lowest, i.e. 8 K in theLa–Fe–Co–Si-based AMR and 5 K in the AMR consisting of $Pr_{0.65}Sr_{0.35}MnO_3$ material.

In 2009 another French group, a company called CoolTech Applications, presented their first magnetic reciprocating device, which was presented in the article of Bour et al. [26]. The experimental device consisted of a Halbach permanent-magnet assembly providing a magnetic field of 1.1 T. The AMR was made of 0.6-mm Gd plates with a spacing of 0.1 mm. The heat-transfer fluid used in the device was a freezing depressant, because of its good corrosion protection. With such an experimental device, the highest no-load temperature span achieved was approximately 16.1 K.

Based on the experience from the first prototype, CoolTech Applications built their second reciprocating prototype in 2011. Operating characteristics are presented in Table 7.6.

Table 7.5 Reciprocating device from the Grenoble Electrical Engineering Laboratory

Name and address of institute	Grenoble Electrical Engineering Laboratory
Name of contact person/email	Afef Kedous-Lebouc/Afef.lebouc@g2elab.grenoble-inp.fr
Year of production	2008
Type	Linear
Maximum frequency	N/A
Maximum cooling power	Currently unload system
Maximum temperature span	10 K
Type of AMR	Parallel plates, powder, sphere, composite, multilayer, etc.
MC material(s)	The tested materials: Gd plates with variable thickness, La–Fe–Co–Si in single and four layers, $PrSrMnO_3$
Type of magnets	Sm–Co Halbach Cylinder: 0.8 T

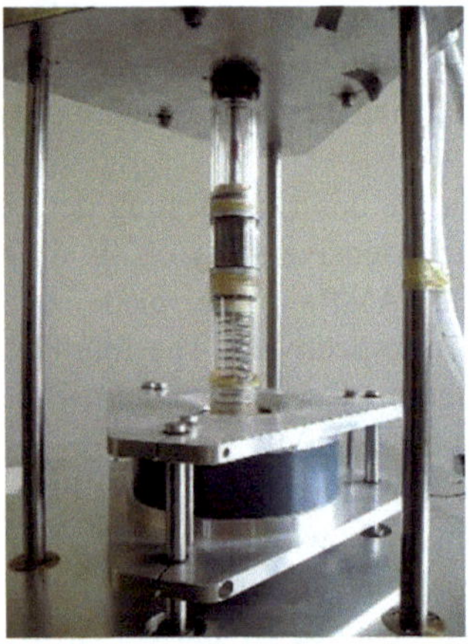

Photography of the main parts of the device

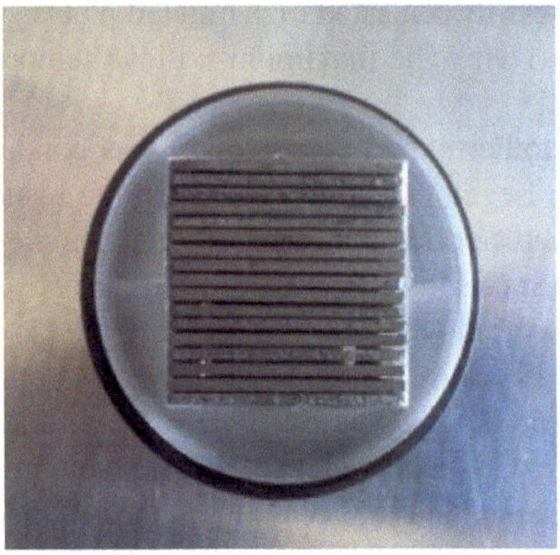

Regenerator and its holder

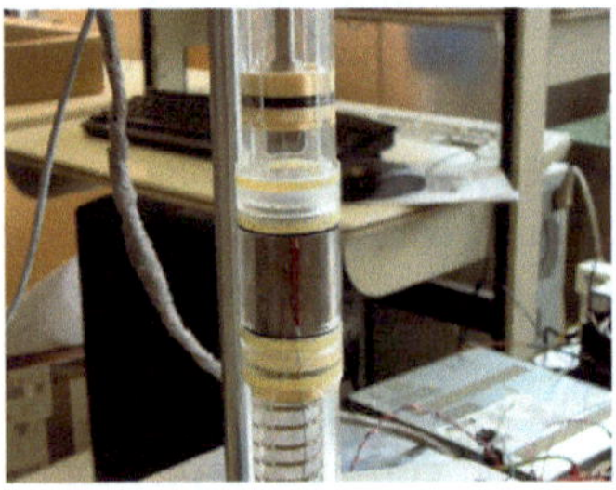

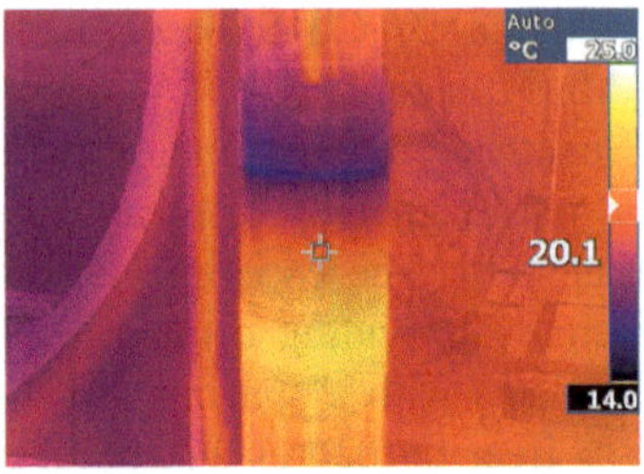

IR Camera monitoring of the temperature and the span creation

Table 7.6 Second reciprocating device from Cooltech Applications

Name and address of institute	Cooltech Applications-5 Impasse Antoine IMBS 67810 Holtzheim-France
Name of contact person/email	Muller Christian/c.muller@cooltechapplications.com
Year of production	2011
Type	Reciprocating 2 stages
Maximum frequency	1.5 Hz
Maximum cooling power	150 W
Maximum temperature span	38 K
Type of AMR	Parallel plates
MC material(s)	Gd & Gd–Tb
Type of magnets	Nd–Fe–B, B = 1.27 T
R200 prototype (www.cooltech-applications.com)	

7.1.6 Danish Prototypes

The first magnetocaloric testing device from the Technical University of Denmark (Table 7.7) was presented in 2008 by Bahl et al. [27]. The AMR was constructed of 0.9-mm-thick Gd plates with a spacing of 0.8 mm between two plates. The total mass of the AMR was 92 g. A magnetic field of up to 1.4 T was provided by an electromagnet. In the initial experimental studies, a number of different influential parameters were varied and tested, such as fluid displacement, cycle periods,

Table 7.7 Reciprocating device from the Technical University of Denmark

Name and address of institute	Department of Energy Conversion and Storage, Technical University of Denmark, Frederiksborgvej 399, DK-4000 Roskilde, Denmark
Name of contact person/email	Christian Bahl/chrb@dtu.dk
Year of production	2007
Type	Reciprocating
Maximum frequency	0.25 Hz
Maximum cooling power	1.5 W
Maximum temperature span	10.2 K (with Gd), 5.8 K (with $La_{0.67}Ca_{0.26}Sr_{0.07}Mn_{1.05}O_3$),
	8.5 K (with 2 $La(Fe,Co,Si)_{13}$)
Type of AMR	Packed bed, parallel plates
MC material(s)	Gd, $La_{0.67}Ca_{0.26}Sr_{0.07}Mn_{1.05}O_3$, $La(Fe,Co,Si)_{13}$
Type of magnets	Single Halbach cylinder
	LakeShore 7407 electromagnet

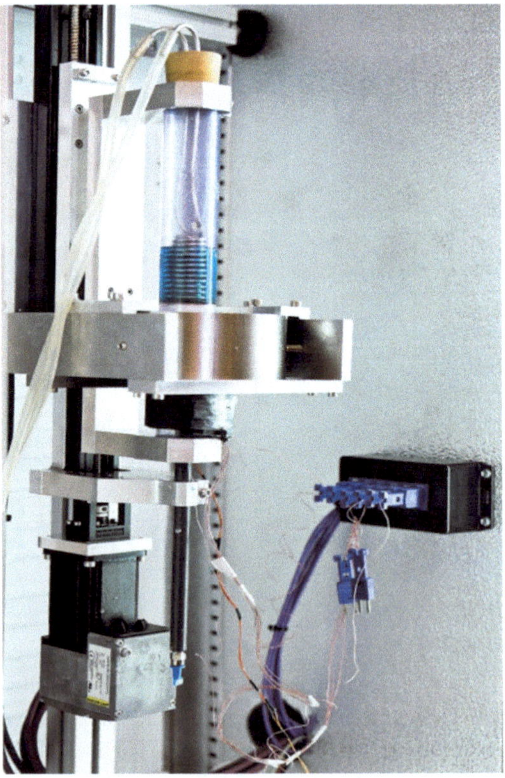

magnetic fields and heat-transfer fluids. These are all fundamental and important parameters, which may give a useful and systematic study for the further research and design of magnetic refrigeration prototypes.

In 2009 Engelbrecht et al. [28] presented the performance of two AMRs: a flat-plate Gd-based AMR and a two-layered AMR consisting of $La(Fe,Co,Si)_{13}$ plates. The device on which the experiments were carried out was the same as in Bahl et al. [27], with the exception that in the last case, instead of the electromagnet a permanent-magnet Halbach assembly was applied. The magnet assembly provided a magnetic field of 1.03 T. The heat-transfer fluid was a mixture of 75 % water and 25 % of automotive anti-freeze. This anti-freeze was used for two purposes: it acted as a corrosion inhibitor and as a fluid with a freezing point below 0 °C. The Gd and $La(Fe,Co,Si)_{13}$ plates were 0.9 mm thick with a spacing of 0.5 mm. The two layers of the $La(Fe,Co,Si)_{13}$ material had Curie temperatures of approximately 276 and 289 K. The maximum no-load temperature span of the Gd-based AMR was about 9 K, whereas in the two-layered AMR a temperature span of approximately 6.5 K was obtained. The reason for the smaller temperature span was the large "temperature gap" between the two Curie temperatures.

Two years later (in 2011) Engelbrecht et al. [29] presented the results of another study on the experimental device described above [28]. Here, five different magnetocaloric materials were tested. These materials were Gd, three different $La(Fe,Co,Si)_{13}$ materials and a ceramic magnetocaloric material, $La_{0.67}Ca_{0.26}Sr_{0.07}Mn_{1.05}O_3$ (LCSM). The Curie temperatures of the $La(Fe,Co,Si)_{13}$ materials were 276, 286 and 289 K, respectively. The Curie temperature of the LCSM was 296 K. The magnetocaloric plates were 0.9 mm thick, with the exception of the LCSM plates, which had a thickness of 0.3 mm. In the experiment, five different AMR configurations were tested: three AMRs with a single material (Gd, LCSM or $La(Fe,Co,SI)_{13}$ with T_c = 289 K), two AMRs with two-layers of $La(Fe,Co,Si)_{13}$ material, first with $T_c = 276$ K, $T_C = 289$ K and second with $T_c = 286$ K, $T_C = 289$ K. The highest no-load temperature span was achieved with the Gd-based AMR (10.2 K). The experiments on other two AMRs with a single material led to a no-load temperature span of 7.9 K for ($La(Fe,Co,Si)_{13}$, $T_c = 289$ K), and 5.9 K of temperature span for LCSM. The two-layered AMR which had magnetocaloric materials with Curie temperatures closer to each other (286, 289 K) performed 8.5 K of a temperature span, whereas the temperature span in another two-layered AMR was 6.5 K.

7.1.7 Slovenian Prototypes

The University of Ljubljana presented its first reciprocating AMR testing device (Table 7.8) in 2012 in the papers by Tušek et al. [30–32].

The device had a static AMR bed and a moving Nd–Fe–B magnet assembly, with a magnetic field of 1.15 T. The maximum operating frequency was 1 Hz. The heat-transfer fluid was a 70/30 % mixture of distilled water and anti-freeze

Table 7.8 Reciprocating device from the University of Ljubljana

Name and address of institute	Laboratory for Refrigeration and District Energy, Faculty of Mechanical Engineering, University of Ljubljana, Askerceva 6, 1000 Ljubljana, Slovenia
Name of contact person/email	Andrej Kitanovski/andrej.kitanovski@fs.uni-lj.si
Year of production	2011
Type	Reciprocating
Maximum frequency	~0.9 Hz
Maximum cooling power	7 W (with Gd), 6.2 W (with La–Fe–Co–Si)
Maximum temperature span	~23.5 K (with Gd), ~20 K (with La–Fe–Co–Si)
Type of AMR	Parallel plates, packed bed (spheres, particles, powder)
MC material(s)	Gd, La–Fe–Co–Si
Type of magnets	Reciprocating Nd–Fe–B permanent magnet, 1.15 T

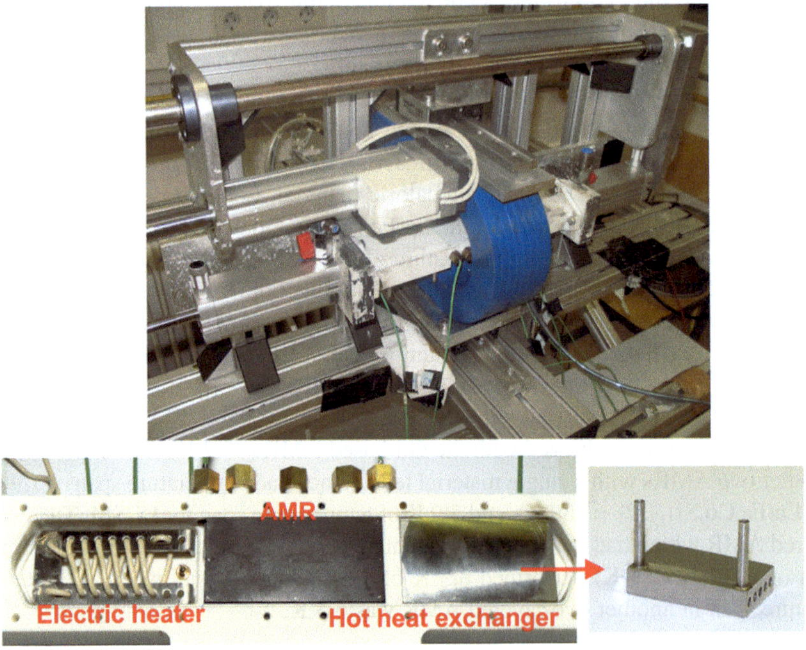

(ethylene glycol). In the first experimental studies, six different AMR geometries were tested, consisting of gadolinium. These were two AMRs with Gd plates having a thickness of 0.25 mm. In the first AMR, the spacing between plates was 0.25 mm and the total mass of Gd was 130.9 g, whereas in the second one, the spacing was 0.1 mm, with the total mass of Gd being 176.3 g. Both AMRs had plates positioned parallel to the magnetic field. Another AMR was built from gadolinium, with the plate thickness of 0.25 mm and a spacing of 0.25 mm, and the total mass of gadolinium was 142.7 g. However, this AMR had plates positioned with the normal to the main surface being perpendicular to the magnetic field.

Further experiments were performed on three AMRs that consisted of packed-bed gadolinium particles. One AMR comprised small Gd cylinders with a diameter of 2.5 mm, a length of 4 mm and a total mass of 120.5 g. The other AMR was filled with 135 g of spheres, having a diameter in the range 0.35–0.5 mm. In the third AMR, 93 g of Gd powder was used. The size of the powder particles was also in the range 0.35–0.5 mm.

Initially, no-load temperature-span tests were conducted on parallel plate AMRs. The best no-load temperature span of 19.8 K was achieved using an AMR with parallel plates that had a smaller spacing. In the AMR with parallel plates that have a wider spacing, the no-load temperature span was 16 K. The no-load temperature span for the AMR with plates positioned with the normal to the main surface being perpendicular to the magnetic field was 14 K due to the higher demagnetization effect.

The best performance of the packed-bed AMRs was obtained with spheres (15.5 K). In the powder-based AMR 7 K of no-load temperature span was obtained. The worst performance was demonstrated by the AMR consisting of cylinders (temperature span of only 4 K). The poor performance of this AMR was mainly due to a too large diameter and length of magnetocaloric cylinders, resulting in an inefficient heat transfer.

Tests on the cooling power were only performed for the best of the evaluated AMRs (parallel plates with a smaller spacing). In this case, the zero-temperature span-specific cooling power was 39.7 W kg^{-1}. At 13.5 K of temperature span the device produced 11.3 W kg^{-1} of specific cooling power.

In the next experimental study from 2014, other magnetocaloric materials were tested in the same experimental device as in Tušek et al. [30–32] and compared to the best performing Gd parallel plate AMR [32]. Three different AMRs were built, with two, four and seven layers of the La–Fe–Co–Si material [33]. The Curie temperatures of two-layered AMR were 291.2 and 296.8 K, respectively. The Curie temperatures of the four-layered AMR were 291.2, 296.8, 303 and 308 K, respectively. The seven-layered AMR had additional materials to the AMR with four layers. Two materials were added at the cold end, having their Curie temperatures at 280.8 and 283.8 K, and additional material was added to the hot end with a Curie temperature of 312 K. The plate thickness of all seven magnetocaloric materials was 0.5 mm, with a spacing of 0.2 mm. The total mass of each AMR was 144 g, regardless of the number of layers. The highest no-load temperature span was similar for the seven-layered, as well as for the four-layered, AMR, and it was approximately 20 K. The two-layered AMR's performance was slightly lower, with a no-load temperature span of about 16 K. The specific cooling power of the seven-layered and four-layered AMR was approximately 15 W kg^{-1} at a temperature span of 8 K. The two-layered AMR preformed approximately 11 W kg^{-1} of specific cooling power at 5 K. Furthermore, the influence of the hot-side (heat sink) temperature was also investigated. In this manner, the highest no-load temperature span of the device was achieved by Gd parallel plate AMR. It was approximately 23.5 K at the heat sink temperature of about 300 K.

7.1.8 Italian Prototypes

Researchers from the University of Genova, Italy, presented their reciprocating magnetocaloric device in 2009 in the article of Tagliafico et al. [34]. They designed a system with two AMRs which moved alternately in and out of the 1.5 T magnetic field provided by a Nd–Fe–B magnet assembly. The operating frequency was rather low, i.e. below 0.2 Hz. The two AMRs consisted of Gd powder with an average diameter of the powder particles equal to 0.3 mm. The heat-transfer fluid was a mixture of water and anti-freeze. Unfortunately, no information regarding the performance of this device was given. In 2013, a new prototype was built based on the experience gained from the first one. This was presented in the article by Tagliafico et al. [35]. The magnet assembly in this device was a Nd–Fe–B magnet assembly with a 1.55 T magnetic field. Also in this case, two AMRs were designed to move alternately in and out of the magnet assembly. Both AMRs were filled with Gd plates. The thickness of plates was 0.8 mm and the total mass of both of the AMRs was 388 g. The heat-transfer fluid was a mixture of 50 % water and 50 % ethanol. The device performed a 5 K temperature span under no-load conditions.

7.1.9 Swiss Prototypes

The University of Applied Sciences of Western Switzerland designed and built an experimental magnetocaloric device (Table 7.9) that was presented in the article of Balli et al. [36] in 2012. A Nd–Fe–B Halbach magnet assembly was used with a magnetic field of 1.45 T. Two AMRs were alternately entering and exiting the magnetic field. These AMRs were filled with Gd plates having a thickness of 1 mm. The total mass of AMRs was 400 g. The heat-transfer fluid was a silicon oil. The highest no-load temperature span was 22 K. Researchers also reported that after certain improvements were made to the device, a no-load temperature span above 30 K had been reached [36, 37].

7.1.10 Korean Prototypes

In 2009 two researchers, Kim and Jeong [38], from Korea Advanced Institute of Science and Technology presented their reciprocating experimental magnetocaloric device. The AMR, which consisted of Gd particles with a size between 0.325 and 0.5 mm, was moving alternately through a magnetic field of 1.58 T, provided with a static Nd–Fe–B Halbach magnet assembly. The maximum frequency of operation was 1 Hz. The heat-transfer fluid was helium. The largest no-load temperature span obtained was 16 K.

Table 7.9 Reciprocating device from University of Applied Sciences of Western Switzerland

Name and address of institute	HES-SO—Yverdon-les-Bains, Clean Cooling Systems SA
Name of contact person/email	Osmann Sari/osmann.sari@heig-vd.ch
	Fouad Rahali/fouad.rahali@clean-cooling-systems.com
Year of production	2009, 2011, 2013
Type	Reciprocating and rotary
Maximum frequency	0.5 Hz
Maximum cooling power	70 W
Maximum temperature span	30 K
Type of AMR	Parallel plates
MC material(s)	Gd, La–Fe–Co–Si, others
Type of magnets	Permanent magnet

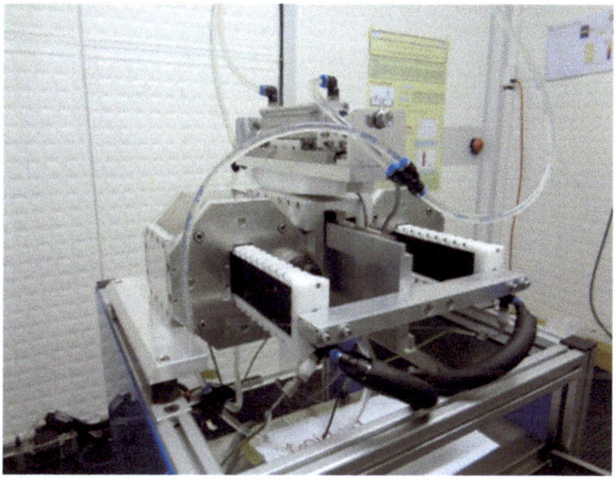

7.1.11 Brazilian Prototypes

First reciprocating magnetocaloric device in Brazil was built at the Federal University of Santa Catarina (Table 7.10) and presented in 2011 by Trevizoli et al. [39].

A Nd–Fe–B Halbach magnet assembly was used to provide the magnetic field of 1.65 T. The AMR consisted of Gd plates with a thickness of 0.85 mm, a spacing of 0.1 mm and a total mass of 195.4 g. The device operated at a maximum frequency of 0.14 Hz. The heat-transfer fluid was water. The no-load temperature span was 4.45 K and the zero-temperature-span specific cooling power was approximately 20 W kg^{-1}.

Table 7.10 Reciprocating device from the Federal University of Santa Catarina

Name and address of institute	POLO—Laboratory of Refrigeration and Thermophysics, Department of Mechanical Engineering, Federal University of Santa Catarina, 88.040-900 Florianópolis, SC, BRAZIL
Name of contact person/email	Prof. Jader R. Barbosa Jr/jrb@polo.ufsc.br
Year of production	2007
Type	Reciprocating
Maximum frequency	0.14 Hz
Maximum cooling power	3.9 W
Maximum temperature span	4.45 K
Type of AMR	Parallel plates
MC material(s)	Gd
Type of magnets	Halbach array—C shape

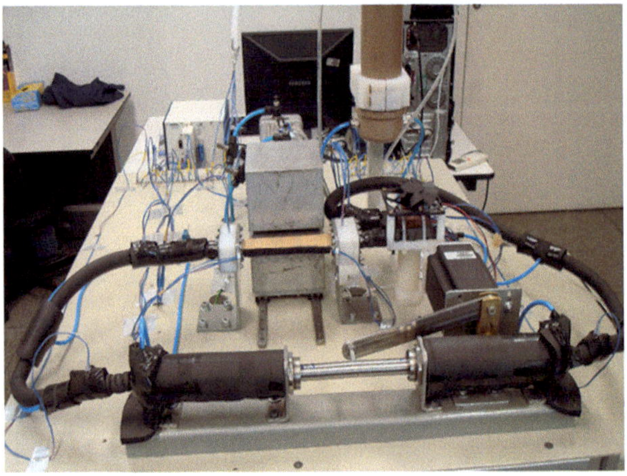

7.1.12 Polish Prototypes

Wroclaw University of Technology built their first prototype (Table 7.11) in 2014 and this was presented by Czernuszewicz et al. [40].

The Halbach permanent-magnet assembly was designed to provide a magnetic field of 1 T. The AMR was filled with 30 g of Gd particles, whose size varied from 2 to 5 mm. The heat-transfer fluid was based on ethylene glycol. Under no-load conditions, the device obtained 2 K of temperature span at an operating frequency of 0.025 Hz.

Table 7.12 presents the second reciprocating prototype from Wroclaw University of Technology. It is an interesting design, since it is a device that works on the principles of the magnetocaloric as well as the barocaloric effect. Some of the characteristics of this device are presented in the Table 7.12.

Table 7.11 Reciprocating device from Wroclaw University of Technology

Name and address of institute	Wroclaw University of Technology, Wybrzeze Wyspianskiego 27, 50-370 Wroclaw, Poland
Name of contact person/email	Daniel Lewandowski/daniel.lewandowski@pwr.edu.pl
Year of production	2010–2013
Type	Linear
Maximum frequency	1 Hz
Maximum cooling power	N/A
Maximum temperature span	6 K
Type of AMR	Packed bed spheres, particles
MC material(s)	Pure gadolinium
Type of magnets	Permanent (Halbach array)

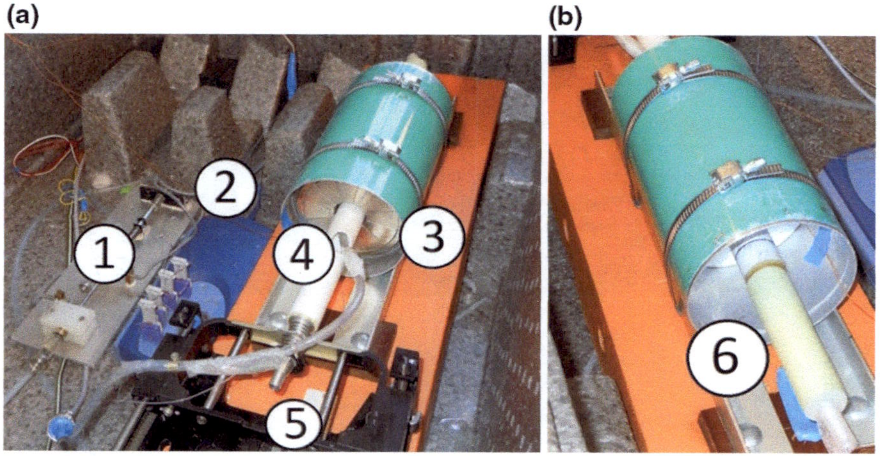

Picture of magnetocaloric cooling system: (a) front and (b) back. *1*—hot heat exchanger, *2*—temperature measurement system, *3*—magnets, Halbach array, *4*—main moving part with MC bed, *5*—movement system, *6*—cold heat exchanger

Table 7.12 Second reciprocating device from Wroclaw University of Technology, combination of magnetocaloric and barocaloric effect

Name and address of institute	Wroclaw University of Technology, Wybrzeze Wyspianskiego 27, 50-370 Wroclaw, Poland
Name of contact person/email	Daniel Lewandowski/daniel.lewandowski@pwr.edu.pl
Year of production	2013–2014
Type	Linear (combination of magnetocaloric and barocaloric)
Maximum frequency	1 Hz
Maximum cooling power	N/A
Maximum temperature span	N/A
Type of AMR	Parallel pipes, plates
MC material(s)	Ni−Mn−X (X = In, Sn, Ga)
Type of magnets	Permanent (Halbach array)

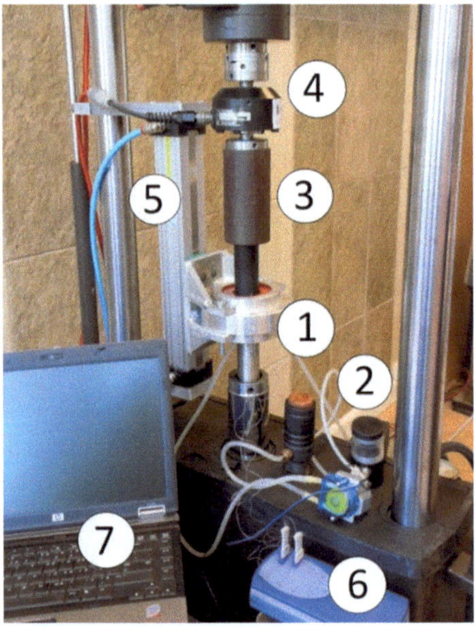

Magneto-barocaloric prototype of heat pump. *1*—Halbach array with MBC material bed inside, *2*—system of heat-transfer fluid flow, pump, HHEX, CHEX, *3*—system of pressure and stress induction, *4*—force measurement cell, *5*—pneumatic system of Halbach array movement, *6*—temperature-measurement system, *7*— controller

7.1.13 Spanish Prototypes

In Spain, the first reciprocating experimental magnetocaloric device was built at the University of Coruna, in 2013. Its performance was presented by Gomez et al. [41]. The magnetic field of 1 T was provided a Nd–Fe–B permanent-magnet assembly. The device had two AMRs filled with Gd plates. The thickness of each plate was

Table 7.13 Reciprocating (linear) magnetocaloric devices built to date, listed with their general characteristics

	Authors, reference	Year	AMR type, materials, mass	Magnet assembly type and magnetic field strength	Heat transfer fluid	Frequency	Temperature span	Specific cooling power
1	Brown [1]	1976	Gd plates, 1 mm, Stirling-like cycle	Superconducting, 7 T	80 % water, 20 % ethyl alcohol	N/A	47 K	No-load
2	Green et al. [3]	1990	1/3 Tb, 1/3 Gd/Tb, 1/3 Gd rolled ribbons, 500 g	Superconducting, 7 T	Nitrogen gas	0.0143 Hz	24 K	No-load
							20.5 K	4 W kg^{-1}
							5 K	10 W kg^{-1}
3	Zimm et al. [4]	1998	Gd spherical particles, 0.15–0.3 mm, 3,000 g	Superconducting, 1.5 and 5 T	Water	N/A	9 K (5 T)	200 W kg^{-1}
							7 K (1.5 T)	70 W kg^{-1}
4	Blumenfeld et al. [5]	2002	Gd spheres 0.2 mm, 64.6 g	Superconducting, 1.7 T	Water	0.017 Hz	23 K	No-load
							18 K	12.4 W kg^{-1}
						0.034 Hz	13.5 K	46.4 W kg^{-1}
5	Shassere et al. [6]	2012	Gd sheets, 1 mm, spacing 0.5 mm, 475 g	Electro, 0.75 T	Water + ethylene glycol	0.17 Hz	5.6 K	8.8 W kg^{-1}
6	Rowe and Barclay [7]	2002	Gd packed bed, 125 g	Nb–Ti superconducting, 2 T	Helium	0.2–1.2 Hz	13 K	No-load
			Gd packed bed, 250 g				18 K	
			Gd packed bed, 375 g				20 K	
	Richard et al. [8]	2004	Gd flakes, 180 g				16 K	No-load
							14 K	5.6 W kg^{-1}
			Gd flakes, 90 g, Gd–Tb flakes, 80 g				20 K	No-load
							14 K	11.8 W kg^{-1}
	Rowe and Tura [9]	2006	Gd, Gd–Tb, Gd–Er, 270 g				44 K (1.5 T)	No-load
							50 K (2 T)	No-load
	Arnold et al. [10]	2011	Gd particles, 66 g, Gd–Er particles, 70 g	Nb–Ti superconducting <5 T			33 K (2 T)	No-load
							0 K (2 T)	110.3 W kg^{-1}
							59 K (5 T)	No-load
							0 K (5 T)	274.1 W kg^{-1}
7	Hirano et al. [11]	2002	Gd spheres, 0.3 mm, 2,200 g	Superconducting, <4 T	Water	0.167 Hz	23 K (4 T)	45.5 W kg^{-1}
							11 K (1 T)	18.2 W kg^{-1}
8	Kawanami [12, 13]	2005, 2007	Gd packed bed	Permanent, 0.9 T	Air	N/A	N/A	N/A
			Gd particles, 0.6 mm, 62 g	Permanent, 1 T	Distilled water	0.35 Hz	12 K	No-load

(continued)

Table 7.13 (continued)

	Authors, reference	Year	AMR type, materials, mass	Magnet assembly type and magnetic field strength	Heat transfer fluid	Frequency	Temperature span	Specific cooling power
9	Nakamura et al. [14]	2008	Gd spheres, 33.4 g	Permanent, 2 T	Air	N/A	7 K	No-load
					Water		21 K	
10	Hirano et al. [15]	2009	La–Fe–Si spheres, 1.6 mm, 12 g	Permanent, 2.3 T	Air	N/A	3.1 K	No-load
			La–Fe–Si spheres, 1 mm, 13 g				3.7 K	
			La–Fe–Si spheres, 2 mm, 15 g				2.7 K	
			Gd spheres, 0.6 mm, 33 g				4.2 K	
11	Toshiba Co., Table 7.3	2006	Gd, Gd alloys, La–Fe–Si–H	Nd–Fe–B permanent	N/A	1 Hz	46 K	No-load
12	Yu et al., Gao et al. [16, 18]	2005	Gd particles, 0.3 mm, 930 g	Electro magnet, 2.18 T	Water	N/A	3 K	20.2 W kg^{-1}
			Gd particles, 0.56 mm, 1,109 g					16.1 W kg^{-1}
			Gd–Si–Ge particles, 0.3–0.75 mm, 1,213 g					8.5 W kg^{-1}
13	Lu et al. [17]	2005	Gd, 0.2 mm; Gd–Si–Ge, 0.2 mm; Gd–Si–Ge–Ga, 0.2 mm	Halbach permanent, 1.4 T	Water	N/A	23 K	No-load
							5 K	40 W (no mass reported)
14	Yao et al. [19]	2006	Gd packed bed, 1167.4 g	Permanent, 1.5 T	Helium	1 Hz	42 K	No-load
							18.2 K	43.9 W kg^{-1}
15	Huang et al. [20]	2006	Gd, La–Fe–Co–Si–B, La–Fe–Co–Si, 0.2–0.5 mm, 950 g	Nd–Fe–B Halbach permanent, 1.5 T	N/A	0.178 Hz	18 K	No-load
							10 K	13.3 W kg^{-1}
							5 K	21 W kg^{-1}
	Cheng et al. [22]	2013	La–Fe–Co–Si–B Particles, 0.42–0.85 mm, 580 g		Na-based	0.9 Hz	15.3 K	No-load
			La–Fe–Co–Si				12.7 K	
			Gd				8 K	
16	Zheng et al. [21]	2009	Gd packed bed, Ericsson cycle	Nd–Fe–B Halbach permanent, 1.5 T	N/A	N/A	N/A	N/A
17	Clot et al. [23]	2003	Gd plates, 1 mm, 0.15 mm spacing, 223 g	Halbach permanent, 0.8 T	Water	0.42 Hz	7 K	No-load
							4 K	39,5 W kg^{-1}

(continued)

Table 7.13 (continued)

	Authors, reference	Year	AMR type, materials, mass	Magnet assembly type and magnetic field strength	Heat transfer fluid	Frequency	Temperature span	Specific cooling power
18	Dupuis et al. [24]	2009	Gd plates, 1 mm	Halbach permanent, 0.8 T	N/A	1 Hz	7.8 K	No-load
	Legait et al. [25]	2014	Gd plates, 1 mm, 0.3 mm spacing, 190 g				11.5 K	
			Pr–Sr–Mn–O plates, 1 mm, spacing 1 mm, 91 g				5 K	
			Single La–Fe–Co–Si plates, 1 mm, spacing 1 mm, 150 g				8 K	
			Layered La–Fe–Co–Si plates, 1 mm, spacing 1 mm, 150 g				10.5 K	
19	Bour et al. [26]	2009	Gd plates, 0.6 mm, spacing 0.1 mm	Halbach permanent, 1.1 T	Freezing depressant	N/A	16.1 K	No-load
20	Cooltech, Table 7.6	2009	Gd and Gd–Tb plates	Nd–Fe–B permanent, 1.27 T	N/A	1.5 Hz (max)	38 K (max)	150 W (max, no mass reported)
21	Bahl et al. [27]	2008	Gd plates, 0.9 mm, spacing 0.8 mm, 92 g	Electro, 1.4 T	Demineralized water + 10 % ethanol	0.055 and 0.084 Hz	6.9 K	No-load
					ethylene glycol		6.4 K	
					propylene glycol		6.2 K	
					olive oil		6 K	
22	Engelbrecht et al. [28]	2009	Gd plates, 0.9 mm, spacing 0.5 mm	Halbach permanent, 1.03 T	75 % water + 25 % freezing depressant	N/A	9 K	No-load
			La–Fe–Co–Si plates, 0.9 mm, spacing 0.5 mm				6.5 K	
	Engelbrecht et al. [29]	2011	Gd — Plates 0.9 mm		N/A	N/A	10.2 K	No-load
			Single La–Fe–Co–Si — Plates 0.9 mm				7.9 K	
			2× Layered La-Fe–Co–Si — Plates 0.9 mm				6.5 K, 8.5 K	
			LCSM plates, 0.3 mm				5.9 K	

(continued)

Table 7.13 (continued)

	Authors, reference	Year	AMR type, materials, mass		Magnet assembly type and magnetic field strength	Heat transfer fluid	Frequency	Temperature span	Specific cooling power
23	Tušek et al. [30–32]	2012	Gd	Plates 0.25 mm, spacing 0.25 mm, 130,9 g	Nd–Fe–B permanent, 1.15 T	70 % distilled water + 30 % anti-freeze	0.15–0.45 Hz	16 K	No-load
				Plates 0.25 mm, spacing 0.1 mm, 176.3 g				23.5 K	No-load
								13.5 K	11.3 W kg^{-1}
								0 K	39.7 W kg^{-1}
				Plates 0.25 mm, spacing 0.25 mm, 142.7 g (perpendicular to mag. field)				14 K	No-load
				Cylinders, $d = 2.5$ mm, $l = 4$ mm, 120.5 g				4 K	
				Spheres, 0.35–0.5 mm, 135 g				15.5 K	
				Powder, 0.35–0.5 mm, 93 g				7 K	
	Tušek et al. [33]	2014	2-layered	La–Fe–Co–Si plates, 0.5 mm, spacing 0.2 mm, 144 g				16 K	No-load
								5 K	11 W kg^{-1}
			4-layered					20 K	No-load
								8 K	15 W kg^{-1}
			7-layered					20 K	No-load
								8 K	15 W kg^{-1}

(continued)

Table 7.13 (continued)

	Authors, reference	Year	AMR type, materials, mass	Magnet assembly type and magnetic field strength	Heat transfer fluid	Frequency	Temperature span	Specific cooling power
24	Tagliafico et al. [34]	2009	Gd powder, 0.3 mm	Nd–Fe–B permanent, 1.5 T	Water + anti-freeze	0.2 Hz	N/A	N/A
	Tagliafico et al. [35]	2013	Gd plates, 08 mm, 360 g		50 % water + 50 % ethanol	N/A	5 K	No-load
25	Balli et al. [36, 37]	2012	Gd plates, 1 mm, 400 g	Nd–Fe–B Halbach permanent, 1.45 T	Silicon oil	N/A	30 K	No-load
26	Kim and Jeong [38]	2009	Gd particles, 0.325–0.5 mm	Nd–Fe–B Halbach permanent, 1.58 T	Helium	1 Hz	16 K	No-load
27	Trevizoli et al. [39]	2011	Gd plates, 0.85 mm, spacing 0.1 mm, 195.4 g	Nd–Fe–B Halbach permanent, 1.65 T	Water	0.14 Hz	4.45 K	No-load
		2011					0 K	20 W kg^{-1}
28	Czernuszewicz et al. [40]	2014	Gd particles, 2–5 mm, 30 g	Nd–Fe–B Halbach permanent, 1 T	Ethylene glycol	0.025 Hz	2 K	No-load
29	Wroclaw Uni, Table 7.12	2014	NiMnX (X = In, Sn, Ga) parallel pipes, plates	Combination of MC and barocaloric	N/A	1 Hz	N/A	N/A
30	Gomez et al. [41]	2013	Gd plates, 0.5 mm, spacing 0.25 mm, 180 g	Nd–Fe–B Halbach permanent, 1 T	Distilled water	N/A	3.5 K	No-load
							0 K	16.7 W kg^{-1}

0.5 mm and the spacing between the plates was 0.25 mm. The total mass of both AMRs together was 180 g. The heat-transfer fluid was distilled water. At no-load, the temperature span was 3.5 K and the specific cooling power at zero temperature span was 16.7 W kg^{-1}.

7.1.14 Conclusion

In this section, we have presented all the reciprocating magnetocaloric devices, as experimental prototypes, which have been built to date. Altogether, 29 devices were built and tested. It is clear that the operating frequencies are limited to 1 Hz and below, which also affects the specific cooling powers. At certain significant temperature spans, e.g., 20 K, the maximum specific cooling powers are in the range of approximately 50 W kg^{-1}. Such specific cooling powers are not sufficient to be commercially viable. From this perspective it is clear that reciprocating magnetocaloric devices are not suitable for real refrigeration/heat-pump applications.

However, reciprocating magnetocaloric devices are mostly useful as the AMR testing devices. Reciprocating devices are usually designed in such a manner that different AMRs may be easily interchangeable. This allows for a fast experimental testing of different AMR configurations. Different geometries and magnetocaloric materials can be tested, as well as different thermodynamic AMR cycles (e.g., Brayton, Ericsson, Stirling). Also, a combination of different ferroic technologies can be investigated as, for example, in a magneto-baro caloric device. All these aspects are extremely important for the further design of rotary magnetic refrigerators or heat pumps.

At the end of this section, all the reciprocating devices are collected in Table 7.13 with their general geometric as well as operating characteristics.

7.2 Rotary Prototypes

7.2.1 USA Prototypes

According to Kirol and Dacus [42] the first-ever rotary magnetic refrigeration prototype was built in Idaho National Engineering Laboratory in 1987. However, the authors stated that the device was being built at that time, thus no photography of the prototype was available as the proof. The prototype consisted of a rotating AMR and a stationary magnet assembly. The rotating AMR was in the form of a disc and consisted of parallel Gd plates. The circular plates were 0.076 mm thick with a spacing of 0.126 mm. The total mass of the whole AMR was 270 g. The heat-transfer fluid used in the device was water. The stationary Nd–Fe–B magnet assembly consisted of four high-field (0.9 T) areas. In this manner, the rotating AMR made four thermodynamic cycles per revolution. The thermodynamic cycles were hybrid ones (Brayton–Ericsson cycles). Furthermore, the magnetic field was applied parallel to the plates in order to minimize the demagnetization effect of the magnetocaloric material. Unfortunately, no performance regarding device's temperature spans or cooling powers was reported. The researchers mainly focused on certain design issues, such as problems with dynamic seals and the corresponding frictional heat generation, taking care to obtain acceptable magnetic field profiles and reducing the complexity of the overall system design.

The second US rotary magnetic prototype was built by Astronautics Corporation of America in collaboration with Tohoku University from Japan and was presented by Zimm et al. [43, 44] at the First International Conference on Magnetic Refrigeration at Room Temperature in Montreaux, Switzerland in 2005. The prototype consisted of a rotating disc with AMRs called "the Wheel" and a stationary Nd–Fe–B magnet assembly. "The Wheel" had three separated sectors. Each sector had one inlet/outlet to/from the hot heat exchanger at each end of the sector. The inlet/outlet to/from the cold heat exchanger was the only one common in the middle of the sector. In this manner, "the Wheel" actually comprised six AMRs. The AMRs were rotating through one stationary 1.5 T magnetic field area. The heat-transfer fluid was water. Two experiments employing two different AMR configurations were carried out. In the first experiment the AMR beds were filled with Gd spherical particles with diameters from 0.425 to 0.5 mm. The mass of the magnetocaloric material was not reported. Such an AMR configuration could produce 18 K of no-load temperature span at an operating frequency of 4 Hz. When a cooling load of 15 W was applied, the temperature span decreased to 14 K. At 0.5 K of temperature span the device could produce 44 W of cooling power. The second experiment was carried out with a changed configuration of the AMRs. The AMR beds were filled with a layer of Gd–Er spherical particles (0.25–0.355 mm in diameter) and a layer of Gd spherical particles (0.425–0.5 mm in diameter). Also, in this case the mass of the magnetocaloric material was not reported. The experiments were again carried out at the operating frequency of 4 Hz. The no-load temperature span achieved was in this case 25 K, whereas at 14 and 0.5 K the device could produce 27 and 41 W of cooling power, respectively.

The second rotary magnetic prototype built by Astronautics Corporation of America was presented in 2007 by Zimm et al. [45]. This device consisted of 12 stationary AMR beds positioned in a ring arrangement and a rotating, modified, Halbach Nd–Fe–B permanent-magnet assembly. The magnet assembly consisted of two high-field (1.5 T) and two low-field areas. The AMR beds were filled with Gd plates. The geometry, regarding the thickness and spacing of the plates, was not reported; however, the total mass of all the beds was 916 g. The experiments were carried out at an operating frequency of 2 Hz with the hot-side temperature fixed at 298 K. In this manner, the no-load temperature span was 11.5 K. At temperature spans of 8 and 4 K, the device could produce 76.4 and 158.3 W kg^{-1} of specific cooling power, respectively. The zero-temperature span specific cooling power was 240.2 W kg^{-1}.

The third prototype built by Astronautics Corporation of America was presented in 2014 by Jacobs et al. [46] (Table 7.14).

Table 7.14 Third rotary magnetic prototype from Astronautics Corporation of America

Name and address of institute	Astronautics Corporation of America, Astronautics Technology Center, 718 Walsh Rd, Madison WI 53714
Name of contact person/email	Dr. Steven L. Russek/s.russek@astronautics.com
Year of production	2013
Type	Rotary
Maximum frequency	4 Hz
Maximum cooling power	3042 W
Maximum temperature span	18 K
Type of AMR	Packed bed
MC material(s)	$La(Fe_{0.885}Si_{0.115})_{13}H_y$
Type of magnets	Permanent, 1.44 T

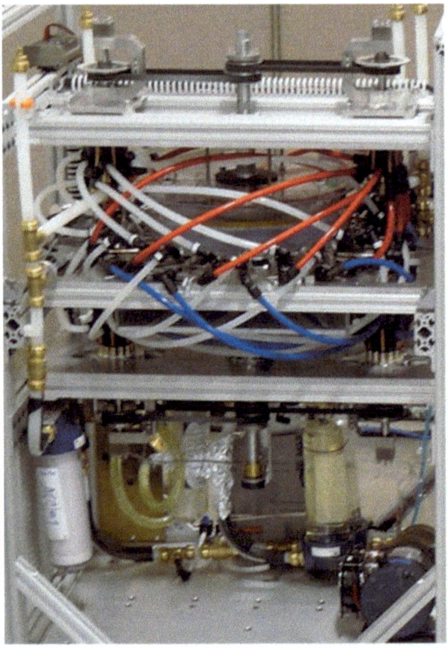

The device used the same magnet assembly and valve system as in their second prototype [45]. Twelve AMR beds were arranged circumferentially. Each AMR bed was filled with six layers of different La–Fe–Si–H spherical particles with diameters of 0.177–0.246 mm. The total mass of all the beds was 1,520 g. The six different La–Fe–Si–H magnetocaloric materials had six different Curie temperatures. These were at 303.5, 306.2, 309.1, 311.6, 313.7 and 316 K, respectively. The heat-transfer fluid used in the device was water mixed with a small amount of anti-corrosion agent.

The device's performance tests were carried out at an operating frequency of 4 Hz and the hot-side temperature was fixed at 317 K. The zero-temperature span-specific cooling power was 2,001 W kg^{-1}, while at the temperature span of 12 K the prototype could produce 1,375 W kg^{-1}, achieving COP = 1.6.

In 2005 researchers from George Washington University and the National Institute of Standards and Technology presented a rotary magnetocaloric device in the article by Shir et al. [47] (Table 7.15). The AMR consisted of a Gd powder bed. A 2 T permanent-magnet assembly was applied. The heat-transfer fluid was helium. The gas flow was managed by a piston–cylinder displacer. The maximum no-load temperature span achieved was approximately 5 K.

Table 7.16 presents a magnetic prototype from General Electrics which was built in 2012. It consists of two dual Halbach magnet assemblies for multistage AMR beds testing. The heat-transfer fluid is controlled by linear piston pumping system.

7.2.2 Spanish Prototypes

The first Spanish rotary magnetic refrigeration prototype was built by the Polytechnical University of Catalonia in collaboration with the Independent University of Barcelona and the University of Barcelona. It was presented in 2000 by Bohigas et al. [48]. A photograph of the prototype and some additional characteristics are also presented in Table 7.17.

The device comprised a rotating AMR disc and a stationary Nd–Fe–B magnet assembly. The magnet assembly had one high-field area where 0.3 T of magnetic field could be induced. The AMR disc consisted of a plastic wheel with a Gd thin ribbon attached to the wheel's lateral surface. Two disc geometries were designed, one with a diameter of 110 mm and the other with a diameter of 75 mm. Both discs had a width of 0.8 mm. The magnetocaloric material was immersed in a container filled with olive oil as the heat-transfer fluid.

The device could produce a no-load temperature span of 1.6 K at an operating frequency of 0.33 Hz. In the next phase, the magnet assembly was improved, by providing a magnetic field of 0.95 T. After the improvement was made a no-load temperature span of 5 K was achieved.

7.2.3 Japanese Prototypes

The first Japanese rotary magnetic prototype was built by the Tokyo Institute of Technology in collaboration with Chubu Electric Power Co. Inc. and presented in 2005 at the First International Conference on Magnetic Refrigeration at Room Temperature in Montreaux, Switzerland by Okamura et al. [49]. The prototype consisted of four AMR beds and a rotating Nd–Fe–B magnet assembly. The magnet assembly had two high-field areas (0.77 T) that were rotating over the four AMR beds.

Table 7.15 Prototype built by George Washington University and the National Institute of Standards and Technology (Farhad Shir next to the device)

Name and address of institute	The George Washington University
Name of contact person/email	Farhad Shir/Farhadshir@aol.com
Year of production	2004
Type	Rotary
Maximum frequency	N/A
Maximum cooling power	N/A
Maximum temperature span	5 K
Type of AMR	Packed bed
MC material(s)	Gadolinium
Type of magnets	"Magic ring" magnet

The AMR beds were filled with four different Gd-based materials. These were $Gd_{0.91}Y_{0.09}$, $Gd_{0.84}Dy_{0.16}$, $Gd_{0.87}Dy_{0.13}$ and $Gd_{0.89}Dy_{0.11}$ spheres with a diameter of 0.6 mm. The total mass of all four AMRs was 1,000 g. The heat-transfer fluid was water. All the performance tests were carried out at an operating frequency of 3.33 Hz. Tests with four different hot-side temperatures were conducted (278, 283, 288, 293 K). At the hot-side temperature of 278 K the specific cooling powers of 10 and 50 W kg^{-1} were achieved at temperature spans of 4 and 0.5 K, respectively. When the hot-side

Table 7.16 Rotary prototype from General Electrics

Name and address of institute	GE Appliance Park GE Appliance park AP35-1301, Louisville KY, USA, 40204	
Name of contact person/email	Michael Benedict David Beers	Michael.benedict@ge.com David.Beers2@ge.com
Year of production	2012	
Type	Reciprocating	
Maximum frequency	4 Hz	
Maximum cooling power	100 W	
Maximum temperature span	50 K	
Type of AMR	Multistage beds, different AMR structures	
MC material(s)	$La(Fe_{0.885}Si_{0.115})_{13}H_y$	
Type of magnets	Two dual Halbach arrays	

temperature was increased to 283 K, the specific cooling powers were 30 W kg^{-1} (at 4.5 K) and 60 W kg^{-1} (at 0.5 K). The increase in the hot-side temperature to 288 K led to specific cooling powers of 10 W kg^{-1} (at 4.5 K) and 40 W kg^{-1} (at 0.5 K). Finally, the hot-side temperature was set to 293 K, which affected the specific cooling powers to decrease to 10 W kg^{-1} (at 1 K) and 18 W kg^{-1} (at 0.5 K).

The second prototype built by Tokyo Institute of Technology was presented in 2007 by Okamura et al. [50] (Table 7.18). It was actually an upgrade of the first prototype [49]. The most important upgrades were made to the magnet assembly. The structure of the iron yoke was changed in order to reduce the Joule heating in

Table 7.17 Spanish rotary magnetic prototype

Name and address of institute	[1]Institut de Ciència de Materials de Barcelona (ICMAB-CSIC), Campus UAB, 08193 Bellaterra, Spain
	[2]Departament de Física i Enginyeria Nuclear, Universitat Politècnica de Catalunya, Diagonal 647, 08028 Barcelona, Catalonia, Spain
	[3]Departament de Física Fonamental, Universitat de Barcelona, Martí i Franquès 1, 08028 Barcelona, Catalonia, Spain
Name of contact person/email	Elies Molins[1]/elies.molins@icmab.es, Bohigas X.[2], Tejada J.[3]
Year of production	1996
Type	Rotary
Maximum frequency	<1 Hz
Maximum cooling power	N/A
Maximum temperature span	5 K
Type of AMR	Any
MC material(s)	Gd
Type of magnets	Permanent (Nd–Fe–B), 0.3 and 0.95 T configurations

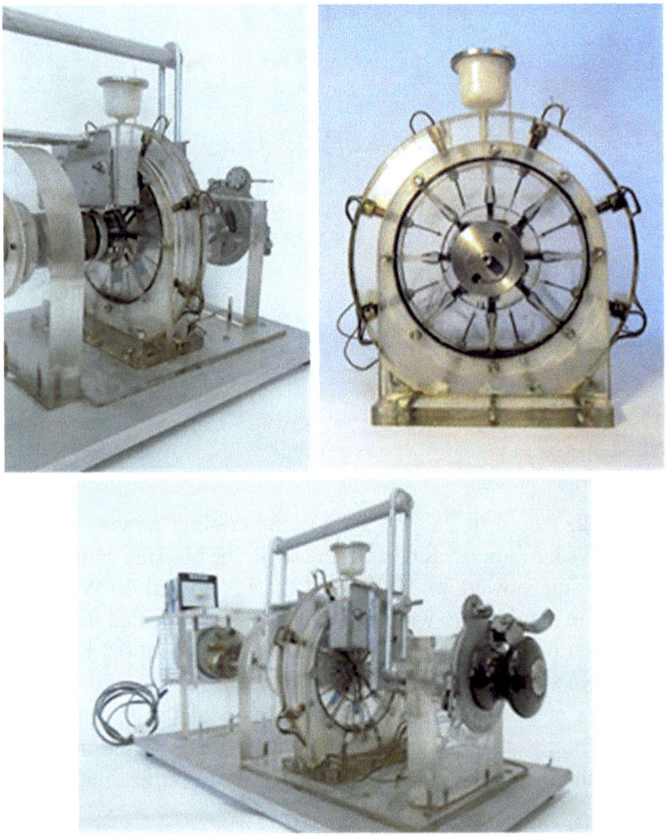

Table 7.18 Second rotary prototype from Tokyo Institute of Technology and Chubu Electric Power Co. Inc

Name and address of institute	Chubu Electric Power Co., Inc.
Name of contact person/email	Naoki Hirano/Hirano.Naoki@chuden.co.jp
Year of production	2007
Type	Rotary
Maximum frequency	1 Hz
Maximum cooling power	540 W
Maximum temperature span	20 K
Type of AMR	Packed bed
MC material(s)	Gd and Gd alloys
Type of magnets	Permanent, 1.1 T

the yoke caused by eddy currents via the magnet rotation. Furthermore, the magnets were replaced by stronger ones to induce a higher magnetic field of 1.1 T. Also, the AMR duct configurations were changed along with the piping diameter to increase the flow rate and decrease the pressure drop.

The four AMR beds were filled with 4,000 g of Gd spheres with a diameter of 0.5 mm. The operating frequency was set to 1 Hz and the hot-side temperature to 294 K. At a temperature span of 5.2 K, the device could produce 37.5 W kg^{-1} of specific cooling power with a COP of 0.5, while at 0.2 K of temperature span the specific cooling power was 135 W kg^{-1} and the COP was 1.8.

Tables 7.19, 7.20, 7.21 and 7.22 present the operating characteristics and pictures of four additional Japanese rotary magnetic refrigeration prototypes built between years 2009 and 2013 at Hokkaido Research Institute, Kobe University, Railway Technical Research Institute and Sanden Co. Inc., respectively.

The prototype from Hokkaido Research Institute (Table 7.19) achieved a maximum cooling power of 150 W and a maximum temperature span of 5.5 K.

The prototype from Kobe University (Table 7.20) could achieve a maximum temperature span of 5 K and a maximum cooling power of 10 W. The magnet assembly consisted of permanent magnets that could induce a magnetic field of 0.6 T.

The device built at the Railway Technical Research Institute (Table 7.21) consisted of a Nd–Fe–B Halbach magnet assembly and could achieve a maximum temperature span 7.6 K, while the maximum cooling power was 150 W.

The prototype built at the Sanden Co. Inc. (Table 7.22) was composed of a permanent-magnet assembly that could induce a magnetic field of 0.85 T. The maximum temperature span was 13.8 K, while the specific cooling power was 157.1 W kg^{-1}.

7.2.4 Swiss Prototypes

The first Swiss rotary magnetic refrigeration prototype was built at the University of Applied Sciences of Western Switzerland and presented in 2006 [51]. The prototype was built as a rotary magnetic demonstrator for air conditioning. The device

Table 7.19 Rotary magnetic prototype from Hokkaido Research Institute

Name and address of institute	Hokkaido Research Institute, N19 W11 Kita-ku, Sapporo, JAPAN
Name of contact person/email	Shigeki Hirano/shigeki-hirano@hro.or.jp
Year of production	2009 (modified in 2013)
Type	Rotary
Maximum frequency	4.5 Hz
Maximum cooling power	150 W
Maximum temperature span	5.5 K
Type of AMR	Packed bed
MC material(s)	Gd and Gd alloys
type of magnets	Permanent (Nd–Fe–B)

consisted of a rotating AMR ring and a stationary permanent-magnet assembly. The AMR ring consisted of Gd particle beds. The magnet assembly could induce a magnetic field of 0.8 T. The heat-transfer fluid was air. The researchers from the University of Applied Sciences of Western Switzerland exhibited the prototype at the industrial fair in Hannover and won first prize at the Swiss Technology Awards 2006.

Table 7.20 Rotary magnetic prototype from Kobe University

Name and address of institute	Kobe University, 1-1 Rokkodai-cho, Nada-ku, Kobe, 657-8501 Japan
Name of contact person/email	Tsuyoshi Kawanami/kawanami@mech.kobe-u.ac.jp
Year of production	2011
Type	Rotary
Maximum frequency	1 Hz
Maximum cooling power	10 W
Maximum temperature span	5 K
Type of AMR	Packed bed
MC material(s)	Gd
Type of magnets	Permanent, 0.6 T

7.2.5 French Prototypes

Tables 7.23 and 7.24 present characteristics and photographs of first and second rotary magnetic prototypes built by Cooltech Applications, while Table 7.25 presents the rotary prototype from the Grenoble Electrical Engineering Laboratory.

The first French rotary magnetic device was built by HVAC Department of National Institute of Applied Sciences in Strasbourg and the company Cooltech Applications. It was presented in 2005 and 2006 by Vasile and Müller [52, 53]. Unfortunately, no data regarding the prototype's performance was given at that point. However, the design of different prototype elements is thoroughly discussed.

Table 7.21 Rotary magnetic prototype from Railway Technical Research Institute

Name and address of institute	Railway Technical Research Institute
Name of contact person/email	Yoshiki Miyazaki/miyazaki.yoshiki.23@rtri.or.jp
Year of production	2011
Type	Rotary
Maximum frequency	1.3 Hz
Maximum cooling power	150 W
Maximum temperature span	7.6 K
Type of AMR	Packed bed
MC material(s)	Gd, La–Fe–Co–Si
Type of magnets	Halbach-arrayed Nd–Fe–B magnet

The device consisted of eight stationary AMR beds and a rotating modified Halbach magnet assembly. The magnet assembly had four 2.4 T magnetic field areas. A special feature of the device is the hydraulic connections of the AMR beds. Eight AMR beds were separated in two parts. Each four beds were hydraulically connected in series. In this manner, fluid could flow through all four beds between the heat exchangers. Furthermore, an important focus was also given to the reduction of thermal losses related to the hot and cold fluid circuits. However, the conclusion was that the researchers needed to conduct additional analyses of the heat transfer in regenerators and to study different magnetocaloric materials before the initial tests of the prototype performance were carried out.

Table 7.22 Rotary magnetic prototype from Sanden Co. Inc

Name and address of institute	Sanden Co., Inc.
Name of contact person/email	pe_sanchoru@sanden.co.jp
Year of production	2013
Type	Rotary
Maximum frequency	3 Hz
Maximum cooling power	200 W
Maximum temperature span	13.8 K
Type of AMR	Packed bed
MC material(s)	Gd, Gd alloy, 1,273 g
Type of magnets	Permanent, 0.85 T

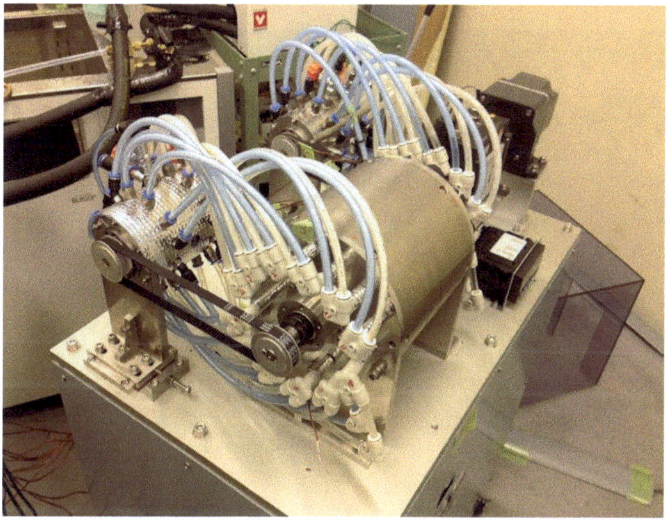

7.2.6 Canadian Prototypes

The first Canadian rotary magnetic refrigeration prototype was built at the University of Victoria and presented in 2007 at the Second International Conference on Magnetic Refrigeration at Room Temperature in Portorož, Slovenia by Tura and Rowe [54]. The device consisted of two stationary AMRs and two Nd–Fe–B Halbach cylindrical magnet assemblies. Both cylindrical magnet assemblies had the outer rings stationary, while the inner ring rotated. Such a magnet assembly could induce a maximum magnetic field of 1.47 T. The AMRs cycled in the opposite phases, which allowed for operating frequencies up to 4 Hz. The fluid flow oscillated by a fluid displacer with a crank mechanism. The heat-transfer fluid was water. The device had two hot heat exchangers for each AMR, while the cold heat exchanger was shared by both AMRs. Both AMR beds were filled with Gd crushed particles with a diameter of 0.6 mm. The total mass of both AMRs was 122 g.

Table 7.23 First rotary magnetic prototype from Cooltech Applications

Name and address of institute	Cooltech Applications-5 Impasse Antoine IMBS 67810 HOLTZHEIM-France
Name of contact person/email	Christian Muller/c.muller@cooltech-applications.com
Year of production	2013
Type	Rotary, 2 stages
Maximum frequency	4 Hz
Maximum cooling power	120 W
Maximum temperature span	42 K
Type of AMR	Packed bed
MC material(s)	Gd and Gd–Tb
Type of magnets	Permanent Nd–Fe–B, 0.98 T

During performance testing the temperature of the hot side was set to 302.5 K. At the operating frequency of 2 Hz, the no-load temperature span was 13 K. Furthermore, the researchers discussed issues regarding high pressure drops, which they assigned to the AMR's geometry.

Later in 2009 an upgraded version of the 2007 prototype [54] was introduced by Tura and Rowe [55, 56]. A number of improvements were made. New heat exchangers were designed and bleed valves were introduced to the cold heat exchanger. Furthermore, a new fluid-flow system was implemented. The hydraulic lines were minimized and the tube diameters were increased to decrease the pressure drop. Moreover, air evacuation prior to fluid filling and then circulating the fluid to expel air bubbles through bleed valves was introduced. After a sufficient time, the inlet and outlet valves were closed and the hydraulic lines were pressurized to

Table 7.24 Second rotary magnetic prototype from Cooltech Applications

Name and address of institute	Cooltech Applications-5 Impasse Antoine IMBS 67810 HOLTZHEIM-France
Name of contact person/email	Christian Muller/c.muller@cooltech-applications.com
Year of production	2014
Type	Rotary, 2 stages
Maximum frequency	6 Hz
Maximum cooling power	>300 W
Maximum temperature span	38 K
Type of AMR	Parallel plates
MC material(s)	Gd and Gd–Er/La–Fe–Si
Type of magnets	Permanent Nd–Fe–B, B_{max} = 1.17 T

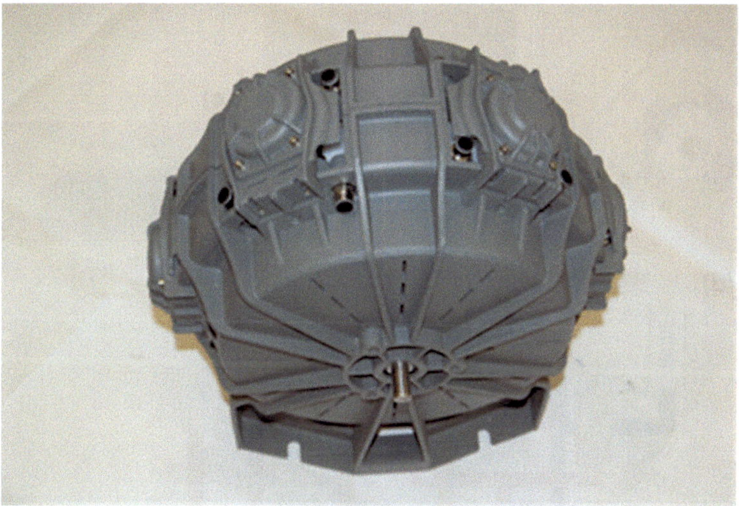

approximately 3 bar. The two AMR beds were filled with Gd spheres of 0.3 mm in diameter. The total mass of both AMRs was 110 g. The upgraded prototype could then achieve 29 K of no-load temperature span. Operating at a frequency of 1.4 Hz, the device could produce 454.5 W kg^{-1} of specific cooling power at a temperature span of 2.5 K. The COP under these operating conditions was 1.6.

The second Canadian prototype was also built at the University of Victoria and presented in 2014 by Arnold et al. [57]. This was a second-generation prototype, which was an upgrade of the 2009 version [55, 56]. A special novelty was the implementation of a triple Halbach magnet assembly. Such a triple Halbach magnet array consisted of three concentric cylinders. The inner cylinder was stationary, while the intermediate and outer ones were rotating in the opposite directions. The opposite rotation of the outer cylinders was important to create a sinusoidal magnetic field wave form with a stationary field direction. Such a magnet assembly had

Table 7.25 Rotary magnetic prototype from Grenoble Electrical Engineering Laboratory

Name and address of institute	Grenoble Electrical Engineering Laboratory
Name of contact person/email	Afef Kedous-Lebouc/Afef.lebouc@g2elab.grenoble-inp.fr
Year of production	2008
Type	Rotary
Maximum frequency	0.5 Hz
Maximum cooling power	60 W
Maximum temperature span	10 K
Type of AMR	Parallel plates
MC material(s)	Gd
Type of magnets	Nd–Fe–B

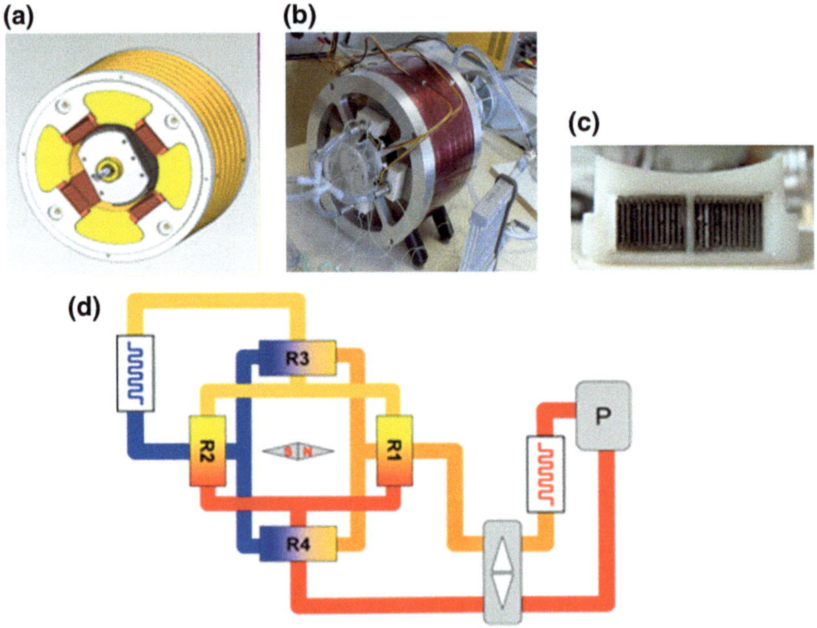

(a) Design of the prototype, (b) Photo of the prototype, (c) Regenerator, (d) Hydraulic circuit management

a peak magnetic field of 1.54 T and it was designed in a manner so that the inner cylinder induced half of the total high-field intensity, whilst the outer cylinders each induced one-quarter of the high field. Furthermore, a novel check-valve configuration was implemented to minimize the fluid's dead volumes. The heat-transfer fluid was a mixture of 80 % water and 20 % glycol. The two AMR beds were filled with Gd spherical particles with an average diameter of 0.5 mm. However, for the initial test the total amount of Gd was 420 g, while later 650 g of Gd was applied. The maximum no-load temperature span when using 650 g of Gd was 33 K, operating at 0.8 Hz, while when using 420 g of Gd the maximum no-load

Table 7.26 First rotary magnetic prototype from the University of Victoria

Name and address of institute	Mechanical Engineering, University of Victoria
Name of contact person/email	Andrew Rowe/arowe@uvic.ca
Year of production	2006
Type	Rotary
Maximum frequency	6 Hz
Maximum cooling power	50 W
Maximum temperature span	29 K
Type of AMR	Spheres, plates, particulate, micro-channel
MC material(s)	Gd, Gd_xTb_{1-x}, Gd_xEr_{1-x}, MnFeAsP, MnFeSiP
Type of magnets	Dual Halbach, 0.1–1.45 T

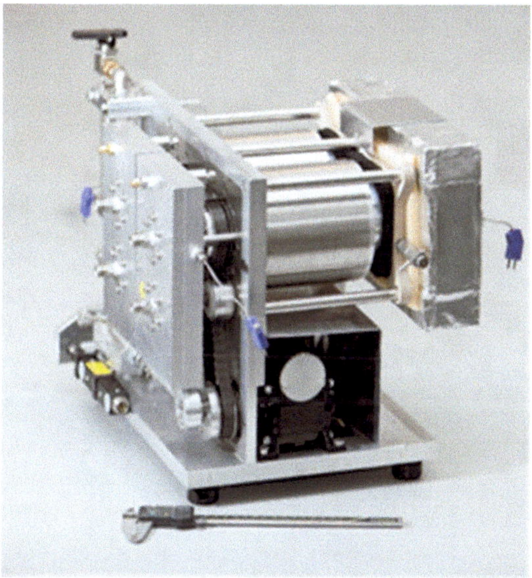

temperature span was approximately 27 K, operating at 0.5 Hz. Moreover, operating at a frequency of 0.8 Hz and a temperature span of 15 K, the device could produce a specific cooling power of 77 W kg^{-1} (with 650 g of Gd).

Both prototypes from the University of Victoria are also presented in Tables 7.26 and 7.27 with some additional information regarding the performance along with photographs of each device.

7.2.7 Chinese Prototypes

The first rotary prototype made in China was built at the School of Materials Science and Engineering, Sichuan University and presented in 2007 by Chen et al. [58].

Table 7.27 Second rotary magnetic prototype from the University of Victoria

Name and address of institute	Mechanical Engineering, University of Victoria
Name of contact person/email	Andrew Rowe/arowe@uvic.ca
Year of production	2012
Type	Rotary
Maximum frequency	2 Hz
Maximum cooling power	80 W
Maximum temperature span	33 K
Type of AMR	Spheres, plates, particulate, microchannel
MC material(s)	Gd, Mn–Fe–As–P
Type of magnets	Triple Halbach, 0.1–1.45 T

The device consisted of a rotating AMR wheel and a stationary Nd–Fe–B Halbach magnet assembly. The magnet assembly had two 1.5 T magnetic field areas. The AMR wheel consisted of 36 sectors, which were filled with Gd particles of 0.5 mm in diameter. The total mass of the magnetocaloric material was 1,000 g. The heat-transfer fluid was water. The maximum no-load temperature span was approximately 11.5 K, while at approximately 6.8 K of temperature span the device could produce 40 W kg^{-1} of specific cooling power. A schematic of the prototype and some additional characteristics are also presented in Table 7.28. Furthermore, in 2005, a second magnetic prototype was built at Sichuan University. The photograph of the second magnetic device and some characteristics are presented in Table 7.29.

The second prototype was built at the Technical Institute of Physics and Chemistry from the Chinese Academy of Sciences and was presented in 2013 by

Table 7.28 First rotary magnetic prototype from Sichuan University

Name and address of institute	School of Materials Science and Engineering, Sichuan University, 24 South Section 1, Yihuan Road, Chengdu 610065, China
Name of contact person/email	Yongbai Tang/tyb1973@163.com
Year of production	2002
Type	Rotary
Maximum frequency	0.2 Hz
Maximum cooling power	N/A
Maximum temperature span	6.2 K
Type of AMR	Parallel plates
MC material(s)	Gd plates
Type of magnets	Permanent, 0.78 T

He et al. [59]. This was a hybrid device, which employed active magnetic regeneration as well as Stirling gas regenerative refrigeration effect. The magnet assembly was a concentric Halbach cylinder, where the inner cylinder rotated and could induce a magnetic field of 1.5 T. The stationary AMR consisted of 1-mm-thick Gd sheets with a total mass of 198 g. The Stirling part of the device consisted of compression and expansion cylinders each fitted with a piston. The heat-transfer fluid was helium. The prototype operated with a frequency of 1.5 Hz and the hot-side temperature was set to 298 K. The main focus of the tests was to study the optimal phase angle between the compression piston of the Stirling part and the magnet position of the AMR part. The phase angle was given as a peak difference between the compression volume and the magnetic field. It indicated the angular relationship between the magnetic field and the compression volume. When the test was carried out operating only Stirling gas refrigeration without the process of active magnetic regeneration, the no-load temperature span achieved was 18.2 K. When 30.3 W kg^{-1} of specific cooling load was applied the temperature span decreased to 12 K. Then, also, the active magnetic regeneration was applied with variations in the phase angle. At a phase angle of 0° the no-load temperature span increased to 19 K. Applying 30.3 W kg^{-1} of specific cooling load also showed an increase in the temperature span, increasing to 12.4 K (in comparison to Stirling process). The highest no-load temperature span was achieved at a phase angle of 90° and it was 21.5 K. Applying 30.3 W kg^{-1} under the same conditions also increased the temperature span to 14.9 K. Increasing the phase angle over 90° resulted in a lower performance of the device.

Tables 7.30 and 7.31 present four rotary prototypes from Baotou Research Institute of Rare Earths. The first two prototypes (Table 7.30) were built in 2006 and 2010. Both prototypes were similar in their design. The magnet assembly consisted of two stationary 1.25 T Nd–Fe–B magnets. The rotating AMR cylinder consisted of eight compartments filled with Gd spheres of 0.4–0.6 mm in diameter. The mass of each compartment was 180 g. The maximum specific cooling power was 34.7 W kg^{-1}, while the maximum temperature span was 18 K.

Table 7.29 Second Chinese rotary magnetic prototype from Sichuan University

Name and address of institute	School of Materials Science and Engineering, Sichuan University, 24 South Section 1, Yihuan Road, Chengdu 610065, China
Name of contact person/email	Yongbai Tang/tyb1973@163.com
Year of production	2005
Type	Rotary
Maximum frequency	0.2 Hz
Maximum cooling power	70 W
Maximum temperature span	11.5 K
Type of AMR	Packed bed
MC material(s)	Gd spheres
Type of magnets	Permanent, 1.5 T

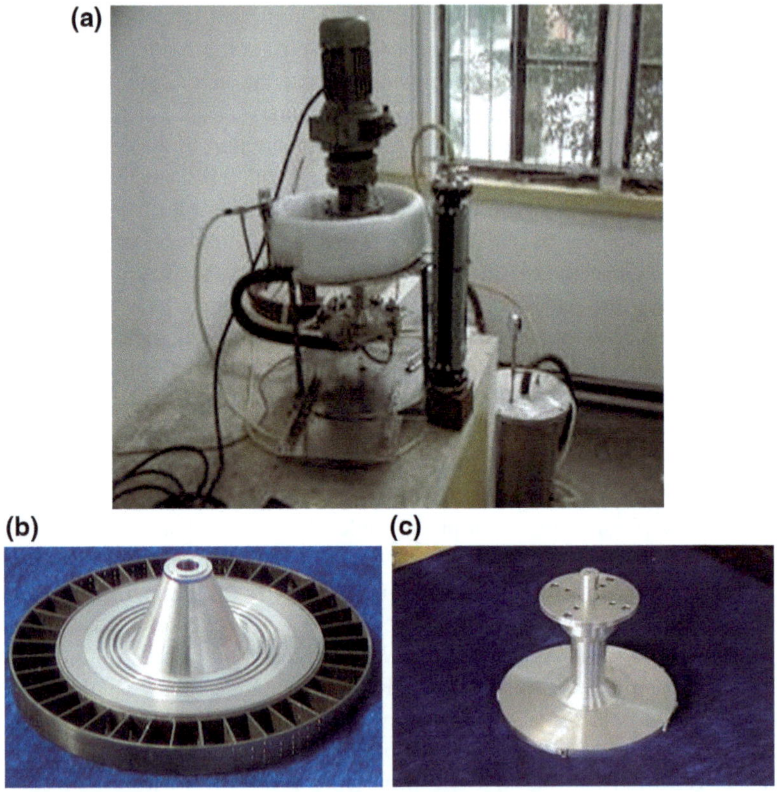

(a) A rotary magnetic refrigerator, (b) A working wheel (AMR bed), (c) A shell of the working wheel

Table 7.30 Third and fourth Chinese rotary magnetic prototypes from Baotou Research Institute of Rare Earths

Name and address of institute	Baotou Research Institute of Rare Earths. Add: No.36, Huanghe Street, RE High-tech Zone, Baotou, Inner Mongolia, P.R. China, 014030
Name of contact person/email	Jiaohong Huang/jiaohongh@sina.cn
Year of production	2006–2010
Type	Rotary
Maximum frequency	0.5 Hz
Maximum cooling power	50 W
Maximum temperature span	18 K
Type of AMR	Packed bed
MC material(s)	Gd spheres, 0.4–0.6 mm, 1,440 g
Type of magnets	2× Permanent Nd–Fe–B, 1.25 T

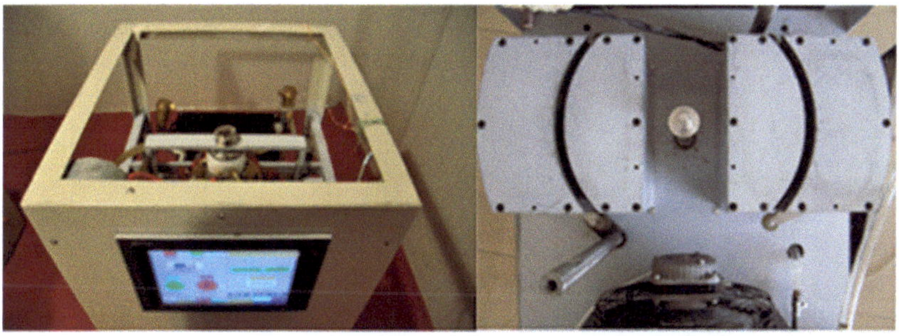

The third and the upgraded fourth rotary prototypes from Baotou Research Institute of Rare Earths (Table 7.31) were built in 2009 and 2012. The magnet assembly was a concentric Halbach cylinder, where the inner cylinder rotated and could induce a magnetic field of 1.4 T. The stationary AMR consisted of Gd and Gd–Tb spheres of 0.3–0.5 mm in diameter. The total mass of all spheres was 1,500 g. When the device operated at the frequency of 0.2 Hz the highest no-load temperature span was 20 K, while the highest specific cooling power was 40 W kg^{-1}.

7.2.8 *Brazilian Prototypes*

The Brazilian rotary magnetic prototype was built in collaboration of The Institute of Physics "Gleb Wataghin", Federal University of Sao Paolo and the National Institute of Metrology and presented in 2009 by Coelho et al. [60].

The device consisted of a rotating AMR wheel and an electromagnet that could induce a 2.3 T of magnetic field. The heat-transfer fluid was ethyl alcohol. Each AMR sector was filled with Gd, with the total mass of all six sectors being 960 g.

Table 7.31 Fifth and sixth Chinese rotary magnetic prototypes from Baotou Research Institute of Rare Earths

Name and address of institute	Baotou Research Institute of Rare Earths. Add: No.36, Huanghe Street, RE High-tech Zone, Baotou, Inner Mongolia, P.R. China, 014030
Name of contact person/email	Jiaohong Huang/jiaohongh@sina.cn
Year of production	2009–2012
Type	Rotary
Maximum frequency	0.2 Hz
Maximum cooling power	60 W
Maximum temperature span	20 K
Type of AMR	Packed bed
MC material(s)	Gd, Gd–Tb spheres, 0.3–0.5 mm, 1,500 g
Type of magnets	Rotating permanent Nd–Fe–B, 1.4 T

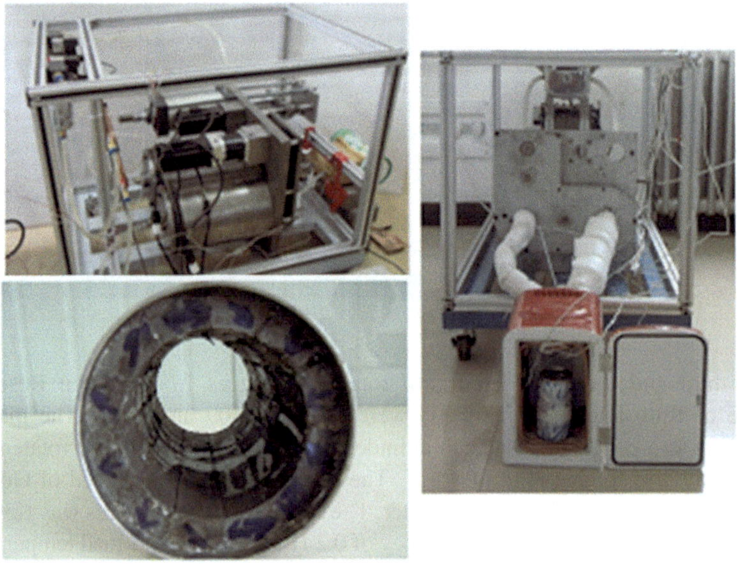

A special feature of the AMRs was the fabrication of the Gd to make a regenerator. A block of Gd was cut with a 0.3-mm-thick wire to get 1-mm-thick plates that were still attached to the base of the bulk. Then similar parallel cuts were made, however, at 90° in relation to the first set. In this way, 11-mm-high pins (still attached to the base) with a 1 mm^2 square cross-section were produced. After this a 0.5-mm epoxy plate was attached to the free surface of the pins. Then the Gd base was cut using a spark-erosion technique. The free surface of the cut pins was also covered with a 0.5-mm thick layer of epoxy. In this way, the rigid structure of the Gd regenerator was obtained. The performance tests were carried out at an operating frequency of 0.5 Hz. The maximum no-load temperature span was 11 K.

Table 7.32 Design plan of the new rotary magnetic prototype from the Federal University of Santa Catarina

Name and address of institute	POLO—Laboratory of Refrigeration and Thermophysics, Department of Mechanical Engineering, Federal University of Santa Catarina, 88.040-900 Florianópolis, SC, Brazil
Name of contact person/email	Prof. Jader R. Barbosa Jr/jrb@polo.ufsc.br
Year of production	2013–2014
Type	Rotary
Maximum frequency	2 Hz
Maximum cooling power	N/A
Maximum temperature span	N/A
Type of AMR	Parallel plates, packed bed, pins
MC material(s)	Gd
Type of magnets	Nested Halbach cylinder

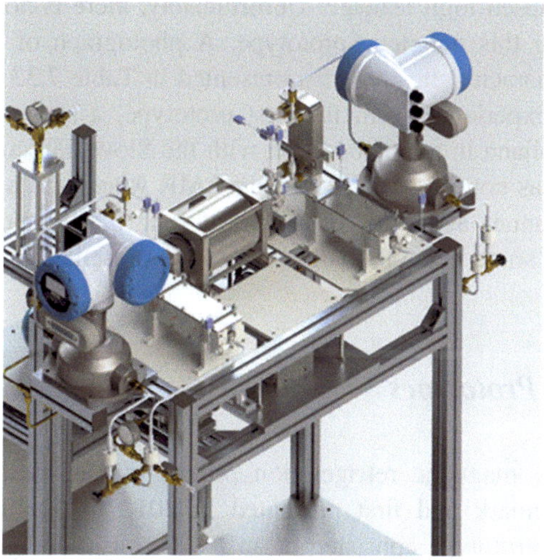

Table 7.32 presents a 3D model of the second Brazilian rotary prototype from the Federal University of Santa Catarina. It is planned to be built in 2014.

7.2.9 Slovenian Prototypes

The first Slovenian rotary magnetic prototype was built at the Faculty of Mechanical Engineering, University of Ljubljana and was presented in 2009 by Tušek et al. [61, 62]. However, it has to be pointed out that the actual development and design of the prototype was mainly the work of Dr. Alen Šarlah, and the authors of this book and his colleagues acknowledge his contribution in the research on magnetic refrigeration at University of Ljubljana, especially in the early years before 2009. Furthermore, a special acknowledgement goes also to the Centre for Element and Structure Modelling (CEMEK) from University of Ljubljana. Researchers from LAMEK contributed greatly to the design of the prototypes built at University of Ljubljana.

The prototype consisted of a rotating AMR drum and a stationary magnet assembly. The magnet assembly consisted of four Nd–Fe–B magnets that could induce a magnetic field of 0.98 T. The AMR drum was divided into 34 sectors that were filled with Gd plates. The plate thickness was 0.25 mm and the total mass was 600 g. The heat-transfer fluid was distilled water. The main goal of the prototype design was to make the parts from low-cost materials with high quality standards and under general tool-making principles, using standard machining methods. However, the prototype faced some issues regarding its high weight and the sealing of the dynamic valve system of the rotating AMR drum. Too-tight sealing caused heat generation due to the friction of the AMR's drum rotation. On the other hand, to loose sealing caused high leakage. Unfortunately, there is no performance data available regarding this magnetic prototype. A photograph of the prototype and some additional characteristics are also presented in Table 7.33.

Based on the experience from the first prototype, a second device from the University of Ljubljana in a collaboration with the Slovenian company SMM was built in 2012. It was composed of a rotating AMR wheel and a Nd–Fe–B magnet assembly. The magnet assembly consisted of four 0.8 T magnetic field areas. A photograph and some general characteristics are presented in Table 7.34.

7.2.10 Danish Prototypes

The Danish rotary magnetic refrigeration prototype was built at the Technical University of Denmark and first presented in 2012 by Engelbrech et al. [63] (Table 7.35). The prototype consisted of an AMR ring rotating in the gap of two concentric cylindrical Nd–Fe–B Halbach magnets. The magnet assembly had four poles with four low-field areas separating them. Such a magnet assembly induced the peak magnetic field of 1.24 T, with an average magnetic field of 0.9 T. The AMR ring had 24 separated beds, which were 250 mm in length. The initial idea was to fill the beds with parallel plates, hence the long AMR beds. However, in reality the beds were filled with Gd spheres of 0.25–0.8 mm in diameter. To reduce the pressure drop the beds were only filled 100 mm in length, which corresponded to 2,800 g of magnetocaloric material. The heat-transfer fluid was a mixture of 75 % deionized water and 25 % commercial ethylene glycol with anti-corrosion inhibitors. The hot-side temperature varied between 300.8 and 301.4 K. The maximum no-load temperature span was 25.4 K, operating at a frequency of 2 Hz. At a temperature span of 21 K and an operating frequency of 2 Hz, the device could produce 35.7 W kg^{-1} of specific cooling power. At a temperature span of 0.3 K and a frequency of 1.8 Hz, 360.7 W kg^{-1} of specific cooling power was produced. When 142.8 W kg^{-1} of

Table 7.33 First rotary magnetic prototype from the University of Ljubljana

Name and address of institute	Laboratory for Refrigeration and District Energy, Faculty of Mechanical Engineering, University of Ljubljana, Askerceva 6, 1000 Ljubljana, Slovenia
Name of contact person/email	Andrej Kitanovski/andrej.kitanovski@fs.uni-lj.si
Year of production	2007
Type	Rotary
Maximum frequency	4 Hz
Maximum cooling power	N/A
Maximum temperature span	N/A
Type of AMR	Parallel plates
MC material(s)	Gd
Type of magnets	Permanent Nd–Fe–B, 0.05–0.98 T

specific cooling load was applied, operating at 1 Hz, the temperature span achieved was 8.9 K. Under these operating conditions the COP was 1.8, while the Carnot COP was 32.4, which corresponded to 5.6 % of the Carnot efficiency. When the specific cooling load was decreased to 71.4 W kg^{-1}, the temperature span increased to 15.4 K. At this point the COP and Carnot COP decreased to 0.8 and 18.3, respectively. This corresponded to a Carnot efficiency of 4.4 %. When operating at frequencies over 2 Hz a gradual decrease in the device performance was reported.

Table 7.34 Second rotary magnetic prototype from the University of Ljubljana

Name and address of institute	Laboratory for Refrigeration and District Energy, Faculty of Mechanical Engineering, University of Ljubljana, Askerceva 6, 1000 Ljubljana, Slovenia
Name of contact person/email	Prof. Alojz Poredoš/alojz.poredos@fs.uni-lj.si
Year of production	2012
Type	Rotary
Maximum frequency	4 Hz
Maximum cooling power	N/A
Maximum temperature span	N/A
Type of AMR	Packed bed, parallel plates
MC material(s)	Gd
Type of magnets	Permanent Nd–Fe–B, 0.8 T

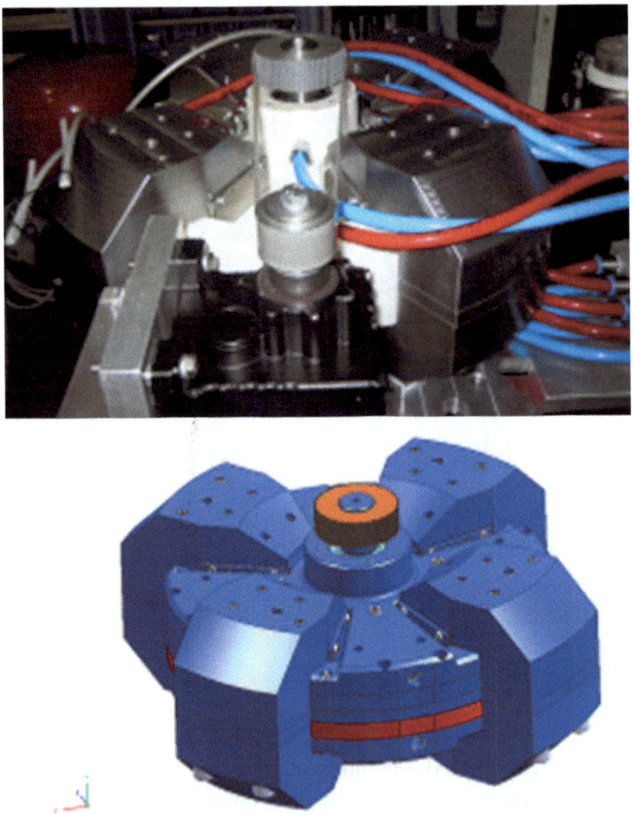

Figure originally published in Tušek et al. [30, Ventil 18: p. 461]; published with kind permission of (c) [Ventil, University of Ljubljana, Faculty of Mechanical Engineering]. All Rights Reserved

Table 7.35 Rotary magnetic prototype from Technical University of Denmark

Name and address of institute	Department of Energy Conversion and Storage, Technical University of Denmark, Frederiksborgvej 399, DK-4000 Roskilde, Denmark
Name of contact person/email	Christian Bahl/chrb@dtu.dk
Year of production	2010
Type	Rotary
Maximum frequency	10 Hz
Maximum cooling power	1012 W
Maximum temperature span	25.4 K
Type of AMR	Packed bed
MC material(s)	Gd (0.25–0.8 mm, 2,800 g)
Type of magnets	Modified Concentric double Halbach cylinder

The decrease in performance is mainly due to the friction in the AMR rotation and the poorer heat transfer at higher operating frequencies.

In 2013 Lozano et al. [64] presented further performance studies on this prototype [63]. They studied the influence of different parasitic losses (in filter, tubes, flow-head and regenerator) in relation to different temperatures at the cold side. Furthermore, they also focused on the influence of the hot reservoir temperature of the hot side and the volumetric flow rate. When operating at a frequency of 1.5 Hz, a hot-side temperature of 297.7 K and a specific cooling load of 142.8 W kg^{-1}, the temperature span was 12.9 K with a Carnot efficiency of 5.6 %.

In 2014, additional studies on the performance of the described prototype were presented by Lozano et al. [65]. They studied the impact of high operating frequencies up to 10 Hz and volumetric flow rates up to 600 l h^{-1}. The temperature of the hot side was fixed at 296 K. Operating at a frequency of 2 Hz the zero-temperature span specific cooling power was 223.2 W kg^{-1}. When the specific cooling load was decreased to 142.8 W kg^{-1}, the temperature span increased to 11.5 K. At 35.7 W kg^{-1}, the temperature span further increased to 18 K. When the operating frequency was increased to 4 Hz, the device showed similar performance as at 2 Hz. Operating at frequency of 6 Hz led to a decrease in the zero-temperature span specific cooling power of 187.5 W kg^{-1}. When the specific cooling load was set to 35.7 W kg^{-1} the temperature span increased to 17 K. Operating at a frequency of 10 Hz further decreased the specific cooling power to 134 W kg^{-1} at a temperature span of 1 K. When 35.7 W kg^{-1} of specific cooling load was applied the temperature span was 14.5 K. The decrease in the device performance with a higher operating frequency was mostly related to the heat generation in valves caused by friction between the seals and the flow-head.

In the same year another study on the performance of the Danish prototype was presented by Bahl et al. [66]. Tests on two different AMR configurations were carried out. In the first tests, the AMR beds were filled with Gd spheres of 0.35–0.85 mm in diameter. The length of the AMRs was 50 mm, which corresponded to a total mass of 1,400 g for all 24 beds. This Gd material had a peak entropy change of 3.5 J kg^{-1} K^{-1} at 290.5 K and a magnetic field of 1 T. The hot-side temperature was set to 297 K. At an operating frequency of 1.5 Hz, the no-load temperature span was 13.9 K, while at 10.5 K the specific cooling power was 71.4 W kg^{-1} and at 1.5 K the device could produce 142.8 W kg^{-1} of specific cooling power. When the operating frequency was increased to only 1.75 Hz, the device performance also increased. At a specific cooling load of 142.8 W kg^{-1} the temperature span increased to 10 K, while at 71.4 W kg^{-1} to 13.9 K. In the second tests, the AMR beds were filled with 2,800 g of Gd spheres, which corresponded to the filled bed length of 100 mm. The diameter of the spheres was 0.25–0.8 mm. The peak entropy change of this Gd was slightly lower than in the first example. It peaked at 3.2 J kg^{-1} K^{-1} at a temperature of 288 K and a magnetic field of 1 T. The hot-side temperature in the tests was set to 298 K. When the device was operating at a frequency of 2.25 Hz and a specific cooling load of 71.4 W kg^{-1} was applied, the temperature span was 18.9 K. However, when the specific cooling load was increased to 142.8 W kg^{-1}, the maximum temperature span was 13.8 K and was achieved at a lower operating frequency of 1.5 Hz.

A photograph of the prototype and some additional characteristics are also presented in Table 7.35.

7.2.11 *Italian Prototypes*

The Italian rotary prototype was built in a collaboration between the University of Salerno, University of Naples Federico II and the Canadian University of Victoria. It was presented in 2014 by Aprea et al. [67]. The device consisted of eight static AMR beds and a rotating magnet assembly. The magnet assembly was a Nd–Fe–B Halbach array structure with two high-field and two low-field areas. The high-field areas peaked with 1.25 T and had an average magnetic field of 1.1 T. Eight AMR beds were filled with 1,200 g of Gd spheres with a diameter of 0.4–0.5 mm. The heat-transfer fluid was demineralized water with anti-corrosion inhibitors. The temperature of the hot side was set to 298 K, while the actual temperature of the ambient varied between 293 and 296 K. The initial studies showed that the device could achieve a no-load temperature span of 13.5 K, operating at a frequency of 0.72 Hz.

7.2.12 German Prototypes

A rotary magnetic prototype has still not been built in Germany. However, through personal correspondence with Professor Dr. Oliver Gutfleisch from TU Darmstadt,

Table 7.36 Design plan of a future rotary prototype from TU Darmstadt, TU Dresden and IFW Dresden

Name and address of institute	Joint BMBF-Project "MagKal" Participants: Technical University of Darmstadt, Technical University of Dresden, IFW Dresden, all Germany
Name of contact person/email	Prof. Dr. Oliver Gutfleisch/gutfleisch@fm.tu-darmstadt.de
Year of production	2014
Type	Rotary
Maximum frequency	4 Hz
Maximum cooling power	100 W (predicted)
Maximum temperature span	6 K (predicted)
Type of AMR	Parallel plates
MC material(s)	Gd (1 mm)
Type of magnets	Nd–Fe–B, 415 kJ m^{-3}

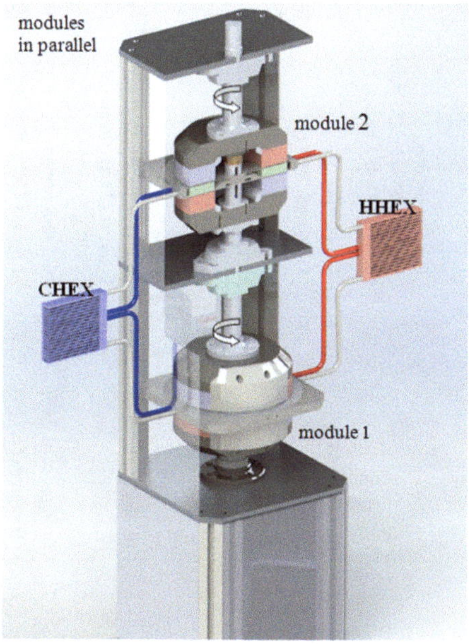

Table 7.37 Rotary magnetocaloric devices built to date, listed with their general characteristics

	Authors, reference	Year	AMR type, materials, mass	Magnet assembly type and magnetic field strength	Heat transfer fluid	Frequency	Temperature span	Specific cooling power
1	Kirol and Dacus [42]	1987	Gd plates, 0.076 mm, 0.126 mm spacing, 270 g	Static Nd–Fe–B, 0.9 T	Water	N/A	N/A	N/A
2	Zimm et al. [43, 44]	2005	Gd particles, 0.425–0.5 mm	Static permanent, 1.5 T	Water	4 Hz	18 K	no-load
							0.5 K	44 W (no mass reported)
			Gd–Er particles, 0.25–0.355 mm, Gd particles, 0.425–0.5 mm				25 K	No-load
							0.5	41 W (no mass reported)
3	Shir et al. [47]	2005	Gd powder	Permanent, 2 T	Helium	N/A	5 K	No-load
4	Zimm et al. [45]	2007	Gd plates, 916 g	Rotating Halbach Nd–Fe–B, 1.5 T	N/A	2 Hz	11.5 K	No-load
							8 K	76.4 W kg^{-1}
							0 K	240.2 W kg^{-1}
5	GE, Table 7.16	2012	N/A	Two dual Halbach	N/A	4 Hz (max)	50 K (max)	100 W (max, no mass reported)
6	Jacobs et al. [46]	2014	Layered La-Fe–Si-H particles, 0.177–0.246 mm, 1520 g	Rotating Halbach Nd–Fe–B, 1.5 T	Water + anti-corrosion agent	4 Hz	12 K	1,375 W kg^{-1}
							0 K	2,001 W kg^{-1}
7	Bohigas et al. [48]	2000	Gd thin ribbon	Static Nd–Fe–B, 0.95 T	Olive oil	0.33 Hz	5 K	No-load
8	Okamura et al. [49]	2005	Layered Gd-Y, Gd–Dy spheres, 0.6 mm, 1000 g	Rotating Nd–Fe–B, 0.77 T	Water	3.33 Hz	1 K	10 W kg^{-1}
9	Okamura et al. [50]	2007	Gd spheres, 0.5 mm, 4000 g	Rotating Nd–Fe–B, 1.1 T	Water	3.33 Hz	5.2 K	37.5 W kg^{-1}
10	Hokakaido Ins., Table 7.19	2009	Gd, Gd alloy packed bed	Permanent (neodymium)	N/A	45 Hz (max)	5.5 K (max)	150 W (max, no mass reported)
11	Kobe Uni., Table 7.20	2011	Gd packed bed	Permanent, 0.6 T	N/A	1 Hz (max)	5 K (max)	10 W (max, no mass reported)

(continued)

Table 7.37 (continued)

	Authors, reference	Year	AMR type, materials, mass	Magnet assembly type and magnetic field strength	Heat transfer fluid	Frequency	Temperature span	Specific cooling power
12	Railway Tech., Table 7.21	2011	Gd, La-Fe-Co-Si packed bed	Halbach Nd–Fe–B	N/A	1.3 Hz (max)	7.6 K (max)	150 W (max, no mass reported)
13	Sanden Co., Table 7.22	2013	Gd, Gd alloy packed bed, 1273 g	Permanent. 0.85 T	N/A	3 Hz (max)	13.8 K (max)	157.1 W (max, no mass reported)
14	West. Switzerland Uni. [51]	2006	Gd particles	Static permanent, 0.8 T	Air	N/A	N/A	N/A
15	Vasille and Müller [52, 53], Cooltech, Table 7.23	2005	Gd, Gd–Tb	Rotating Halbach Nd–Fe–B, 0.98 T	N/A	4 Hz (max)	42 K (max)	120 W (max, no mass reported)
16	Cooltech, Table 7.24	2014	Gd, Gd–Er, La-Fe–Si	Permanent Nd–Fe–B, 1.17 T	N/A	6 Hz (max)	38 K (max)	300 W (max, no mass reported)
17	Grenoble Lab, Table 7.25	2008	Gd parallel plates	Nd–Fe–B	N/A	0.5 Hz (max)	10 K (max)	60 W (max, no mass reported)
18	Tura and Rowe [54–56]	2007, 2009	Gd crushed particles, 0.6 mm, 122 g	Rotating Nd–Fe–B dual Halbach, 1.47 T peak	Water	2 Hz	13 K	No-load
			Gd spheres, 0.3 mm, 110 g			1.4 Hz	29 K	No-load
							2.5 K	454.5 W kg^{-1}
19	Arnold et al. [57]	2014	Gd spherical particles, 0.5 mm, 650 g	Rotating Nd–Fe–B triple Halbach, 1.54 T peak	80 % water + 20 % glycol	0.8 Hz	33 K	No-load
							15 K	77 W kg^{-1}
20	Chen et al. [58], Table 7.28	2002	Gd particles, 0.5 mm, 1,000 g	Static Nd–Fe–B Halbach, 1.5 T	Water	N/A	11.5 K	No-load
							6.8 K	40 W kg^{-1}
21	Sichuan Uni, Table 7.29	2005	Gd spheres	Permanent, 1.5 T	N/A	0.2 Hz (max)	11.5 K (max)	70 W (max, no mass reported)

(continued)

Table 7.37 (continued)

	Authors, reference	Year	AMR type, materials, mass	Magnet assembly type and magnetic field strength	Heat transfer fluid	Frequency	Temperature span	Specific cooling power
22	He et al. [59]	2013	Gd sheets, 1 mm, 198 g, + Stirling regenerative refrigeration	Rotating Halbach, 1.5 T	Helium	1.5 Hz	21.5 K	No-load
							14.9 K	30.3 W kg^{-1}
23, 24	BRIRE, Table 7.30	2006, 2010	Rotating AMR, Gd spheres, 0.4–0.6 mm, 1440 g	Nd–Fe–B, 1.25 T	N/A	0.5 Hz (max)	18 K (max)	34.7 W kg^{-1} (max)
25, 26	BRIRE, Table 7.31	2009, 2012	Gd, Gd–Tb spheres, 0.3–0.5 mm, 1500 g	Rotating Halbach, 1.4 T	N/A	0.2 Hz	20 K (max)	40 W kg^{-1} (max)
25	Coelho et al. [60]	2009	Rotating AMR, Gd pins, 960 g	Electro, 2.3 T	Ethyl alcohol	0.5 Hz	11 K	No-load
27[a]	POLO, Table 7.32	2014	Gd, parallel plates, packed bed, pins	Halbach	N/A	2 Hz (max)	N/A	N/A
28	Tušek et al. [61, 62]	2007	Gd plates, 0.25 mm, 600 g	Static Nd–Fe–B, 0.98 T	Distilled water	4 Hz (max)	N/A	N/A
29	Ljubljana Uni., Table 7.34	2012	Gd, packed bed, parallel plates	Static Nd–Fe–B, 0.8 T	70 % distilled water + 30 % anti-freeze	4 Hz (max)	N/A	N/A
30	Engelbrecht et al. [63]	2012	Gd spheres, 0.25–0.8 mm, 2,800 g	Static Nd–Fe–B Halbach, 1.24 T peak, 0.9 T average	75 % deionized water + 25 % ethylene glycol with anti-corrosion inhibitors	2 Hz	25.4 K	No-load
						1.8 Hz	0.3 K	360.7 W kg^{-1}
	Lozano et al. [64, 65]	2013, 2014				1.5 Hz	12.9 K	142.8 W kg^{-1}
						2 Hz	18 K	35.7 W kg^{-1}
						6 Hz	17 K	
						10 Hz	14.5	
	Bahl et al. [66]	2014				2.25 Hz	18.9 K	71.4 W kg^{-1}
						1.5 Hz	13.8 K	142.8 W kg^{-1}
31	Aprea et al. [67]	2014	Gd spheres, 0.4–0.5 mm, 1,200 g	Rotating Nd–Fe–B Halbach, 1.25 T peak, 1.1 T average	Demineralized water + anti-corrosion inhibitors	0.72 Hz	13.5 K	No-load
32[a]	TU Darmstadt, Table 7.36	2014	Gd plates, 1 mm	Nd–Fe–B, 415 kJ m^{-3}	N/A	4 Hz (max)	6 K (max)	100 W (max, no mass reported)

[a] Not built yet, predictions only

the authors of this book have obtained the design of a 3D model of the first rotary magnetic prototype. It is planned to be built towards the end of 2014. Some information regarding prototype characteristics and 3D model are presented in Table 7.36.

7.3 Conclusion

This section on rotary prototypes presents most of the rotary magnetic devices built so far. There have been more than 20 different rotary devices built in the past 30 years. However, since there are a vast number of different design issues, it is of course hard to address all of them. As one can see, the problems in rotary magnetic refrigeration prototype design are not only in the AMR and magnet assembly design, but also in the peripheral components, such as fluid-flow circuits, valve systems, heat exchangers, etc. Regarding the AMR and magnet assemblies, the choice of the magnetocaloric material, its geometry and the choice of the driving system for rotating the AMR/magnets are also of great importance. It has been noted that all the components may seriously affect the efficiency of the device. However, gradual improvements in the magnetic system components can be seen and there is still a lot more room for optimization and development.

All the rotary devices are collected in Table 7.37, together with their general geometric as well as operating characteristics.

References

1. Brown GV (1976) Magnetic heat pumping near room temperature. J Appl Phys 47 (8):3673–3680
2. Yu B, Liu M, Egolf PW et al (2010) A review of magnetic refrigerator and heat pump prototypes built before the year. Int J Refrig 33:1029–1060
3. Green G, Chafe J (1990) A gadolinium-terbium active regenerator. Adv Cryog Eng 35: 1165−1174
4. Zimm C, Jastrab A, Sternberg A et al (1998) Description and performance of a near-room temperature magnetic refrigerator. Adv Cryog Eng 43:1759–1766
5. Blumenfeld PE, Prenger FC, Sternberg A (2002) High temperature superconducting magnetic refrigeration. In: Advances in cryogenic engineering: proceedings of the cryogenic engineering conference, vol 47
6. Shassere B, Abdelaziz O, Evans B et al (2012) Thermal imaging of active magnetic regenerator MCE materials during operation. In: Fifth IIF-IIR international conference on magnetic refrigeration at room temperature, Thermag V, Grenoble, France, pp 565–572
7. Rowe AM, Barclay JA (2002) Design of an active magnetic regenerator test apparatus. In: Advances in cryogenic engineering: proceedings of the cryogenic engineering conference, vol 47
8. Richard MA, Rowe AM, Chahine R (2004) Magnetic refrigeration: single and multimaterial active magnetic regenerator experiments. J Appl Phys 95(4):2146–2150
9. Rowe A, Tura A (2006) Experimental investigation of a three-material layered active magnetic regenerator. Int J Refrig 29:1286–1293

10. Arnold DS, Tura A, Rowe A (2011) Experimental analysis of a two-material active magnetic regenerator. Int J Refrig 34:178–191
11. Hirano N, Nagaya S, Takahashi M et al (2002) Development of magnetic refrigerator for room temperature application. In: Advances in cryogenic engineering: proceedings of the cryogenic engineering conference, vol 47
12. Kawanami T, Chiba K, Ikegawa M et al (2005) Optimization of a magnetic refrigerator at room temperature for air cooling systems. In: First IIF-IIR international conference on magnetic refrigeration at room temperature, Thermag I, Montreux, Switzerland, pp 275–282
13. Kawanami T (2007) Heat transfer characteristics and cooling performance of an active magnetic regenerator. In: Second IIF-IIR international conference on magnetic refrigeration at room temperature, Thermag II, Portoroz, Slovenia, pp 23–34
14. Nakamura K, Kawanami T, Hirano S et al (2008) Improvement of room temperature magnetic refrigerator using air as heat transfer fluid. In: Thermal issues in emerging technologies, ThETA 2
15. Hirano S, Kawanami T, Nakamura K et al (2009) A development of spherical-shaped magnetocaloric materials using power coating method. In: Third IIF-IIR international conference on magnetic refrigeration at room temperature, Thermag I, Des Moines, Iowa, USA, pp 457–464
16. Yu BF, Gao Q, Wang CF et al (2005) Experimental investigation on refrigeration performance of a reciprocating active magnetic regenerator of room temperature magnetic refrigeration. In: First IIF-IIR international conference on magnetic refrigeration at room temperature, Thermag I, Montreux, Switzerland, pp 375–391
17. Lu DW, Xu XN, Wu HB et al (2005) A permanent magnet magneto-refrigerator study using Gd/Gd-Si-Ge/Gd-Si-Ge-Ga alloys. In: First IIF-IIR international conference on magnetic refrigeration at room temperature, Thermag I, Montreux, Switzerland, pp 291–296
18. Gao Q, Yu BF, Wang CF et al (2006) Experimental investigation on refrigeration performance of a reciprocating active magnetic regenerator of room temperature magnetic refrigeration. Int J Refrig 29:1274–1285
19. Yao GH, Gong MQ, Wu JF (2006) Experimental study on the performance of a room temperature magnetic refrigerator using permanent magnets. Int J Refrig 29:1267–1273
20. Huang J, Liu J, Jin P et al (2006) Development of permanent magnetic refrigerator at room temperature. Rare Metals 25(Spec Issue):641
21. Zheng ZG, Yu HY, Zhong XC et al (2009) Design and performance study of the active magnetic refrigerator for room-temperature application. Int J Refrig 32:78–86
22. Cheng J, Liu G, Huang J et al (2013) Refrigeration effect of $La(FeCoSi)_{13}B_{0.25}$ compounds and gadolinium metal in reciprocating magnetic refrigerator. J Rare Earths 31(12):1163
23. Clot P, Viallet D, Allab F et al (2003) A magnet-based device for active magnetic regenerative refrigeration. IEEE Trans Magn 39(5):3349−3351
24. Dupuis C, Vialle AJ, Legait U et al (2009) New investigations in magnetic refrigeration device, AMR cycle and refrigerant bed performance evaluation. In: Third IIF-IIR international conference on magnetic refrigeration at room temperature, Thermag I, Des Moines, Iowa, USA, pp 437–442
25. Legait U, Guillou F, Kedous-Lebouc A et al (2014) An experimental comparison of four magnetocaloric regenrators using three different materials. Int J Refrig 37:147–155
26. Bour S, Hamm JL, Michot H et al (2009) Experimental and numerical analysis of a recirpocating room temperature active magnetic regenerator. In: Third IIF-IIR international conference on magnetic refrigeration at room temperature, Thermag I, Des Moines, Iowa, USA, 2009, pp 415–424
27. Bahl CRH, Petersen TF, Pryds N et al (2008) A versatile magnetic refrigeration test device. Rev Sci Instrum 79:1−7
28. Engelbrecht K, Jensen JB, Bahl CRH et al (2009) Experiments on a modular magnetic refrigeration device. In: Third IIF-IIR international conference on magnetic refrigeration at room temperature, Thermag I, Des Moines, Iowa, USA, pp 431–436

29. Engelbrecht K, Bahl CRH, Nielsen KK (2011) Experimental results for a magnetic refrigerator using three different types of magnetocaloric material regenerators. Int J Refrig 34:1132–1140
30. Tušek J, Kitanovski A, Flisar A et al (2012) Active magnetic regenerator (AMR) experimental device. In: Fifth IIF-IIR international conference on magnetic refrigeration at room temperature, Thermag V, Grenoble, France, pp 429–436
31. Tušek J (2012) Thermomagnetic properties of active magnetic regenerators. Ph.D. thesis, University of Ljubljana, Faculty of Mechanical Engineering, Ljubljana
32. Tušek J, Kitanovski A, Zupan S et al (2013) A comprehensive experimental analysis of gadolinium active magnetic regenerators. Appl Therm Eng 53:57–66
33. Tušek J, Kitanovski A, Tomc U et al (2014) Experimental comparison of multi-layered La-fe-Co-Si and single-layered Gd active magnetic regenerators for use in a room-temperature magnetic refrigerator. Int J Refrig 37:117–126
34. Tagliafico LA, Scarpa F, Tagliafico G et al (2009) Design and assembly of a linear reciprocating magnetic refrigerator. In: Third IIF-IIR international conference on magnetic refrigeration at room temperature, Thermag I, Des Moines, Iowa, USA, pp 425–430
35. Tagliafico LA, Scarpa F, Valsuani F et al (2013) Preliminary experimental results from a linear reciprocating magnetic refrigerator prototype. Appl Therm Eng 52:492–497
36. Balli M, Sari O, Mahmed C et al (2012) A pre-industrial magnetic cooling system for room temperature application. Appl Energy 98:556–561
37. Sari O, Balli M (2014) From conventional to magnetic refrigerator technology. Int J Refrig 37:8–15
38. Kim Y, Jeong S (2009) Investigation on the room temperature active magnetic regenerative refrigerator with permanent magnet array. In: Third IIF-IIR international conference on magnetic refrigeration at room temperature, Thermag I, Des Moines, Iowa, USA, pp 393–400
39. Trevizoli PV, Barbosa JR Jr, Ferreira RTS (2011) Experimental evaluation of a Gd-based linear reciprocating active magnetic regenerator test apparatus. Int J Refrig 34:1518–1526
40. Czernuszewicz A, Kaleta J, Krolewicz M et al (2014) A test stand to study the possibility of using magnetocaloric materials for refrigerators. Int J Refrig 37:72–77
41. Gomez JR, Garcia RF, Carril JC et al (2013) Experimental analysis of a reciprocating magnetic refrigeration prototype. Int J Refrig 36:1388–1398
42. Kirol LD, Dacus MW (1987) Rotary recuperative magnetic heat pump. In: Cryogenic engineering conference, St. Charles, IL, USA, pp 757–765
43. Zimm C, Boeder A, Chell J et al (2005) Design and performance of a permanent magnet rotary refrigerator. In: First IIF-IIR international conference on magnetic refrigeration at room temperature, Thermag I, Montreux, Switzerland, pp 367–376
44. Zimm C, Boeder A, Chell J et al (2006) Design and performance of a permanent-magnet rotary refrigerator. Int J Refrig 29:1302–1306
45. Zimm C, Auringer J, Boeder A et al (2007) Design and initial perfomance of a magnetic refrigerator with rotating permanent magnet. In: Second IIF-IIR international conference on magnetic refrigeration at room temperature, Thermag II, Portoroz, Slovenia, pp 341–347
46. Jacobs S, Auringer J, Boeder A et al (2014) The performance of a large-scale rotary magnetic refrigerator. Int J Refrig 37:84–91
47. Shir F, Bennet LH, Mavriplis C (2005) Transient response in magnetocaloric regeneration. IEEE Trans Magn 41(6):2129−2133
48. Bohigas X, Molins E, Roig A et al (2000) Room-temperature magnetic refrigerator using permanent magnets. IEEE Trans Magn 36(3):538−544
49. Okamura T, Yamada K, Hirano N et al (2005) Performance of a room-temperature rotary magnetic refrigerator. In: First IIF-IIR international conference on magnetic refrigeration at room temperature, Thermag I, Montreaux, Switzerland, pp 319–324
50. Okamura T, Rachi R, Hirano N et al (2007) Improvement of 100 W class room temperature magnetic refrigerator. In: Second IIF-IIR international conference on magnetic refrigeration at room temperature, Thermag II, Portoroz, Slovenia, pp 377–382
51. Yu B, Liu M, Egolf PW et al (2010) A review of magnetic refrigerator and heat pump prototypes built before the year 2010. Int J Refrig 33:1029–1060

52. Vasile C, Muller C (2005) A new system for a magnetocaloric refrigerator. In: First IIF-IIR international conference on magnetic refrigeration at room temperature, Thermag I, Montreux, Switzerland, pp 357–366
53. Vasile C, Muller C (2006) Innovative design of a magnetocaloric system. Int J Refrig 29:1318–1326
54. Tura A, Rowe A (2007) Design and testing of a permanent magnet magnetic refrigerator. In: Second IIF-IIR international conference on magnetic refrigeration at room temperature, Thermag II, Portoroz, Slovenia, pp 363–370
55. Tura A, Rowe A (2009) Progress in the characterization and optimization of a permanent magnet magnetic refrigerator. In: Third IIF-IIR international conference on magnetic refrigeration at room temperature, Thermag III, Des Moines, Iowa, USA, pp 387–392
56. Tura A, Rowe A (2011) Permanent magnet magnetic refrigerator design and experimental characterization. Int J Refrig 34:628–639
57. Arnold DS, Tura A, Ruebsaat-Trott A et al (2014) Design improvements of a permanent magnet active magnetic refrigerator. Int J Refrig 37:99–105
58. Chen YG, Tang YB, Wang BM et al (2007) A permanent magnet rotary magnetic refrigeration. In: Second IIF-IIR international conference on magnetic refrigeration at room temperature, Thermag II, Portoroz, Slovenia, pp 309–313
59. He XN, Gong MQ, Zhang H et al (2013) Design and performance of a room-temperature hybrid magnetic refrigerator combined with Stirling gas refrigeration effect. Int J Refrig 36 (5):1465–1471
60. Coelho AA, Gama S, Magnus A et al (2009) Prototype of a Gd-based rotating magnetic refrigerator for work around room temperature. In: Third IIF-IIR international conference on magnetic refrigeration at room temperature, Thermag III, Des Moines, Iowa, USA, 2009, pp 381–386
61. Tušek J, Zupan S, Šarlah A et al (2009) Development of a rotary magnetic refrigerator. In: Third IIF-IIR international conference on magnetic refrigeration at room temperature, Thermag III, Des Moines, Iowa, USA, pp 409–414
62. Tušek J, Zupan S, Šarlah A et al (2010) Development of a rotary magnetic refrigerator. Int J Refrig 33(2):294–300
63. Engelbrecht K, Eriksen D, Bahl CRH et al (2012) Experimental results for a novel rotary active magnetic regenerator. Int J Refrig 35(6):1498–1505
64. Lozano JA, Engelbrecht K, Bahl CRH et al (2013) Performance analysis of a rotary active magnetic refrigerator. Appl Energy 111:669–680
65. Lozano JA, Engelbrecht K, Bahl CRH et al (2014) Experimental and numerical results of a high frequency rotating active magnetic refrigerator. Int J Refrig 37:92–98
66. Bahl CRH, Engelbrecht K, Eriksen D et al (2014) Development and experimental results from 1 kW AMR. Int J Refrig 37:78–83
67. Aprea C, Greco A, Maiorino A et al (2014) Initial experimental results from a rotary permanent magnet magnetic refrigerator. Int J Refrig 43:111–122

Chapter 8
Design Issues and Future Perspectives for Magnetocaloric Energy Conversion

This chapter provides information about the design characteristics for particular types of magnetic refrigerators and heat pumps. The sections therefore indicate the different categories to which a particular magnetocaloric device can belong. For these we show the different design configurations. These are related to linear devices that are based on the linear movement of a magnetic field source or magnetocaloric material; rotary devices, which are based on the rotation of the magnetic field source or the rotation of the magnetocaloric material; and static magnetocaloric devices. Some of the configurations were already applied in research. However, we also present a few new solutions, which are based on our research experiences and which might be applied in future studies and related devices. We again address the importance of the thermal diode mechanism in combination with the active magnetic regeneration (AMR) principle. In addition, a note on power generation is added, with a brief review of the existing work in this particular domain. By pointing out the most successful design approaches, this chapter also serves as a future guideline on the magnetic refrigeration and heat pumping.

Since the magnetocaloric devices will, in the following decade, most probably start to penetrate some market niches, it is important, before starting, to address the standardization (or classification) issue relating to magnetocaloric energy conversion. Only a few studies have been performed in this particular and very important domain. However, the research community in the field of magnetocaloric energy conversion decided to enter such a standardization process.

For instance, for devices, Scarpa et al. in 2012 [1] proposed a classification method for room-temperature magnetic refrigerators. They proposed a classification of magnetic refrigerators using 12 different criteria, marked by the numbers 1–12. Here, we denote these; however, we marked in bold a slight addition and changes that we think should be included or modified in such a classification.

(1) Device type **(0 for single-stage cycle without regeneration, 1 for passive regeneration, 2 for active regeneration)**,
(2) magnet type (0 for permanent, 1 for electromagnet),
(3) type of permanent magnets **(0 for simple magnets, 1 for 2D magnets with Halbach principle, 2 for 3D magnets with Halbach principle)**,

A. Kitanovski et al., *Magnetocaloric Energy Conversion*,
Green Energy and Technology, DOI 10.1007/978-3-319-08741-2_8

(4) magnetic field application (0 by immersion of magnetocaloric material, 1 by fields composition of moving magnets, 2 by switching on/off electromagnets),
(5) number of layers in magnetocaloric regenerators (number of different magnetocaloric material layers with different Curie temperatures),
(6) magnetocaloric regenerators structure (0 for ordered: calibrated spheres, plates, wires, sheets, honeycomb; 1 for disordered: random spheres, powder, fibres, porous matrix; 2 for magnetocaloric fluid: ferrofluid, particle suspension; **3 for magnetocaloric material with thermal diodes**),
(7) working fluid (0 for liquid, 1 for gas, **2 for any phase-change fluid**),
(8) pumping system (0 for bi-directional; 1 for uni-directional),
(9) relative motion between the magnetocaloric material and magnet (0 for no motion, 1 for motion),
(10) relative motion between the magnetocaloric material and magnet (0 for static magnetocaloric material, 1 for moving magnetocaloric material),
(11) relative motion between the magnetocaloric material and magnet (0 for linear, 1 for rotational, **2 for static device**) and
(12) relative motion between the magnetocaloric material and magnet (0 for discontinuous bi-directional, 1 for discontinuous uni-directional; 2 for continuous).

We also propose to use the following characters in front of the number of the classification, as follows: **R** for refrigerators, freezers, chillers; **H** for heat pumps; **AC** for air conditioners and **P** for magnetocaloric power generators. In the following text we use the classification mentioned above for all types of refrigeration magnetocaloric devices. In cases that all or at least two options of a certain classification can be used, we simply leave the brackets and the number of such a classification.

8.1 Linear AMR Magnetocaloric Devices

In this chapter we do not focus on the principles where a single AMR is used, because this, especially at low frequencies of operation, does not provide a continuous cooling process and requires higher work input. This is especially the case because the two forces on the magnetocaloric material during the magnetization or demagnetization process are not counter-balanced.

In Fig. 8.1, a linear system is shown in which the magnet assembly is moved over the two AMRs, so all the time one AMR is magnetized and the other is demagnetized. Of course, this only holds for the case when we neglect the time required for the movement of a magnet. In this particular case the linear device means that the linear movement of the magnet assembly will be provided over the first AMR shown in Fig. 8.1a. Since the magnetocaloric material will heat due to the positive change of the magnetic field, by using the AMR principle, the fluid will flow from the CHEX (heat source heat exchanger) through the AMR to the HHEX (heat sink heat exchanger). The fluid flow in this case is driven by a bi-directional

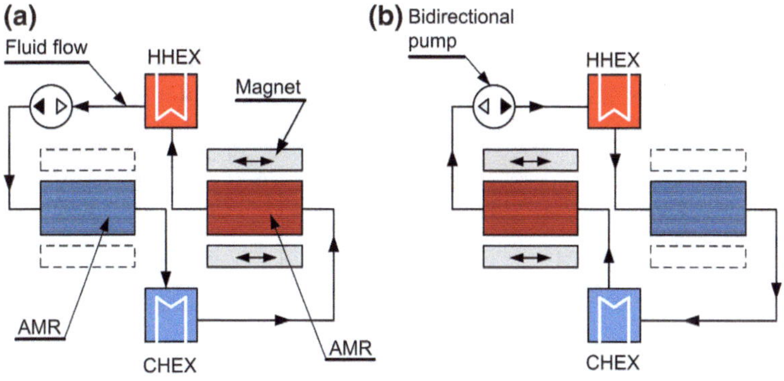

Fig. 8.1 A magnetocaloric device with a linear movement of the magnet assembly over two static AMRs with the application of a bi-directional pump (**a** *left* AMR demagnetized, *right* AMR magnetized; **b** *left* AMR magnetized, *right* AMR demagnetized), classification no: R20(3)1(5)(6) 001000

pump. This will enable the flow of the fluid from the HHEX through the demagnetized AMR. There, the fluid will cool down and it will continue to flow to the CHEX. In the case of Fig. 8.1b, the first cycle (shown in Fig. 8.1a) will actually be repeated, and this will be provided by the movement of the magnetic field source from the first to the second AMR. The previously warm AMR will thus cool down due to demagnetization and the previously cold AMR will heat up due to magnetization. In this particular case the bi-directional pump will switch the flow direction and thus the fluid flow for the case in Fig. 8.1b will be provided in the opposite direction with respect to the case of Fig. 8.1a. This kind of system may be considered as simple since it does not necessarily need switching valves. Switching is provided by the bi-directional pump. Since the oscillating flow is also provided through the CHEX and HHEX heat exchangers we have to keep in mind that in this particular case they somehow also act as the passive regenerators, which can slightly influence their design.

It is simple to show the case in Fig. 8.1 for the rotating magnet assembly. In this particular case the reader has to observe Fig. 8.1 as she (he) would look at it from the rotating centre, thus in the radial direction. Figure 8.2 shows such an example. In the particular case of Fig. 8.2 we show four AMRs to be put together in a device. Therefore, changing the position of the magnets from a magnetized AMR to a demagnetized AMR will require switching of the bi-directional pump. Note that the example in Fig. 8.2 is only illustrative. More information about rotary devices is given in the subsequent text.

Note that in any case, the bi-directional pump should be as close as possible to both heat exchangers, CHEX and HHEX, respectively. This also holds for the AMR. The reason to reduce such a distance to the minimum possible is in order to avoid a too large dead volume. In the worst case, the fluid particle from the AMR will actually not enter the heat exchanger or vice versa. This, of course, means that the system will not operate.

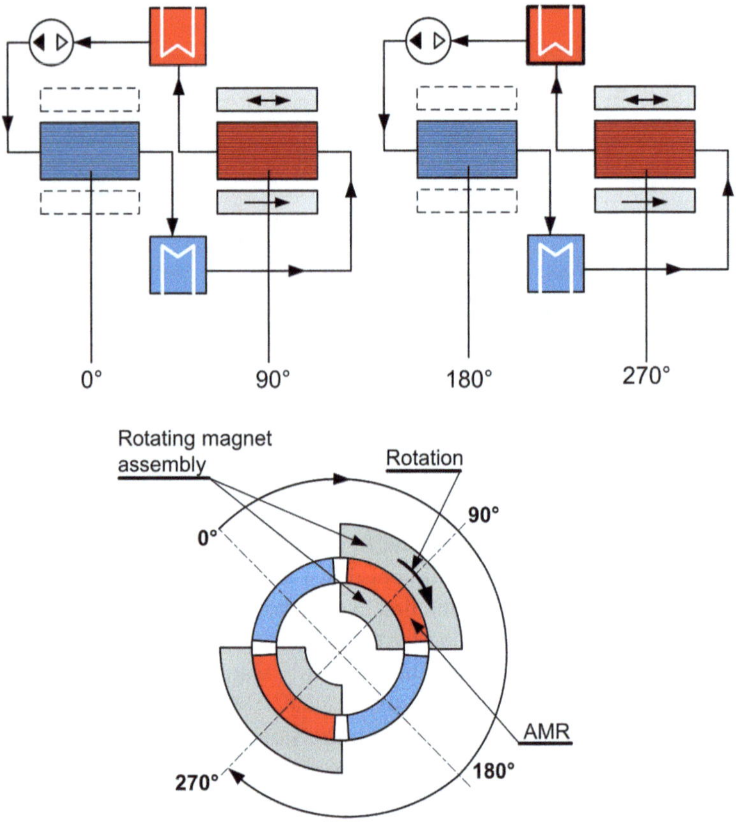

Fig. 8.2 Example of rotary movement for the magnetocaloric device, based on the same principle as the linear device in Fig. 8.1, classification no: R20(3)1(5)(6)00101(12)

When the uni-directional pump is applied, then the system in Fig. 8.1 requires a valve system, which switches the fluid flow in order to provide the oscillatory flow of the fluid through the AMRs. In this particular case, the system in Fig. 8.1 has to be transformed into the one presented in Fig. 8.3. In the case of uni-directional pump the valve system should be as close as possible with regard to the AMR and the heat exchangers. This is due to the same reason of avoiding or minimizing the dead volume.

Note also that the fluid flow through the heat sink HHEX and the heat source CHEX is in this case always in the same direction, so no passive regeneration occurs in the heat exchangers. The system in Fig. 8.3, however, is more complex. The switching-valve system can be stationary or rotary. A smart design can also lead to a single, but complex, valve system. By using the Roman numbers from I to XI in Fig. 8.3 we denote the different processes that the fluid experiences.

Let us take, for example, Fig. 8.3a. The fluid flows from I through the CHEX. Therefore, it absorbs heat from the cooled environment or media and reaches state

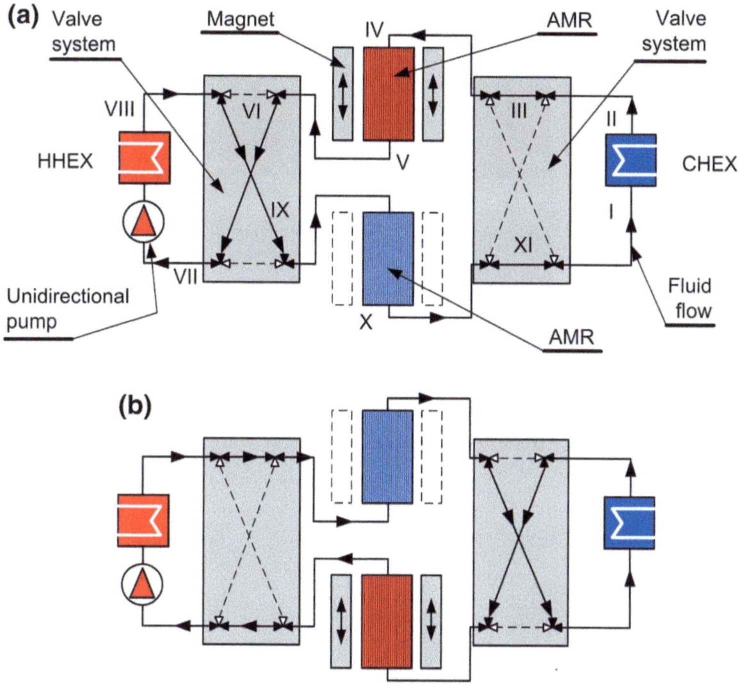

Fig. 8.3 A magnetocaloric device with a linear movement of the magnet assembly over two static AMRs with the application of the switching-valve system and the uni-directional pump (**a** *upper* AMR magnetized, *lower* AMR demagnetized; **b** *upper* AMR demagnetized, *lower* AMR magnetized), classification no: R20(3)1(5)(6)011001

II. From II it enters the switching-valve system III, which guides the fluid to the AMR, which is in the magnetized state IV. There the fluid absorbs heat from the AMR V and passes to the second valve system VI, where it is redirected to the pump VII and HHEX. In HHEX it rejects heat to the environment VIII.

Then the fluid again enters the switching-valve system IX, which guides the fluid to another AMR, which is in the demagnetized state. The fluid cools down X and enters the valve system XI. When the magnets move from the position in Fig. 8.3a to the position in Fig. 8.3b, the valve system switches the directions of the fluid flow and thus provides a quasi-continuous production of cold. We denote it as "quasi" because of the fact that during the movements of the magnets, especially if the AMR Brayton-like process is applied, this will require a certain time period.

If the magnetocaloric material moves instead of the magnetic field source (Fig. 8.4), this can be provided either with a static permanent magnet assembly or with an electric coil. In the latter case, it of course makes sense to avoid any motion (with "on-off" operation of the electromagnet). Note that the first magnetic refrigerators based on superconducting solenoids applied the motion of the magnetocaloric material in and out of the superconducting solenoid.

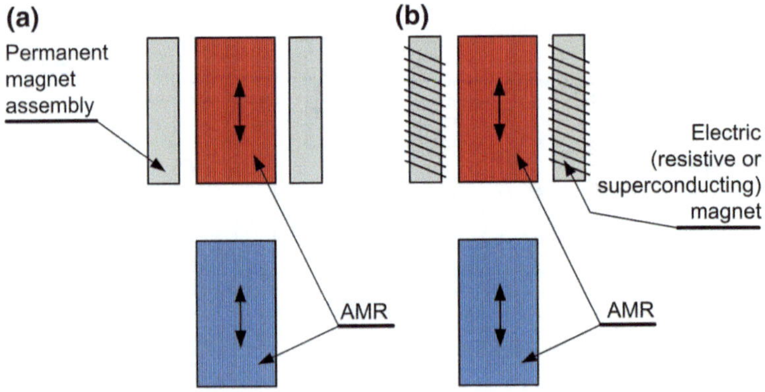

Fig. 8.4 **a** The linear operating principle with moving magnetocaloric regenerators, classification no: R20(3)0(5)(6)0(8)1001, **b** the linear operating principle with a static magnetic field source, classification no: R21(3)2(5)(6)000

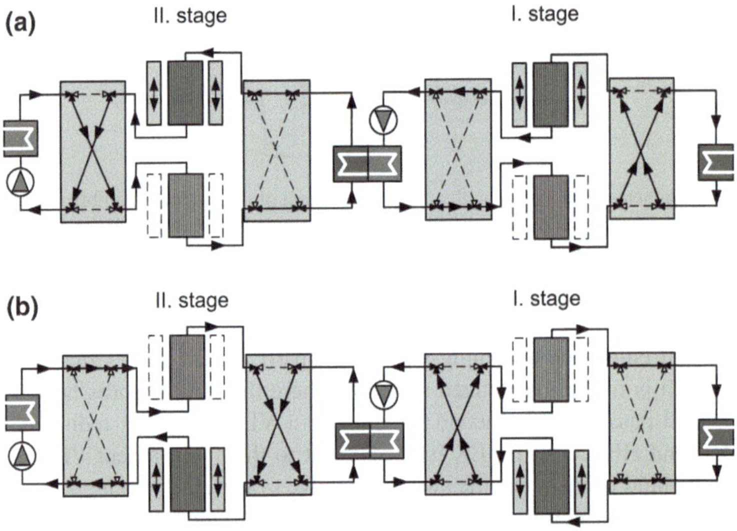

Fig. 8.5 A cascade magnetocaloric device based on the linear operating principle with moving magnetic field sources (**a** *upper* AMRs magnetized, *lower* AMRs demagnetized; **b** *upper* AMRs demagnetized, *lower* AMRs magnetized)

There is a possibility that the single magnetic refrigeration unit is connected in a cascade arrangement with the second magnetic refrigeration unit (Fig. 8.5).

This would be similar to a conventional cascade refrigeration system (e.g., double, triple or multiple stage). However, if each of the magnetic refrigeration units provides an AMR cycle, then we will show that the cascading does not make much sense, unless the geometric or other restrictions demand this. Figure 8.5 shows an example of the connection of two AMR-based magnetic refrigeration units. This is

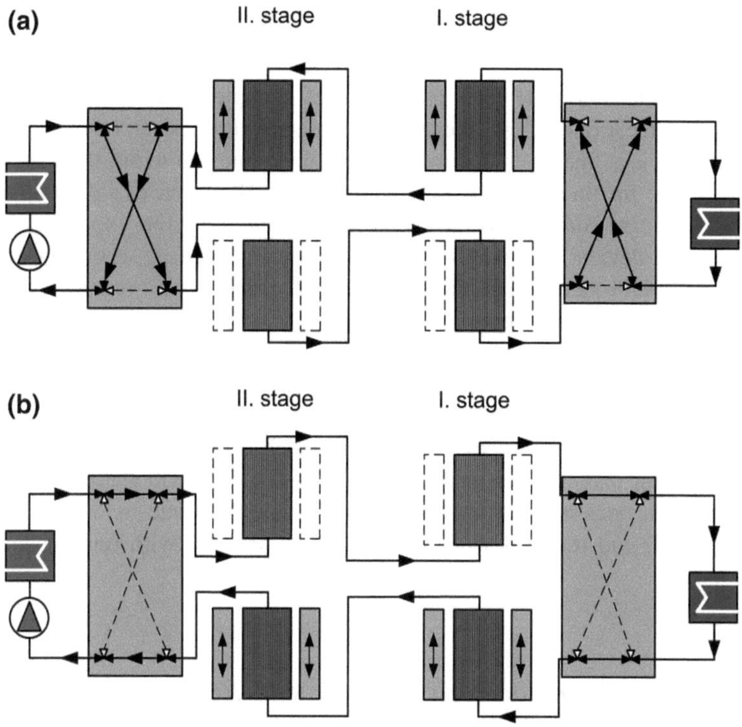

Fig. 8.6 A simplified cascade magnetocaloric device based on the linear operating principle with moving magnetic field sources (**a** *upper* AMRs magnetized, *lower* AMRs demagnetized; **b** *upper* AMRs demagnetized, *lower* AMRs magnetized)

done in a similar way as in conventional double-stage refrigerators, where the condenser of the first unit is coupled with the evaporator of the second unit.

As one can see, the transfer of heat between the lower and upper stages is usually provided by the counter-flow heat exchanger. In the ideal case, the inlet temperature on the warm side would equal the inlet temperature at the cold side. If one removes the intermediate heat exchanger then a new configuration can be obtained, as shown in Fig. 8.6.

As can be seen from Fig. 8.6, when a cascade system is required to produce a larger temperature span between the heat source and the heat sink, but the AMR principle is applied, then the intermediate heat exchangers between the two stages are actually not required. The reason for this is the fact that the AMRs already provide the regeneration of the heat and therefore the temperature span. From the point of view of design and also efficiency, it makes sense to actually have a single and longer AMR instead of two shorter AMRs, such as shown in the case of Fig. 8.6. In this way one can avoid any potential dead volume, heat transfer and heat losses (gains) and other losses, related, e.g. to additional viscous losses, magnetic flux leakage, etc.

8.2 Rotary AMR Magnetocaloric Devices

Rotary magnetocaloric devices present more complex and advanced systems compared to linear ones. Also, the efficiency of such devices can be expected to be higher for devices that operate continuously, since one can avoid the losses associated with the inertia and discontinuous operation. However, in a discontinuously rotating device no substantial advantage in efficiency can be expected over the linear type of device. This holds true, especially for devices, where the magnetocaloric material rotates, since these devices are in most cases related to losses in the valve distribution system and its friction or leakage. However, in most rotary devices, we can certainly expect the advantage of compactness or higher power density for such a device, compared to a linear device.

In this section we provide some brief information about the different arrangements of magnetic field sources and AMRs, which also depend on the rotation of these two main parts of the magnetocaloric device. We will therefore generally divide these devices into: rotary devices with static magnetic field sources and a rotating AMR, and rotary devices with a static AMR and with rotating magnetic field sources.

8.2.1 Rotary Magnetocaloric Devices with Rotating AMRs

These devices can be further divided according to the fluid flow through the AMR as follows:

- Axial fluid-flow devices,
- Radial fluid-flow devices,
- Azimuth fluid-flow devices.

8.2.1.1 Axial Rotary Magnetocaloric Devices with Rotating AMRs

Figure 8.7 shows an example of a two-pole rotary magnetic device in which the working fluid flows through the AMRs in the axial direction (direction of the observer in Fig. 8.7, parallel to the axis of rotation). The principle of the operation is also explained in Fig. 8.7 (right). In this particular case we do not show the fluid switching-valve system, but the reader should note that this, in the case of the unidirectional pump, will not differ much from the one presented in Fig. 8.3.

According to Fig. 8.7 (right), the coaxial cylinder consisting of four AMRs is rotating through the two high and two low magnetic field regions. We consider in this case a continuously rotating and operating device. This means that not only is the continuous rotation provided, but also that the continuous fluid flow is provided to the magnetocaloric material, so in any of time periods such a device produces cold.

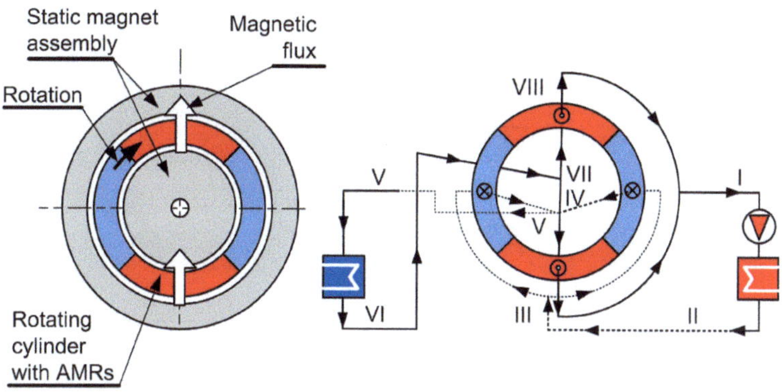

Fig. 8.7 An example of the operating principle of an axial rotary magnetocaloric device, classification no: R20(3)0(5)(6)011112

The fluid flows continuously through the two interconnected and oppositely positioned AMRs, since these operate at the same temperature level. The warm working fluid leaves the AMRs in the magnetized state and enters the HHEX at position I. There it rejects the heat to environment II and flows to the fluid distribution valve (not shown), which divides fluid flow III and guides it to the AMRs in the demagnetized state. Note that the entering of the working fluid into the AMRs in the magnetized state and the demagnetized state has to be provided on opposite sides of the coaxial AMR ring (to provide a counter-flow direction). The working fluid cools down in both demagnetized AMRs and flows to the valve system (flow divider) IV, which directs the fluid flow to the CHEX V. In the CHEX the working fluid heats VI since it absorbs heat from the cooled environment and flows back to valve system VII, which directs the fluid flow to both magnetized AMRs. In these AMRs the working fluid absorbs heat VIII and flows towards the CHEX.

The magnetic field source in axial rotating devices can be very different. For instance, in Fig. 8.8, two different cases are presented in which the magnet assembly does not necessarily cover the whole region of the AMR's coaxial ring. Therefore, we refer to it as the opened magnet assembly.

In Fig. 8.8a, the coaxial ring consists of eight separate AMRs, which can continuously or discontinuously rotate through the magnetic field. A discontinuous rotation in this case would require fast rotation for 180° (note that in the case of Fig. 8.7, a discontinuous rotation of 90° would be required). The fast rotation is desired to provide the adiabatic conditions, as are, for instance, required for the AMR Brayton-like cycle.

Continuous rotation, however, is also possible, but this can increase the complexity of the valve-distribution system. This holds especially for the case when separate AMRs are rotating and when the AMR passes from the low to high magnetic field region and vice versa.

This can be partially solved by applying other thermodynamic cycles than the AMR Brayton-like one. Whereas in Fig. 8.8a the magnetic flux direction is

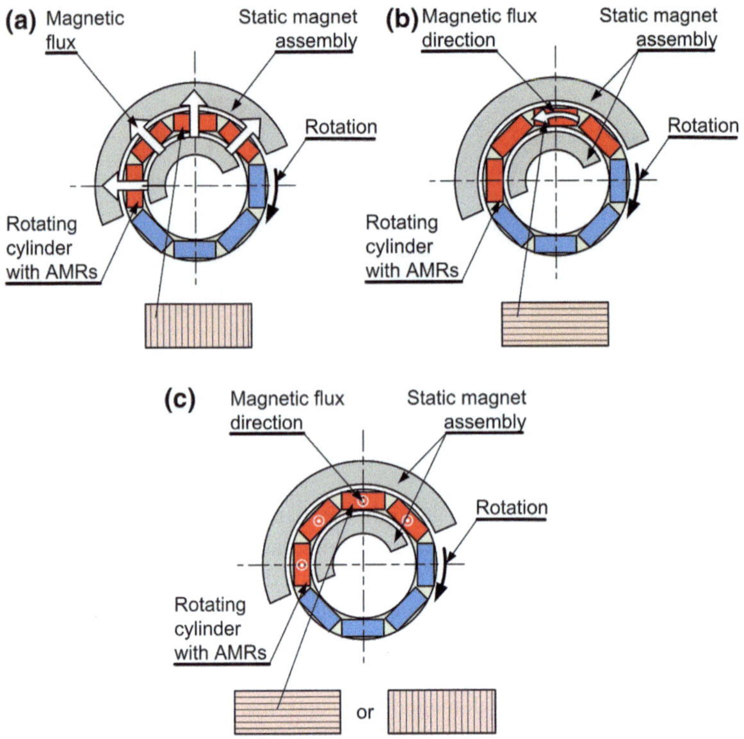

Fig. 8.8 Axial rotary magnetocaloric device principle with an opened magnet assembly. The figure shows three different applications of the magnetic flux direction (**a** magnetic field in radial direction; **b** magnetic field in azimuth direction; **c** magnetic field in axial direction), classification no: R20(3)0(5)(6)0(8)111(12)

provided in the radial direction, note that the arrangement of magnets can be made also in the azimuth direction (Fig. 8.8b) or the axial direction (Fig. 8.8c).

The rotation of the coaxial structure with the AMRs can be continuous or discontinuous. In the latter case, the stepper motor can, for instance, provide such a rotation. The stepper motor comprises an incremental drive digital actuator and is driven in fixed angular steps. The digital signal is used to drive the motor and for every time sequence given by the digital pulse the rotor of the motor will rotate by a specific number of degrees in rotation. The faster the motor turns, the lower will be the torque. This is another reason why low torque is desired. Another factor is the number of poles. If the stepper motor performs two revolutions per second (which is actually a very slow rotation speed, also for stepper motors) for a desired torque, then the frequency of operation of 8 Hz of the magnetic refrigerator will require a 4-pole magnetic field source. The stepper motors, depending on their type, can operate with 50–500 steps per revolution and provide between 1,000 and 10,000 steps per second. However, as an example, 20 Nm will be provided at the rotation by 500 steps per second, and 5 Nm at the rotation speed of 1,500 steps. Compared

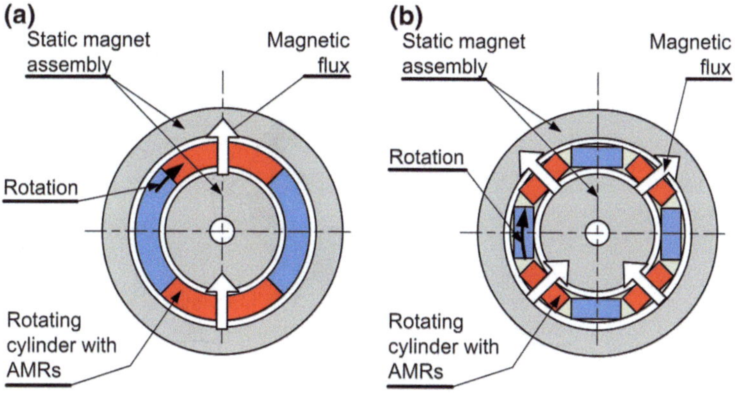

Fig. 8.9 Axial rotary magnetocaloric device principle with the closed magnet assembly, **a** two-pole magnet assembly, classification no: R20(3)0(5)(6)01111(12), **b** four-pole magnet assembly, classification no: R20(3)0(5)(6)01111(12)

to other permanent motor drives, stepper motors are usually not very energy efficient; therefore, we need to take important care about this issue.

Closed magnet assemblies generally provide better "efficiency" of the magnets, as well as offer better shielding of the magnetic flux to the environment. Two such examples are shown in Fig. 8.9.

As can be seen from Figs. 8.9 and 8.10, different number of poles can be applied in the system. Furthermore, the coaxial rotating ring may consist fully of magnetocaloric material (one rotating AMR), or may consist of separated AMRs. In the latter case, especially if the magnetic field distribution in the high and low field is not homogeneous, it makes sense to apply the stepper motor and the fast rotation of the coaxial ring. This issue however depends also on the type of the thermodynamic cycle. The difference between applying the full coaxial ring of AMRs or separated AMRs is mostly in the rotation principle (continuous or discontinuous) or in the valve distribution system. The latter needs to provide or prevent certain fluid flows for the AMRs, which are in transition from the high to low magnetic field region or vice versa. Other issues relate to the costs of the construction of the AMR.

Axial rotary devices may also apply multiple magnet assemblies, as shown in Fig. 8.10. However, as noted before, the closed magnet assembly will provide better usage of the magnetic energy by reducing the leakage of the magnetic flux.

Another arrangement of magnet assemblies is shown in Fig. 8.11. Both solutions a and b, respectively, show two-pole arrangements. However, Fig. 8.11a shows another, alternative solution for the guidance of the magnetic flux.

In Fig. 8.11a, the whole magnet assembly contributes to the guidance and concentration of the magnetic flux, whereas in the case of Fig. 8.11b, only a portion of the magnet assembly is actually active. Other parts (where AMRs in the demagnetized state are positioned) serve mostly for shielding. In a comparison of both cases, Fig. 8.11a provides a more efficient solution (with regard to the intensity

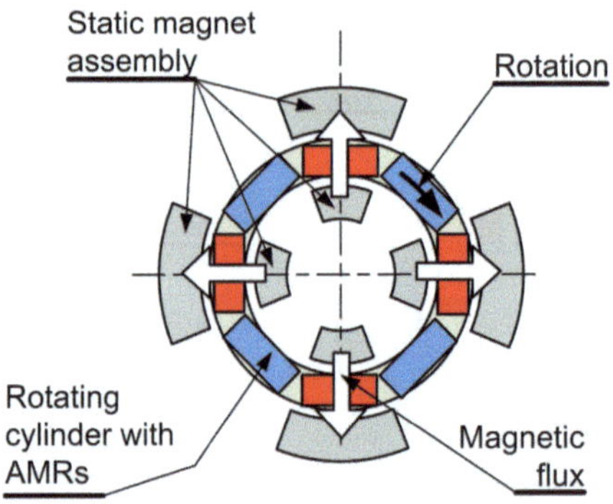

Fig. 8.10 Four-pole opened magnet assembly for the axial rotary magnetocaloric device principle with eight beds of AMR, classification no: R20(3)0(5)(6)0(8)111(12)

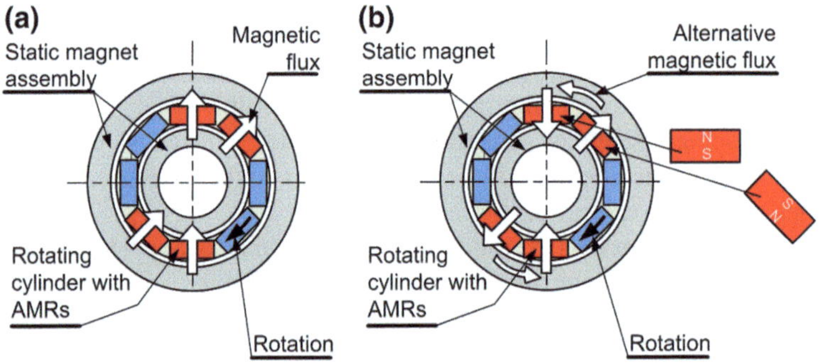

Fig. 8.11 **a** Two-pole closed magnet assembly for the axial rotary magnetocaloric device principle with eight beds of AMR, classification no: R20(3)0(5)(6)0(8)111(12), **b** two-pole closed magnet assembly for the axial rotary magnetocaloric device principle with eight beds of AMR a simple permanent magnet configuration no: R20(3)0(5)(6)0(8)111(12)

of the magnetic flux density and consequently the efficiency of a device), whereas the case in Fig. 8.11b provides a cheaper solution, which will in most cases also lead to a lower magnetic flux density in the high magnetic field region.

8.2.1.2 Radial Rotary Magnetocaloric Devices with Rotating AMRs

In radial devices, the working fluid is usually applied from the central axis of the rotation, from where it is distributed by the valve system. Figure 8.12 shows an example of applying the "opened" magnet assembly. This is also a more common option for radial devices compared to a closed magnet assembly. However, imagine that the magnet assembly is constructed as a 3D magnet assembly (where the AMR is sandwiched between the upper and lower blocks of the magnet assembly). In this case a closed arrangement can also be constructed. According to Fig. 8.12, the magnet assembly has to provide the magnetic flux in the direction of the observer of the system (direction of the central rotating axis of the coaxial ring with AMRs). The fluid flow is therefore usually directed from the central axis in a radial direction through the AMRs or vice versa, from the outer diameter towards the centre of the device. Similar as in the case of Fig. 8.7, we can describe the operation of such an architecture.

According to Fig. 8.12, the fluid flows continuously through two, oppositely positioned and interconnected AMRs, since these operate at the same temperature level. Note, also, in this case that the entering of the working fluid in the AMRs in the magnetized state and demagnetized state has to be provided on opposite sides of the coaxial AMR ring in the radial direction. The warm working fluid that is leaving the AMRs in the magnetized state enters the HHEX at position I. There it rejects heat II to the environment and flows to the fluid distribution valve (not shown), which divides fluid flow III and guides it in a radial direction through the AMRs in the demagnetized state. The working fluid cools down in both demagnetized AMRs and flows to the valve system (flow divider) IV, which directs the fluid flow to the CHEX V. In the CHEX the working fluid heats VI since it absorbs heat from the cooled environment and flows back to the valve system VII, which directs the fluid flow in the radial direction to both magnetized AMRs. In those AMRs, the working fluid absorbs the heat VIII and flows towards the CHEX. Also, in this case, the rotation of the coaxial structure with the AMRs can be continuous or discontinuous.

Figure 8.13 shows two ways of applying the magnetic flux direction through the AMR. In the first case (Fig. 8.13a), the parallel plates of the AMR have to be arranged in the direction of the magnetic flux, i.e. in the radial direction. In the case of Fig. 8.13b, the magnetic flux direction is parallel to the axis of rotation. Therefore, the parallel plates in the AMR have to be arranged in the same direction with the magnetic flux.

As can be seen in this case, the length of the AMR is more restricted in radial devices than in axial ones. Because of this the difference between the outer and inner diameters of such a ring has to be sufficiently long (e.g. from about 50 to

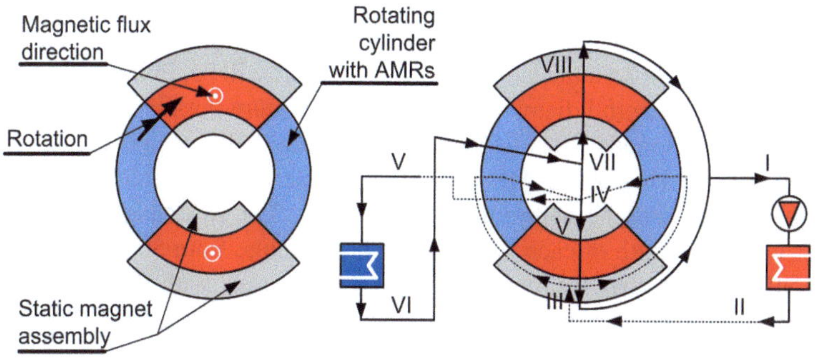

Fig. 8.12 An opened magnet assembly of the rotary magnetocaloric device with the radial fluid flow through AMRs, classification no: R20(3)0(5)(6)01111(12)

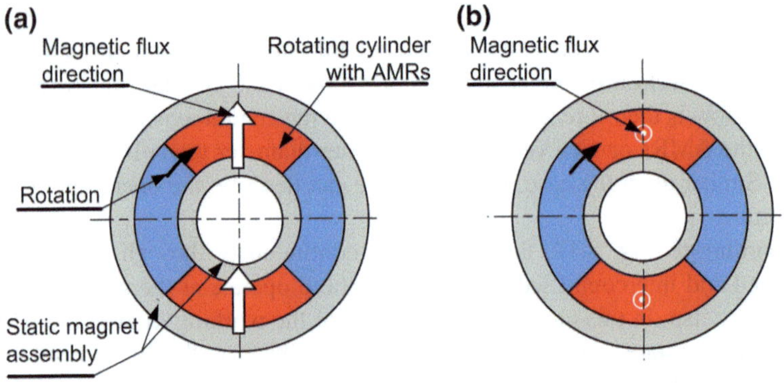

Fig. 8.13 Different magnetic flux directions will require different arrangements of AMR plates as shown in cases (**a**) and (**b**), classification no: R20(3)0(5)(6)0(8)111(12)

150 mm, depending on the type of the working fluid, the AMR structure, the magnetocaloric material, the frequency of the operation and of course the desired temperature span). This, however, in some cases will not be sufficient to provide the desired temperature span. However, there are solutions in which for instance the fluid flows forwards and backwards in the radial direction through the U-shaped AMR. Another solution is when the fluid passes the AMR in the radial direction; however, it also moves in a kind of zig-zag arrangement in the azimuth direction.

Also in radial devices, the number of poles (high magnetic field regions) can vary. Therefore, a similar example as shown in Figs. 8.10 and 8.11 can be applied for radial devices as well.

8.2.1.3 Azimuth Rotary Magnetocaloric Devices with Rotating AMRs

The magnetocaloric material in this case is attached to a rotating ring. It is important to separate the AMRs in order to provide pipe connections and therefore enable the azimuth flow of the working fluid (Fig. 8.14). What kind of central diameter of such an AMR structure (ring) has to be provided? For the two-pole magnet, four AMRs can be applied. Since the space that allows the entering and exiting of the working fluid has to be comprised, the length of a single AMR will be, in this case, about 0.5 of the value of the central diameter (where the maximum possible length of each of four AMRs corresponds to π d/4, where d represents the central diameter of the AMR ring) of the ring with AMRs. In the four-pole magnet (eight AMRs), half of this value (0.25) will be approximately available. Let us consider the four-pole magnet with the length of each AMR being in the range from 50 to 150 mm. Then this would require that the outer diameter of the ring with eight AMRs (four magnetized and four demagnetized) is in the range from about 200 to 600 mm (for the length of a single AMR being L = 0.25 d). This simple example shows the design restrictions of devices based on the azimuth flow of the working fluid. Furthermore, one should also take care about finding the proper fluid flow solution

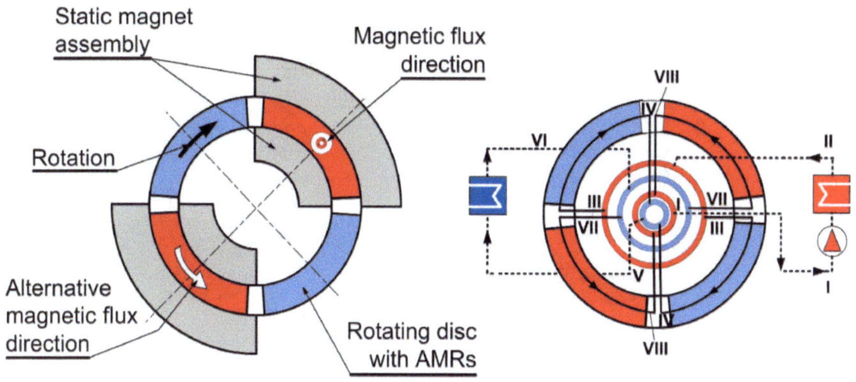

Fig. 8.14 An opened magnet assembly of a rotary magnetocaloric device with the azimuth fluid flow through AMRs, classification no: R20(3)0(5)(6)01111(12)

for the case when the AMR is in transition from the high-field to the low-field region and vice versa. Furthermore the discontinuous rotation would certainly lead to better performance of a device.

In a rotating disc with AMRs and with the azimuth flow of the working fluid, usually a central valve system will serve for the switching and the distribution of the fluid. In Fig. 8.14 (right), a simple example is shown, where we denote using the Roman numbers from I to VIII the particular process with regard to the fluid flow. The case of Fig. 8.14 (right) shows the application of the uni-directional pump I, which pumps the fluid through the HHEX. There the fluid rejects the heat to the environment and flows further to the valve-distribution system. From this the fluid is redirected to two AMRs being in the demagnetized state III. When the working fluid passes the demagnetized AMRs, it cools down IV and then re-enters the valve system in which it is redistributed again to V. However, this time it flows toward the CHEX, where it absorbs heat VI from the refrigerated environment. The working fluid flows back to the valve system, which guides it now towards the AMRs in the magnetized state VII. By passing the magnetized AMRs the working fluid heats up VIII and continues the flow to the valve system.

In Fig. 8.15a, the AMRs are shaped as rectangular inserts. This simplifies their production. However, one should take care of the magnetic flux density distribution in such AMRs. In Fig. 8.15b another example is shown in which the AMR consists of parallel discs, whose shape is reminiscent of a Tesla turbine (magnets are positioned below and above such discs—in Fig. 8.15b only the lower part of magnet assembly is shown). This kind of arrangement looks interesting. However, there is the question of a sufficient heat-transfer surface provided by the magnetocaloric material.

Also in azimuth fluid-flow devices, as this holds for axial and radial devices, the number of poles (high magnetic field regions) can vary. Therefore, a similar example as that shown in Figs. 8.10 and 8.11 can be applied for azimuth devices as well.

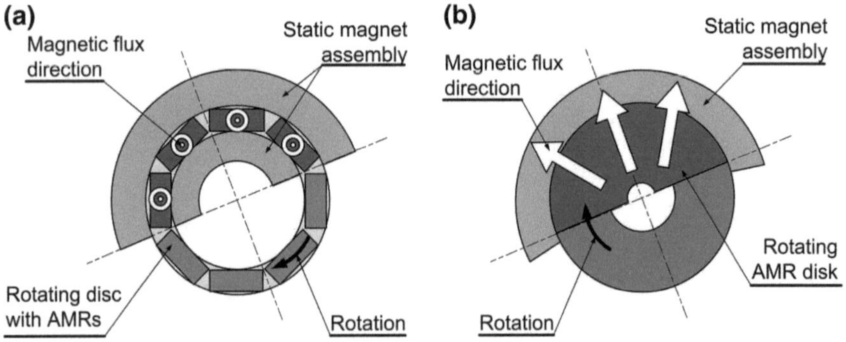

Fig. 8.15 An opened magnet assembly of a rotary magnetocaloric device with the azimuth fluid flow through the AMRs, **a** the rotating disc with eight AMRs, classification no: R20(3)0(5)(6)0(8) 111(12), **b** the rotating AMR discs, classification no: R20(3)0(5)(6)0(8)111(12)

8.2.2 Rotary AMR Magnetocaloric Devices with Rotating Magnetic Field Sources

The rotating magnetic field source has advantages over static sources because of the fact that one can avoid complex valve systems and related losses. These are still required in the case of applications of uni-directional pumps. However, conventional valves can be applied as well. Figure 8.16 shows an example of the application of a rotary magnet system in which two rotating magnets are applied over two static AMRs. This is done in order to provide a quasi-continuous production of cold. In the case of Fig. 8.16a, a bi-directional pump is used. The working fluid flows to both AMRs, one being magnetized and the other being demagnetized. Because of the rotary magnet system, the magnets will provide a kind of ON-OFF operation for the magnetization or demagnetization of the AMRs. Therefore, as shown in Fig. 8.16b, the previously magnetized AMR (from Fig. 8.16a) will be demagnetized and the previously demagnetized AMR will become magnetized. In order to provide the regenerative process, the bi-directional pump should switch its operation from the case shown in Fig. 8.16a to the case shown in Fig. 8.16b. As noted before, since the oscillatory fluid flow is provided also through the heat exchangers, we should pay attention to the passive regeneration in those heat exchangers.

The application of the bi-directional pumping system provides an advantage over the uni-directional pumping system because in the first no valves are actually required. However, the potential losses due to the dead volume should be as small as possible. The fluid simply oscillates with the help of the bi-directional pump, which is synchronized with the rotation of the magnetic field source (synchronized with the magnetization and demagnetization). The time period between the switching from magnetized to demagnetized state and vice versa is not desired, since it restricts the power density. In the case of the AMR Brayton-like refrigeration cycle, the pump will not operate during the processes of magnetization and demagnetization.

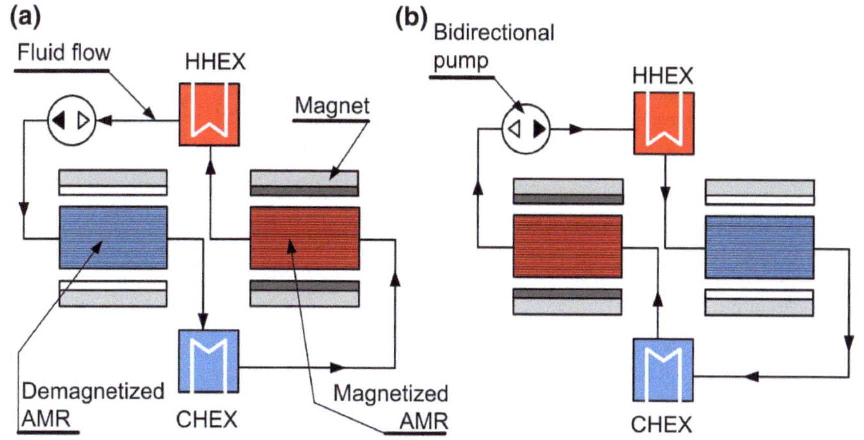

Fig. 8.16 A magnetocaloric device with a rotary movement of the magnet assembly over two static AMRs with the application of the bi-directional pump (**a** *left* AMR demagnetized, *right* AMR magnetized; **b** *left* AMR magnetized, *right* AMR demagnetized), classification no: R20(3)1(5)(6)00101(12)

Figure 8.17 shows the example of a rotary magnetic field source with the application of a uni-directional pump. In this case the fluid flow through the CHEX and HHEX will be provided, and always in the same direction. However, a valve system has to be applied. This provides the switching and the distribution of the fluid flow. Let us first consider Fig. 8.17a, in which different processes that relate to the fluid flow are denoted using the Roman numbers from I to XI. The fluid enters the CHEX at I, where it absorbs heat from the refrigerated environment II. Then it passes to the valve system III, which directs the fluid to the AMR, which is in the magnetized state IV. By passing through the magnetized AMR the working fluid warms V and leaves in the direction of another valve system VI. There it is directed to the HHEX VII. By passing the HHEX, the working fluid rejects heat to the environment and flows back to the valve system, where it is directed towards the AMR in the demagnetized state IX. By passing the demagnetized AMR the working fluid cools down X and enters the valve system XI, from which it is directed to the CHEX. Since the magnetic field source is rotating, in the next phase, the previously magnetized AMR will become demagnetized and the previously demagnetized AMR will become magnetized, respectively (Fig. 8.17b). Note that in the case of the AMR Brayton-like cycle, the time that the system switches from the case of Fig. 8.17a to the case of Fig. 8.17b should be as small as possible.

In the case of Fig. 8.17b, the working fluid still enters the CHEX at the same position as in the case of Fig. 8.17a. However, now the valve system instead of the bi-directional pump serves for switching of the directions of the fluid flow.

The rotation of the magnetic field source can be provided by different architectures of the magnet assembly. For instance, the magnets can be arranged as rotating bars (shown in Fig. 8.18a) or the magnet can consist of the stationary part

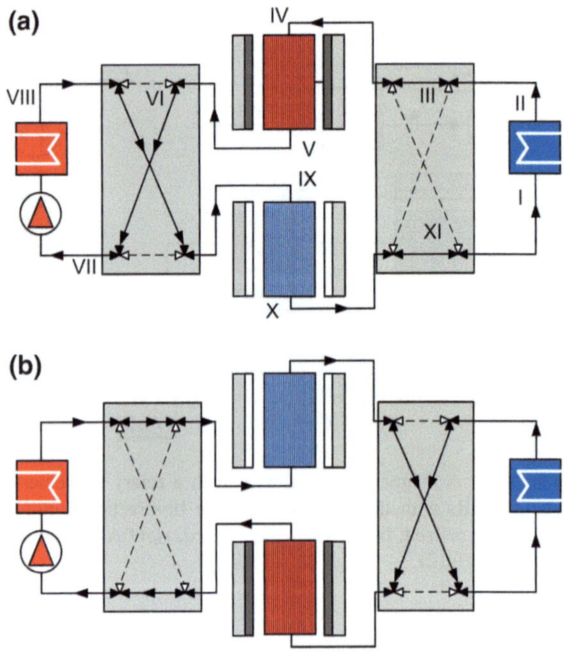

Fig. 8.17 A magnetocaloric device with a rotary movement of the magnet assembly over two static AMRs with the application of the switching valve system and the uni-directional pump (**a** *upper* AMR magnetized, *lower* AMR demagnetized; **b** *upper* AMR demagnetized, *lower* AMR magnetized), classification no: R20(3)1(5)(6)01101(12)

(stator) and the rotary part (rotor) (Fig. 8.18b). The simplest case for the latter is to apply a double or triple Halbach structure. However, this will not be the most cost-effective solution. A more effective solution can be provided by the rotation of the magnetic field source, such as, for example, shown in Fig. 8.19.

According to Fig. 8.19 (left), a four-pole rotating magnet provides for the magnetization and demagnetization of eight AMR beds. Such a magnet assembly can be simply designed to be longer in the case of axial fluid flow (parallel to the rotation axis), when a longer AMR is required. This certainly represents a simpler construction over applications with a radial or azimuth fluid flow through the AMR. This is especially so if the up-scaling of devices is required. In Fig. 8.19 (right) another example is shown in which the two-pole magnet is applied. In this case we show two different examples of the magnetization of AMRs. However, in most cases such an arrangement will provide a smaller magnetic flux density than in the case of Fig. 8.19 (left).

Another very important issue that needs to be addressed in the case of rotating magnetic field sources is to follow the design and knowledge from permanent magnet motors. Such an example is shown in the two cases of Fig. 8.20.

In the cases a–c for Fig. 8.20, we can follow the knowledge on the design of permanent magnet motors in order to provide an efficient solution for the magnetic flux distribution as well as to minimize the torque. The magnetic flux density in such a particular case of multiple pole magnets will be, in most cases, low. Therefore, compared to permanent magnet motors, the design architecture should be modified for application of magnetic refrigeration.

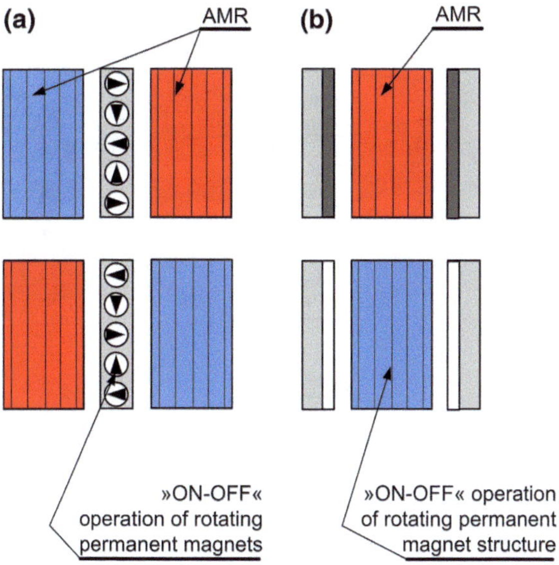

Fig. 8.18 Application of special magnet assemblies in a rotary device with rotating magnetic field sources (**a** application of rotating permanent magnet bars; **b** application of the rotating magnet assembly), classification no: R20(3)1(5)(6)0(8)101(12)

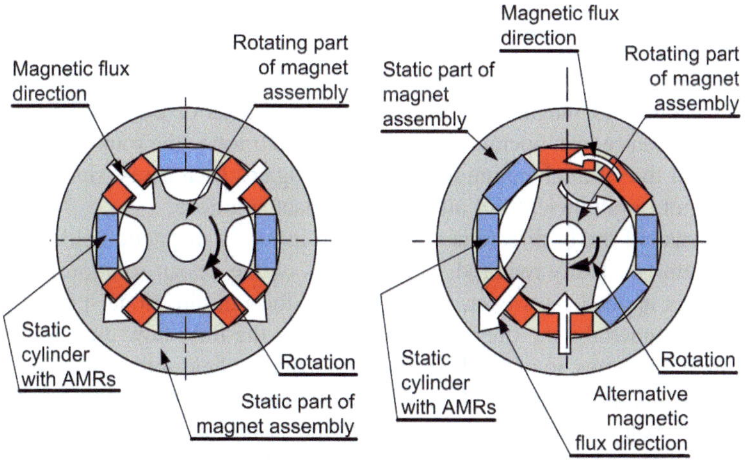

Fig. 8.19 Application of a closed magnet assembly with the four- (*left*) and two-pole (*right*) rotating magnetic field source, classification no: R20(3)1(5)(6)0(8)101(12)

The fourth case presented in Fig. 8.20d has not yet been shown to the scientific community. This is despite the fact that this issue was already a part of some scientific debates, for instance between our group and the group from DTU Risoe institute from Denmark. Namely, it is possible to construct a motor, which at the

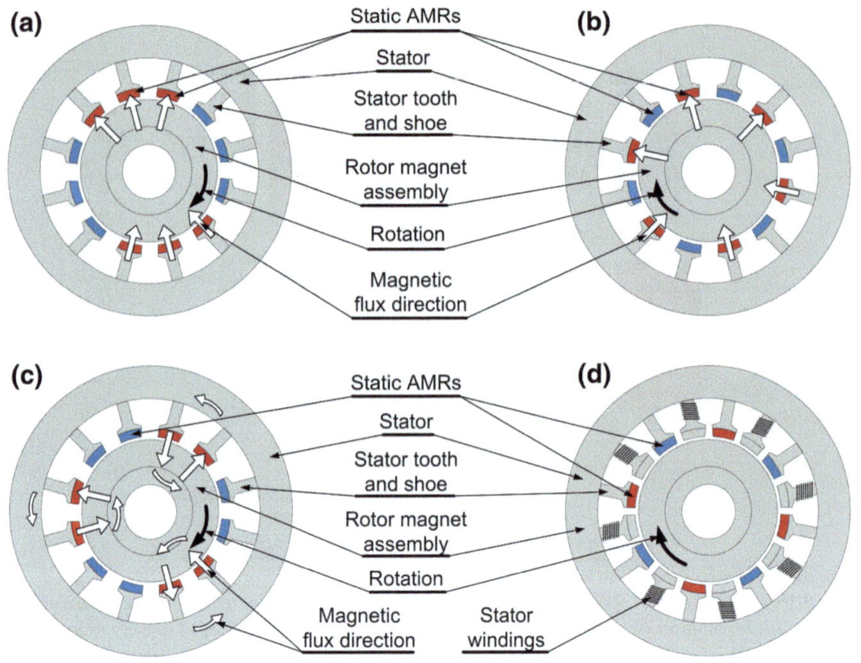

Fig. 8.20 **a–c** Application of the principle of a permanent-magnet rotor, classification no: R20(3)1(5)(6)0(8)101(12), **d** application of the principle of the hybrid motor-magnetizer for the purpose of magnetocaloric energy conversion, classification no: R20(3)1(5)(6)0(8)101(12)

same time provides magnetization and demagnetization of the AMRs. A stepper motor, for instance, can operate with a single pole to rotate the rotor. Therefore, we propose that the research community also investigates such cases, since they could represent not only efficient but also very compact solutions.

In the case of the bi-directional pump, also in rotary magnetic field sources, no valve system is actually required. This depends on the design solution for how the fluid flow is applied. However, in the case of the uni-directional pump, a valve system will undertake the operation for switching the fluid flow direction.

Figure 8.21 shows an example of a rotary magnetic field source system, where a bi-directional pump is applied. In this particular case we show two pairs of CHEX and HHEX. Note that this is only because of the easier illustration. Otherwise, such system can consist of a single CHEX and a single HHEX, which will most probably take the position at the front and at the back of the magnet assembly (with regard to the observer of Fig. 8.21).

Although the application of the axial fluid flow through the AMRs seems to be the best and simplest option, we will focus here on other solutions. Therefore, we show in Fig. 8.21 an example that regards the azimuth flow through the AMRs for a two-pole rotating magnetic field source. A similar situation can be considered for a multiple-poles rotating magnet assembly.

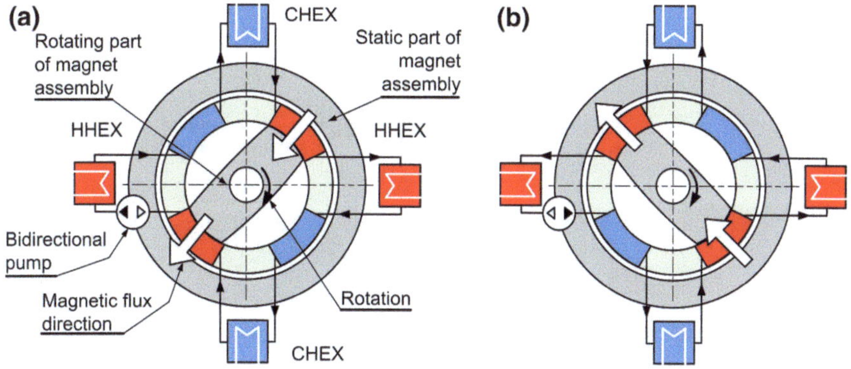

Fig. 8.21 Application of the closed, two-pole rotating magnetic field source with the application of the fluid flow in the azimuth direction and with the use of the bi-directional pump (**a** and **b** denote different direction of the fluid flow and different positions of the two pole rotating magnet, shifted for the angle of 90°), classification no: R20(3)1(5)(6)00101(12)

This is shown in Fig. 8.22, which relates to a four-pole magnetic field source. Also in this case, the CHEX and HHEX are shown only because of the illustration. Note again that only one of each type of heat exchangers (CHEX or HHEX, respectively) is actually required.

Another option in rotating magnetic field sources is to apply a uni-directional pump and a valve system that serves for the fluid-flow distribution and the switching of the fluid-flow direction. In Fig. 8.23, the system consists of two pairs of CHEX and HHEX as well as two pairs of valve systems. These correspond to each of the pairs of heat exchangers. Note that this case is shown only for illustration purposes. Therefore, a single CHEX with a single valve system can be

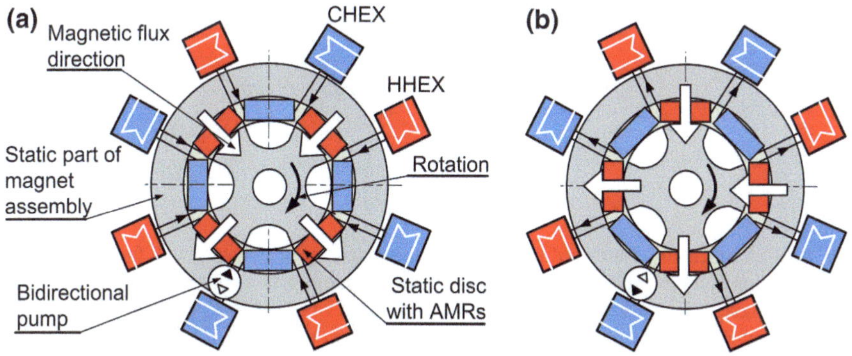

Fig. 8.22 Application of a closed, rotating four-pole magnetic field source, with the use of the bi-directional pump and the azimuth flow of the working fluid (**a** and **b** denote different direction of the fluid flow in AMRs and different positions of the four pole rotating magnet, shifted for the angle of 45°), classification no: R20(3)1(5)(6)00101(12)

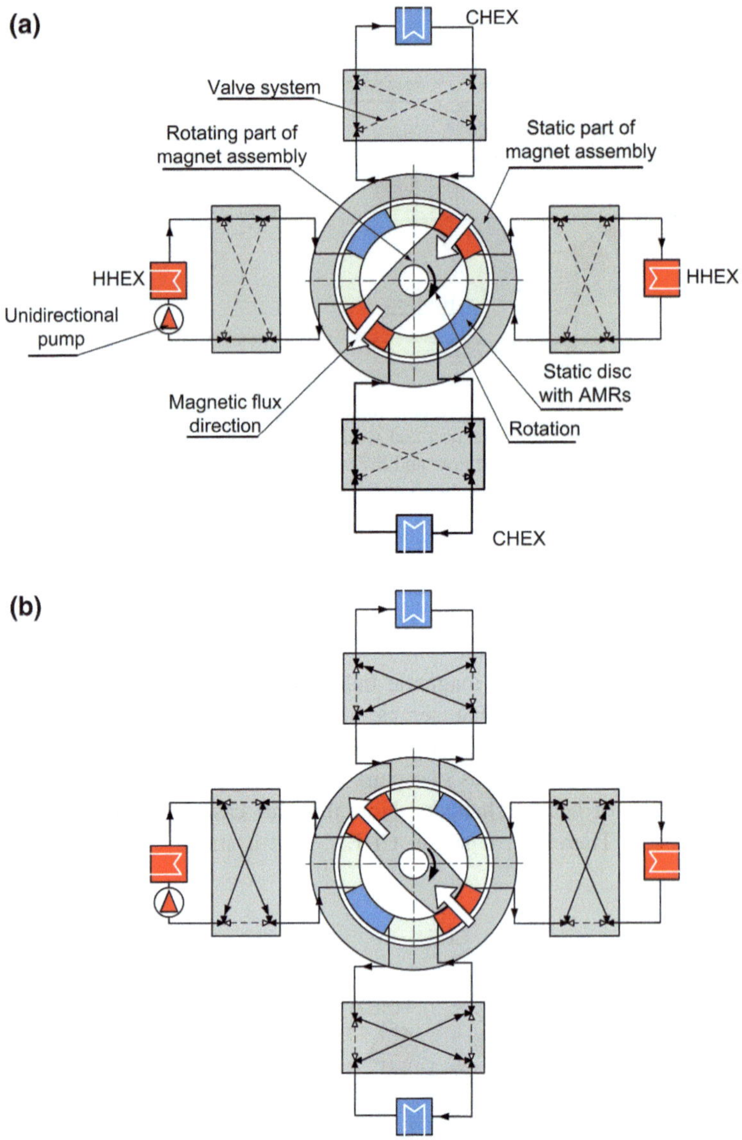

Fig. 8.23 Application of a closed two-pole rotating magnetic field source with the azimuth fluid flow direction and the application of the uni-directional pump with a corresponding valve system (**a** and **b** denote different direction of the fluid flow in AMRs and different positions of the two pole rotating magnet, shifted for the angle of 90°), classification no: R20(3)1(5)(6)01101(12)

applied as well. This also holds true for the HHEX and the corresponding valve system. Both the valve systems can be, however, also integrated into a single, more complex valve system (not shown).

8.3 Static AMR Magnetocaloric Devices

Currently, there exists no practical solution for the application of a permanent-magnet assembly that would apply static switching of the high and low magnetic field region. This could be performed by temperature gradients, but the risk of demagnetizing the magnet as well as the slow heat transfer process associated with such a principle will not lead to a practical solution.

There are, however, ideas that have been recently published by Bali et al. [2]. They found that the rotation of a single crystal of the anisotropic $HoMn_2O_5$ alloy in a constant field of 7 T led to rather high isothermal entropy change. Despite the fact that this kind of solution is actually related again to the movement of the magnetocaloric material (rotation for a certain angle with respect to the magnetic flux direction), the working principle applies the anisotropy of the magnetocaloric material and does not really change in the external parameters.

In this section we therefore focus only on the application of resistive and superconducting magnets. An example of such a system with two static magnetic field sources and two static AMRs is shown in Fig. 8.24.

Imagine the application of the bi-directional pump. An ON-OFF operation of the magnetic field source provides magnetization and demagnetization of AMRs. To this process the bi-directional pump is synchronized, which in the case of the AMR Brayton-like cycle switches immediately after the process of magnetization or demagnetization is performed. A major advantage of such a system is that, besides the pump system and related fluid flow, there is no need for moving parts. However,

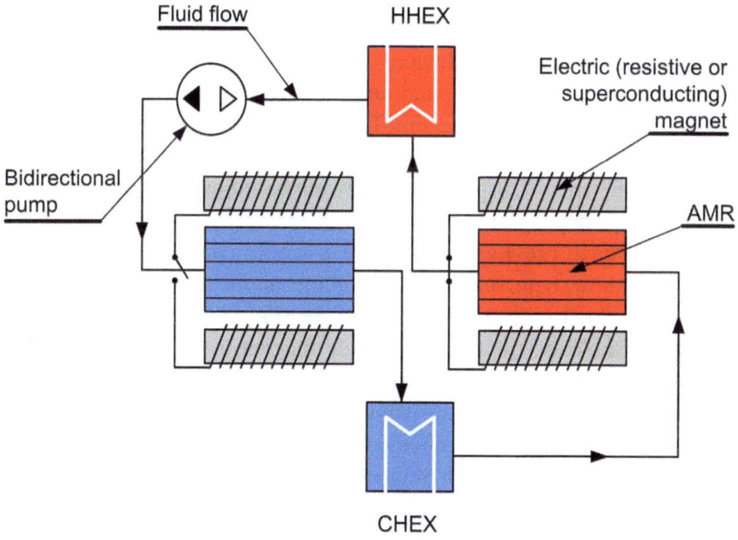

Fig. 8.24 Application of an ON-OFF operation for the static magnetic field source over the static AMRs, classification no: R21(3)2(5)(6)00002

as noted already before the resistive magnets, especially at higher magnetic flux densities, will require in most cases also cooling associated with the Joule heating due to the electrical resistance in the coil. Despite this, if such a system is built, one has to keep control of the electrical circuit in order to provide fast switching of the magnet. The Joule heating can be avoided by the application of superconducting magnets. However, for fast switching it makes sense to apply in both cases a tandem system of two magnets, interconnected into the serial electrical circuit and the single cryocooler. In this manner, the electric current and the associated magnetic energy could be shifted from one part of the system to another part. There is no evidence that such a tandem system has been ever built for the purpose of the magnetocaloric energy conversion.

8.4 AMR Devices with Thermal Diode Mechanisms

The application of the AMR principle is restricted by the frequency of the operation, associated mostly with the convective heat-transfer between the magnetocaloric material and the working fluid. For the 40 K temperature difference between the heat source and the heat sink, and at least the same efficiency of the device as that of the compressor-based refrigerator, the future maximum frequency of the operation of the advanced AMR with the working fluid based on water and additives, will most probably not exceed 5 Hz. In the case of the application of liquid metals, we could consider the limiting frequency to be 10 Hz (see Sect. 4.6). However, this might not represent a sufficient power density of device. Therefore, the research community should also focus on developments of the AMR devices with the application of thermal diode mechanisms. Depending on the particular mechanism or the design of the device, the frequency of the operation can be very high, even 100 Hz. Our experimental and numerical analyses show that the maximum performance when applying thin-film Peltier modules, for the thermal diode mechanism, will be reached at a frequency of about 25–30 Hz. However, other thermal diode mechanisms, especially if they are based on a micro-scale design, can boost such a frequency to be higher.

Although we have previously dedicated a whole chapter to the thermal diode mechanism we show here one example of the application of the thermal diode mechanism, combined with the AMR principle (Fig. 8.25).

In Fig. 8.25, a closed, rotating, four-pole magnet assembly provides the magnetization and demagnetization processes to the magnetocaloric material. This is in contrast to other rotary systems consisting of a tiny plate and not a porous structure. The magnetocaloric material is embodied between two thermal diode mechanisms. We show in Fig. 8.25 eight such segments. When the magnetic field is rotating, then in the case of segments that are under the high magnetic field region (magnetization), all the upper thermal diode mechanisms in four of such segments are switched on, providing the heat flux to be transferred to the upper micro-heat exchanger. In these segments, the lower thermal diode mechanisms have to be

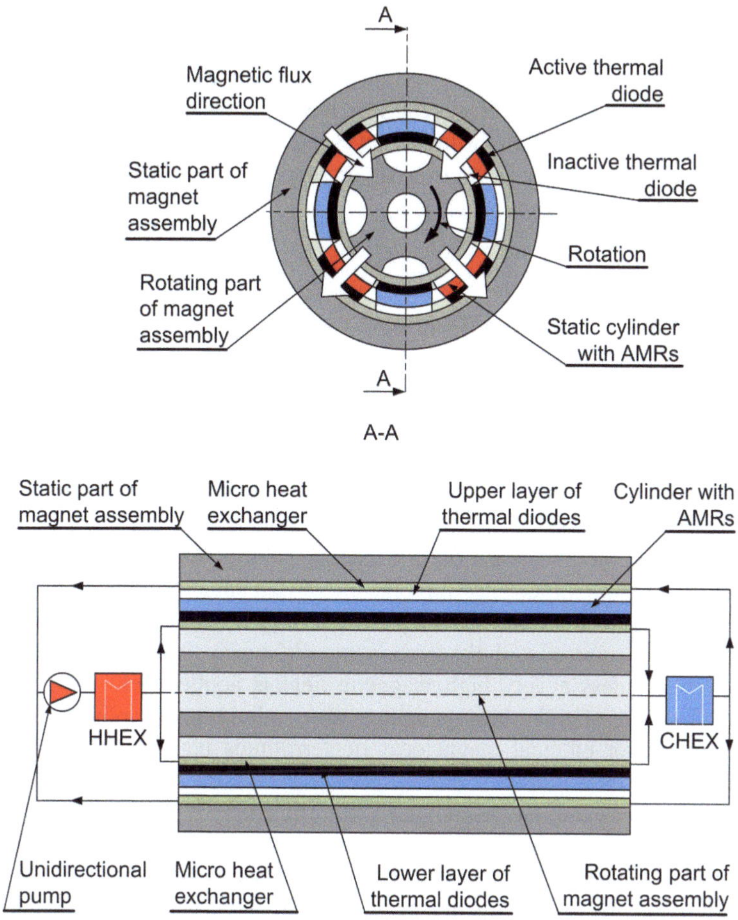

Fig. 8.25 Application of the AMR principle combined with a thermal diode mechanism, classification no: R2011(5)3(7)11012

inactive, thus preventing the heat to flow from magnetized magnetocaloric material to the lower micro-heat exchanger. We show in Fig. 8.25 an additional heat source CHEX and a heat-sink heat exchanger HHEX, respectively. However, this might not be necessary, since the fluid flow in micro-heat exchangers (the "upper" micro-heat exchanger represents the heat sink and the "lower" micro-heat exchanger represents the heat source) apply the uni-directional flow of the working fluid. Since the frequency of the operation of such a system is fast, the heat is actually continuously transferred (pulsated) from the heat source to the demagnetized magnetocaloric material and from the magnetized magnetocaloric material to the heat sink. Additionally, a counter working fluid-flow direction in combination with the operation of the rotating magnetic field source and synchronized thermal diode mechanisms provides heat regeneration along the magnetocaloric material (parallel to the axis of the rotation).

This is why we refer to such a system as an AMR with the application of the thermal diode mechanism. We encourage the research community to seriously consider this kind of solution for future developments of the magnetocaloric energy conversion. In the chapter on thermal diode mechanisms (Chap. 6) we have provided sufficient information, which may serve as the basis for such developments.

8.5 Devices with Magnetocaloric Fluids

The idea of applying magnetocaloric fluids as the refrigerant is certainly very attractive. The architectures of devices that can be applied in this particular case can be found in the chapter on magnetocaloric fluids. However, the major obstacle of magnetocaloric fluids is in the content of the magnetocaloric material in base (nanofluids—ferrofluids) or the carrier fluid (magnetocaloric particle suspensions). In the first, the thermal mass of the magnetocaloric material is relatively low compared to the base fluid. Therefore, the effective adiabatic temperature change of the magnetocaloric fluid will be very low. If one accounts for irreversible heat transfer that is associated with the temperature difference, then it becomes clear that such devices, unless a very high magnetic field is applied, will not represent a solution for refrigeration, heat pumping, air-conditioning or power generation. Other solution for efficient magnetocaloric ferrofluids regards the application of liquid metals as base fluids.

If, on the other hand, the magnetocaloric particle suspension is applied (magnetocaloric magnetorheological fluid), this will be associated with highly non-Newtonian rheological behaviour, which does not depend only on the particle size and their volume fraction, but also on the magnetic field and associated with the yield stress.

Magnetocaloric fluids can be successfully applied in thermal management as self-running systems (similar to heat pipes). If they are applied for refrigeration, heat pumping or power generation, rather high magnetic field will be required.

Note that magnetorheological fluids have a good potential to be applied in magnetocaloric devices with solid magnetocaloric material. They can be used in valve system, or as the actuator in a special pump system. These fluids can be also successfully applied as dynamic seals. Moreover, magnetorheologic fluids can serve as thermal diode mechanisms.

The authors ask the reader to see the chapter on magnetocaloric fluids (Chap. 5). There we show design concepts for magnetocaloric energy conversion with magnetocaloric fluids. Moreover, different potential applications of magnetic fluids in the magnetocaloric energy conversion are discussed.

8.6 A Note on Magnetocaloric Power Generation

Imagine a realistic system with a low-enthalpy heat source with a temperature of 120 °C and a heat-sink temperature of 30 °C (i.e. such can be used by Organic Rankine Cycle—ORC). A 90 K temperature difference: This of course is not much for some other technologies, but the present knowledge in magnetocaloric energy conversion does not provide an efficient solution to develop a market-ready application. This kind of development will actually demand a longer period than magnetic refrigeration or magnetic heat pumping. Therefore, despite the fact that magnetocaloric power generation can represent an important future power-generation application, the high-temperature difference that is required simply represents at the moment a too large obstacle. This, of course, is not a statement to discourage future developments. However, it makes sense that the new emerging technology first enters market domains in which it can become well established. Despite this fact and because of studies that show good future potential, we have made below a short review on the existing work in this particular domain.

8.6.1 How to Perform Magnetocaloric Power Generation?

The answer is given by the temperature dependency of the magnetization of the magnetocaloric material. For instance, if the magnetocaloric material is in the ferromagnetic state, it will be attracted towards the higher magnetic field. Imagine now an opened thermodynamic cycle, where a disc of magnetocaloric material is partially placed in the high-magnetic-field region and partially in the low-magnetic-field region. Of course, if no temperature gradient is applied, the disc will stay at rest. Now let us introduce the heat source and the heat sink. The heat sink will be provided in the low (or no field) field region. However, the heat source will be provided in the region before the disc exits the magnet field. Therefore, the magnetocaloric disc will have a temperature gradient in the magnetic field. It will be less attracted to the magnetic field at its exit, and more attracted to the magnetic field region at its entrance. This will create the rotation of the disc.

8.6.2 Review of Magnetocaloric Power Generation

The first to start investigation of the phenomenon of magnetocaloric power generation were Tesla [3] and Edison [4]. Much later, in the 1950s, researchers were analysing the idea of magnetocaloric power generation by applying magnetocaloric suspensions as working fluids [5, 6]. However, some work was additionally done by Brillouin and Iskenderian [7], Van der Voort [8], Chilowsky [9] and Elliott [10], but no evidence of actual prototypes from that time exists.

However, Murakami and Nemoto [11] published their work on experiments and considerations on the behaviour of thermomagnetic motors in 1972. Later, in the 1980s, Kirol [12, 13] and Salomon [14] investigated magnetocaloric power generators with solid working materials.

The idea of magnetocaloric power generation was later rediscovered by the research group from the University of Applied Sciences of Western Switzerland, guided by Egolf. As a result, a comprehensive study was performed for the Swiss Federal Office of Energy in 2008 [15]. Later, a number of publications were produced by the same group (see Diebold et al. [16], Egolf et al. [17], Vuarnoz et al. [18–20]).

In the period 2006–2014, also the group under the guidance of Brück from Delft University (starting at the University of Amsterdam) was very active in finding magnetocaloric materials that are suitable for magnetocaloric power generation. They published a report on these developments as a result of collaboration with the company BASF (see examples in Brück et al. [21], Brück [22]).

In 2014 the same group published an article on a small proof-of-the concept device [23].

Valuable work was also performed by Takahashi et al. [24, 25], who published work on analyses of thermomagnetic engines in 2004 and 2006.

It is important to mention the work on thermomagnetic wheels, which represents a special case of magnetic energy harvesting and which have been published by Karle in 2000 [26], Palmy in 2007 [27], Palmy and Egolf [28] and Alves et al. [29].

Some additional and newer publications on magnetocaloric power generation can be found in Cleveland and Liang [30], Alves et al. [31], Ferreira et al. [32], and Trapanese et al. [33].

8.7 Future Perspectives and Guidelines for Magnetocaloric Energy Conversion

According to our experiences and knowledge we put here some guidelines for the future developments of magnetocaloric energy conversion. We believe that in the future, magnetocaloric energy conversion devices will consist of static magnetocaloric material and moving, rotating or switching magnetic fields. Why not rotation of the magnetocaloric material? All the present experiences with such devices show a large problem associated with the dynamic sealing of such systems. Therefore, a number of researchers doing such configurations experienced internal or even external leakage of the working fluid or very large friction due to the dynamic seal. On the other hand, if an efficient valve for such a system is designed, it will certainly represent an expensive part of the device. It is therefore easier to develop a system in which the magnetocaloric material stays at rest, and therefore no dynamic seal is required, unless a rotating valve system is also applied in this case.

In the following text we present some major design issues that need to be solved in the future. We, however, will not address new permanent-magnet materials or new magnetocaloric materials, because this would be too speculative.

8.7.1 Active Magnetic Regeneration AMR (Conventional Principle)

The application of the principle of AMR will certainly be improved in the future by using some new processes of manufacturing for magnetocaloric regenerators (very fine, ordered structures with large heat-transfer surfaces and a layered magnetocaloric material), or by applying some advanced working fluids with a higher thermal conductivity than that of water and with low viscosity.

For instance, if the feature of the AMR (which strongly depends on the working fluid) is the following: porosity from 30 to 45 %; wall thickness of the structure 30–100 microns; fluid voids from 30 to 100 microns (depends on the working fluid) with no flow maldistribution, then the magnetocaloric device will approach a frequency of the operation of 5 Hz and simultaneously similar or better efficiency as the compressor refrigerator with a temperature span of about 40 K (see also Sect. 4.4).

These features can be achieved because of the very high heat-transfer surface as well as the ordered structure, by which high viscous losses will be avoided, especially at higher frequencies of operation.

In order to perform this, better manufacturability and processing knowledge with respect to magnetocaloric materials is required (see Sect. 4.7). One should also not forget the mechanical properties which will allow such tiny structures to sustain mechanical stresses.

8.7.2 Active Magnetic Regeneration with Thermal Diodes

The upper limit of the frequency for operation of the AMR will also limit market applications, whether because of the low efficiency at large temperature spans, or low efficiency at higher power density. Therefore, in order to boost the frequency of the operation from, e.g. 5–25, 50 Hz or even 100 Hz, the conventional AMR principle certainly cannot be applied. The only at present known mechanism that can solve this problem is the application of thermal diodes. Note that the principle of thermal diodes also solves other existing problems for the magnetocaloric energy conversion technology. Namely, the solution of thermal diode, with embodied magnetocaloric material, enhances the heat-transfer rate and simultaneously also provides a solution for fluid connections and fluid-flow dividers, and thus also prevents internal fluid leakage and avoids high viscous losses. Furthermore, it can create a significant impact of improvements for the technology by achieving the following solutions:

- the magnetocaloric "regenerator" can be shaped in a simple plate, foil or tape, not a complex porous structure, and the small mass of magnetocaloric material is compensated by the very high frequency,
- smaller fluid-flow viscous losses and larger heat-transfer surface—the fluid flow can be performed through an expanded surface of a heat exchanger, which is not made of magnetocaloric material and other materials can be processed in much better and more precise ways (e.g. Cu and similar materials). Thus frictional losses can be decreased and the heat-transfer area with the fluid flow can be enhanced,
- prevented losses due to the fluid oscillatory or bi-directional flow (fluid switching), since no fluid switching is required anymore and no "dead" volume of the fluid occurs. Namely, the working fluid always flows in the same direction,
- prevented internal fluid leakage and prevented friction by dynamic seals and valves—by the introduction of the proposed concept the rotation of the magnetic field source enables the use of "static" valves (if any), so neither mechanical friction nor the related fluid leakage is expected. The thermal diode mechanisms actually overtake the role of the valve system.

8.7.3 Magnet Assembly and Related Motor Drive

In the case of superconducting or electric resistive magnets, a motor is not required. The system can simply operate with an ON-OFF operation. However, it would make sense to perform a design of a tandem magnet, so the electric current could switch from one to system another. As we have noted already before, very high magnetic field changes (e.g. above 3 T) will require application of a rather different magnetic thermodynamic cycle than the AMR Brayton-like one.

In the case of permanent magnets, one way to develop magnet assemblies is that they consist of multiple poles in similar way as permanent-magnet motors. Such a structure will most probably be embodied with a shield (and yoke), and therefore no additional need for shielding will be required. Anyway, a good magnet design for the magnetocaloric device should lead to the minimum possible leakage of the magnetic flux into the environment. The multiple-poles feature can be applied for AMR principles as well as principles based on a thermal diode mechanism. One should pay attention to the torque as well as eddy currents, which might be produced as well. The torque issue is also very important for the selection of the motor, and therefore one should try to keep it as low as possible. We also did a note on solving the cogging torque, related to rotating magnet designs (see the chapter on magnetic field sources). One of the solutions for preventing eddy currents is by using laminates of material with electrical insulation or by choosing materials with good magnetic but low electric conductivity.

Motor cores are, for instance, made of stacks of thin, coated steel sheets, where steel comprises additives such as silicon for a higher electrical resistance. Note that due to the relatively high electrical conductivity of Nd–Fe–B magnets, the eddy-current losses can be significant, especially in devices with a high pole number or a high rotational speed. This can also lead to heating of the magnet, which causes additional problems of lower magnetization (see, e.g. Zhu et al. [34]).

Depending on the operation of a device, the rotating magnet device can apply a continuous permanent magnet motor or a stepper motor where in the latter a discrete number of steps is used (e.g. 200 per revolution, see e.g. Hughes [35]). This kind of motor will precisely shift the positions from the magnetized to the demagnetized zone and vice versa. A note in this chapter was also given for the combination of the motor drive and the magnet assembly. This kind of system has not been yet developed, but it makes sense to enter such developments as well.

8.7.4 Pumping and Valve System

Two types of pumping systems can be applied in future magnetocaloric devices, uni-directional and bi-directional pumps. For the first there is no need for some deeper explanation, except for the fact that heating from the pump motor should not be transferred to the working fluid. One solution is also to place such a pump in the vicinity of the HHEX and therefore to reject heat from the pump to the environment. Note also that the pump losses are not negligible so the selection of a pump with good efficiency is very important. A larger problem is with the bi-directional pump. An electric drive of the two pistons in order to provide the oscillating movement of fluid is one solution. However, our tendency is to drive such a system by, e.g. 20 Hz (for 10 Hz operation frequency of the magnetocaloric device). A gear pump is another solution, but it is rather expensive. The peristaltic pump also represents a solution, but it is inefficient and causes pulsations of flow. Another way is to apply membrane pumps. The issue of the cost and energy efficient bi-directional pumping system was not fully solved so far. Note that in the case of high frequency and very efficient AMR, the fluid flow particle will in some cases make only about 25 % of the full length of the AMR. If the AMR is short, e.g. 50 mm, this will mean that the fluid particle will oscillate in the path of ±12.5 mm.

Since we propose a static AMR for future applications, we also consider a static valve system for the application of the uni-directional pump. In this case, one should pay attention of the dead volume in which the working fluid is captured between the valve system and the AMR. The same holds true also for the heat-sink (HHEX) and the heat source (CHEX) heat exchangers, which should be positioned as close as possible to the AMR.

All the dynamic valves to date were developed for the purpose of being used in magnetic refrigeration technology. However, in static valve systems, besides applying some traditional valves, these can be newly developed for the application.

Because of the varying magnetic field it makes sense to think about a field-actuated valve system. One such example has been proposed by Eriksen et al. [36].

Another way is to apply a ferrofluid or better, a magnetorheologic fluid, which activates/de-activates in the presence/absence of a magnetic field. Therefore, to learn more about the potential design of such a system the reader is directed here to the literature which we denote in the chapter on magnetocaloric fluids, where a large part of the chapter is dedicated to the rheology of magnetic fluid and magnetic particle suspensions.

Note again that in the application of thermal diode mechanisms a uni-directional pump is required and actually no valve system is needed.

8.7.5 Working Fluid

Most of the applications apply water with freezing depressant, which in some cases acts also as the inhibitor to prevent corrosion of the magnetocaloric material. However, other corrosion inhibitors may serve as well. The most important characteristics of the working fluid are the low viscosity and high thermal conductivity. Note that application of nanofluids will not lead to a better solution than that with water, despite the slightly higher thermal conductivity. This is due to the fact that they cause larger viscous losses, which result not only in increased power consumption of the pump, but also heat generation in the fluid due to higher viscosity and related friction at the wall. Our experimental and theoretical work with the liquid metal Galinstan, however, has shown that liquid metals possess excellent properties, which may be successfully applied in AMRs and allow higher operating frequencies than water-based systems (see also Sect. 4.6 for details). However, the cost of liquid metals based on Ga, In, Sn is very high. Despite this our analyses have shown that the high cost of the liquid metal will be compensated by the substantially reduced cost of magnets. Other solutions that may apply toxic mercury or NaK could be applied as well. In this case, however, there is a question about their use in usual applications. So one can expect such applications will be applied in special market niches as well as environments.

There is no evidence that (phase-change) refrigerants, with condensation and evaporation, have been applied in magnetocaloric devices. This is possible, but not in applications where such a fluid is oscillating with high frequency. Most probably such fluids can be applied as secondary refrigerants or as the working fluid in the case of the application of thermal diode mechanisms, where no oscillation of the working fluid is required.

8.7.6 Power Generation

As already mentioned in the section on power generation, the major problem of performing such processes is in need for a very large number of layers of magnetocaloric material (in the case of La- or Mn-based materials, for low-enthalpy sources with temperature at about 120–160 °C, this means about 20–25 layers of materials with different Curie temperatures). This is why we propose the development of these systems after the market applications in the field of magnetic refrigeration, heat pumping and air-conditioning.

8.7.7 General Characteristics of Future Magnetocaloric Devices

Table 8.1 shows some of the predicted features that future magnetocaloric devices will have. As shown in Table 8.1, we predict that future magnetocaloric devices will apply permanent magnets up to a heating or cooling power of 300 kW. Above this range, the application of superconducting magnets makes sense. Therefore, large-scale chillers and heat pumps with cooling or heating power above 300 kW and up to a few MW will be based on superconducting magnets. Also, in power generation, it is expected that micro units will not be developed, but rather larger superconducting systems operating with a low enthalpy source of heat. Note, however, that the operation with low-enthalpy heat sources, even if the power generator operates with the Carnot efficiency (100 % exergy efficiency), such a device will have a thermodynamic efficiency of the heat-to-power conversion in the range from 10 to 30 %. Therefore, a power unit which converts 1 MW of heat with the thermodynamic efficiency of 10 % and the Carnot efficiency of 50 % will generate 50 kW of mechanical power or electricity. Note also that the permanent magnets, especially Nd–Fe–B are quite sensitive to high temperatures. With regard to the frequency of the operation, one can expect for the 40 K temperature difference and the efficiency of at least of that of compressor units, that devices with the AMR principle will reach 5 Hz. In the case of AMR principles with some

Table 8.1 Predicted general features of future magnetocaloric devices depending on the magnetic field source

Target market/ target of design	Small scale refrigerators	Small scale air conditioners	Chillers/ heat pumps <300 kW	Large-scale heating and/or cooling >300 kW	Power generation >50 kW_{el}
Permanent magnets	✓	✓	✓		
Superconducting magnets				✓	✓

Table 8.2 Predicted general features of future magnetocaloric devices depending on the frequency of the operation (number of thermodynamic cycles per unit of time)

Target market/target of design	Frequency				
	1 Hz	5 Hz	10 Hz	20 Hz	50 Hz
AMR	✓	✓			
Thermal diode mechanism			✓	✓	✓
AMR with liquid metal		✓	✓		

advanced working fluids, e.g. liquid metals, the frequency could be increased up to 10 Hz. All the higher frequencies of operation will demand the application of a thermal diode or similar mechanisms.

Table 8.1 shows the potential future magnetocaloric devices related to the magnetic field source. It is based on present knowledge with respect to magnetocaloric energy conversion. As can be seen from Table 8.1, we can expect that magnetic refrigerators, magnetic air conditioners, magnetic heat pumps and magnetic chillers, whose cooling or heating power will be in a range below 300 kW, will most probably apply permanent magnets. However, for large-scale units, including the power generation, a superconducting magnetic field source will be applied.

Another example is shown in Table 8.2. This shows the frequency dependence of three different applications that relate to a magnetocaloric material, its geometry and operation: basic AMR principle, the AMR principle with advanced working fluids (e.g., with liquid metals) and the principle of AMR combined with the thermal diode mechanism. This table predicts magnetocaloric energy conversion that enables high efficiencies and certain temperature spans that are comparable with conventional compressor-based refrigeration.

As can be seen from Table 8.2, the basic AMR principle with water-based fluids or similar will most probably not bridge the frequency of the operation of 5 Hz. Of course, a high frequency with such devices will be possible, but their efficiency and also the restricted temperature span will represent too large an obstacle with respect to conventional devices. The solution with liquid metals and the AMR principle is actually more cost-effective and more energy efficient solution than that of the application of conventional fluids.

Thus we may expect such solutions to enable doubling of the power at the same efficiency with respect to solutions of AMR with conventional fluids, for instance, water. For high frequencies of operation above 10 Hz, the present knowledge shows that only solutions with thermal diode mechanisms can lead to such applications. These, in contrast to other AMR principles, can also apply conventional refrigerants and can even operate in a similar manner as heat pipes. Note that the application of a thermal diode mechanism will not be perfect for frequencies of operation below, e.g. 5 or 10 Hz. We hope again that the research community will be encouraged by this book to enter into design of magnetocaloric devices with thermal diode mechanisms.

References

1. Scarpa F, Tagliafico G, Tagliafico LA (2012) Classification proposal for room temperature magnetic refrigerators. Int J Refrig 35:453–458
2. Bali M, Jandl S, Fournier P et al (2014) Anisotropy-enhanced giant reversible rotating magnetocaloric effect in $HoMn_2O_5$ single crystals. Appl Phys Lett 104:232402
3. Tesla N (1890) Pyromagneto-electric generator. US Patent 428.057
4. Edison TA (1892) Pyromagnetic generator. US Patent 476.983
5. Resler EL Jr, Rosensweig RE (1964) Magnetocaloric power. AIAA J 8(2):1418–1422
6. Resler EL Jr, Rosensweig RE (1967) Regenerative thermomagnetic power. J Eng Power 89:399–406
7. Brillouin L, Iskenderian HP (1948) Thermomagnetic generator. Federal Telecommunication Laboratories, Nutley, pp 300–313
8. Van der Voort E (1969) Ideal magnetocaloric conversion. Appl Sci Res 20:98–114
9. Chilowsky C (1952) Thermomagnetic generator and refrigerator. US Patent 2.619.603
10. Elliott JF (1959) Thermomagnetic generator. J Appl Phys 11(30):1774–1777
11. Murakami K, Nemoto M (1972) Some experiments and considerations on the behavior of thermomagnetic motors. IEEE Trans Mag 8:387–398
12. Kirol LD, Mills JI (1984) Thermomagnetic generator. In: Proceedings of intersociety energy conversion engineering conference, vol 3, San Francisco, USA, 19 Aug 1984, pp 1361–1368
13. Kirol LD, Mills JI (1984) Numerical analysis of thermomagnetic generators. J Appl Phys 56 (3):824–828
14. Salomon D (1989) Thermomagnetic mechanical heat engines. J Appl Phys 65(9):3687–3693
15. Kitanovski A, Diebold M, Vuarnoz D et al (2008) Applications of magnetic power production and its assessment: a feasibility study-final report, Bern: Swiss Federal Office of Energy, SFOE contract and project number: 101776/152190, p 33
16. Diebold M, Kitanovski A, Vuarnoz D et al (2007) Force, torque and energy of machines with porous magnetisable wheels. In: Proceedings of second workshop on magnetostrictive materials and magnetic refrigeration, Vienna, Austria, 13–15 Aug 2007, pp 15–16
17. Egolf PW, Kitanovski A, Diebold M et al (2009) Magnetic power conversion with machines containing full or porous wheel heat exchangers. J Magn Magn Mater 321(7):758–762
18. Vuarnoz D, Kitanovski A, Diebold M et al (2009) Feasibility study of a power conversion system utilizing industrial waste heat, Yverdon: HES-SO 2009; final report project 19144
19. Vuarnoz D, Kitanovski A, Gonin C et al (2012) Thermodynamic and exergy efficiencies of magnetocaloric energy conversion utilising industrial waste heat. Int J Exergy 10(4):365–378
20. Vuarnoz D, Kitanovski A, Gonin C et al (2012) Quantitative feasibility study of magnetocaloric energy conversion utilizing industrial waste heat. Appl Energ 100:229–237
21. Brück E, Tegus O, Buschow K (2006) Novel magnetocaloric materials: not only for cooling applications. In: Proceedings of IEEE international magnetics conference, INTERMAG 2006, p 85
22. Brück E (2012) Magnetocaloric materials not only for cooling applications FOM-BASF IPP I18 2008–2012, industrial partnership programme I18, final report, p 27
23. Christiaansen T, Brück E (2014) Proof-of-concept static thermomagnetic generator experimental device. Metall Mater Trans E 1e:36–40
24. Takahashi Y, Matsuzawa T, Nishikawa M (2004) Fundamental performance of the disc-type thermomagnetic engine. Electr Eng Jpn 148(4):26–33
25. Takahashi Y, Yamamoto K, Nishikawa M (2006) Fundamental performance of triple magnetic circuit type cylindrical thermomagnetic engine. Electr Eng Jpn 154(4):68–74
26. Karle A (2000) The thermomagnetic Curie-motor for the conversion of heat into mechanical energy. Int J Therm Sci 40:834–842
27. Palmy C (2007) A thermo-magnetic wheel. Europhys News 38(3):32–34

28. Palmy C, Egolf PW (2007) Levitation and acceleration of a thermo-magnetic wheel. In: Proceedings of the second international conference on magnetic refrigeration of the international institute of refrigeration, IIF/IIR in Paris, pp 299–307
29. Alves CS, Colman FC, Foleiss GL et al (2012) Simulation of solar Curie wheel using NiFe alloy and Gd. Int J Refrig 37:215–222
30. Cleveland M, Liang H (2012) Magnetocaloric piezoelectric composites for energy harvesting. Smart Mater Struct 21:047002
31. Alves CS, Colman FC, Foleiss GL et al (2013) Numerical simulation and design of a thermomagnetic motor. Appl Therm Eng 61:616–622
32. Ferreira LDR, Bessa CVX, da Silva I et al (2014) A heat transfer study aiming optimization of magnetic heat exchangers of thermomagnetic motors. Int J Refrig 37:209–214
33. Trapanese M, Viola A, Franzitta V (2012) Design and experimental test of a thermomagnetic motor. AASRI Proc 2:199–204
34. Zhu ZQ, Ng K, Schofield N et al (2004) Improved analytical modelling of rotor eddy current loss in brushless machines equipped with surface-mounted permanent magnets. Proc Inst Elect Eng-Elect Power Appl 151(6):641–650
35. Hughes A (2013) Electric motors and drives: fundamentals, types and applications, 4th edn. Newnes, Oxford, p 440
36. Eriksen D, Bahl C, Pryds N et al (2012) A novel magnetic valve using room temperature magnetocaloric materials. In: Proceedings of the fifth IIF-IIR international conference on magnetic refrigeration at room temperature: Thermag V, Grenoble, France, pp 389–396

Chapter 9
Economic Aspects of the Magnetocaloric Energy Conversion

In the past decade, several studies have been performed to evaluate the economic potential and the cost of magnetic refrigerators and magnetic heat pumps. These studies were mostly based on theoretical predictions, by taking into account estimations of the costs for a particular part of the magnetic refrigerator, as well as by taking into account the operating costs of a particular device. Therefore, they can serve for a rough estimation and as a guideline for the design characteristics that can affect different costs. Namely, the cost estimation of an emerging technology is rather difficult, especially if it is compared with some mature refrigeration technology. Therefore, one has to take into account the potential mass production, as well as the optimization of the technology, which, together with different scenarios, can predict future developments.

In this chapter we address the cost issues that relate to magnetic field sources and magnetocaloric materials. We further divide magnetic field sources into permanent-magnet structures, electric resistive magnets and superconducting magnets. Note that the costs of a magnetic field source, and partially also the magnetocaloric material, at the moment represent the major cost contribution and also the major economic obstacle of magnetic refrigeration technology.

In addition, this chapter provides a review of the different economic studies covering magnetic refrigeration, including some ecological aspects that are related to the carbon footprint and lifecycle analyses (LCA).

9.1 A Brief Discussion About the Market and the Costs of Nd–Fe–B Permanent Magnets

Since most of the magnetic field sources in magnetic refrigeration are based on Nd–Fe–B magnets, it is important to address some serious issues, which may drastically influence not only a particular device, but the whole domain of magnetic refrigeration, or more generally, the magnetocaloric energy conversion.

First of all, one has to be aware that most of the rare earth neodymium (Nd) comes from China, and according to the estimation of Benecki [1], at least 80 % of

A. Kitanovski et al., *Magnetocaloric Energy Conversion*,
Green Energy and Technology, DOI 10.1007/978-3-319-08741-2_9

Table 9.1 The growth of Nd–Fe–B magnet market production [7]

Production by Country/Region	Metric tons × 1000	
	2012	2015
China	50	65
EU	1	1
Japan	10	8
USA	0	2
Others	2	2
Total	63	78

global Nd–Fe–B magnet production will be in China in 2015 (see also Table 9.1). In other words, the market for Nd will be very dependent on this fact (unless the production of magnets is based in China). Since each country tries to be energy independent, material source dependence should be avoided as much as possible. Any restriction in terms of quotas or a variation in prices can drastically influence not only the magnetic refrigeration technology, but also many new, renewable-energy-related technologies based on permanent magnets (see also Bradsher [2] and Qi [3]).

In 2013, the average global price for sintered Nd–Fe–B material (52 MGOe) was about 30 eurokg^{-1}, whereas in 2011 the price of the same material was about twice as high [4]. A rapid decrease in Chinese prices (which differ from global prices) for Nd–Fe–B magnets in 2012 and 2013 can be seen in Fig. 9.1 (note that these are of course lower than export-related prices).

The following example is used to address how important it is to minimize the costs of magnets or magnet assemblies. For this purpose an example of a magnet assembly, which our laboratory ordered from a commercial supplier in 2010, is shown in Fig. 9.2 [8]. This magnet was designed in our laboratory. The cost of 2,085 euro were related only to the European industry manufacturing costs and the costs of materials (soft iron + Nd–Fe–B 52 MGOe). The ratio Λ_{cool}/P_{field} is approximately 0.15, which is actually a rather good magnet configuration. Of course, the price of 2,085 euro for a single magnet assembly is too high to be considered in a real economic evaluation.

Let us now consider the application of two different AMRs, consisting of magnetocaloric materials, Gd and La–Fe–Co–Si (Table 9.2). For these we assume approximately the same Curie temperature of T = 292 K. This will be taken into consideration as the refrigeration temperature T_R. We assume that the magnetic field change corresponds to 1 T. We also assume a perfect AMR geometry (e.g. with a very fine order tiny structure) as well as that AMR occupies all the available volume in the high magnetic field region. The average porosity of such an AMR can be considered 30 %, therefore, the volume of each AMR will be 84 cm^3. The maximum refrigeration capacity of each material is defined according to the definition given in the chapter on thermodynamics (Chap. 4). In Table 9.3 the maximum cooling power that corresponds to the different frequencies of operation (the number of thermodynamic cycles per unit of time) is shown. For this we do not

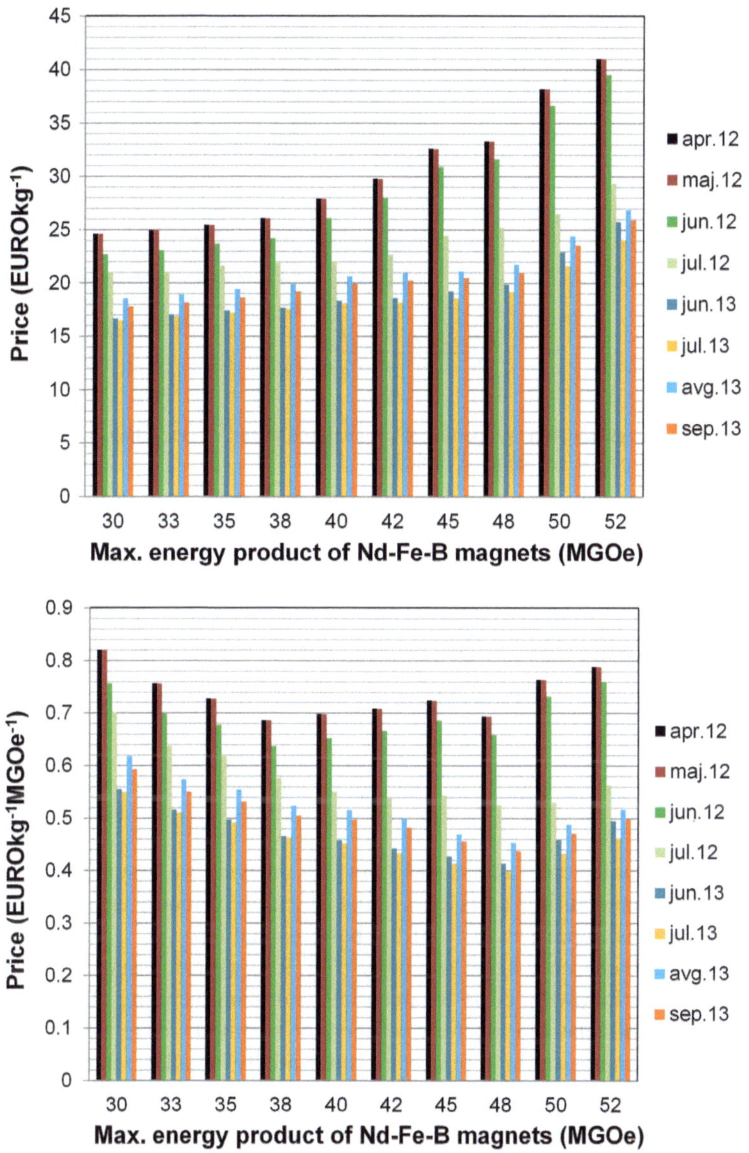

Fig. 9.1 Chinese prices of Nd–Fe–B magnets in 2012 and 2013 (data taken from [5, 6])

consider how such a cycle is performed. We also assume that the manufacturing costs of the AMRs equals 60 eurokg^{-1} for La–Fe–Co–Si AMR and 150 eurokg^{-1} for the Gd-based AMR. Note that the cost of Gd on the market, when buying a small quantity of plates of, e.g. 0.2 mm, can exceed 1,000 euro/kg. Another issue that increases the price is, of course, the purity of the Gd. Therefore, we take 150 eurokg^{-1} as being optimistic and valid for mass production of high-purity

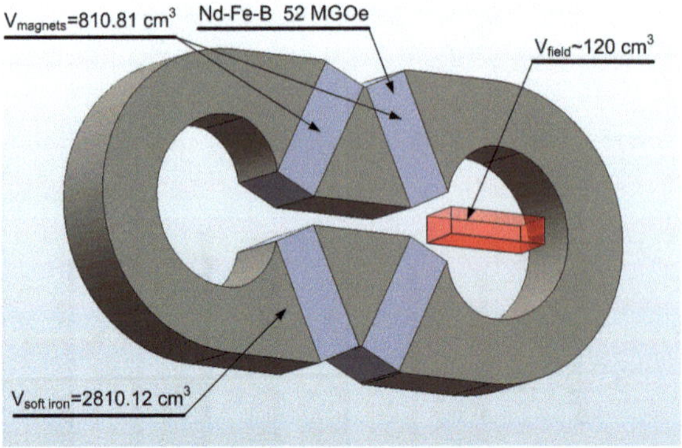

Fig. 9.2 Permanent-magnet assembly that provides 1 T of magnetic flux density in a volume of 120 cm^3 (see also [8])

Table 9.2 The magnetocaloric properties of the analysed materials (see Chap. 4 for the definition of the maximum specific refrigeration capacity)

Material	ΔT_{ad} (K) [9]	Δs_{is} ($Jkg^{-1}K^{-1}$) [9]	q_{Rmax} (Jkg^{-1})
Gd	3.3	3.1	910
LaFeCoSi	2.2	5.3	1,553

Table 9.3 The theoretical maximum cooling power (W) for the selected case at T_R = 292 K

	Frequency (Hz)			
Material	1	2	5	10
Gd (W)	604	1,208	3,019	6,039
LaFeCoSi (W)	1,031	2,061	5,153	10,306

gadolinium manufactured as the AMR. In this analysis we assume that the mass production manufacturing costs of magnet assembly (labour, energy and manufacturing devices) represent 25 % of the basic capital cost.

Now, for the chosen magnet assembly (Fig. 9.3) we can assume the mass production and the following costs for materials and manufacturing (Table 9.4).

The maximum relative costs for a refrigerator's compressor relates to very small units (large compressor unit represent substantially smaller relative costs than smaller units). Despite this a 50-W compressor's production costs are approximately 12 euros or 0.24 euros/Wcooling power. So, what level of theoretical costs for the magnet assembly is acceptable? We assume again the most ideal case, which considers that the costs of the compressor are equivalent only to the magnet assembly with the magnetocaloric material, by neglecting all the other components of the magnetic refrigerator. By combining the results in Tables 9.3 and 9.4 we can now express the specific costs in euro per watt of cooling power, as shown in Fig. 9.3.

Despite the fact that Fig. 9.3 shows that a frequency of operation higher than 2 Hz is required to obtain the same specific costs as for the compressor, note that the magnet assembly with the magnetocaloric material will account for up to 60–80 % of the total costs of the magnetic refrigerator [10]. Therefore, in order to have a

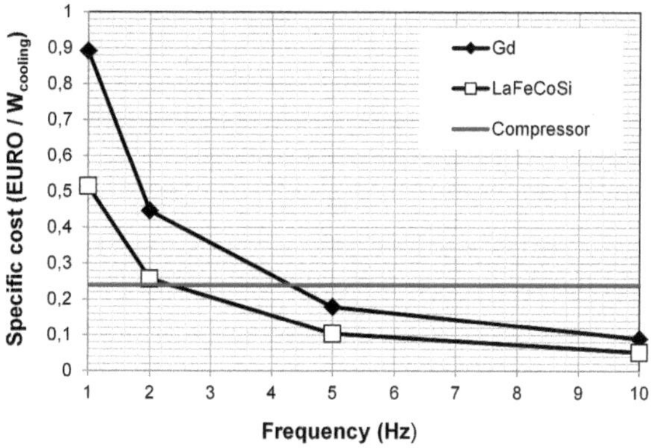

Fig. 9.3 Comparison of the specific costs for a 50-W_c compressor and the equivalent ideal magnetic refrigerator

Table 9.4 The mass-production cost estimation of the magnet assembly in Fig. 9.3

Material	$Euro kg^{-1}$	Mass (kg)	COST (Euro)	Manufacturing cost of assembly	Total cost (Euro)
NdFeB	40	6	240	+25 %	300
Soft iron	5	22.2	111	+25 %	139
AMR LaFeCoSi	60	0.60	36	/	36
					475
NdFeB	40	6	240	+25 %	300
Soft iron	5	22.2	111	+25 %	139
AMR Gd	150	0.66	99	/	99
					538

comparison with the whole magnetocaloric device, a frequency of operation of above 3 Hz will actually be required. At this frequency, the magnetic refrigerator will have to operate with an efficiency at least as high as a compressor refrigerator, as well as with an equivalent temperature span between the heat source and heat sink (i.e. 40 K or more). Note again that the 40 K temperature span with the predicted same or higher efficiency of the magnetic refrigerator represents an ideal case. This also holds true for the frequency of operation. For such an ideal case the AMR should actually consist of a layered magnetocaloric material and produced as a structure with very tiny plates (a honeycomb or similar ordered structure) with thickness of, e.g. 30–100 μ, porosity of about 30–45 %, and with a similar thickness of equally sized fluid voids. For frequencies above 5 Hz, using water as the heat transfer fluid in such an AMR will also not be a good option.

Note also that any complication related to the magnet structure will increase the costs. In Fig. 9.2, the magnet assembly applies simple rectangular prisms of permanent magnets. If we wish to make such magnets more complex, the first question is of course why is this required? For instance, we have already shown in the chapter on AMRs (Chap. 4) that the homogeneity of the magnetic field is perhaps

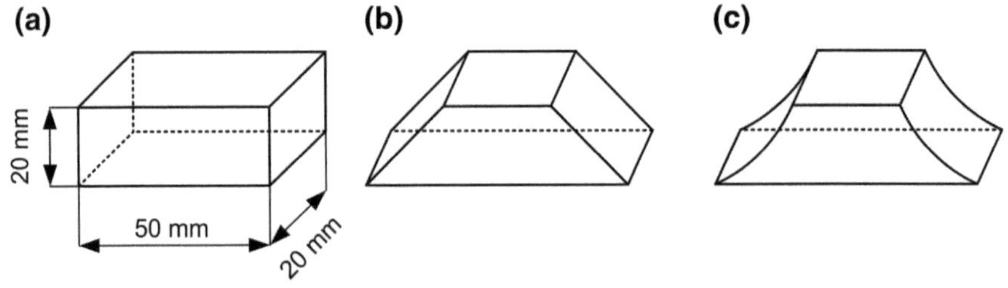

Fig. 9.4 Comparison of different permanent-magnet structures

not the best solution, especially if we want to perform thermodynamic cycles other than the AMR Brayton-like cycle.

Moreover, any complexity of the permanent-magnet structures will also lead to complexity of the assembly process, which will drastically increase the costs. Figure 9.4 has been constructed according to our communication with Barcza from Vacuumschmelze GmbH & Co [11]. If the basic prism (a) in Fig. 9.4 costs 100 units of money, then it will follow for (b) and (c) that the cost will represent 110 units. Any additional complexity, e.g. towards asymmetric structures, will increase the costs. This will be further evident for the costs that concern the assembly process.

Since at present no Nd–Fe–B equivalent alternative permanent magnetic field sources are available, we should try to decrease the mass of the permanent-magnet material as much as possible. This can be done either with a smart design of the magnet assembly, or by a substantial increase in the frequency of the operation (number of thermodynamic cycles per unit of time).

9.2 A Brief Discussion on the Market and the Costs of Superconducting Magnets

Despite superconducting magnets having already been applied for the purpose of magnetic refrigeration, there is no evidence that any of these superconducting magnets has already been designed for the purpose of magnetic refrigeration at room temperature. Therefore, no cost evaluation for superconducting magnetic refrigerators or chillers can be found in the literature, except that made for the Swiss Federal Office of Energy in 2010 by Egolf et al. [12]. We will come back to that study in the subsequent text.

Note that in the following decades people will witness the very rapid development of different sectors, which are related to the application of superconducting materials, such as for instance those used in electrical equipment (i.e. cables, transformers, fault current limiters), rotating machines (motors and generators) as well as superconducting magnetic energy storage [13].

Because of these facts we would like here again to encourage the industry related to superconducting applications to enter the development of large-scale superconducting magnetic chillers or heat pumps.

Most of the work that concerns the evaluation of costs for superconducting magnets regards the superconducting storage of electric energy (SMES). These systems store electrical energy in the magnetized space, with most of these devices applying toroidal and cylindrical solenoids. The maximum energy stored per unit volume at a certain magnetic field is calculated using simple mathematical relations. The procedure to perform an economic analysis on superconducting magnets was shown by Egolf et al. [12].

Following the magnetic energy storage, the maximum energy stored in the magnetized space (gap) of the superconducting magnet can be defined as [12]:

$$\mathrm{d}w = \frac{\mu_0}{2} d(H_0)^2 = \frac{1}{2\mu_0} d(B_0)^2 \tag{9.1}$$

where H_0 represents the applied magnetic field in the empty gap of the superconducting coil. Since the empty magnetized space can be considered as a material with linear magnetic characteristics, a change in the stored energy in a certain space can be defined as [12]:

$$\mathrm{d}W = \frac{1}{2\mu_0} \int_V d(B_0)^2 \mathrm{d}V \tag{9.2}$$

If the magnetic flux density is constant, then it follows from Eq. (9.2) that:

$$W = \frac{(B_0)^2}{2\mu_0} V \tag{9.3}$$

With the application of Eq. (9.3) it is now possible to show the variation of the magnetic energy depending on the magnetized volume and the magnetic flux density (Fig. 9.5).

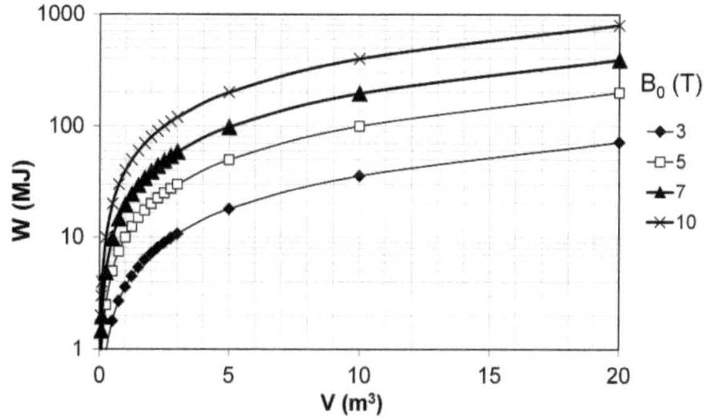

Fig. 9.5 The maximum energy stored per unit volume for the magnetized space

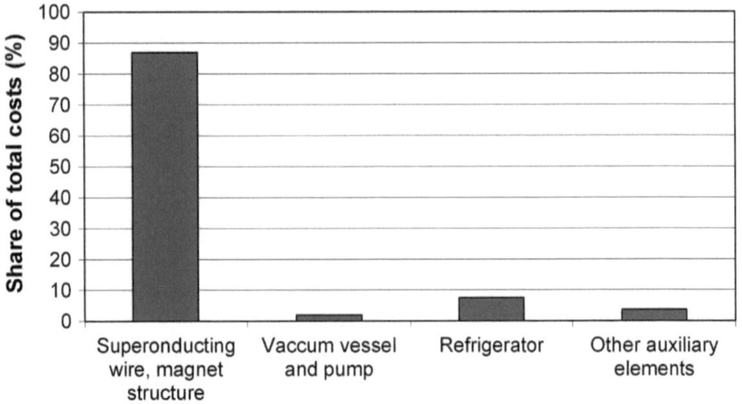

Fig. 9.6 The share of the total costs for the SMES (data taken from Luongo et al. [14])

The major cost contribution to the superconducting magnetic energy storage system comes from the superconducting coil (material and manufacturing). This can also be seen in Fig. 9.6, which shows the cost breakdown presented by Luongo in 2001 [14]. In this case the superconducting magnet has a toroidal structure for storage of 535 MJ, with peak magnetic flux density of 8.1 T. As can be seen from Fig. 9.6, the majority of the cost contribution of the system is caused by the superconducting magnet with its structure and wires.

Note that the operating costs of the cryogenic cooling system are actually very low compared to the capital costs, and also the total efficiency of the superconducting magnetic storage system will be very high and in the range from 85 to 90 % or even higher, which shows that the cryogenic system, especially for larger systems, will not have a drastic effect on the overall efficiency.

In the article of Zhu et al. [15], a preliminary study of Superconducting Magnetic Energy Storage (SMES) system design and a cost analysis for the power-grid application was presented. They also showed the costs of such systems calculated on the basis of the commercial price of superconductors. Table 9.5 shows the cost estimation for different SMES. The costs in Table 9.5 relate only to superconductors.

Table 9.5 Specification and costs of different scales of SMES systems [15]

	1.2 kJ (solenoid)	1.6 MJ (solenoid)	1.3 MJ (toroid)	1 GJ (toroid)
Inner radius (mm)	45	200	80	340
Outer radius (mm)	68	280	120	860
Height (mm)	24	120	/	/
Total length of superconducting tape (km)	0.17	12	5	533
COST (US$)	1.4×10^4	1.02×10^6	4.25×10^5	4.5×10^7
Unit price (US$J^{-1})	1.4	0.64	0.33	0.045

Another comparison of different toroidal SMES with different magnetic flux densities and different configurations of magnetized spaces was performed by Lee et al. [16] in 2009. More information about the cost analyses of SMES systems can be found in Egolf et al. [12], Schenung et al. [17, 18], Seeber et al. [19] and Lubell and Lue [20].

In 2008, Green and Strauss published a comprehensive analysis on the costs for low-temperature superconducting (LTS) magnets [21]. In the same year, Green also published an article on the costs of helium refrigerators and coolers for superconducting devices [22]. Based on their work, Green and Strauss provided fitting functions, which may serve for the cost estimation of the LTS magnets:

$$\mathrm{COST}\,(\mathrm{Mil.\,US\$}) = 0.92 \cdot W^{0.6} \tag{9.4}$$

where W is the stored magnetic energy in MJ. Equation (9.4) was also expressed in terms of the volume and the magnetic flux density as

$$\mathrm{COST}\,(\mathrm{Mil.\,US\$}) = 0.8 \cdot (B_0 \cdot V)^{0.6} \quad B_0 V \left(\mathrm{T\ m^3}\right) \tag{9.5}$$

These cost-fitting lines (least squared fits) are valid for all magnets, solenoid magnets and toroid magnets. For solenoid magnets only, Green and Strauss [21] provided the following relation:

$$\mathrm{COST}\,(\mathrm{Mil.\,US\$}) = 0.95 \cdot W^{0.67} \quad W(\mathrm{MJ}) \tag{9.6}$$

or:

$$\mathrm{COST}\,(\mathrm{Mil.\,US\$}) = 0.55 \cdot (B_0 \cdot V)^{0.67} \quad B_0 V \left(\mathrm{T\ m^3}\right) \tag{9.7}$$

For toroid magnets only, Green and Strauss [21] provided the following relation:

$$\mathrm{COST}\,(\mathrm{Mil.\,US\$}) = 2.04 \cdot W^{0.5} \quad W(\mathrm{MJ}) \tag{9.8}$$

or:

$$\mathrm{COST}\,(\mathrm{Mil.\,US\$}) = 2.01 \cdot (B_0 \cdot V)^{0.5} \quad B_0 V \left(\mathrm{T\ m^3}\right) \tag{9.9}$$

Figure 9.7 shows a comparison of costs for LTS superconducting magnets. The calculation was based on Eqs. (9.7 and 9.9).

Additional information was also given for the costs of the cryogenic system, since these were not accounted for in Eqs. (9.4–9.9). The capital costs for helium refrigerators (0.01–30 kW cooling at 4.5 K) were defined by Green [22] to be

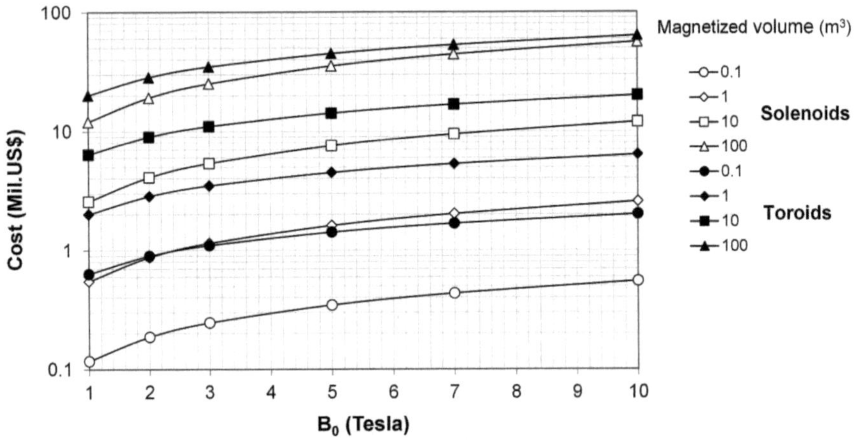

Fig. 9.7 The cost of solenoidal and toroidal superconducting magnets

$$\mathrm{COST\,(Mil.\,US\$)} = 2.6 \cdot \left(\dot{Q}_R\right)^{0.63} \quad \dot{Q}_R(\mathrm{kW}) \tag{9.10}$$

where the refrigeration power was defined as the design refrigeration power in 0, normalized to the temperature of 4.5 K. A similar equation was given for small cryocoolers (0.1–10 W of cooling at 4.2 K) [22]:

$$\mathrm{COST\,(1000 \times US\$)} = 37 \cdot \left(\dot{Q}_R\right)^{0.38} \quad \dot{Q}_R(\mathrm{W}) \tag{9.11}$$

where the refrigeration power was defined as the design refrigeration power in Watts, normalized to the temperature of 4.2 K. Figure 9.8 (left) shows the cost of helium refrigerators (cold boxes and compressors) for refrigeration at 4.5 K. The calculation was based on Eq. (9.10). Figure 9.8 (right) shows the capital cost of

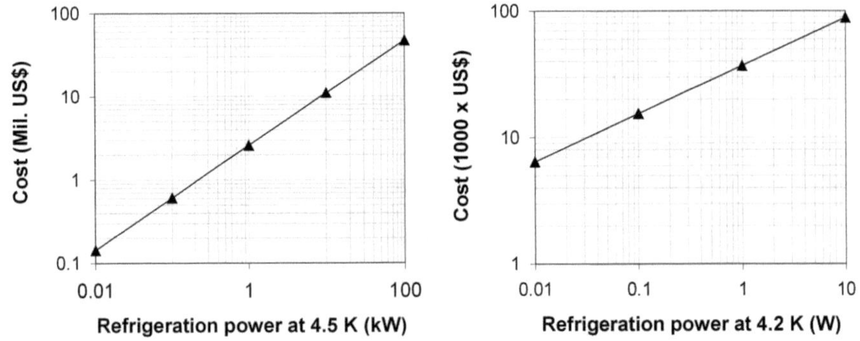

Fig. 9.8 The cost of helium refrigerators (*left* large-scale helium refrigerators; *right* small-scale helium refrigerators)

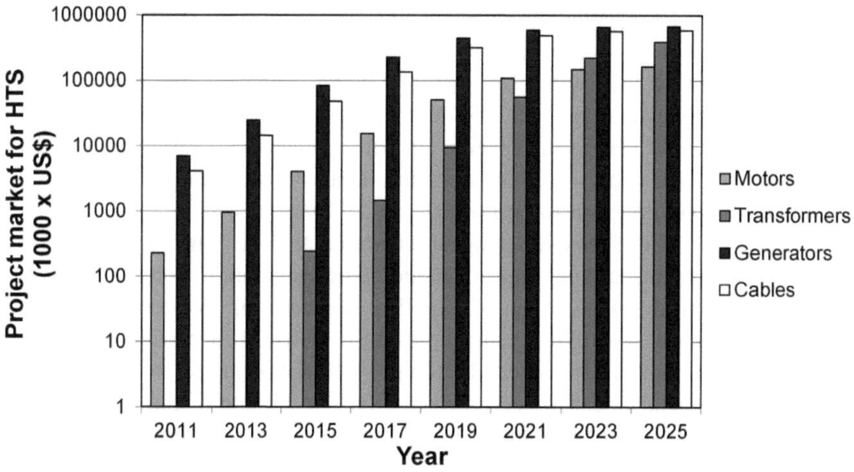

Fig. 9.9 The market projection for high-temperature superconductor applications (data taken from Mulholland et al. [23])

small helium coolers (k$) for refrigeration at 4.2 K (W) [22]. The calculation was based on Eq. (9.11).

Our search for market analyses on superconducting applications led to rather old reports. For instance, in the report by Mulholland et al. [23] the authors did an analysis of future prices and markets for high-temperature superconductors (HTS). Despite this being a rather old report from 2001, it is important to address the trends of production as well as the costs associated with the applications of HTS. For the first, Fig. 9.9 shows a projection of the market for high-temperature superconductor applications. For the second, the authors have stated that researchers will substantially increase the current carrying capacity of HTS wires, and by also taking into account the rapid decrease of HTS wire costs with an increased length; this will substantially reduce the investment costs by 2020.

9.3 Review of Cost Analyses for Magnetocaloric Energy Conversion

Some of the earliest cost analyses on magnetic refrigeration were performed by Russek and Zimm in 2006 [24], where they focused on the application of air conditioners. They used a simple approach in which they defined the specific costs of the magnetocaloric material and the permanent-magnet material per cooling power of the device. Their study revealed that when using Gd or similar material at a cost of 20 US$kg^{-1}, for an operating frequency of 10 Hz for the device, the costs for such a material will be 1.5 US$(kWcooling)$^{-1}$. They estimated the cost of the $La(Fe_{1-x}Si_x)_{13}$ material to be 8 US$kg^{-1}, which led to the specific cost of such a material being 0.1 US$(Wcooling)$^{-1}$ for a frequency of operation of 10 Hz. For the case of Gd, but in this case the evaluation was focused on $Nd_2Fe_{14}B$-type permanent magnets (1.4 T) with a cost of 27 US$kg^{-1}, the specific cost of such a magnet assembly would be 10.2 US$(Wcooling)$^{-1}$. In the case of $La(Fe_{1-x}Si_x)_{13}$ material, this would lead to the specific cost of the magnet assembly being 2.2 US $(Wcooling)$^{-1}$. No analyses of the operating characteristics and efficiency of the magnetic air conditioner were performed.

Most of the comprehensive technical economic analyses on magnetic refrigeration, magnetic heat pumping and magnetic power generation were performed by the group from the University of Applied Sciences of Western Switzerland (USAWS), who did several different studies for the Swiss Federal Office of Energy in the period 2004–2010 (these can be found in the following references [25–29]). Note that these studies can be accessed free on the website of the Swiss Federal Office of Energy. In the subsequent text we present some of the most important findings of these studies.

In 2006, a feasibility study on magnetic heat pumps was published by Egolf et al. [25]. The authors did a technical-economic analyses for an 8-kW, ground-source, rotary magnetic heat pump with the temperature of the heat source and the heat sink (on the water side) being 0 and 35 °C, respectively. The aim of the analysis was to compare such a heat pump with a conventional compressor-based heat pump for the same efficiency COP = 6.5 and for the same operating parameters.

For this purpose the following material costs and masses were estimated (see also Fig. 9.10):

- magnetocaloric material (125 eurokg^{-1})
- permanent magnet material (60 eurokg^{-1})
- soft iron (45 eurokg^{-1})
- other material costs (10 eurokg^{-1})

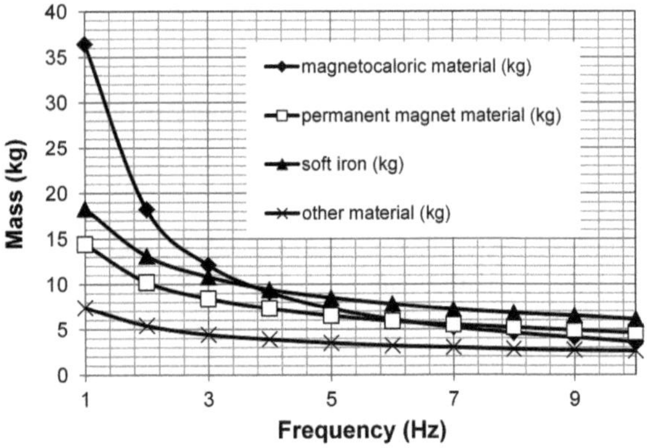

Fig. 9.10 Masses depending on the frequency of the operation (data taken from Egolf et al. [25])

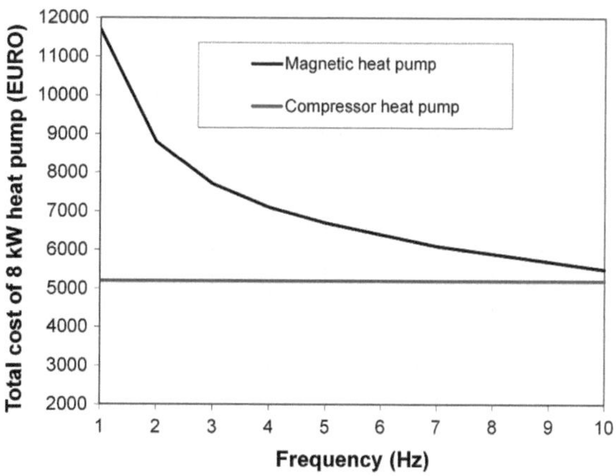

Fig. 9.11 The manufacturing costs for the 8-kW heat source magnetic heat pump and the equivalent compressor-based heat pump (data taken from Egolf et al. [25])

Figure 9.11 was made according to the results of the study by Egolf et al. [25]. In Fig. 9.11 the total manufacturing costs for an 8-kW, ground-source, magnetic heat pump are shown.

Note that the mass of the magnet assembly was underestimated in the study, and therefore also the costs for the magnet assembly. To be on the safe side the authors took rather high costs for the other parts of the device (i.e. motor, pumping system, valves, pipework, housing, etc.). Despite this the results shown in Fig. 9.11 are very important. Although they do not show exact costs, they demonstrate the importance of the frequency of the operation of the magnetocaloric device. Since the costs of the evaluated magnetic heat pump in Egolf et al. [25] did not reach the cost of the compressor-based heat pump, the authors emphasize that by taking into account the advantage of potentially better efficiency and also operating costs, the magnetic heat pump, operating with a frequency of 5 Hz, will be economically competitive with a compressor-based heat pump. The issue of high efficiency, besides low operating costs, defines the efficiency label. This fact must not be missed.

In the study on the applications of magnetic refrigeration (Kitanovski et al. [26], see also Kitanovski and Egolf [30]) the authors showed the most important and feasible market domains for magnetic refrigeration, i.e. household refrigeration, heat pumps, wine/beverage refrigeration, small medicine refrigeration appliances, room air conditioners, water cooled chillers, large-scale superconducting chillers, etc. Based on this study, some further economic evaluations have been performed by the same group.

In the case of magnetocaloric power generation, Kitanovski et al. [27] and Egolf et al. [31] emphasized the use of low-enthalpy heat sources. Furthermore, they addressed the importance of energy harvesting by applying superconducting magnets. The same group also performed a comprehensive technical-economic feasibility study of a magnetocaloric power-conversion system for the utilization of waste heat [32]. In subsequent publications by Vuarnoz et al. in 2012 [33, 34] the authors found a payback period of 4.9 and 2.4 years (for electricity price 0.1 and 0.2

CHFkWh^{-1}) for the best-case scenario and large-scale magnetocaloric power generation of 15 MW.

An additional study was performed in 2010 by Egolf et al. [29]. This time central magnetic chiller applications were studied (see also Kitanovski et al. [35]). For the purpose of the study the authors designed three different magnet assemblies for rotary magnetic chillers, with three different magnetic fields (1, 1.5 and 2 T, respectively). The design of such a magnet assembly can also be seen in the chapter on magnetic field sources (Chap. 3), Fig. 3.33. Since the design was such that it enabled up-scaling of the magnet assembly for different applications (related to cooling power), this also provided a good basis for further technical and economic evaluation. In all the analyses the coldest point of the magnetocaloric material was defined to be 7 °C and the hottest point was defined to be 32 °C (i.e. 25 K difference between the heat source and the heat sink). These two temperatures were compared with the temperatures of the chilled and the cooling fluid at the outlet of the evaporator and the outlet of the condenser, respectively. For these data, a COP = 5.5 for the compressor chiller was defined as the minimum required value for the magnetic chiller. This coefficient of performance was also the basis on which the further economic analysis was performed. For the purpose of the study, the thickness of the magnetocaloric plates in the regenerator was one of the main parameters. This was varied from 100 to 400 microns. Also, the volume fraction of the magnetocaloric material in the regenerator was varied from 30 to 60 %. Two working fluids were evaluated, the 20 % ethanol water solution and the liquid metal Galinstan (see Sect. 4.6). The latter was chosen because of its good heat transfer characteristics.

Figure 9.12 was constructed using data from Egolf et al. [29]. It shows the manufacturing cost ratio between a magnetic chiller based on the AMR principle, and an equivalent compressor chiller for two different working fluids. In the case of water-ethanol, this corresponds to a range of cooling powers up to 8 kW. In the case of Galinstan, this corresponds to cooling powers up to 30 kW. Note that for the

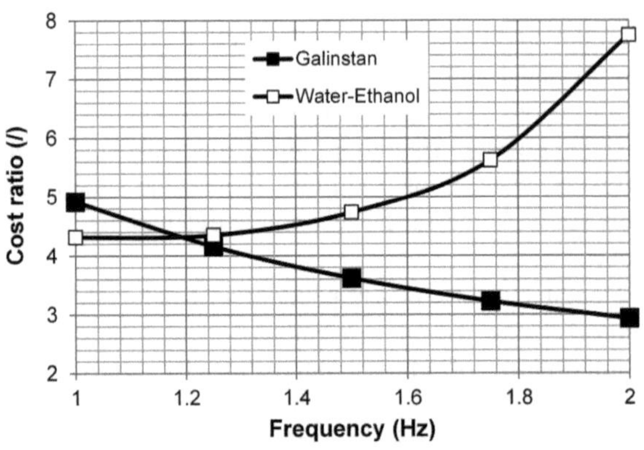

Fig. 9.12 The ratio between the manufacturing costs for the AMR-based magnetic chiller with permanent magnets and the equivalent compressor chiller for two different heat-transfer fluids (calculation performed on the basis of data from Egolf et al. [29])

purpose of the comparison in Fig. 9.12, the best results for the magnetic chiller were taken (60 % volume fraction, thickness of the magnetocaloric material 100 microns). The results reveal that the most important issue to be addressed for future research is to target substantial improvements in the frequency of operation, which are very strongly related to the AMR geometry and heat transfer.

The authors have further emphasized [29, 35] that devices based on permanent magnets and the AMR principle have a certain restriction (upper limit) for cooling power up to 200 kW, which is related to the geometry of the permanent magnet. Namely, for a large cooling power the dimensions of the permanent-magnet assemblies can be such that this can lead to substantially higher manufacturing costs. The authors added an important note [29, 35] that the problems associated with cost, power density, compactness and efficiency may be solved in the future; however, with new approaches, which are for instance based on the application of advanced working fluids, new heat transfer principles with the application of thermal diodes, advanced design approaches, etc.

In addition, Egolf et al. [29] did a comprehensive analysis on the feasibility of a superconducting rotary magnetic chiller. The study revealed that when magnetic chillers are compared to conventional ones, this shows an advantage for both efficiency and cost. Namely, the cost ratio between the manufacturing costs of the superconducting chiller versus compressor chillers were estimated to be from 0.55 to 0.75. However, the minimum cooling power for such devices should be not less than 1 MW.

In 2011, a more accurate model for characterization of AMR's operation was applied for an investigation of magnetic superconducting chillers by Kitanovski et al. [36]. Since this was reported only internally we present here some of the results of the calculations that were performed at the University of Ljubljana. For the purpose of the study the supply and return temperatures for the chilled fluid were taken to be 7/12 °C and for the cooling fluid 27/32 °C, respectively. Such a temperature difference at the heat source or heat sink is relatively low, i.e. about 5–6 K. A difference of 5–6 K can also be obtained with many different magnetocaloric materials where the process of adiabatic magnetization (demagnetization) is performed for a magnetic field change of less than 3 T, without the need for very high magnetic fields. This is also why the analyses in the study considered magnetic field changes of superconducting magnets being 1.5, 2, and 3 T.

In the study it was assumed that the rotary superconducting chiller (i.e. the rotation of the magnetic field source or the rotation of the AMRs) performs the AMR Brayton-like cycle.

The numerical analysis of the characteristics of the operation at given parameters was based on a dynamic simulation model, developed and described by Tušek et al. in 2011 [37]. The following parameters were considered in the analysis:

- Heat-transfer fluid: water;
- Return temperature of the chilled fluid 12 °C;
- Supply temperature of the cooling fluid 27 °C;
- Frequency of the operation <5 Hz;

Table 9.6 Geometry of the regenerator with a 60 % volume fraction of magnetocaloric material [36]

Spheres diameter d = 0.25 mm	L = 5 cm	L = 10 cm	L = 15 cm
Mass of MCM material (kg)	298.43	596.85	895.28
Parallel plates thickness s = 0.25 mm	L = 20 cm	L = 30 cm	L = 40 cm
Mass of MCM material (kg)	1193.7	1790.55	2387.4

- MC material: ideally layered La (Fe, Si, H) material;
- Magnetocaloric structure: Coaxial cylinder with volume fraction 60 % (see Table 9.6);
- Magnetic field change: 1.5, 2 and 3 T.

Figure 9.13 shows the maximum cooling power and the corresponding COP as a function of frequency for the magnetic field change of 1.5 T for the AMR consisting of spheres (top) and for the AMR consisting of plates (bottom). As can be seen from Fig. 9.13 (top), the maximum cooling power may be obtained for AMRs with lengths between 10 and 15 cm. It is further evident that the increase in the length of the regenerator from 10 to 15 cm leads to a decrease in the cooling power at frequencies above 3 Hz. Also, the COP is very low at frequencies above 3 Hz and cannot be comparable with that of compressor chillers. As can be also seen from Fig. 9.13 (top), the highest COP is obtained for the shortest packed-bed AMRs (for a given temperature span).

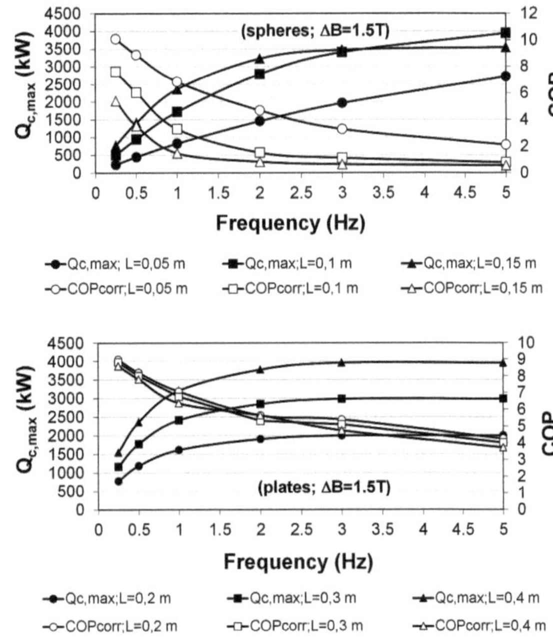

Fig. 9.13 The maximum cooling power and the corresponding COP as a function of frequency for the AMR with spheres (*top*) and for the AMR with plates (*bottom*) under a magnetic field change of $\mu_0 \Delta H_0 = 1.5$ T [36]

The advantage of the application of AMRs consisting of parallel plates can be seen in Fig. 9.13 (bottom). Since the application of spheres leads to large heat transfer surfaces, it could be expected that such geometry produces higher cooling power than that of the parallel plates, which is actually true. Note that with regard to Figs. 9.13 and 9.14 the masses of the evaluated packed-bed AMRs were significantly smaller than those of the parallel-plate AMRs. However, for the same coefficient of performance COP the AMR with parallel plates enables greater cooling power compared to the AMR with packed beds of spheres. The length of the magnetocaloric regenerator also plays a crucial role. In the case of a packed bed of spheres, the viscous losses, as well as the related heat generation inside the structure, have a large impact on both the efficiency and the COP. This is especially true for long regenerators. The impact of the AMR's geometry is also explained in Sect. 4.4.

If the magnetic field change is increased to 3 T, this will increase the cooling power and efficiency (Fig. 9.14).

By following the results on the characteristics of the operation of the AMR, Kitanovski et al. [36] also performed an economic analysis. This was based on the economic evaluation that was already presented in Egolf et al. [29]. The economic analyses considered, besides the cost of the whole superconducting magnet system, the cost of the magnetocaloric material. The production cost for a La(Fe,Si,H) regenerator was estimated to be 60 eurokg^{-1}. Other costs were neglected.

The results in Figs. 9.15 and 9.16 show that the geometry of the AMR has an important influence on the cost of the magnetic chiller. Note that for the same COP

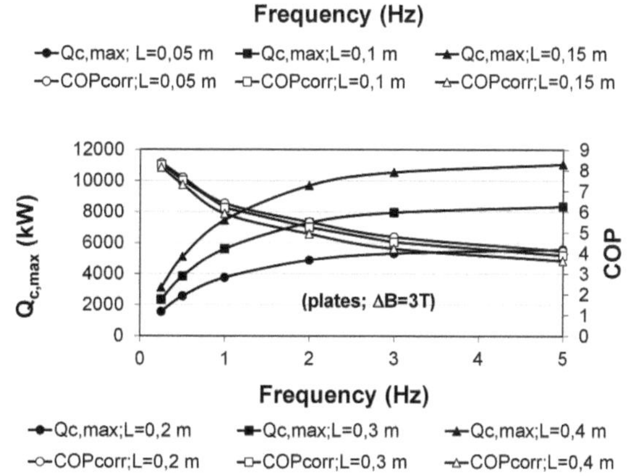

Fig. 9.14 The maximum cooling power and the corresponding COP as a function of frequency for the AMR with spheres (*top*) and for the AMR with plates (*bottom*) under a magnetic field change of $\mu_0 \Delta H_0 = 3$ T [36]

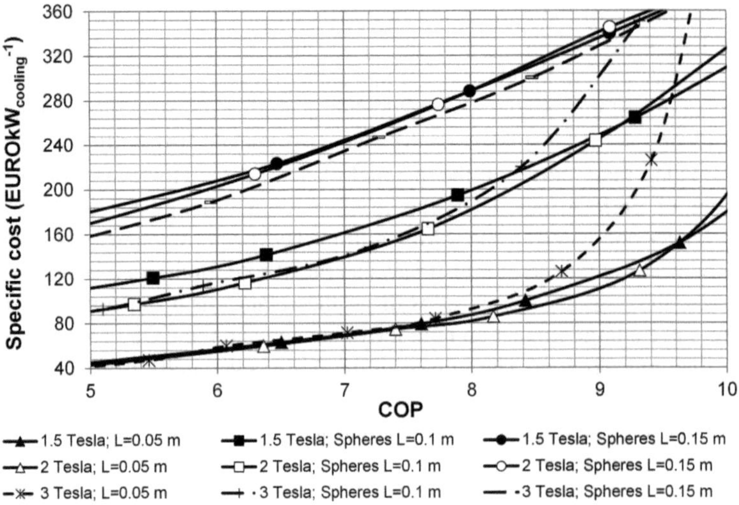

Fig. 9.15 The specific cost of the superconducting magnetic chiller with packed-bed AMRs in the cooling power range from 1 to 5 MW (data recalculated from Kitanovski et al. [36])

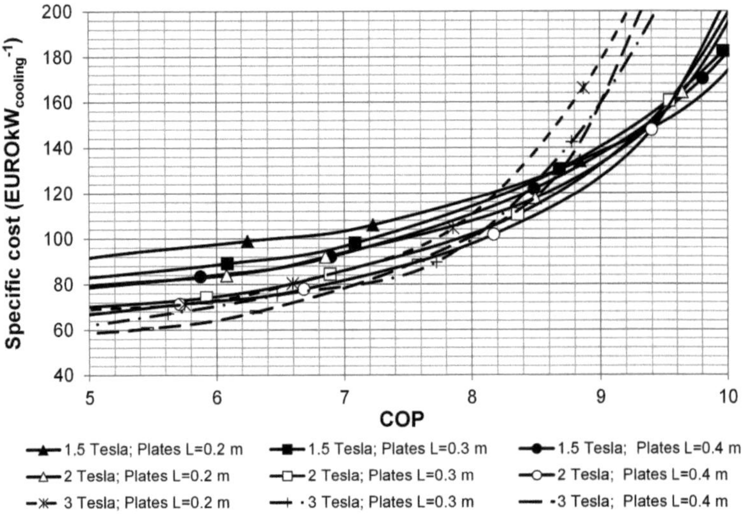

Fig. 9.16 The specific cost of the superconducting magnetic chiller with parallel plate AMRs in the cooling power range from 1 to 8 MW (data recalculated from Kitanovski et al. [36])

and magnetic field, the cost of a device with packed beds of spheres decreases with a decreasing length of the regenerator. In the case of parallel plates, the cost of the device decreases with an increasing length of the parallel plates. The reason for this is in the heat transfer efficiency, as well as the viscous losses, which strongly depend on the geometry of the AMR and the working fluid.

For both the spheres and the parallel plates the cost decreases with an increasing magnetic field change from 1.5 to 2 T. The increase in the COP will lead to a higher relative cost (due to the decrease of the cooling power with the increase of the COP). For the case of longer regenerators (i.e. $L = 15$ cm for packed beds and all the considered lengths for parallel plates), slightly better results can be obtained with a magnetic field change of 3 T compared to a 2 T magnetic field change. However, shorter regenerators lead to higher relative costs for the 3 T magnetic field, compared to the 2 T field. Moreover, at high magnetic field changes (e.g. 3 T) and for longer regenerators with spheres, there is a substantial increase in the relative cost at higher COPs.

In 2011, another comprehensive economic evaluation was performed by Bjørk et al. [38]. Furthermore, the authors defined an expression for the mass of the magnet and the magnetocaloric material that is required for a magnetic refrigerator. This was determined by numerical modelling for both parallel plate and packed-sphere-bed regenerators as a function of the temperature span and cooling power. For the purpose of the study the magnetocaloric material gadolinium was considered to have a constant adiabatic temperature change with infinitely linearly graded properties (i.e. ideally layered).

They introduced a new figure of merit for the permanent magnet assembly, defined as [38]:

$$M^* = \left(\frac{\mu_0 H}{B_{\text{rem}}}\right)^2 \frac{V_{\text{field}}}{V_{\text{mag}}} \tag{9.12}$$

where V_{field} is the volume of the air gap where there is a constant applied magnetic field, $\mu_0 H$, and V_{mag} is the volume of permanent magnets which have a remanence magnetic flux density B_{rem}.

The authors further stated that the maximum value of $M^* = 0.25$ and denoted such a magnet assembly as M_{25}. Based on Eq. (9.12) the authors provided another expression that holds for devices in which the magnetocaloric material is continuously utilized:

$$m_{\text{mag}} = \frac{1}{2}\left(\frac{\mu_0 H}{B_{\text{rem}}}\right)^2 \frac{m_{\text{mc}}\rho_{\text{mag}}}{(1-\varepsilon)\rho_{\text{mc}} M^*} \tag{9.13}$$

In Eq. (9.13) the m_{mc} represents the total mass of the magnetocaloric material, the term $(1 - \varepsilon)$ represents the volume fraction of the magnetocaloric material in the AMR, and ρ_{mag} and ρ_{mc} represent the densities of the permanent magnets and the magnetocaloric material, respectively. In an economic analyses, Bjørk et al. [38] considered the cost of the magnet material to be 40 US$/kg and the cost for the magnetocaloric material to be 20 US$/kg. The manufacturing costs of the magnet assembly and the manufacturing costs of the regenerator were not included, nor were the costs related to the motor drive, regulation and control, and other auxiliaries. However, the aim of the study was to optimize (find the minimum) the costs of the permanent magnet material as well as that of the AMR, since these represent the major contribution to the costs of a device. Table 9.7 presents the parameters that have been applied in the analyses of Bjørk et al. [38] for AMRs based on spheres and on parallel plates.

The Δx in Table 9.7 represents the fluid strokes length, which describes the fraction of fluid that is displaced with respect to all the volume of fluid in the AMR. The τ_{rel} represents the relative cycle time, which is the ratio between the time used for the magnetization or demagnetization of the AMR and the time used for the fluid displacement.

The analysis was performed in such a manner that for each set of process parameters the cooling power was defined. Based on this, the mass of magnetocaloric material was defined as a function of the magnetic flux density for a desired

Table 9.7 Parameters used in the analysis of Bjørk et al. [38]

Spheres									
Δx (%)									
70	90	110	135	150	180	215			
Frequency of operation (Hz)									
1	2	4	10						
τ_{rel} (/)									
0.1	0.25	0.5							
Sphere diameter (mm)									
0.1	0.25	0.5							
Maximal magnetic flux density (T)									
0.4	0.6	0.8	1	1.2	1.4	1.6			
Temperature of the cold end of AMR T_{cold} (K)									
268	272	276	278	280	284	288	292	296	298
Parallel plates									
Δx (%)									
40	50	60	70	80	90				
Frequency of operation (Hz)									
0.167	0.33	1	2	4					
τ_{rel} (/)									
0.25	0.5								
Height of the fluid channel (mm)									
0.1	0.25	0.5							
Thickness of the plate (mm)									
0.1	0.25	0.5							
Maximal magnetic flux density (T)									
0.4	0.6	0.8	1	1.2	1.4	1.6			
Temperature of the cold end of AMR T_{cold} (K)									
268	272	276	278	280	284	288	292	296	298

cooling power and temperature span. There it was assumed that the cooling power is directly proportional to the mass of the magnetocaloric material. The authors used Eq. (9.13) and for each temperature span and magnetic field calculated the combined cost of the magnet and the magnetocaloric material. This was repeated for all the parameters. Then the lowest cost was chosen for the presentation.

Figure 9.17 shows the minimum costs for a Gd-based AMR consisting of a packed bed of spheres and parallel plates, respectively. The figure was produced based on data from Bjørk et al. [38]. The costs are presented as a function of the cooling power. Additionally, the costs for an ideally layered AMR with the properties of gadolinium are shown. All the cases correspond to a temperature span of 30 K.

The authors also presented the masses that corresponded to a particular cooling power and temperature span. However, they did not consider the coefficient of performance at which the temperature span and the particular cooling power can be obtained.

In 2011, Rowe also presented his work on the configuration and performance analysis of magnetic refrigerators [39]. He based his economic evaluation on an exergy analysis and proposed a model for such an evaluation. In the model he presented the specific costs per watt of cooling as a function of the specific exergetic cooling power, which he defined as

$$\mu = \frac{f_R}{B_0}\left(\frac{E}{V_{\mathrm{mc}}}\right) \tag{9.14}$$

which has units of watt of exergy of cooling per litre of magnetocaloric material and per magnetic flux density in Tesla and where E represents the exergy of the cooling energy:

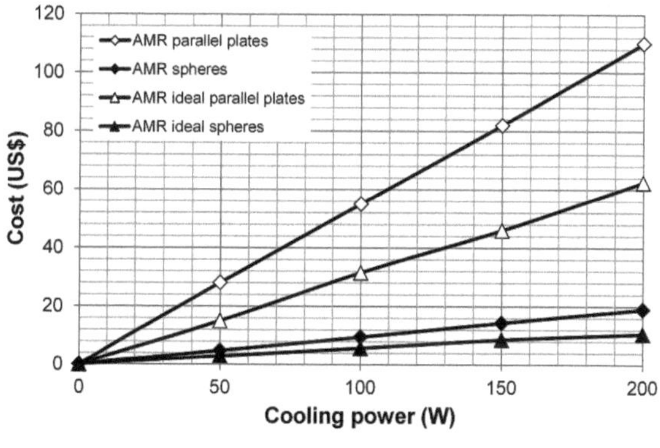

Fig. 9.17 The minimum combined cost of permanent-magnet material and magnetocaloric material for a parallel-plate and packed-bed magnetic refrigeration system with a magnetocaloric material of gadolinium for a temperature span of 30 K (data taken from Bjørk et al. [38]). Ideal represents an ideally layered AMR with the properties of Gd

$$E = Q_c\left(\frac{T_H}{T_C} - 1\right) \tag{9.15}$$

and B_0 represents the applied magnetic flux density in units of Tesla, V_{mc} is the volume of magnetocaloric material and f_R is the frequency at which the magnetocaloric material is cycled.

The device frequency f_D was defined as the rate at which the alignment of magnets and regenerator is repeated. The relation between the magnetocaloric material's frequency and the device frequency was given with the following relation:

$$f_R = ab \cdot f_D \tag{9.16}$$

where a represents the fraction of refrigerant seen by a single high-field region and b represents the number of high-field regions.

As an example, in the article he was investigating the economics of a small magnetic refrigerator with a cooling power of 70 W, operating with a heat source temperature of 7.4 °C and a heat sink temperature of 54.4 °C (temperature span of 47 K). The cost of the magnetic refrigerator was based on the costs of the permanent-magnet material (40 US\$kg^{-1}) and the costs of the magnetocaloric material (20 US\$kg^{-1}). No other costs were taken into account, since the magnets and the magnetocaloric material were assumed to represent the major cost contribution. In order to compare with the conventional refrigeration technology, Rowe [39] considered the cost of the compressor unit as 0.5 US\$/Wcooling (for COP = 1.6) and 1.9 US\$/Wcooling (for COP = 2.6).

The results of the study reveal that for the considered application, in order to be applied on the market, the specific exergetic power must be between 400 and 1,000 W(lT)$^{-1}$ if the COP is in the range 1.6–2.6, with the cost of the cooling being less than 2 US\$(Wcooling)$^{-1}$. However, Rowe [39] emphasized that if the COP of the magnetic refrigerator can be higher, then a higher device cost and a lower specific exergetic power will be acceptable.

A year earlier, in 2010, Roudaut et al. [40] carried out a cost analysis and related comparison of existing prototype devices. For the purpose of the study the costs of the Nd–Fe–B magnets, gadolinium (as the magnetocaloric material), Fe and FeCoV were considered to be 100, 130, 1 and 80 eurokg^{-1}, respectively. The study revealed the following specific costs (Table 9.8) of a device per cooling power (see

Table 9.8 Devices for which the cost estimation was performed by Roudaut et al. [40]

Authors	Specific cost (euroWcooling^{-1})
Bjørk et al. [41]	3.16
Tura and Rowe [42]	16.4
Lee et al. [43]	5.35
Zheng et al. [44]	2.46
Okamura et al. [45, 46]	1.69
Zimm et al. [47]	4.56

chapter on magnetic field sources and the chapter on prototypes for more information about the devices).

The authors of that study emphasized that all the calculations were run on non-optimized geometries, as well as that some information about the structures was lacking, and therefore the results did not necessarily reflect the potential of the original design of a particular device.

In 2014 another economic evaluation was made by Tura and Rowe for a concentric Halbach cylinder [48]. In the study the authors took into consideration the costs of the magnetocaloric regenerator to be 150 US$kg^{-1}, and the cost of the Nd–Fe–B magnets to be 42 US$kg^{-1}. Their first estimation was based on an analysis of the operating characteristics of their nested (double) Halbach structure, for which they found that the cost at a 2 Hz frequency of operation ranged from a minimum 10 US$W^{-1} to a maximum of 21 US$W^{-1} for a zero temperature span and it grows exponentially as the temperature span approaches the maximum and the cooling power goes to 0 W. The authors were then further searching for the optimized parameters of the geometry, utilization and frequency in order to minimize the operating and capital costs. The resulting costs of the magnets varied from 100 to 800 US$ and the costs of the AMR varied from 50 to 400 US$ for the cooling powers from 100 to 400 W. Finally, the optimized costs of cooling (US $(Wcooling)$^{-1}$) varied from 0.6 to 1.05 US$(Wcooling)$^{-1}$ for the COP from 2 to 4.5, respectively, and for cooling powers in the range from 0 to 400 W.

In most of the analyses shown in this chapter the authors have shown the economic potential of magnetocaloric energy conversion. As can be seen, this will strongly depend on the costs of permanent magnets, and then the costs of the magnetocaloric material, which both represent high cost fraction of the device, e.g. 60–80 %. Furthermore, the frequency of operation is the parameter that will for a certain temperature span between the heat source and heat sink as well as the desired efficiency, define the power of the magnetocaloric device and related specific cost. This, however, mostly depends on the efficiency of the heat transfer between the working fluid and the magnetocaloric material.

9.4 A Note on Economic Analyses for Magnetocaloric Energy Conversion

As can be seen from the review of existing economic analyses a large number of these analyses do not consider all the costs that are present in the capital costs of the magnetocaloric device. The correct economic evaluation must consider mass production and the following costs:

- Manufacturing costs of the permanent-magnet assembly (material costs, labour costs, manufacturing costs);
- Manufacturing costs of the magnetocaloric regenerator and its housing (material costs, manufacturing costs);

- Driving system (motor and related costs);
- Pump and pipework costs;
- Costs of valves, sensors, control and regulation system;
- Cost of heat exchangers;
- Housing costs.

When compared to existing technologies, the costs should not be defined on the selling prices of those (which include margins), but one should consider the manufacturing costs.

Furthermore, the two technologies are comparable if they operate at the same temperature span as well as the efficiency. When the efficiency of the magnetocaloric energy conversion device is higher than the comparable conventional technology for the same operating parameters of the heating, cooling or electric power (heat pumps; refrigerators, chillers, air conditioners; power generators, respectively) and the same temperature conditions, this can be then be added to the analyses that also concern the operating costs. However, based on our experience, one should note that investors usually do not concern themselves about operating costs, but only investments. Therefore, in such a case it is helpful if additionally the efficiency labelling is provided, because it may influence the added value of the product.

Moreover, an additional lifecycle assessment (LCA) is required. For conventional refrigeration technologies there are a large number of such studies (see as an example Horie [49], Vendrusculo et al. [50]).

To date, the only LCA on magnetic refrigeration at room temperature was performed by Monfared et al. [51], a team representing experts in conventional compressor-based refrigeration. They found that magnetic refrigeration has a larger environmental impact than conventional compressor-based refrigeration due to the use of rare-earth materials in the magnets, and that recycling processes for such materials are important.

An important note by authors of this book for future LCAs suggests that these should be performed by a team that has good knowledge and experience of both comparable technologies. And in such an analysis one has to take care when comparing the emerging technology (with a lot of potential of improvements) with the mature technology. Namely, a lot of different technological issues, which today represent an obstacle, can in the future of magnetic refrigeration have a simple solution. Therefore, different scenarios have to be created, which take into account particular improvements. One such example is the substantial increase in the power density of a device.

Additional search on global sources of materials and their future availability (see as an example Du and Graedel [52], Graf and Held [53], de Boer and Lammertsma [54], Walters et al. [55] and Moss et al. [56]), including recycling (e.g. Binnemans et al. [57] and Liu et al. [58]) should support such studies.

References

1. Benecki WT (2013) The permanent magnet market–2015. In: Presentation on magnetics 2013 conference, Orlando, Florida, USA, 7–8 Feb 2013
2. Bradsher K (2011) Prices of rare earth metals declining sharply. New York Times, 16 Nov
3. Qi L, Yap CW (2014) China to appeal WTO rare-earth ruling group backed US, EU, Japan in complaint accusing beijing of violating trade rules. New York Times, 17 April 2014
4. http://www.researchinchina.com/Htmls/Report/2013/6714.html
5. Smith L (2012) The rare earth market, a monthly analysis for the rare earth industry. Core Consultants & Metal Pages, Aug 2013, p 15
6. Smith L (2013) The rare earth market, a monthly analysis for the rare earth industry. Core Consultants & Metal Pages, Nov 2013, p 14
7. Benecki WT (2011) Permanent magnets 2010–2020. In: Magnetics 2011 conference, San Antonio, Texas, USA, 1 March 2011
8. Tušek J, Kitanovski A, Zupan S et al (2013) A comprehensive experimental analysis of gadolinium active magnetic regenerators. Int J Refrig 53(1):57–66
9. Bjørk R, Bahl CHR, Katter M (2010) Magnetocaloric properties of $LaFe_{13-x-y}Co_xSi_y$ and commercial grade Gd. J Magn Magn Mater 322(24):3882–3888
10. Kitanovski A, Poredoš A, Tušek J et al (2010) Industrial magnetic dryer and magnetic precision water cooling system: final report of the feasibility study. University of Ljubljana, Faculty of Mechanical Engineering, p 57
11. Personal communication with Barcza A and Katter M from Vacuumschmelze GmbH & Co, 2011 and 2014
12. Egolf PW, Kitanovski A, Gonin C (2010) Central magnetic cooling and refrigeration machines (Chiller) and their assessment, a feasibility study. Final report, Swiss Federal Office of Energy, SFOE contract and project number: 102873/153691, March 2010, p 34
13. Wolsky AM (2013) HTS from precommercial to commercial: a roadmap to future use of higher-temperature superconductors by the power sector. International Energy Agency, p 146
14. Luongo CA (2001) Optimization of toroidal superconducting magnet storage systems. Physica C 354:110–114
15. Zhu J, Zhang H, Yuan W et al (2013) Design and cost estimation of superconducting magnetic energy storage (SMES) systems for power grids. In: The 2013 IEEE power and energy society general meeting (PES), 21–25 July, pp 1–5
16. Lee SY, Kwak SY, Seo JH et al (2009) Optimal design of HTS magnets for a modular toroid-type 2.5 MJ SMES using multi-grouped particle swarm optimization. Phys Supercond 469 (15–20):1789–1793
17. Schoenung SM, Meier WR, Fagaly RL et al (1993) Design, performance and cost characteristics of high temperature superconducting magnetic storage. IEEE Trans Energy Convers 8(1):33–39
18. Schoenung SM, Bieri RL, Meier WR et al (1993) Cost savings and prospects for applications of micro superconducting magnetic energy storage (SMES) using high temperature superconductors. IEEE Trans Appl Supercond 3(1):200–203
19. Seeber B (ed) (1998) Handbook of applied superconductivity, vol 1. IOP Publishing, Bristol
20. Lubell MS, Lue JW (1995) Structure and cost scaling for intermediate size superconducting magnetic energy storage (SMES) systems. IEEE Trans Energy Convers 5(2):345–349
21. Green MA, Strauss BP (2008) The cost of superconducting magnets as a function of stored energy and design magnetic induction times the field volume. IEEE Trans Appl Supercond 18 (2):248–251
22. Green MA (2008) The cost of helium refrigerators and coolers for superconducting devices as a function of cooling at 4 K. In: Advances in cryogenic engineering: transactions of the cryogenic engineering conference-CEC, vol 52. AIP Conference Proceedings, vol 985, pp 872–878

23. Mulholland J, Sheahen TP, McConnell B (2001) Analysis of future prices and markets for high temperature superconductors. Work for US Department of Energy, Sept 2001, p 100. http://www.ornl.gov/sci/htsc/pubs.html#reports
24. Russek SL, Zimm CB (2006) Potential for cost effective magnetocaloric air conditioning systems. Int J Refrig 29:1366–1373
25. Egolf PW, Gendre F, Kitanovski A et al (2006) Machbarkeitsstudie für magnetische Wärmepumpen: Anwendungen in der Schweiz–Schlussbericht. Bern: Swiss Federal Office of Energy, SFOE contract and project number: 100873/151017, p 69
26. Kitanovski A, Diebold M, Vuarnoz D et al (2008) Applications of magnetic refrigeration and its assessment: a feasibility study-final report. Bern: Swiss Federal Office of Energy, SFOE contract and project number: 101776/152191, p 42
27. Kitanovski A, Diebold M, Vuarnoz D et al (2008) Applications of magnetic "Power Production" and its assessment: a feasibility study-final report. Bern: Swiss Federal Office of Energy, SFOE contract and project number: 101776/152190, p 33
28. Egolf PW, Kitanovski A, Gonin C et al (2010) Magnetic heat pump with ground heat source. Optimized prototype-Final report, SFOE contract and project number: 152928/100873, p 22
29. Egolf PW, Kitanovski A, Gonin C (2010) Central magnetic cooling and refrigeration machines (Chiller) and their assessment. A feasibility study-final report, SFOE contract and project number: 102873/153691, p 34
30. Kitanovski A, Egolf PW (2009) Application of magnetic refrigeration and its assessment. J Magn Magn Mater 321(7):777–781
31. Egolf PW, Kitanovski A, Diebold M et al (2009) Magnetic power conversion with machines containing full or porous wheel heat exchangers. J Magn Magn Mater 321(7):758–762
32. Vuarnoz D, Kitanovski A, Diebold M et al (2009) Feasibility study of a power conversion system utilizing industrial waste heat. Yverdon: HES-SO 2009. Final report project 19144
33. Vuarnoz D, Kitanovski A, Gonin C et al (2012) Thermodynamic and exergy efficiencies of magnetocaloric energy conversion utilising industrial waste heat. Int J Exergy 10(4):365–378
34. Vuarnoz D, Kitanovski A, Gonin C et al (2012) Quantitative feasibility study of magnetocaloric energy conversion utilizing industrial waste heat. Appl Energy 100:229–237
35. Kitanovski A, Egolf PW, Poredoš A (2012) Rotary magnetic chillers with permanent magnets. Int J Refrig 35(4):1055–1066
36. Kitanovski A, Egolf PW, Tušek J et al (2011) Superconducting magnetic chillers for cooling and air conditioning applications. Internal Report, University of Ljubljana in collaboration with P.W. Egolf from University of Applied Sciences of Western Switzerland, p 13
37. Tušek J, Kitanovski A, Prebil I et al (2011) Dynamic operation of an active magnetic regenerator (AMR): numerical optimization of a packed-bed AMR. Int J Refrig 34 (6):1507–1517
38. Bjørk R, Smith A, Bahl CRH et al (2011) Determining the minimum mass and cost of a magnetic refrigerator. Int J Refrig 34:1805–1816
39. Rowe A (2011) Configuration and performance analysis of magnetic refrigerators. Int J Refrig 34(1):168–177
40. Roudaut J, Yonnet J-P, Kedous-Lebouc A et al (2010) Permanent magnet sources for magnetic refrigeration. In: REPM'10—Proceedings of the 21st workshop on rare-earth permanent magnets and their applications, pp 86–89
41. Bjørk R, Bahl CRH, Smith A et al (2008) Optimization and improvement of Halbach cylinder design. J Appl Phys 104(1):013910
42. Tura A, Rowe A (2007) Design and testing of a permanent magnet magnetic refrigerator. In: Proceedings of the second IIF-IIR international conference on magnetic refrigeration at room temperature, pp 363–370
43. Lee SJ, Kenkel JM, Pecharsky VK et al (2002) Permanent magnet array for the magnetic refrigerator. J Appl Phys 91(10):8894–8896
44. Zheng ZG, Yu HY, Zhong XC et al (2009) Design and performance study of the active magnetic refrigerator for room-temperature application. Int J Refrig 32(1):78–86

45. Okamura T, Yamada K, Hirano N et al (2006) Performance of a room-temperature rotary magnetic refrigerator. Int J Refrig 29(8):1327–1331
46. Okamura T, Rachi R, Hirano N et al (2007) Improvement of 100 W class room temperature magnetic refrigerator. In: Proceeding of the second IIF-IIR international conference on magnetic refrigeration at room temperature, pp 377–382
47. Zimm C, Auringer J, Boeder A et al (2007) Design and initial performance of a magnetic refrigerator with a rotating permanent magnet. In: Proceedings of the second IIF-IIR international conference on magnetic refrigeration at room temperature, pp 341–347
48. Tura A, Rowe A (2014) Concentric Halbach cylinder magnetic refrigerator cost optimization. Int J Refrig 37:106–116
49. Horie YA (2004) Life cycle optimization of household refrigerator-freezer replacement, Report No. CSS04-13. MSc. thesis, University of Michigan, Aug 2004, p 81
50. Vendrusculo EA, de Castilho Queiroz G, De Martino Jannuzzi G et al (2009) Life cycle cost analysis of energy efficiency design options for refrigerators in Brazil. Energy Effi 2:271–286
51. Monfared B, Furberg R, Palm B (2014) Magnetic vs. vapor-compression household refrigerators: a preliminary comparative life cycle assessment. Int J Refrig 42:69–76
52. Du X, Graedel TE (2011) Global rare earth in-use stocks in NdFeB permanent magnets. J Ind Ecol 15(6):836–843
53. Graf R, Held M (2013) Life cycle assessment of rare earths under consideration of new supply trends. In: The 6th international conference on life cycle management in Gothenburg 2013, p 4
54. de Boer MA, Lammertsma K (2012) Scarcity of rare earth elements, a review commissioned by the Royal Netherlands Chemical Society. University of Amsterdam, 2012, p 27
55. Walters A, Lusty P, Hill A (2011) Rare earth elements. British Geological Survey, Nov 2011, p 54
56. Moss RL, Tzimas E, Kara H et al (2011) Critical metals in strategic energy technologies. Assessing rare metals as supply-chain bottlenecks in low-carbon energy technologies, EUR 24884 EN–2011, p 164
57. Binnemans K, Jones PT, Blanpain B et al (2013) Recycling of rare earths: a critical review. J Clean Prod 51:1–22
58. Liu J, Chinnasamy C (2012) Rare earth magnet recycling. In: presentation at the rare earth elements workshop, EPA-Colorado, May 2012, p 31

Chapter 10
Alternative Caloric Energy Conversions

In this chapter, some alternative, "ferroic", solid-state energy conversion technologies are presented. These are the electrocaloric (pyroelectric), the barocaloric and the elastocaloric energy conversions. In their nature, they are analogous to magnetocaloric energy conversion; however, different external influences are needed to initialize the caloric effect. In the case of electrocaloric energy conversion, this is related to a change in the electric field; in the case of barocalorics, to a change in the hydrostatic pressure and in the case of elastocaloric energy conversion, to a change in the mechanical stress. Each group, to some extent, possesses possible advantages as well as some disadvantages in comparison with magnetocaloric energy conversion. However, since all three alternative solid-state energy conversion technologies are at an early stage of development, it is not yet reasonable to compare them with magnetocaloric energy conversion. It is only time and further research that will show the full potential of these alternative solid-state energy conversion technologies.

This chapter is divided into three sections. In each section, the physical principle behind the discussed ferroic effect will be presented and an overview of existing materials with their physical properties will be made. Furthermore, different possibilities for designing an energy conversion device using these materials will be reviewed (especially for electrocalorics). However, since the technology based on these three effects is at an early stage of development, only a few prototypes of energy conversion devices have been presented.

10.1 Electrocaloric and Pyroelectric Energy Conversion

In this subsection, the electrocaloric and pyroelectric energy conversions are presented and discussed. In general, the electrocaloric energy conversion stands for the heat pumping processes (or refrigeration), whereas the pyroelectric energy conversion stands for the power generation.

The underlying mechanism of the electrocaloric energy conversion is the so-called electrocaloric effect. The electrocaloric effect is expressed as the temperature change of dielectric materials subjected to a varying electric field. To

A. Kitanovski et al., *Magnetocaloric Energy Conversion*,
Green Energy and Technology, DOI 10.1007/978-3-319-08741-2_10

simplify, as the electrocaloric material is subjected to a positive electric field change the material heats up, yet as the electric field is turned off the opposite occurs and the material cools down. Speaking thermodynamically, the electrocaloric effect is analogue to the magnetocaloric effect, though instead of the magnetic field change an electric field change is required to induce the caloric effect in the material. However, the electrocaloric energy conversion has some potential advantages over the magnetocaloric energy conversion, such as higher power density and higher compactness of the energy conversion devices, no dependence on rare-earth materials, no moving parts of the device, operation of the devices with less vibrations, silent operation, etc. Nevertheless, the area of the electrocaloric energy conversion only recently attracted more attention by the scientific community and therefore the development of the electrocaloric energy devices is at its early stage.

When the pyroelectric material is subjected to a variation of the temperature and short circuited, an electrical current named the pyroelectric current will be generated. Therefore, the pyroelectric energy conversion is a way to directly convert thermal energy into electrical energy. The materials used for pyroelectric power generation are of the same group as those used for electrocaloric energy conversion. However, they are exposed to different thermodynamic cycles and operating conditions and therefore the desired characteristics of the materials used for pyroelectric power generation can be, to some extent, different from those used for electrocaloric energy conversion.

In this chapter, the physics behind the electrocaloric and pyroelectric effect are discussed and some materials with the electrocaloric and pyroelectric effect are reviewed. The emphasis, however, is on the first concepts and prototypes of the electrocaloric and pyroelectric energy conversion devices.

10.1.1 Introduction to the Electrocaloric Effect

The electrocaloric effect (ECE) is a physical phenomenon that occurs in some dielectric materials under the influence of a varying electric field. It is expressed as the adiabatic temperature or isothermal entropy change of the material. To explain this phenomenon let us consider a polar dielectric material in which charged particles are initially randomly distributed (Fig. 10.1a).

When the material is exposed to an electric field, an electric force acts on the charged particles. These tend to move or shift in the direction of the field and undergo a transition from the disordered to the ordered state (Fig. 10.1b). A consequence of this orientation is an apparent accumulation of negative charges on the top surface, and positive charges on the bottom surface, of the dielectric material. The amount of charge accumulated per surface area is defined as the polarization. However, this is a simplified explanation of polarization, which is known to be a combination of a number of different effects. For a detailed description of polarization and the properties of dielectric materials, the reader is referred to Kao [1].

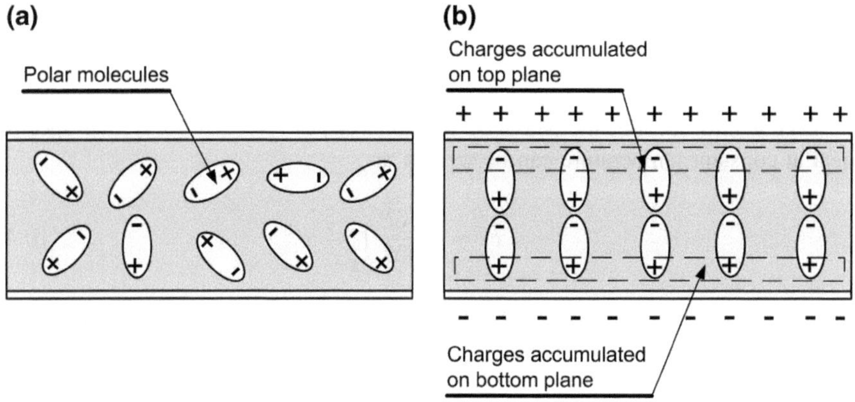

Fig. 10.1 **a** Schematic presentation of the orientation of polar molecules in a dielectric material that is not exposed to an electric field, **b** Schematic presentation of the orientation of polar molecules in a dielectric material under the influence of an electric field

Another consequence of a transition from the disordered to the ordered state is a reduction of the material's dipolar entropy. If the dielectric material is subjected to a rapid change of electric field, so that the heat transfer to the surroundings can be neglected, the process can be considered as adiabatic. In this case the total entropy of the material must remain constant, which then causes the temperature of the electrocaloric material to increase. When the electric field is turned off, the electrocaloric material transits back to the disordered phase and its temperature decreases. If the polarization is performed under isothermal conditions, the dipolar entropy change of the electrocaloric material will lead to an increase in the total entropy of the electrocaloric material (or a decrease in the total entropy due to the depolarization).

The electrocaloric effect is analogous to the magnetocaloric effect. However, instead of a magnetic field change, the electrocaloric effect is induced by an electric field change. If the entropy of the dielectric material is considered as a function of the temperature and the electric field, then the total change of the entropy can be expressed as [2]:

$$\mathrm{d}s(T,E) = \left(\frac{\partial s}{\partial T}\right)_E \mathrm{d}T + \left(\frac{\partial s}{\partial E}\right)_T \mathrm{d}E \tag{10.1}$$

where s and T are the specific entropy and the temperature of the material, respectively, and E is the electric field. Next, the specific heat at constant electric field c_E and the specific heat at constant temperature c_T are defined using the following relations:

$$c_E = T\left(\frac{\partial s}{\partial T}\right)_E \tag{10.2}$$

$$c_T = T\left(\frac{\partial s}{\partial E}\right)_T \tag{10.3}$$

Taking into account the Maxwell relation $(\partial P/\partial T)_E = (\partial s/\partial E)_T$, the specific heat at constant temperature can be written as:

$$c_T = T\left(\frac{\partial P}{\partial T}\right)_E \tag{10.4}$$

where P is the polarization of the electrocaloric material. This expression is then inserted into Eq. (10.1) to obtain:

$$\mathrm{d}s(T,E) = \frac{c_E}{T}\mathrm{d}T + \left(\frac{\partial P}{\partial T}\right)_E \mathrm{d}E \tag{10.5}$$

Let us first consider the case where the change of the electric field occurs under adiabatic conditions. In this case, the total entropy of the electrocaloric material remains constant, and therefore:

$$0 = \frac{c_E}{T}\mathrm{d}T + \left(\frac{\partial P}{\partial T}\right)_E \mathrm{d}E \tag{10.6}$$

From Eq. (10.6), the adiabatic temperature change of the material can be expressed as:

$$\Delta T_{\mathrm{ad}} = -\int_{E_1}^{E_2} \frac{T}{c_E}\left(\frac{\partial P}{\partial T}\right)_E \mathrm{d}E \tag{10.7}$$

If we now consider the isothermal case ($\mathrm{d}T = 0$), the total entropy change of the material can be written as:

$$\mathrm{d}s(T,E) = \left(\frac{\partial P}{\partial T}\right)_E \mathrm{d}E \tag{10.8}$$

After the integration of Eq. (10.8), the expression for the isothermal entropy change of the material is obtained:

$$\Delta s_{\mathrm{ist}} = -\int_{E_1}^{E_2} \left(\frac{\partial P}{\partial T}\right)_E \mathrm{d}E \tag{10.9}$$

It is clear from Eqs. (10.7) and (10.9) that in order to achieve a large electrocaloric effect, the rate of change of the polarization of the electrocaloric material with respect to the temperature should be large. Furthermore, the electrocaloric material should be subjected to a large change in the electric field.

Ferroelectrics are a group of dielectric materials that possess a large rate of change of polarization with respect to the temperature. One of their characteristics is that they possess a spontaneous polarization (polarization at zero electric field) within a certain temperature interval. In this interval, the dielectric material is said to be in a ferroelectric phase. However, there exists a critical temperature, referred to as the Curie temperature, above which the spontaneous polarization vanishes and the material is said to be in a paraelectric phase. The transformation of a dielectric material from the ferroelectric phase to the paraelectric phase is called the phase transition. With regard to this phase transition, ferroelectrics can be divided into first-order and second-order phase transition materials. The characteristic of a first-order phase transition is a sudden drop in the polarization around the Curie temperature (Fig. 10.2a). On the other hand, ferroelectric materials with a second-order phase transition undergo a continuous change of polarization from the ferroelectric to the paraelectric phase (Fig. 10.2b). In both cases, a relatively large change in the polarization occurs in the vicinity of the Curie temperature. Therefore, in this temperature range, a relatively large electrocaloric effect is expected.

An interesting group of the dielectric materials are the so-called relaxor ferroelectrics, which can be either polymers or ceramics. They usually possess no spontaneous polarization; however, they undergo a phase transition under the influence of an electric field (field-induced phase transition) [3]. As in the case of common ferroelectric materials, relaxor ferroelectric materials can either possess a first- or second-order phase transition. A special characteristic of relaxor ferroelectrics is that the maximum value of the electrocaloric effect is usually observed over a broad temperature interval (up to a few tens of degrees kelvin). On the other hand, the maximum electrocaloric effect of common ferroelectrics is observed over a much narrower temperature interval, usually in the range of a few degrees kelvin. Detailed information about the electrocaloric effect in relaxor ferroelectrics can be found in Pirc et al. [3].

Besides the magnitude of the polarization change as a function of temperature, the electrocaloric effect is defined by the magnitude of the electric field change. In general, a larger electric field change induces a larger electrocaloric effect, though this is only true until the saturation point is reached. At the saturation point, the dipoles are perfectly aligned along the field and by further increasing the field, the dipolar entropy of the dielectric material remains the same [4]. However, in most cases, the dielectric material breaks down before the electric field is increased up to the saturation point [4].

The equations for the adiabatic temperature change (Eq. 10.7) and the isothermal entropy change (Eq. 10.9) for an electrocaloric material give a good insight into the main factors that influence the magnitude of the electrocaloric effect. However, it is important to stress that these expressions do not always correctly predict the values of the electrocaloric effect [2, 5]. The reason for this is in the hysteresis behaviour as the ferroelectric materials have two equilibrium states for every field applied

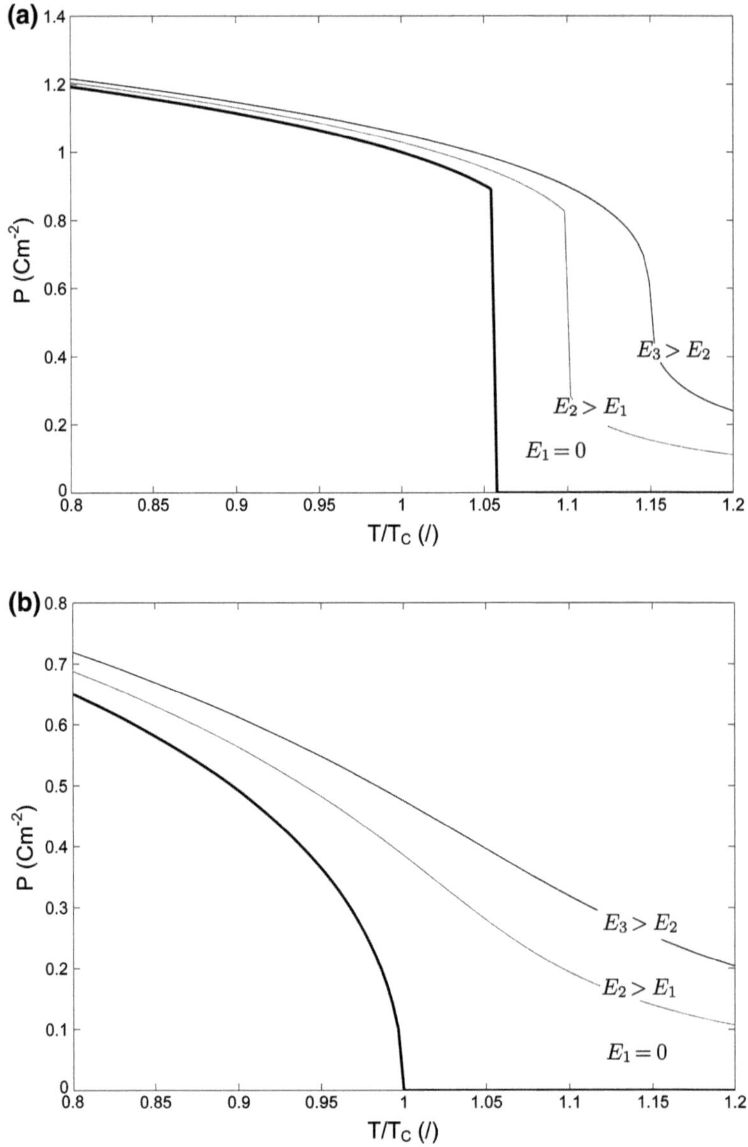

Fig. 10.2 **a** First-order phase transition of a ferroelectric material, **b** Second-order phase transition of a ferroelectric material

and therefore a thermodynamic treatment is not the most appropriate [2]. In some cases, in order to determine the electrocaloric effect more accurately and to confirm the values obtained by indirect method, other methods, such as direct measurements, should be used [2, 5].

10.1.2 Electrocaloric Materials

The first report of a material exhibiting the electrocaloric effect was in 1930, when Kobeko and Kurchatov [6] measured the electrocaloric effect in Rochell salt. However, they did not report any values from the measurements. Later, in 1963, Wiseman and Kuebler [7] repeated the experiment and measured an adiabatic temperature change in a range of few millikelvins. During the 1960s and 1970s, bulk materials exhibiting the electrocaloric effect were intensively studied, but then the research was gradually abandoned since the values of the electrocaloric effect did not exceed 1 K [4]. An important milestone came in 2006 when Mischenko et al. [8] reported the so-called giant electrocaloric effect in $PbZr_{0.95}Ti_{0.05}O_3$ (PZT) ceramic thin films. Using indirect measurements, they demonstrated that the material undergoes an adiabatic temperature change of 12 K for an electric field change of 48 MVm^{-1}. After 2006 many researchers reported the discovery of new materials, with the electrocaloric effect, including several ceramics and some polymers [4, 6, 9].

In this section, the two largest groups, i.e. ceramic and polymer electrocaloric materials, are further discussed. The general characteristics are outlined and for each group a set of materials together with their electrocaloric properties are presented. The main parameter that characterizes an electrocaloric material is the adiabatic temperature change. In most cases, this is obtained by indirect or direct measurements. In the case of an indirect method, the polarization curves for different electric fields at different temperatures are measured. The adiabatic temperature change and isothermal entropy change of the material are then calculated using Eqs. (10.7) and (10.9). Since indirect measurements do not always provide accurate results [2, 5, 10], direct methods, such as differential scanning calorimetry [11] and high-resolution calorimetry [12], should be used to determine the electrocaloric effect. For a detailed description of measuring the electrocaloric effect, the reader is referred to Correia et al. [13].

Besides the adiabatic temperature change, the isothermal entropy change gives important information about the properties of an electrocaloric material. Furthermore, the maximum specific cooling capacity, which is explained in detail in Chap. 4, can be used to characterize the cooling performance of the electrocaloric material. It can be calculated as:

$$q_{R\max} = \frac{(2\,T_R + \Delta T_{\mathrm{ad}})\,\Delta s_{\mathrm{ist}}}{2} \tag{10.10}$$

where T_R denotes the lowest temperature of the electrocaloric material in the cycle. In cases where the ΔT_{ad} at a specific temperature was reported, T_R can be expressed as:

$$T_R = T_{\mathrm{ref}} - \Delta T_{\mathrm{ad}} \tag{10.11}$$

where T_{ref} is the temperature at which the electrocaloric effect was obtained.

10.1.2.1 Electrocaloric Ceramic Materials

Ceramic materials exhibiting the electrocaloric effect can be, based on their thickness, divided into bulk ceramics (with a thickness greater than 100 μm), thick-film ceramics (with a thickness of a few tens of micrometres) and thin-film ceramics (with a thickness of less than 1 μm). Such a classification is also common in other areas where electroceramics are used [14].

The adiabatic temperature change in bulk ceramics is usually of the order of a few kelvins, as can be seen from Table 10.1. The highest adiabatic temperature change of a bulk ceramic material known to the authors of this book was 4.5 K for the $Ba(Zr_{0.2}Ti_{0.8})O_3$ (BZT) ceramic [15].

Thick-film ceramics are, in general, 10–100 times thinner than bulk ceramics. In most cases they are in the form of a multilayer structure, also known as a multilayer capacitor. The largest electrocaloric effect in a multilayer capacitor was reported by Bai et al. [24]. They measured an adiabatic temperature change of 7.1 K in a multilayer capacitor constructed from 63 layers of 3-μm-thick $BaTiO_3$. Kar-Narayan et al. [25, 26] measured the adiabatic temperature change of a very similar, commercially available, multilayer structure; however, they reported an adiabatic temperature change of 0.5 K. Furthermore, Kar-Narayna et al. [27] analysed the cooling powers achievable with multilayer capacitors. Based on theoretical estimations they predicted that an idealized electrocaloric cooling device with multilayer capacitors could achieve a specific cooling power up to 2,875 Wkg^{-1}. One of the few cases where the electrocaloric effect was measured in a thick-film ceramic that was not in the form of a multilayer capacitor was reported by Rožič et al. [12]. They measured an adiabatic temperature change of 1.8 K in a 28-μm-thick PLZT 8/65/35 ceramic film. The electrocaloric properties of some thick-film ceramics are collected in Table 10.2.

Table 10.1 Electrocaloric properties of bulk ceramics

Material	ΔT_{ad} (K)	Δs_{is} ($Jkg^{-1}K^{-1}$)	T_{ref}^a (K)	ΔE (MVm^{-1})	$q_{r,max}$ (Jkg^{-1})	Measurement procedure	References
BNT-0.3BT	−2.1	/	423	5	/	Indirect	[16]
BNT-KN	1.73	/	349	1.7	/	Indirect	[17]
BZT (x = 0.2)	4.5	8	312	14.5	2,478	Direct	[15]
BZT (x = 0.15)	4.2	7.3	342	15	2,481	Direct	[15]
NBT	−0.34	−0.43	413	5	176	Indirect	[18]
NBT-0.08BT	0.19	0.26	370	4	96	Indirect	[18]
PLZT 8/65/35	2.2	/	385	8.8	/	Direct	[19]
PMN	2.6	/	451	9	/	Direct	[12]
PMN	0.11	/	274	1	/	Indirect	[20]
PMN-0.85PT	1.7	/	291	1.6	/	Direct	[21]
PMN-0.3PT	2.7	/	429	9	/	Direct	[19]
PMN-0.08PT	1.3	/	296	1.5	/	Direct	[22]
PZN-0.08PT	0.25	/	453	1.2	/	Direct	[23]

[a] T_{ref}—temperature at which ΔT_{ad}, Δs_{is} and $q_{r,max}$ were obtained

Table 10.2 Electrocaloric properties of thick-film ceramics

Material	ΔT_{ad} (K)	Δs_{is} ($Jkg^{-1}K^{-1}$)	T_{ref}^{a} (K)	ΔE (MVm^{-1})	$q_{r,max}$ (Jkg^{-1})	Measurement procedure	Note	References
$BaTiO_3$	7.1	50	333	80	16,473	Direct	MC[b]	[24]
$BaTiO_3$	0.55	/	350	30	/	Direct	MC[b]	[25]
PST	2.4	/	/	13.8	/	Direct	MC[b]	[28]
PLZT 8/65/35	1.8	1.5	384	6.8	575	Direct	/	[29]

[a] T_{ref}—temperature at which ΔT_{ad}, Δs_{is} and $q_{r,max}$ were obtained
[b] *MC*—multilayer capacitor

Thin-film ceramics have attracted a lot of attention due to the discovery of the so-called giant electrocaloric effect in PZT thin-film ceramics [8]. The thickness of these materials can be as little as a few 100 nm and hence direct measurements are a great challenge. Therefore, the evaluation of the electrocaloric effect in thin-film materials is usually reported on the basis of indirect measurements. For example, in 2013 Peng et al. [30] predicted an electrocaloric effect of 45.3 K in thin-film $Pb_{0.8}Ba_{0.2}ZrO_3$ (PBZ) ceramics obtained by an indirect measurement. Nevertheless, Lu et al. [31] reported on a directly measured electrocaloric effect of 40 K in 400-nm-thick La-doped $Pb(ZrTi)O_3$ (PLZT) relaxor ceramics. Some other thin-film ceramic materials, together with their electrocaloric properties, are presented in Table 10.3. In comparison with bulk and thick-film ceramics, the electrocaloric effect in thin-film ceramics can be one order larger. The main reason for this lies in the higher dielectric strength of thin-film materials, which can be as high as 120 MVm^{-1} [31].

In conclusion, it can be said that all three groups of ceramics have the potential for future cooling or other energy conversion applications. Though bulk ceramics do not exhibit sufficient adiabatic temperature changes to be used in a single-stage cooling device, they could be applied in devices where heat regeneration allows a significant increase of the temperature span [37]. This also holds true for thick-film ceramics, which are usually in the form of a multilayer capacitor. However, they

Table 10.3 Electrocaloric properties of thin-film ceramics

Material	ΔT_{ad} (K)	Δs_{is} ($Jkg^{-1}K^{-1}$)	T_{ref}^{a} (K)	ΔE (MVm^{-1})	$q_{r,max}$ (Jkg^{-1})	Measurement procedure	References
PLZT 8/65/35	40	50	318	120	14,900	Direct	[31]
PBZ	45.3	/	290	59.8	/	Indirect	[30]
PMN-0.35PT	31	/	413	74.7	/	Indirect	[32]
PMN-0.1PT	5	/	348	89.5	/	Indirect	[33]
PMN-0.07PT	9	/	298	72.3	/	Indirect	[34]
PST	−6.2	−6.3	341	77.4	2,129	Indirect	[35]
PZ	11.4	/	508	40	/	Indirect	[36]
PZT	12	/	499	48	/	Indirect	[8]

[a] T_{ref}—temperature, at which ΔT_{ad}, Δs_{is} and $q_{r,max}$ were obtained

could have advantages over bulk ceramics since a lower voltage is needed to induce the electrocaloric effect. On the other hand, thin-film ceramics can, in some cases, exhibit a sufficient electrocaloric effect to be used in a cooling device without heat regeneration. Because of their small thickness, thin-film ceramics possess poor mechanical properties. Furthermore, their thermal mass is very low. Therefore, they should be processed in a multilayer structure that could improve their mechanical properties and increase their thermal mass.

10.1.2.2 Electrocaloric Polymers

Polymers are a very interesting group of electrocaloric materials. They are mainly polyvinylidene fluoride (PVDF)-based copolymers or terpolymers in the form of thick films with their thickness usually ranging between 1 and 15 μm. They normally possess a very high dielectric strength, which can be as high as 350 MVm^{-1}. Therefore, the adiabatic temperature changes of polymer films can be comparable to those achieved in ceramic thin films. For example, Li et al. [38] reported an adiabatic temperature change of 35 K measured in a 10-μm-thick polymer film. The electrocaloric properties of some other polymers are presented in Table 10.4. Another favourable property of polymer films is that they are not as brittle as ceramic materials. Furthermore, the large electrocaloric effect makes polymer-based materials good material candidates for future electrocaloric devices. However, like in the case of thick- and thin-film ceramic materials, they should be further

Table 10.4 Electrocaloric properties of polymers

Material	ΔT_{ad} (K)	Δs_{is} ($Jkg^{-1}K^{-1}$)	T_{ref}^{a} (K)	ΔE (MVm^{-1})	$q_{r,max}$ (Jkg^{-1})	Measurement procedure	References
P(VDF-TrFE 55/45)	12	70	340	120	23,380	Direct	[39]
P(VDF-TrFE 68/32)	2.3	/	489	49	/	Direct	[12]
Irradiated P(VDF-TrFE 68/32)	20	95	306	160	28,120	Direct	[31]
Irradiated P(VDF-TrFE 65/35)	35	160	323	180	48,880	Direct	[38]
P(VDF-TrFE 70/30)	21.2	/	390	300	/	Indirect	[40]
P(VDF-TrFE-CFE) (56.2/36.3/7.6)	21.6	/	350	350	/	Indirect	[40]
annealed P(VDF-TrFE-CFE) (56.2/36.3/7.6)	9	47	318	300	14,735	Indirect	[41]
quenched P(VDF-TrFE-CFE) (56.2/36.3/7.6)	5	35	318	300	11,043	Indirect	[41]
Terpolymer-copolymer blend (90/10)	10	/	300	150	/	Direct	[42]

[a] T_{ref}—temperature at which ΔT_{ad}, Δs_{is} and $q_{r,max}$ were obtained

processed in the form of a multilayer capacitor in order to enhance their mechanical properties and increase the total thermal mass of one electrocaloric element.

In conclusion, it would be hard to say which electrocaloric material is the best choice for future applications. This will probably depend strongly on the desired operating parameters, such as temperature span, cooling (or heating) power and the efficiency of the device. Moreover, a lot of information is still missing for individual materials. Many authors do not report on the entropy change of the materials, which is, together with the adiabatic temperature change, a very important parameter. Furthermore, the influence of some other effects, for example, the effect of the number of polarization/depolarization cycles on the electrocaloric and mechanical properties of the materials, are not well known.

10.1.3 Review of Device Concepts and First Prototypes

In the past 8 years, since the discovery of the giant electrocaloric effect in PZT ceramics in 2006, research in the area of electrocaloric refrigeration and heat pumping has mainly been focused on searching for new materials and investigating their properties. In order to transfer these new discoveries to an application, e.g. an electrocaloric cooling device, it is important that the ideas about how to design an electrocaloric cooling system are investigated and realized. So far, several different concepts have been presented and are discussed in this chapter. However, just a few of these presented concepts have been realized and experimentally verified. This chapter is divided into two parts. The first part is focused on the different concepts associated with electrocaloric cooling devices and a theoretical evaluation of their performance. In the second part, the first electrocaloric prototypes devices are presented.

10.1.3.1 Electrocaloric Cooling Devices: Concepts and a Theoretical Evaluation of the Performance

Electrocaloric cooling devices should be designed in such a way that during or after the polarization, the heat generated due to the electrocaloric effect is effectively transferred from the electrocaloric material to the heat sink. In contrast, during or after the depolarization, the electrocaloric material should absorb the heat from the heat source. Furthermore, no heat should be transferred from the heat source to the heat sink at any time, to avoid unnecessary heat losses (or gains).

For a better understanding, let us consider the simple case illustrated in Fig. 10.3. The electrocaloric material with dimensions w and h is put into direct thermal contact (contact thermal resistance is neglected) with the heat sink and the heat source. If the thickness of the electrocaloric material h is much smaller than the width w, then the ratio between the heat-transfer area and the volume of the electrocaloric material is large. In this case, the electrocaloric material can efficiently exchange heat with the heat sink/source. However, during the polarization (when the material heats up) and

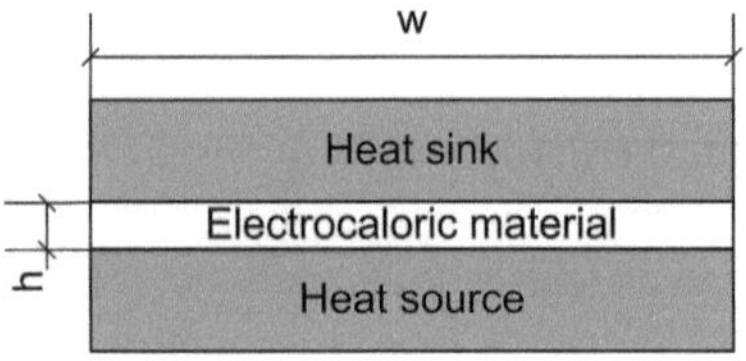

Fig. 10.3 Schematic presentation of an electrocaloric material positioned between a heat sink and a heat source

depolarization (when the material cools down), heat will be, at the same time, transferred in both directions, towards the heat sink and the heat source, respectively. Therefore, only a portion of the heat is actually "pumped" from the heat source to the heat sink and such a device cannot operate efficiently as a refrigerator or a heat pump.

To achieve the efficient operation of an electrocaloric cooling or heat pumping device, the simple system presented in Fig. 10.4 should be considered. The heat sink and heat source are physically separated and no heat can be transferred between them. During the time of polarization (Fig. 10.4a), the electrocaloric material is only in thermal contact with the heat sink and transfers the heat generated, due to the electrocaloric effect, to the heat sink. Next, the electrocaloric material is moved from the heat sink to the heat source (Fig. 10.4b). The electrocaloric material is then depolarized and can absorb the heat from the heat source. Such a device could already work as a refrigerator or a heat pump; however, its performance would probably be very limited. Therefore, other concepts about how to design an electrocaloric refrigeration or heat pumping device should be considered. Those known to the authors of the book are discussed later in this chapter.

Device with an Active Heat Regenerator

The concept of a device with an active heat regenerator was proposed and explained by Barclay and Steyer [43] in 1982 for magnetic refrigeration and heat pumping near room temperature. The concept proved to be one of the best solutions presented so far, and since then the majority of magnetocaloric devices are based on such an active heat regenerator [44]. This idea was first introduced to the field of electrocalorics by Sinyavsky and Brodyansky [45] in 1992. Their experimental

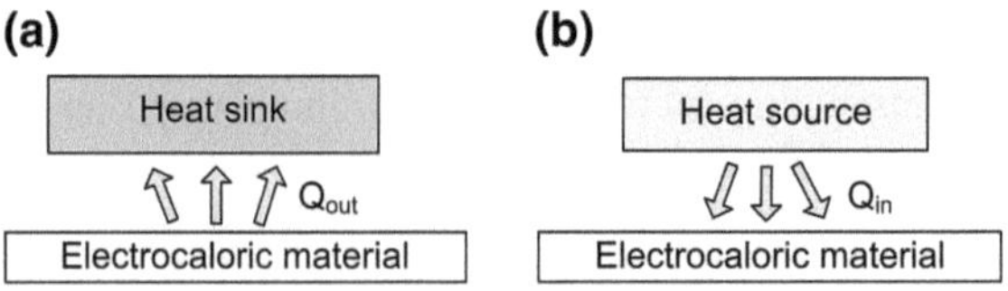

Fig. 10.4 A schematic presentation of a simple electrocaloric refrigeration or a heat pumping system, **a** The electrocaloric material is in an ideal contact with the heat sink, **b** The electrocaloric material is in ideal contact with the heat source

device is described in more detail later in this chapter. Here, a general description of the active electrocaloric heat regenerator (AER) is given. Moreover, the working principle of a refrigeration or heat pumping device based on an AER is described.

An AER is a porous structure of electrocaloric material through which a fluid can flow [46]. An example of an AER is presented in Fig. 10.5. According to Fig. 10.5 the basic element of the AER is a plate consisting of an electrocaloric material, which is coated on both sides with an electrically conductive film. The plates are put on top of each other and the distance holders are inserted in between to create a void for the fluid flow. Simultaneously, the distance holders serve as conductors for the electric current. The AER is put in a housing between the heat sink (the hot heat exchanger) and the heat source (the cold heat exchanger), as presented in Fig. 10.6. The housing is filled with a dielectric fluid and a mechanical system, e.g. a bi-directional pump, is used to move the fluid through the AER. The working of the device can be divided into four main steps. First, the electrocaloric material is polarized and, due to the electrocaloric effect, the electrocaloric material heats up. Next, the fluid is pumped from the heat source to the heat sink. The fluid flows through the voids between the electrocaloric plates and the heat generated due to the electrocaloric effect is transferred from the electrocaloric plates to the fluid. At the same time, part of the fluid is already entering the heat sink, where it can transfer the heat. In the third step, the material is depolarized and therefore it cools down. Now the fluid flows from the heat sink to the heat source, and in this way the fluid is cooled as part of the process. Simultaneously, a part of the fluid is already

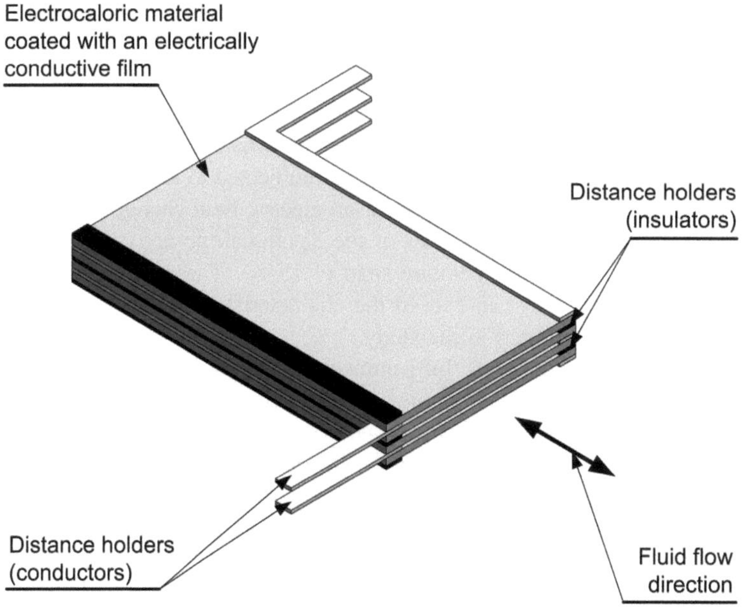

Fig. 10.5 Schematic presentation of an AER

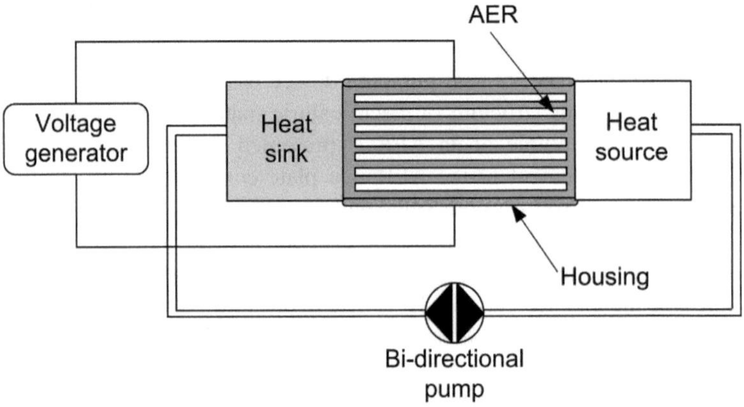

Fig. 10.6 Schematic presentation of a refrigeration or a heat pumping system with an AER

entering the heat source where it can absorb heat. The device operates by repeating those four steps. After a certain number of steps the steady state is reached and a temperature span between the heat sink and heat source is established. An important characteristic of a device with an AER is that the temperature span between the heat sink and the heat source can be a factor greater than the adiabatic temperature change of the electrocaloric material [9].

Ožbolt et al. [47] performed a numerical simulation in order to compare the performance of a magnetocaloric and an electrocaloric active heat regenerator (i.e., the maximum specific cooling power and the COP). The geometry of the regenerator was the same for both cases. In the first case, gadolinium and in the second case a multilayer capacitor constructed from layers of $Pb(Sc_{0.5}Ta_{0.5})O_3$ ceramics and electrodes were defined as the active materials. The magnetic field change in the case of the magnetocaloric active regenerator was set to be 1 T. On the other hand, the electrocaloric active regenerator was subjected to different changes of the electric field. The results revealed that for an electric field change of 77.4 MVm^{-1} the maximum specific cooling power of the electrocaloric active regenerator was around 500 Wkg^{-1} (for a temperature span of 15 K). These results show that the characteristics of the AER can exceed the characteristics of the active magnetocaloric regenerator considered in the study.

In another paper, Guo et al. [48] numerically analysed the performance of a micro-scale electrocaloric cooling device with an active heat regenerator. The regenerator was considered to be composed of P(VDF-TrFE-CFE) terpolymer plates (thickness of 10 μm), with a total length of the regenerator of 2 mm. In order to pump the fluid through the regenerator diaphragm actuators were assumed. Their numerical results indicate that if the system under investigation would be operating at a temperature span of 15 K an efficiency equal to 31 % of the Carnot COP and a cooling-power density of 3 Wcm^{-2} could be achieved.

A fluid flow is, however, not the only medium that could be used to transfer heat from the AER to the heat sink and heat source. Instead of a fluid a matrix made out of a solid material could be used, and such a solution was proposed by Gu et al. [49]. The solid matrix can simply be a set of parallel plates that can fit into the gaps between the electrocaloric plates. The working principle of a device stays the same, only this time, instead of a fluid, a solid matrix is moved in and out of the AER by a mechanical system. Gu et al. [49] analysed the performance of a 10-mm-long AER with a solid-state matrix. The electrocaloric plates were assumed to be made of P (VDF-TrFE-CFE) terpolymer. They predicted that such a device with an AER length of 10 mm could provide 9 Wcm^{-3} of cooling-power density for a 20 K temperature span. Moreover, they estimated that it could reach more than 50 % of the Carnot efficiency.

Devices with Thermal Diodes

In order to control the heat-flux direction from the electrocaloric material in a specific working step of an electrocaloric device (the heat should be transferred to a heat sink during or after the polarization and to the heat source during or after the depolarization) an element, called a thermal diode, could be applied. A detailed description of different thermal diodes and their application in the case of magnetocaloric energy conversion can be found in Chap. 6. Due to the analogy between magnetocaloric and electrocaloric energy conversion, the concepts presented in Chap. 6 could also be applied for electrocaloric refrigeration and heat pumping. In this section, some of the concepts of electrocaloric energy conversion devices with thermal diodes are discussed.

For example, Epstein et al. [50] presented the concept of an electrocaloric device that is constructed from a thin-film multilayer capacitor (constructed from layers of thin-film electrocaloric material and layers of electrically conductive material) sandwiched between two thermal diodes. The device is schematically illustrated in Fig. 10.7. The operation of the device consists of two main steps and two sub-steps. First, the electrocaloric material in the multilayer capacitor is polarized. During the first half of the polarization, the thermal diodes are not active (no heat is transferred from the multilayer capacitor) and the multilayer capacitor heats up. In the second half of the polarization, the thermal diode in the contact with the heat sink is active and the heat is isothermally transferred from the multilayer capacitor to the heat sink (Fig. 10.7a). Next, during the first part of the depolarization, the multilayer capacitor cools down (both thermal diodes are inactive and no heat is transferred). During the second part of the depolarization, the thermal diode that is in contact with the heat sink is activated and the heat can be absorbed by the multilayer capacitor (Fig. 10.7b). This process repeats many times and after a number of repetitions the steady state is achieved. Based on a numerical simulation, the authors estimated that if a thermal diode with a conductivity contrast (defined as the ratio between the thermal resistances of the active and inactive thermal diodes) of

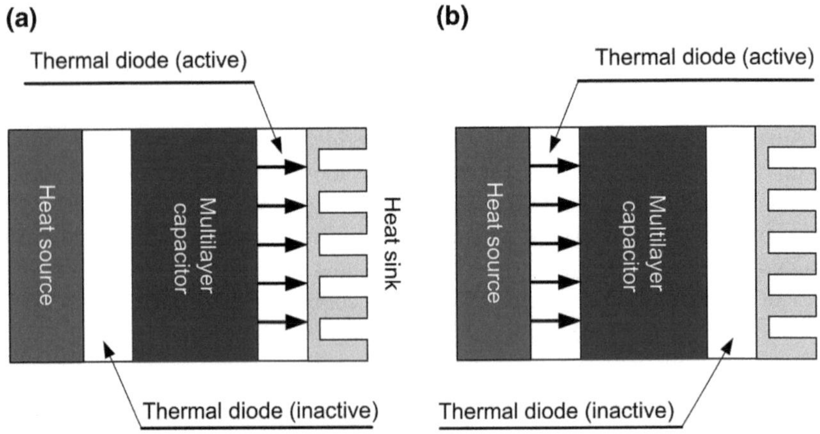

Fig. 10.7 Schematic presentation of the two main steps in the working cycle of an electrocaloric cooling device with heat switches, **a** Thermal diode between the heat sink and the electrocaloric element is active, **b** Thermal diode between the heat source and the electrocaloric element is active [50]

Fig. 10.8 Schematic presentation of a cooling-bridge device

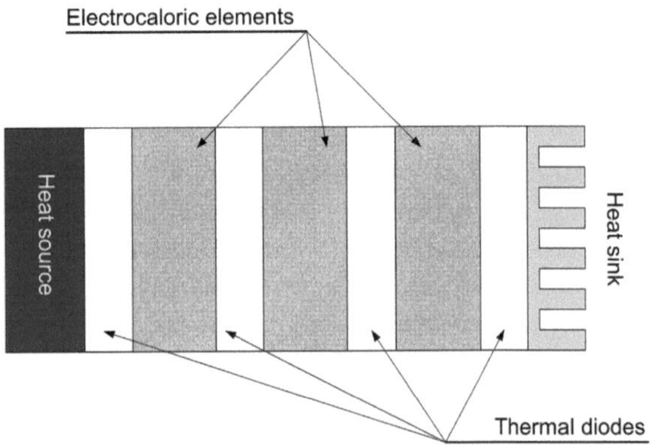

above 100 were to be applied, the exergy efficiency of the device could exceed 66 %. This concept, however, was not experimentally validated.

An upgrade to the system presented above is a device that consists of several layers of electrocaloric elements (multilayer capacitors, thin/thick-film electrocaloric materials or bulk electrocaloric materials) and thermal diodes, as shown in Fig. 10.8. Such a system was proposed by Kutnjak et al. [20]. They named the device "a cooling bridge". No performance analysis, theoretical or experimental, has been found for this particular concept.

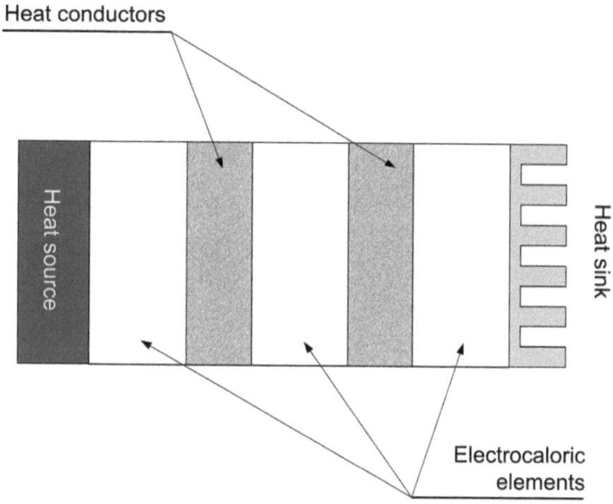

Fig. 10.9 Schematic presentation of the *solid cooling line*

Solid-State Cooling Line

The concept of a solid-state cooling line was presented by Karmanenko et al. in 2007 [51]. A schematic presentation of the device can be seen in Fig. 10.9. The device consists of a total of five elements, i.e. two electrocaloric elements and three heat conductors, and is constructed in a similar way to the cooling-bridge device. However, the cooling-line device does not apply thermal diodes. Its working principle is based on the unsymmetrical pattern of an applied electric field and is explained in detail in Es'kov et al. [52]. With a 1D numerical model it was predicted that the device would be able to achieve a temperature span of 25 K between the heat sink and the heat source at the operating frequency (the number of thermodynamic cycles per unit of time) of 3 Hz. However, the presented concept was not experimentally tested.

10.1.3.2 Electrocaloric Cooling Devices: First Prototypes

In 1992 Sinyavsky and Brodyansky [45] presented the first concept of an electrocaloric cooling device. The electrocaloric material used was a $PbSc_{0.5}T_{a0.5}O_3$ (PST) bulk ceramic in the form of plates with dimensions of $20 \times 10 \times 0.3\ mm^3$. The plates were positioned on top of each other and copper wires were installed between the plates to create a gap for the fluid flow. Furthermore, the wires served as a conductive material for the electrical charge. From the above description it is obvious that the device was actually based on an active electrocaloric heat regenerator. The total length of one regenerator was 55 mm with a cross-section of $10 \times 5\ mm^2$ and had a mass of 35 g. Two regenerators in parallel were inserted into a housing, as can be seen in Fig. 10.10. On one end of the regenerator, a hot heat

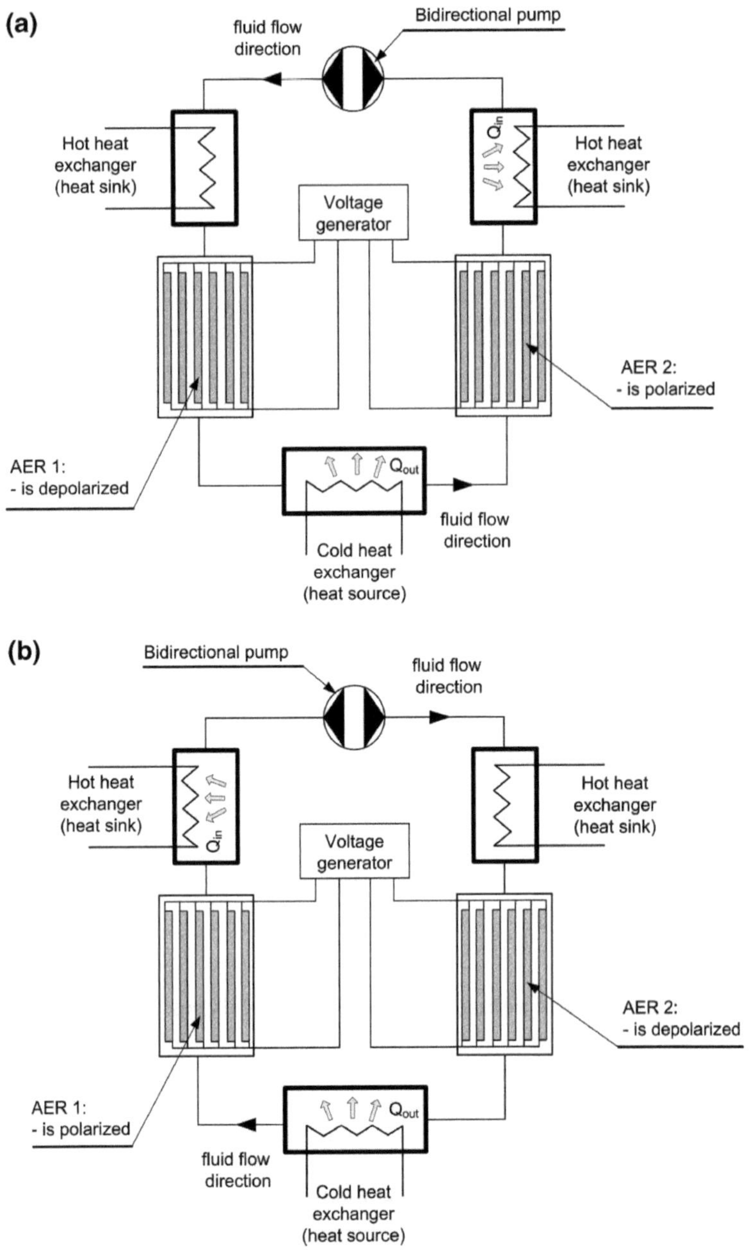

Fig. 10.10 Schematic presentation of an electrocaloric cooling device by Sinyavsky and Brodynasky [45], **a** AER 1 is depolarized, AER 2 is polarized and the fluid is flowing in a *counter-clockwise direction*, **b** AER 1 is polarized, AER 2 is depolarized and the fluid is flowing in *clockwise direction*

exchanger (heat sink) was installed, and on the other end, there was a cold heat exchanger (heat source). The device also included a voltage generator (to create a varying electric field) and a pump (to move the fluid through the regenerator). The synchronized operation of the voltage generator and the pump ensured that after the polarization of the electrocaloric material, the fluid was flowing from the heat source to the heat sink. In contrast, after the depolarization, the fluid was flowing from the heat sink to the heat source. The authors reported a maximum temperature span of 5 K. The latter was achieved when the material was subjected to an electric field change of 6 MVm^{-1} and pentane was used as the working fluid. The authors [45] did not report on the adiabatic temperature change of the PST electrocaloric material used in the experiment.

Over the next 15 years, the amount of research in the field of electrocaloric cooling was in decline. However, 6 years after the discovery of the giant electrocaloric effect, in 2012, Jia et al. [53] presented a new, small-scale electrocaloric cooling device. A schematic presentation of the device is shown in Fig. 10.11. It

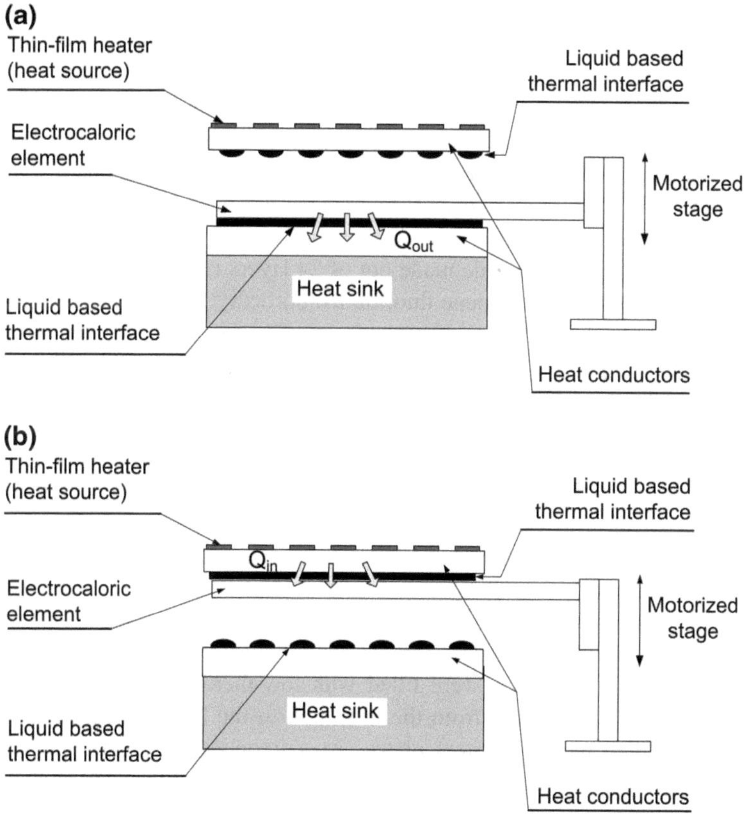

Fig. 10.11 Electrocaloric cooling device with a liquid-based switchable thermal interface, **a** Electrocaloric material is in an ideal thermal contact with the heat sink, **b** Electrocaloric material is in an ideal thermal contact with the heat source (see also Jia et al. [53])

consisted of an electrocaloric $BaTiO_3$ multilayer capacitor and a mechanical drive that was used to move the multilayer capacitor and create an alternating thermal contact with the heat sink and the heat source. The contact surfaces of the heat sink and the heat source were coated with liquid-based thermal interfaces to enhance the heat transfer from the multilayer capacitor to the heat sink/source. The operation of the device can be divided into four main steps. In the first step, the electrocaloric material is polarized and heats up. Next, the multilayer capacitor is moved into physical contact with the heat sink. Therefore, the heat is transferred from the multilayer capacitor to the heat sink by thermal conduction (Fig. 10.11a). In the third step, the material is depolarized and, due to the electrocaloric effect, it cools down. Then the capacitor is moved into contact with the heat source and can absorb the heat (Fig. 10.11b). In this study, a constant Joule heating of 15 mW was induced in a thin-film heater (heat source). The authors reported that at a frequency of 3 Hz (the number of thermodynamic cycles per unit of time) the heat source was cooled down by 1 K with respect to a reference point. To set the reference point, the device was first operated only by applying mechanical motion. The electrocaloric element was not subjected to an electric field change and no electrocaloric effect was induced. In this case, the heat was passively transferred from the heat source to the heat sink.

An interesting prototype device was presented by Gu et al. [54] in 2013. In their paper, they describe a chip-sized electrocaloric cooling device. The concept of the device was already described earlier in this chapter and it is based on an active heat regenerator with a solid-state matrix to transfer the heat from the electrocaloric material to a heat source/sink. As a basic unit of the active electrocaloric heat regenerator, a multilayer module made out of 24 layers of 8-μm-thick high-energy-electron-irradiated poly(vinylidene fluoride-trifluoroethylene) 68/32 % (copolymer) films was used. Each layer of the film was covered with Au electrodes. The layers were then glued together to form a 0.25-mm-thick multilayer module. It was shown that the polymer used to construct the multilayer module can achieve an adiabatic temperature change that is even higher than 20 K in a 160-MVm^{-1} electric field change [31]. However, the adiabatic temperature change of the constructed multilayer module was reported to be much smaller. It was measured to be 2.25 K under 80 MVm^{-1} of electric field change. Two multilayer modules were used in the device, a schematic presentation of which can be seen in Fig. 10.12. The photograph of the device and its main characteristics are presented in Table 10.5. The solid matrix was constructed from stainless-steel plates with a thickness of 0.5 mm. Thin slits were cut in the stainless-steel plates in the direction perpendicular to the mechanical motion. The slits were filled with low-thermal-conductivity epoxy to decrease the rate of heat flux from the heat sink to the heat source. A step-motor drive was used to move the steel plates. At a frequency of 1 Hz and under an electric field change of 100 MVm^{-1}, the maximum temperature span between the heat sink and heat source was measured to be 6.6 K at room temperature.

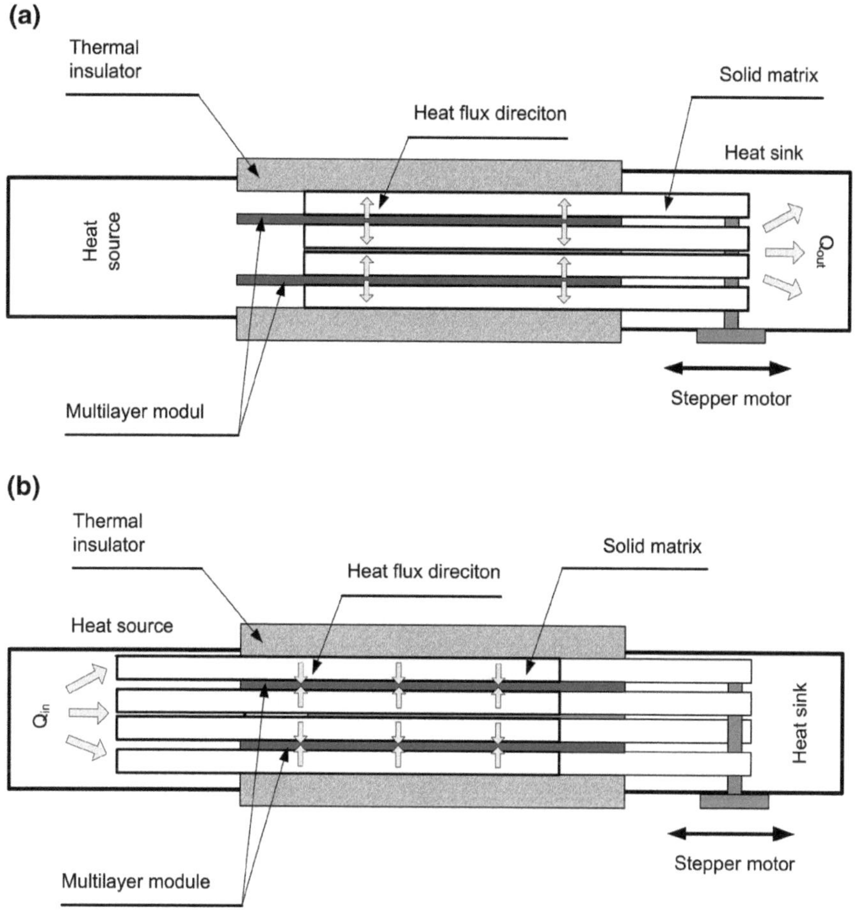

Fig. 10.12 Schematic presentation of an electrocaloric cooling device with a solid matrix, **a** After the polarization, the solid matrix is in thermal contact with the heat sink and the multilayer module rejects heat to the solid matrix, **b** After the depolarization the solid matrix is in thermal contact with the heat source and the multilayer module absorbs heat from the solid matrix (see also Gu et al. [54])

At the end of 2013, Chukka et al. [55] presented a paper describing an electrocaloric device for active cooling. Their investigation was focused on how to reduce the time needed to cool down a chip-sized object to ambient temperature. A schematic presentation of the system under investigation is shown in Fig. 10.13. In the middle of a sandwich-like structure, there is an electrocaloric plate made out of a bulk PMN-70PT single-crystal ceramic material that exhibits a maximum adiabatic temperature change of 2.7 K under an electric field change of 1.2 MVm^{-1}. On each side of the electrocaloric material, a Peltier element (used as a temperature sensor) was placed to form a sandwich-like structure. On the top side of the sandwich-like structure, a copper block was placed and this functioned as a heat

Table 10.5 Chip-sized electrocaloric cooling device and its properties

Name and address of institute	The Pennsylvania State University	
Name of contact person/email	Qiming Zhang	qxz1@psu.edu
Year of production	2012	
Maximum frequency	1 Hz	
Maximum cooling power	N/A	
Maximum temperature span	6.6 K	
EC material	High-energy electron-irradiated P(VDF-TrFE) 68/32	

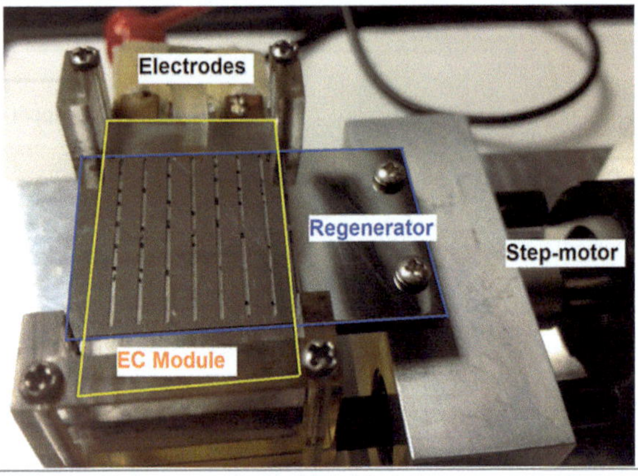

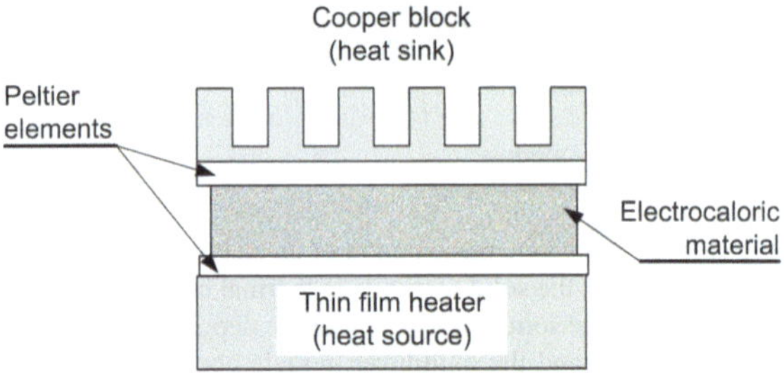

Fig. 10.13 Electrocaloric cooling device by Chukka (see also [55])

sink. On the bottom side, a thin-film heater with 10 W of power (heat source) was positioned. Additional temperature sensors were inserted between the thin-film heater and the electrocaloric material and between the electrocaloric material and the copper block. In the first-case scenario, the device was heated up to 85 °C (by turning on the thin-film heater). Then the heater was turned off and the time needed for the sandwich-like structure to cool down to ambient temperature was measured. In this case, the electrocaloric material was not subjected to a varying electric field. In the second case scenario, the sandwich-like structure was again heated up and after it had reached the desired temperature of 85 °C, the thin-film heater was turned off. However, this time, after the heater was turned off, the electrocaloric material was subjected to a rectangular electric signal (thereby inducing the electrocaloric effect in the electrocaloric material). The electric signal consisted of rectangular pulses with a constant amplitude of 12 $kVcm^{-1}$; however, different pulse repetition rates (the time between two neighbouring electrical pulses) were considered. The results showed that in the second experiment, at a pulse repetition rate of 2.4 s, the sandwiched-like structure cooled down to ambient temperature in approximately 600 s, which was almost twice as fast as in the first experiment.

A team of researchers (also the authors of this book) from the Faculty of Mechanical Engineering, University of Ljubljana and scientists from the Institute "Jožef Stefan" (IJS) developed a small-scale electrocaloric cooling device with an AER [56]. As the electrocaloric material, 0.2-mm-thick $(PbMg_{1/3}Nb_{2/3}O_3)_{0.90}$-$(PbTiO_3)_{0.10}$ bulk ceramic plates, characterized and processed by the team at the IJS were used. The maximum adiabatic temperature change of the ceramic material was measured for an electric field change of 160 MVm^{-1} and accounted for 3.5 K (which is one of the highest ECE values ever measured in bulk ceramics). However, the electrocaloric cooling device operated at much lower electric field changes, the maximum being 50 MVm^{-1} (which corresponded to 0.89 K of adiabatic temperature change). A total of 30 ceramic plates were used, constructing a 60-mm-long AER with a total mass of approximately 9 g. To create a void for the fluid flow, a conductive distance holder with a thickness of 0.1 mm was inserted between the individual ceramic plates. The distance holders were also used to connect the electrodes of the ceramic plates to the electric field generator. The device is schematically presented in Fig. 10.14. The AER was inserted into a housing and a peristaltic pump was used to pump the fluid through the AER. Silicon oil was used as the working fluid. A detailed description of the working principle of an electrocaloric cooling device with an AER was already described in this chapter. The authors reported that a maximum temperature span of 3.3 K (the temperature difference between the heat source and the heat sink) was measured at a frequency of 0.75 Hz and an electric field change of 50 MVm^{-1}. This means that a regeneration factor (the temperature span of the device over the adiabatic temperature change of the electrocaloric material) of 3.7 was achieved. The photograph of the device and its main characteristics are presented in Table 10.6.

Additionally, the teams from the University of Ljubljana and the IJS proposed numerous ideas and solutions about how to improve the performance of such a device and gathered them in a patent application [46].

Table 10.7 shows the main characteristics of the five presented prototypes. It can be concluded that these first prototypes are far from being competitive with magnetocaloric or compressor-vapour cooling devices. Nevertheless, these prototypes serve more as proof-of-concept devices. However, the domain of electrocaloric refrigeration and heat pumping was only recently given more attention by the research community. In Fig. 10.15, the total number of publications related to the topic of electrocaloric refrigeration and heat pumping since 1990 is presented. The data from Fig. 10.15 were obtained from the online database Web of Science. It is clear that the number of publications is increasing exponentially and has more than doubled in the past 3 years. Therefore, an increase in the number and the performance of electrocaloric cooling devices is expected in the future.

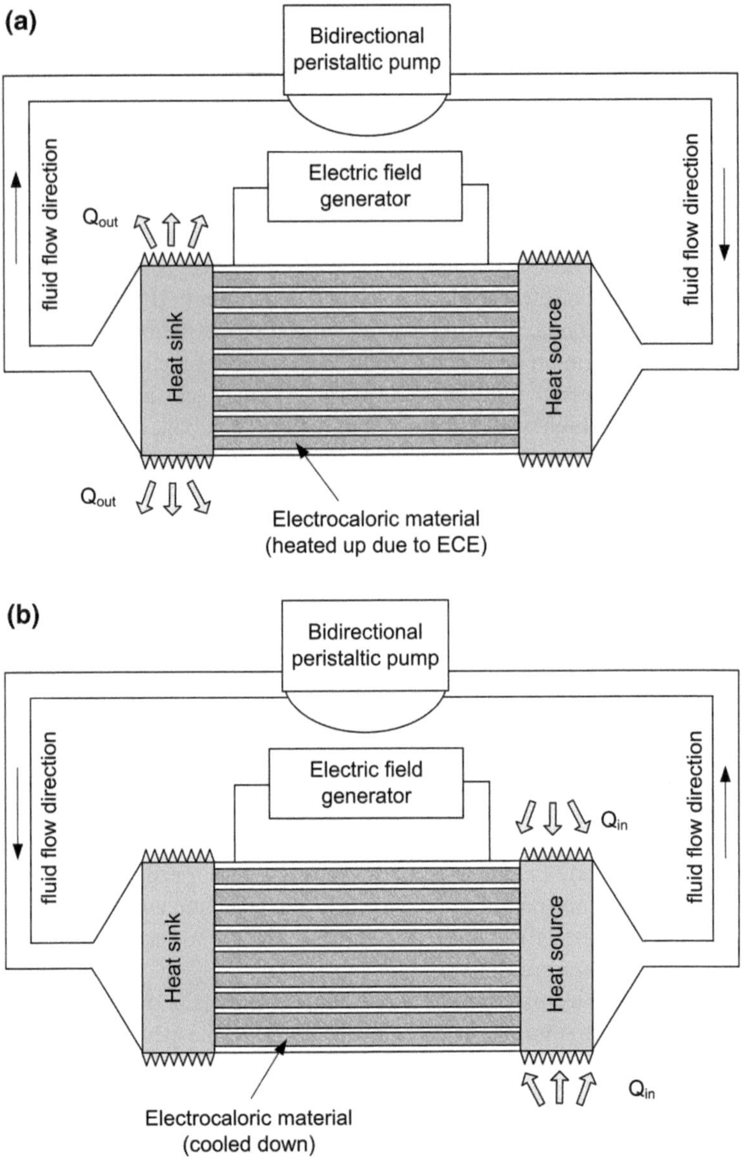

Fig. 10.14 Schematic presentation of a small-scale electrocaloric cooling device with an AER, **a** After the polarization the fluid flows from the heat source to the heat sink, where it rejects the heat to the environment, **b** After the depolarization, the fluid flows from the heat sink to the heat source, where it absorbs heat

Table 10.6 Small-scale electrocaloric cooling device with an AER and its characteristics

Name and address of institute	University of Ljubljana, faculty of mechanical engineering	
Name of contact person/email	Andrej Kitanovski	andrej.kitanovski@fs.uni-lj.si
Year of production	2013	
Maximum frequency	2 Hz	
Maximum cooling power	N/A	
Maximum temperature span	3.3 K	
EC material	$(PbMg_{1/3}Nb_{2/3}O_3)_{0.90}$-$(PbTiO_3)_{0.10}$	

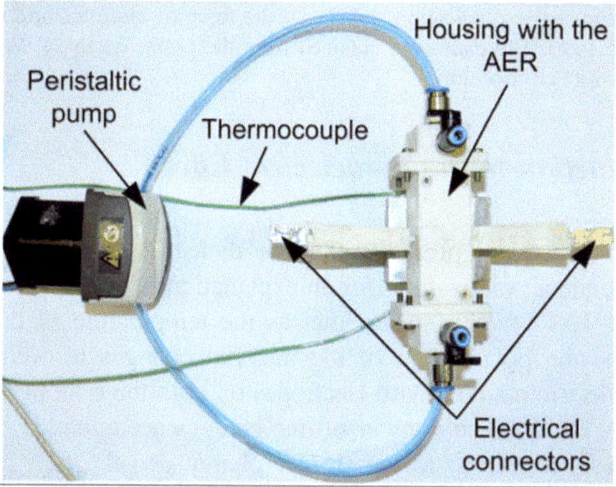

Table 10.7 Electrocaloric cooling devices

Author	Year	Short description	ΔT_{ad}	Temperature span	Evidence of realized prototype[a]	References
Sinyavsky and Brodyansky	1991	Device with active heat regenerator	1.3 K at 3MVm^{-1}	5 K at 6 MVm^{-1}	No	[45]
Jia et al.	2012	Device with liquid-based switchable thermal interfaces	0.5 K at 30 MVm^{-1}	1 K at 30 MVm^{-1}	No	[53]
Gu et al.	2013	Device with a solid state matrix	2.25 K at 80 MVm^{-1}	6.6 K at 100 MVm^{-1}	Yes	[54]
Chukka et al.	2013	Device for active cooling	/	/	Yes	[55]
Plaznik el al.	2014	Small-scale device with active heat regenerator	0.89 K at 5 MVm^{-1}	3.3 K at 5 MVm^{-1}	Yes	[56]

[a] A photograph of the device was considered as evidence (either published in a paper or obtained from the authors by means of personal communication)

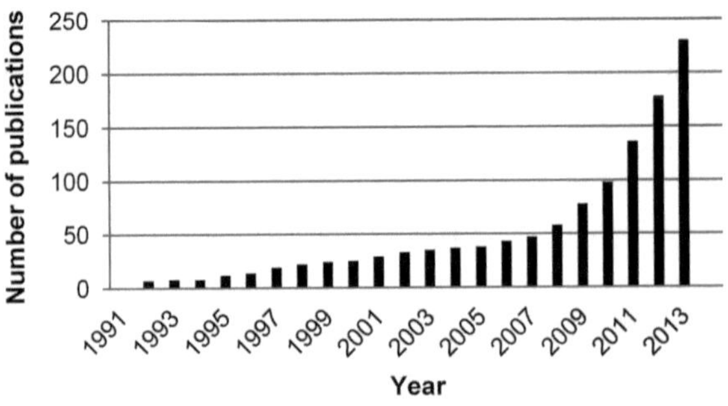

Fig. 10.15 Total number of publications related to the topic of electrocaloric refrigeration and heat pumping since 1990. The data were obtained from the online database Web of Science by searching for the word "electrocaloric"

10.1.4 Introduction to the Pyroelectric Effect

The pyroelectric effect is a property of some dielectric materials that possess a temperature-dependent polarization (for an explanation of polarization the reader is referred to Sect. 10.1.1). This means that as the temperature of the pyroelectric material changes, the polarization of the material changes as well. Let us now consider a pyroelectric material with electrodes on both the bottom and top planes, as shown in Fig. 10.16a. The number of free charges accumulated on the top and bottom electrodes, referred to as the surface charge density, is a function of the polarization of the pyroelectric material and the electric field. In the case of a linear and isotropic material, the surface charge density, also named electric displacement, can be expressed as [1]:

$$\sigma = \varepsilon_0 E + P \tag{10.12}$$

where P is the polarization, E is the electric field and ε_0 is the electric permittivity of vacuum. It is clear from Eq. (10.12) that a change in the polarization of the pyroelectric material (due to a temperature change) will have an effect on the amount of free charges accumulated on the top and bottom electrodes of the pyroelectric material. If the pyroelectric material is further connected to an electrical circuit, the change of the temperature of the pyroelectric material will result in a pyroelectric current [1]. A simplified example of the above description is presented in Fig. 10.16 (see also [57]). The pyroelectric material is initially in a thermal equilibrium (its temperature does not change with time) and since in this case the polarization remains constant, no electrical current will flow (Fig. 10.16a). However, when the pyroelectric element is heated, the spontaneous polarization will decrease and therefore electrons will flow from the negatively charged plate to the positively charged plate, creating an electrical current (Fig. 10.16b). If the material is then cooled back down to its initial temperature, the polarization of the pyroelectric material will decrease and the current will flow in the opposite direction

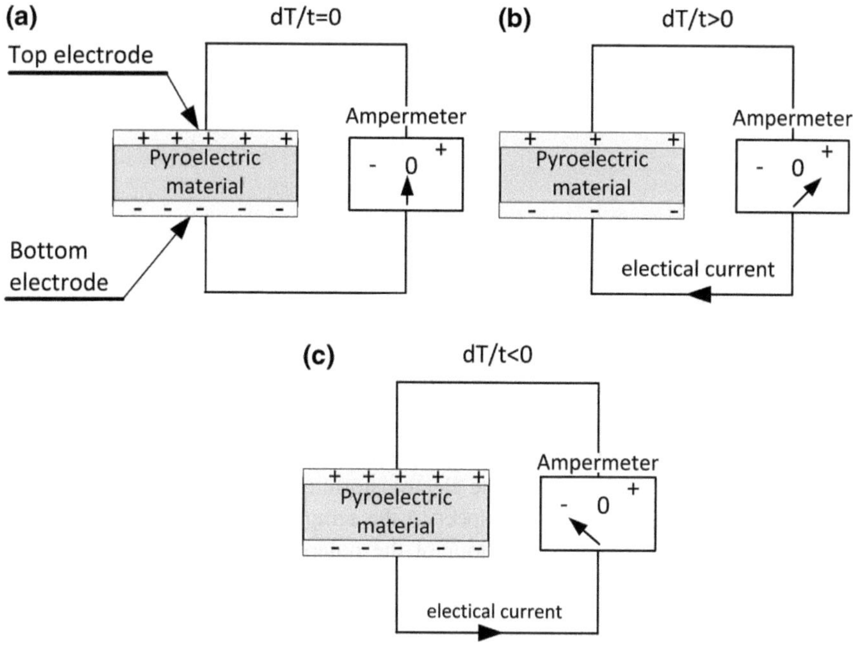

Fig. 10.16 Schematic presentation of a pyroelectric material in an electrical circuit, **a** The pyroelectric material is in a thermal equilibrium, **b** The pyroelectric material is heating up, **c** The pyroelectric material is cooling down

(Fig. 10.16c). To describe how the polarization of a pyroelectric material changes with temperature, a so-called pyroelectric coefficient is used and is defined as [1]:

$$p = \frac{\mathrm{d}\sigma}{\mathrm{d}T} \tag{10.13}$$

where T is the temperature of the pyroelectric material. If the expression for the surface charge density is inserted into Eq. (10.13), the pyroelectric coefficient can be written as:

$$p = \frac{\mathrm{d}(\varepsilon_0 E + P)}{\mathrm{d}T} \tag{10.14}$$

For a constant electric field and with the assumption that the field-induced polarization is much smaller than the spontaneous polarization, the pyroelectric coefficient can be expressed as [1]:

$$p = \left(\frac{\partial P_s}{\partial T}\right)_{E,X} \tag{10.15}$$

where P_s is the spontaneous polarization, X is the elastic stress and E is the electric field. Furthermore, if the pyroelectric material is short-circuited and subjected to a temperature change we can write:

$$I_p = \frac{p \cdot \mathrm{d}T}{\mathrm{d}t} \cdot A \tag{10.16}$$

where A is the surface area of the pyroelectric material and t is the time.

The pyroelectric current generated due to the temperature fluctuations of the pyroelectric material can be harvested using the different thermodynamic cycles and devices that are presented later in this chapter. However, it is already obvious, that in order for a pyroelectric material to convert as much thermal energy as possible into electrical energy, the pyroelectric material should possess a large pyroelectric coefficient. A large pyroelectric coefficient is a property of dielectric materials whose polarization changes intensively by changing their temperature. The same conclusion was already made for the electrocaloric effect, i.e. the larger the rate of change of the polarization with respect to the temperature, the larger will be the electrocaloric effect. Such a behaviour of the polarization can be found in ferroelectric materials around their phase transition (transition from the paraelectric to the ferroelectric phase) temperature. To find out more about the properties of ferroelectric materials, the reader is referred to Sect. 10.1.1.

10.1.5 Pyroelectric Materials for Energy Harvesting

The pyroelectric effect was probably already observed by the Greeks more than 2,400 years ago in the mineral tourmaline; however, the pyroelectric effect was first introduced to the scientific community in 1,717 by the physician and chemist Louis Lemery [57]. Intensive investigations of the properties of pyroelectric materials began in 1960 and during the period 1960–2003 there were more than 8,500 publications [58]. Nowadays, pyroelectric materials can be found in many fields of applications, such as thermal imaging, light, motion and fire detection, etc. In this chapter the focus will be on pyroelectric materials that could be applied for energy harvesting applications. The pyroelectric materials used in the first generation of prototypes for energy harvesting devices can be divided between ceramic and polymer materials. In general, it is desirable for the pyroelectric material to possess a large pyroelectric effect. However, other properties should be taken into account when choosing the appropriate pyroelectric material, such as a high electrical resistivity (to minimize the leakage current through the pyroelectric element), a low heat capacity (to minimize the heat input needed to change the temperature of the pyroelectric material) and a small hysteresis [59, 60]. Furthermore, for some applications, the high dielectric strength of the pyroelectric material is important [61, 62].

Table 10.8 Pyroelectric coefficient of some ceramic and polymer materials

Material	Material type	T[a] (°C)	p ($\mu Cm^{-2}K^{-1}$)	References
$BaTiO_3$	Ceramic	20–47	330	[1]
$Pb(ZrTi)O_3$	Ceramic	20–100	260	[63]
$Pb(Mg_{1/3}Nb_{2/3})O_3$-$30PbTiO_3$	Single-crystal ceramic	190	6,500	[64]
Polyvinylidene fluoride	Polymer	30–65	62	[65]
Poly(vinylidene fluoride-trifluoroethylene) 60/40	Polymer	50	144	[66]

[a] Temperature or temperature range where the pyroelectric coefficient was measured

The ceramic materials used for pyroelectric energy conversion applications are mostly in the form of bulk materials with a thickness of around 100 μm. They usually possess a large pyroelectric coefficient, and the most commonly used ceramic pyroelectric material for energy harvesting is $PbZr_{0.95}Ti_{0.05}O_3$ (PZT). The pyroelectric coefficients of some ceramic materials are listed in Table 10.8.

The polymer materials considered for pyroelectric energy harvesting are normally in the form of thick films, with their thicknesses being a few tens of micrometres. They possess a smaller pyroelectric coefficient in comparison with ceramic materials, but they can be exposed to higher electric fields before any dielectric breakdown occurs and are considerably less expensive than ceramic materials [61]. The most commonly used are polyvinylidene fluoride (PVDF) films and their copolymers, such as poly(vinylidene fluoride-trifluoroethylene) P(VDF-TrFE). The pyroelectric properties of some polymer films are presented in Table 10.8.

From Table 10.8 it is clear that some ceramics, for example the $({}_{0.67}Ba_{0.33}Sr)$ TiO_3 ceramic, possess a very large pyroelectric effect (7,000 $\mu cm^{-2}K^{-1}$). However, it must be noted that such a high value is usually observed only in the vicinity of the phase transition temperature and not over a broader temperature range [64].

10.1.6 Review of Device Concepts and First Prototypes for Pyroelectric Energy Harvesting

In order to harvest the electrical energy from the temperature fluctuations of a pyroelectric material, the material must be a part of a larger system where it undergoes a thermodynamic cycle. Different solutions and ideas about how to design such a system were proposed and are reviewed later in this chapter. In addition, the first prototype devices are presented. The chapter is divided into two sections. In the first section, the background behind the concepts of pyroelectric energy harvesting devices is presented. In the second section, the experimental realization of these concepts is reviewed.

10.1.6.1 Pyroelectric Energy Harvesting: Concepts and a Theoretical Evaluation of the Performance

Pyroelectric energy harvesting devices can be divided into two groups. The property of the devices in the first group is that they do not need any external control and any external electrical power source. They simply take advantage of the change of the spontaneous polarization of the pyroelectric material due to its temperature fluctuation. We will name this group of devices the self-driven pyroelectric energy converters. On the other hand, the devices in the second group work on the basis of the so-called Ericsson or Olson thermodynamic cycle [67]. In order to execute such a cycle, an external voltage source and therefore a power supply is needed.

Self-Driven Pyroelectric Energy Converters

The most straightforward technique to convert the temperature fluctuations of a pyroelectric material into electrical energy is based on the concept already described in Sect. 10.1.4. In this case, the pyroelectric material is cyclically subjected to the temperature fluctuations and, as a consequence, if the pyroelectric material is short-circuited, an electrical current is generated, as stated in Eq. (10.16). A temperature fluctuation in a pyroelectric material can be induced in many different ways, for example, by the changing weather conditions (such as solar radiation [63, 68]) or by the alternating movement of a pyroelectric material from a heat source to a heat sink [69, 70]. The electrical energy can then be harvested using different electrical circuits, thereby subjecting the pyroelectric material to different thermodynamic cycles. Some of them were presented and discussed in detail by Sebald et al. [71]. An important characteristic of such devices is that they do not need any external power supply and could therefore be used instead of batteries for powering sensors, monitoring hot pipes and microcontrollers. Some of the first such prototype devices are presented later in this chapter.

Pyroelectric Energy Harvesters Based on the Ericsson Cycle

In contrast to the self-driven pyroelectric energy converters, a pyroelectric material used in a device working under the pyroelectric Ericsson cycle must be connected to a voltage source. The material is subject to the thermodynamic cycle presented in Fig. 10.17. Let us start with our explanation in the lower left corner of the cycle, where in its initial state the pyroelectric material is subjected to a small electric field (E_{low}) and has the temperature T_{high}.

In the first step (1–2), the temperature of the material is decreased to T_{low} at a constant electric field and thereby the polarization of the material increases. In the next step (2–3), the electric field is isothermally increased to the value E_{high} and the polarization of the material increases even further. Then the material`s temperature is decreased at a constant electric field (3–4) and in the last step the electrical field is

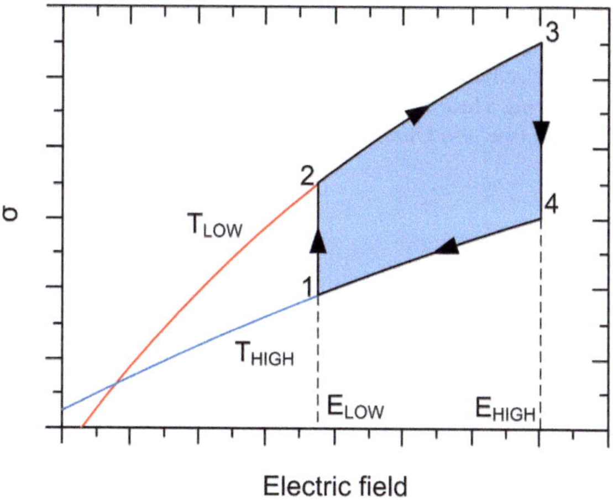

Fig. 10.17 Pyroelectric Ericsson thermodynamic cycle. The *red line* represents electric displacement versus electric field at T_{low} and the *blue line* the electric displacement versus electric field at T_{high} of an arbitrary pyroelectric material

changed to the low value (E_{low}) (4–1). The pyroelectric material is now, thermodynamically speaking, in the same state as at the beginning of the cycle. However, the pyroelectric material had done some work, which is equivalent to the surface area of the enclosed path presented in the σ–E diagram (Fig. 10.17).

Such a cycle was first introduced to the field of pyroelectric energy harvesting by Olsen et al. [67]. For this reason, in the literature, the pyroelectric Ericsson cycle is often simply referred to as the Olsen cycle [72–74]. However, it was already Olsen and his coworkers [67] who concluded that such a device working under the basic Ericsson cycle is energy inefficient. Namely, the heat in such a cycle has to be supplied to the pyroelectric material during the process of isofield polarization (3–4) as well as during the process of isothermal depolarization (4–1). However, a portion of the heat supplied to the process of isofield polarization could be obtained from the process of isofield depolarization. The last process regards regeneration, which should be employed to increase the efficiency of the cycle [75]. The concept of a pyroelectric energy conversion system that employs regeneration was first described by Olsen et al. [75] and is schematically presented in Fig. 10.18. According to Fig. 10.18, the pyroelectric material is initially in thermal contact with the heat source and the heat is transferred from the heat source to the pyroelectric material. Then as the pyroelectric material is moved towards the heat sink, the heat is transferred from the pyroelectric material to the heat regenerator. As the pyroelectric material reaches the heat sink most of the heat has already been transferred to the heat regenerator and only a small portion of the heat is transferred to the heat sink. An analogue process occurs as the pyroelectric material is moved back to the heat source and the heat that was transferred from the pyroelectric material to the regenerator in the previous step can now be used to heat up the pyroelectric material. In this case, the temperature variation of the pyroelectric material is the same as if no regeneration process was employed and therefore the pyroelectric material performs the same amount of work.

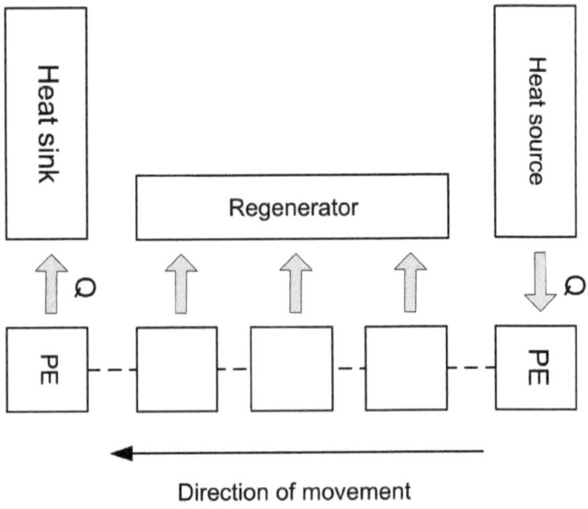

Fig. 10.18 Schematic presentation of the concept of a pyroelectric device employing a heat regeneration process (see also Olsen et al. [75])

However, in the cycle with the heat regeneration, considerably less heat is transferred from the heat source to the material, thus increasing the efficiency of the cycle. Olsen and Brown [75] estimated the efficiency (defined as the ratio between the work performed by the pyroelectric material and heat transferred from the heat source to the pyroelectric material) of an idealized device working under the regenerative pyroelectric Ericsson cycle (neglecting the energy losses such as the Joule heating of the pyroelectric material, the losses due to the hysteresis of the pyroelectric material, the energy input needed to move the pyroelectric material from the heat source to the heat sink, and the other way around, and the heat flow between the heat source and the heat sink by means of thermal conduction). They showed that such a device could reach an efficiency equal to the efficiency of a Carnot engine operating between the same temperatures. However, in practical cases, the devices working under the regenerative pyroelectric Ericsson cycle differ from the concept presented in Fig. 10.18. Furthermore, the efficiencies achieved are considerably lower than those of an ideal Carnot engine working between the same temperatures [62, 67]. For example, Olsen et al. [67] reported that a pyroelectric energy converter working under the Ericsson cycle and employing heat regeneration reached an efficiency of 0.42 %, which was equal to 5.7 % of the Carnot (Exergy) efficiency.

10.1.6.2 Pyroelectric Energy Harvesting: First Prototypes

In this section, the first generation of pyroelectric energy harvesting prototypes is presented. These first prototypes can be divided into two groups. The first group includes the so-called self-driven pyroelectric energy harvesting devices that need no external power supply. The second group relates to the pyroelectric energy harvesting devices based on the Ericsson thermodynamic cycles.

Self-Driven Pyroelectric Energy Harvesters

Cuadras et al. [65] constructed an experiment in which the pyroelectric effect was used to charge a capacitor. A special electrical circuit was developed in order to store the electrical charge in the capacitor (generated due to the pyroelectric effect) during the heating and cooling of the pyroelectric material. The pyroelectric material was heated up from room temperature to 62 °C using a hair dryer and then cooled down, based on free cooling, back to the initial temperature. Moreover, to achieve a higher voltage on the capacitor and to store more energy in the capacitor, two pyroelectric elements (in this case two PZT ceramics with a thickness of 100 μm) were electrically connected in parallel. After a number of cycles (one cycle was defined as a process during which the pyroelectric material heats up and cools back down once), in approximately 1,000 s, a maximum voltage of 31 V was reached on the charged capacitor, with an available energy of 0.5 mJ.

Zhang et al. [68] investigated the possibility of harvesting electrical energy from the temperature fluctuations of a pyroelectric material exposed to solar radiation and changing wind conditions during daytime. The experiment was conducted under laboratory conditions, where solar radiation was simulated using a spotlight with 20 W of power and a centrifugal fan was used to simulate the changing wind conditions. As the pyroelectric material, a 140-μm-thick PZT ceramic in the shape of a disc with a diameter of 24 mm was used. The pyroelectric material was exposed to a constant solar radiation of 1000 Wm^{-2}. On the other hand, the fan was periodically switched on and off, creating different airflow velocities around the pyroelectric material and changing the heat-transfer rate due to the heat convection. The tests lasted for approximately 430 s and every 100 s the fan was turned on for 30 s, thereby cooling down the pyroelectric material for 16 K. Due to the temperature variation of the pyroelectric material, an average electric power density of 4.2 μWcm^{-3} (per volume of the pyroelectric material) was produced.

A similar experiment to the one described above was conducted by Mane et al. [63]. They investigated the possibility of harvesting electrical power by exposing a pyroelectric material to changing levels of radiation. Three different pyroelectric materials, a single-crystal PMN-0.3PT, a commercially available PZT ceramic and a pre-stressed composite PZT ceramic, were considered. A lamp was used as the light source and a rotating disc with an aperture was positioned between the lamp and the sample (the pyroelectric material). By rotating the disc, the sample was periodically exposed to increased levels of radiation, and therefore a temperature variation was induced in the sample. Different cyclic frequencies of the disc were investigated. The results showed that the maximum peak power density of 8.64 μWcm^{-3} (per volume of the pyroelectric material) was achieved in the PMN-30PT single-crystal ceramic material at an angular velocity of 0.64 $rad \cdot s^{-1}$ of the rotating disc and at a heating rate of the pyroelectric material of 8.5 Ks^{-1}. Under the same conditions, a peak power density of 4.48 μWcm^{-3} was measured in the commercially available PZT ceramic and a peak power density of 6.31 μWcm^{-3} in the prestressed composite PZT ceramic. However, the authors did not provide

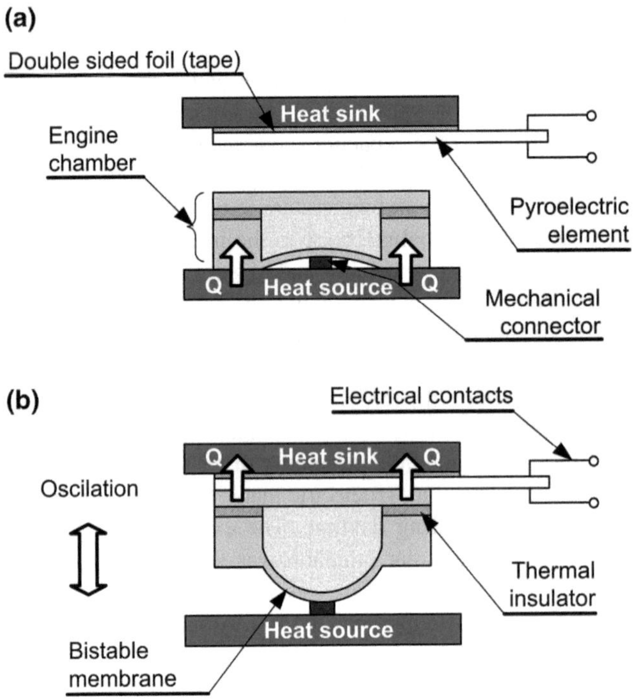

Fig. 10.19 A self-sustaining pyroelectric energy harvester, **a** The engine chamber is in thermal contact with the heat source, **b** The engine chamber is in thermal contact with the heat sink (see also Ravindran et al. [69])

details about the dimensions of the samples and the time-averaged power density achieved. Therefore, a direct comparison with other systems is not possible.

Ravindran et al. [69] presented a prototype of a self-sustaining pyroelectric energy converter (PEG). The device is schematically presented in Fig. 10.19. It consists of a heat sink, a heat source, a pyroelectric material and the so-called engine chamber. In the centre of the engine chamber, there is a sealed cavity filled with a working fluid. The bottom of the sealed cavity is made of a bi-stable membrane. When the engine chamber is in physical contact with the heat source (Fig. 10.19a), the heat is transferred from the heat source to the engine chamber by means of heat conduction. As a result, the working fluid heats up, the pressure in the cavity increases and the bi-stable membrane moves the engine chamber up. Now the engine chamber is in physical contact with the pyroelectric element (attached to the heat sink) and the heat is transferred from the engine chamber through the pyroelectric element to the heat sink (Fig. 10.19b). As a consequence, the pyroelectric element heats up. On the other hand, the engine chamber is cooled down in this process and as a certain temperature it reached it moves back to its original position. Since the engine chamber and the pyroelectric material are now no longer in contact, the pyroelectric element cools down and the cycle can repeat

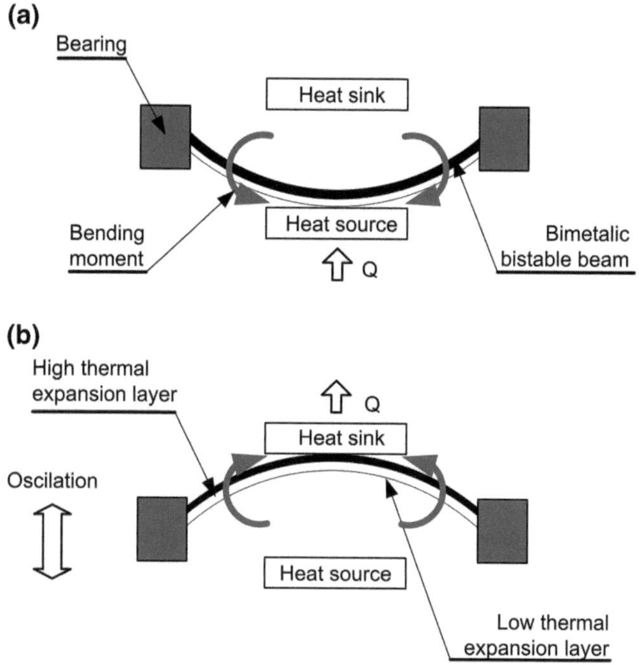

Fig. 10.20 Bimetallic micro heat engine for pyroelectric energy conversion, **a** The pyroelectric material is in thermal contact with the heat source, **b** The pyroelectric material is in thermal contact with the heat sink (see also Ravindran et al. [70])

from the beginning. In the experiment, a 200-μm-thick commercially available piezoceramic Vibrit 1100, which also exhibits pyroelectric properties, was used. The temperature difference between the heat source and the heat sink was varied between 45 and 80 K and the device was operating between a frequency (defined as the number of thermodynamic cycles per unit of time) of 0.1 and 0.42 Hz. A maximum power output of the pyroelectric element of 3.03 μW was measured for a temperature difference of 79.5 K and a frequency of the device equal to 0.42 Hz.

Ravindran et al. [70] presented an additional, similar concept to the one described above. In the centre of the device is a pyroelectric material attached to a bimetallic strip, as seen in Fig. 10.20. On each end the bimetallic strip is mounted on a bearing and placed between a heat sink and a heat source. When the bimetallic strip is in physical contact with the heat source (Fig. 10.20a), the heat is transferred from the heat source to the bimetallic strip by means of heat conduction. As a result, the bimetallic strip heats up, which causes a bending moment in the strip that moves from the heat source to the heat sink (Fig. 10.20b). Now the bimetallic strip is in physical contact with the heat sink and the heat is transferred from the bimetallic strip to the heat sink. In this way, the bimetallic strip cools down and moves back to its original position (Fig. 10.20a). This cycle is repeated many times and a sort of oscillatory movement of the bimetallic strip is established. Since the pyroelectric

Table 10.9 Characteristics of the self-driven pyroelectric energy converters

Authors	Material	ΔT (°C)	Power (μW)	References
Cuadras et al.	PZT	30	0.5	[65]
Zhang et al.	PZT	16	2.7	[68]
Mane et al.	PMN-0.3PT	/	/	[63]
Mane et al.	PZT	/	/	[63]
Ravindran et al.	Piezoceramic vibrit 1100	80	3.0	[69]
Ravindran et al.	PZT	96	4.4	[70]

material is attached to the bimetallic strip, its temperature oscillates as well and a pyroelectric current is generated. To test the concept, a 250-μm-thick PZT ceramic material with an area of 10×10 mm^2 was attached to a bimetal strip (type MS, Rau GmbH) with dimensions $40 \times 6 \times 0.28$ mm^3. Peltier elements were used to simulate the heat source and the heat sink of the device. The experiment was performed for various values of the temperature difference between the heat sink and the heat source. Based on the measured voltage and electrical current, the generated power was calculated. At a temperature difference of 96 K between the heat source and the heat sink and a frequency of 0.2 Hz, a maximum power of 4.4 μW was achieved.

In Table 10.9, the main characteristics of the above-presented, self-driven, pyroelectric, energy harvesting prototypes are presented. The generated electrical power of the devices is in the range of a few μW. For example, this could be enough to power an electronic watch [71]. However, it has to be taken into account that the mass of the pyroelectric materials used in the prototypes presented was relatively low since in almost all the cases only one pyroelectric element was used. A further increase in the generated power could also be achieved by increasing the frequency of the devices. This calls for better heat-transfer mechanisms and the use of thermal diodes.

Pyroelectric Energy Harvesters Based on the Ericsson Cycle

The most common way to realize the pyroelectric Ericsson thermodynamic cycle is by performing a so-called dipping experiment. In this case, two thermal baths, one having the temperature T_{high} and the other the temperature T_{low}, are prepared. The pyroelectric material is further connected to a voltage source and alternatively moved from the hot to the cold thermal bath. Furthermore, the voltage applied to the electrodes of the pyroelectric material is changed so that the material undergoes the pyroelectric Ericsson cycle. The performance of the material is usually expressed as the electrical energy generated per litre of pyroelectric material in one thermodynamic cycle constrained by the predefined temperatures (T_{high}, T_{low}) and the electric fields (E_{high}, E_{low}). Many different authors reported on the pyroelectric material's performance measured using the dipping experiment. The results of some of these experiments are gathered in Table 10.10; however, only the maximum

Table 10.10 Results of the dipping experiment

Material	Material type	T_{low} (°C)	T_{high} (°C)	E_{low} (MVm^{-1})	E_{high} (MVm^{-1})	Work per cycle (JL^{-1})	d[a] (μm)	References
P(VDF-TrFE) 60/40	Polymer	50	100	6.7	27.6	165	25	[60]
P(VDF-TrFE) 60/40	Polymer	55	100	6.7	20	95	25	[60]
P(VDF-TrFE) 60/40	Polymer	55	100	6.7	20	70	25	[60]
PLZT 9/65/35	Bulk ceramic	3	150	0.4	7.5	654	200	[61]
PLZT 8/65/35	Bulk ceramic	25	160	0.2	7.5	888	290	[61]
PLZT 7/65/35	Bulk ceramic	30	200	0.2	7	1,014	190	[61]
PLZT 6/65/35	Bulk ceramic	40	210	0.4	8.5	949	180	[61]
PLZT 5/65/35	Bulk ceramic	40	250	0.4	7.5	799	200	[61]
$0.9Pb(Mg_{1/3}Nb_{2/3})O_3$-$0.1PbTiO_3$	Bulk ceramic	30	80	0	3.5	186	1,000	[77]
$Pb(Zn_{1/3}Nb_{2/3})_{0.95}Ti_{0.4}O_3$	Single-crystal ceramic	100	160	0	2	243	1,000	[78]
$68PbMg_{1/3}Nb_{2/3}O_3$-$32PbTiO_3$	Single-crystal ceramic	80	170	2	9	100	140	[79]
$0.945PbZn_{1/3}Nb_{2/3}O_3$-$0.055\ PbTiO_3$	Single-crystal ceramic	100	190	0	1.2	150	200	[72]

[a] d is the thickness of the pyroelectric material

values of the electrical energy generated per cycle reported by the authors are presented. In general, the electrical energy generated is a function of the temperature difference between the hot and the cold thermal bath and the difference between the applied electrical fields. It is clear that the higher the temperature change of the pyroelectric material and the larger the electric field change [61, 76], the larger will be the amount of generated electrical energy. However, other effects, such as the current leakage, can have a significant impact on the electrical energy generated in such processes [4].

However, the experimental setups for performing the dipping experiment cannot be considered as real pyroelectric energy harvesting devices. The sample is, in most cases, manually moved from the hot to the cold thermal bath and the frequencies of the operation are relatively low. For example, Lee et al. [61] reported that the pyroelectric material was moved from the hot thermal bath to the cold thermal bath every 15–25 s. Nevertheless, the results of the dipping experiment give important information about the pyroelectric material itself.

Cha et al. [80] presented a pyroelectric energy harvesting device that uses liquid-based thermal interfaces and operates under a pyroelectric Ericsson cycle. No regeneration process was employed and the pyroelectric material was subjected to the same thermodynamic cycle as in the case of the dipping experiments. However, the fre quency of the device reached 1 Hz and it worked autonomously. Therefore, the device is not considered as an experiment, the aim of which is to define the maximum energy produced in the pyroelectric element per single thermodynamic cycle (as in the case of the dipping experiment), but rather as a prototype energy conversion device. The device is schematically presented in Fig. 10.21. In the centre of the device was a 5-μm-thick P(VDF-TrFE) 56/44

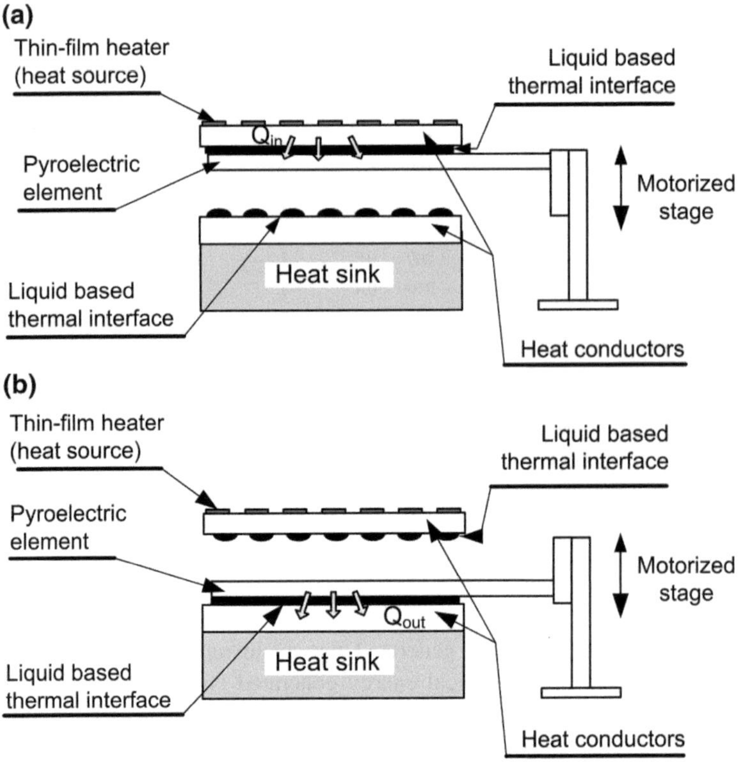

Fig. 10.21 Pyroelectric energy harvesting device using liquid-based switchable thermal interfaces working according to the pyroelectric Ericsson cycle, **a** The pyroelectric material is in thermal contact with the heat source, **b** The pyroelectric material is in thermal contact with the heat sink (see also Cha et al. [80])

copolymer material with pyroelectric properties. The pyroelectric material had electrodes on both sides and was positioned on a glass substrate for reasons of mechanical stability. The glass substrate with the pyroelectric material (pyroelectric element) was mounted on a motorized stage and positioned between the heat sink and the heat source. The contact surfaces of the heat sink and the heat source were covered with liquid-based thermal interfaces to enhance the heat transfer from the pyroelectric element to the heat source/sink. The device works in accordance with the following description. First, the pyroelectric element is moved into physical contact with the heat source and therefore the material heats up (Fig. 10.21a). Next, the electric field is switched to the low value. The pyroelectric element is then moved into physical contact with the heat sink and the heat is transferred from the pyroelectric element to the heat sink by means of heat conduction (Fig. 10.21b). The electric field is then increased and the cycle can be repeated. During the experiment, the pyroelectric current and the applied voltage were measured and based on these measurements the generated electrical power was calculated. The

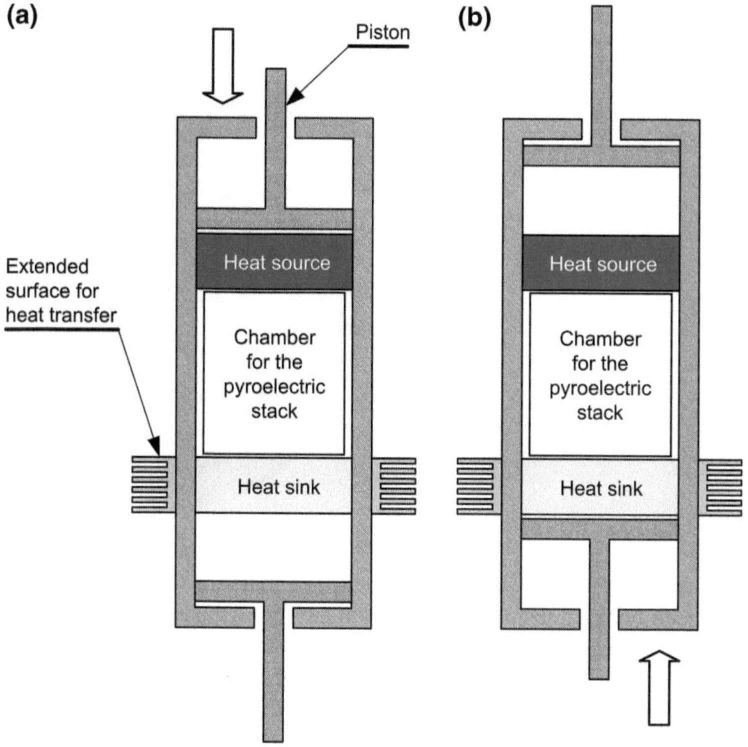

Fig. 10.22 Pyroelectric energy converter that employs regeneration (see also Olsen et al. [67])

effect of the cyclic frequency and the temperature of the heat source on the power density of the device were investigated. The maximum power density output achieved was approximately 135 W/L of the pyroelectric material with the following parameters: a heat sink temperature of 40 °C, a heat source temperature of 100 °C, a low electric field of 20 MVm^{-1}, a high electric field of 50 MVm^{-1} and a frequency of approximately 1 Hz.

In 1981 Olsen et al. [67] suggested that in order to increase the efficiency of a pyroelectric energy converter, a heat regeneration process should be employed. Based on this idea they designed the device that is schematically presented in Fig. 10.22. The device consisted of a stack of pyroelectric elements, an electrical heater (heat source), an aluminium block (heat sink), a Teflon housing and a fluid pumping system. The pyroelectric stack was put in the housing, which was then filled with silicone oil. The silicone oil was pumped from the heat sink to the heat source in an oscillatory manner, creating temperature oscillations in the pyroelectric elements. To pump the fluid, a piston driven by an electrical motor was used. Furthermore, in order to achieve the pyroelectric Ericsson cycle, the applied voltage was controlled as well. The voltage was changed from the low value to the high value after the piston had moved the fluid from the heat source to the heat sink,

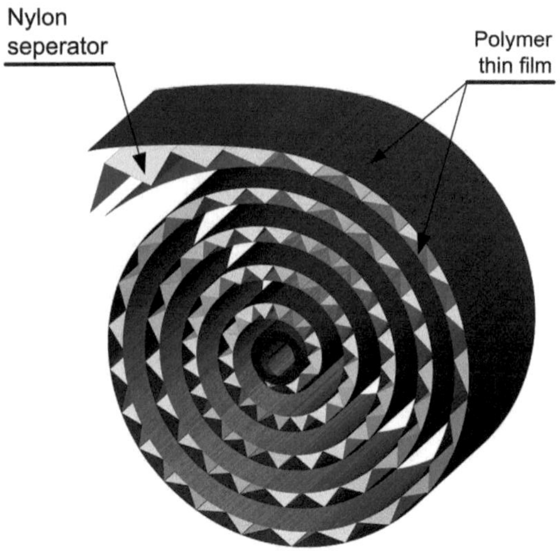

Fig. 10.23 Spiral stack made of layers of polymer and nylon separator (see also Olsen et al. [81])

thereby heating up the stack of the pyroelectric material (Fig. 10.22a). After the fluid had been pumped from the heat sink to the heat source, thereby cooling down the stack of pyroelectric elements, the voltage was decreased from the high value to the low value (Fig. 10.22b). The stack of pyroelectric elements was constructed from $Pb_{0.99}Nb_{0.2}(Zr_{0.68}Sn_{0.25}Ti_{0.07})_{0.93}O_3$ (PZLT) ceramic plates with dimensions of 24 × 24 × 0.25 mm^3. Wires were soldered to the electrodes of each ceramic plate and 24 of such assemblies created a porous stack with a 0.25-mm gap between the pyroelectric elements. The electrically wired pyroelectric stack was connected to a measurement system to measure the generated pyroelectric current and the voltage. Based on the measured current and the voltage, an electrical energy of 0.31 J·$cycle^{-1}$ (130 J·$cycle^{-1}$/L of the pyroelectric material) was calculated. The temperatures of the heat source and the heat sink were 178 and 145 °C, respectively, and the electric field was varied between the high value of 3.2 MVm^{-1} and the low value of 0.4 MVm^{-1}. The frequency of the device was 0.128 Hz and the electrical power output was 40 mW. The heat input of the electrical heater was 9.6 W and, accordingly, the efficiency of the device was 0.42 %.

Olsen et al. [81] performed an additional experiment, similar to the one described above; however, in this case a polymer pyroelectric material P(VDF-TrFE) was used. The polymer material was formed in a spiral stack that was constructed from layers of the polymer pyroelectric material. Between the layers of the polymer material a nylon separator was inserted to create a void for the fluid flow (Fig. 10.23). The results of the experiment showed that in comparison with the PLZT-based pyroelectric stack, the produced electrical power of 30 J·$cycle^{-1}$/L of the pyroelectric material was almost an order smaller. The temperatures of the heat sink and the heat source were 23 and 67 °C, respectively, and the electric field was varied between approximately 22.5 and 53 MVm^{-1}.

Table 10.11 Characteristics of the pyroelectric energy converters working under the pyroelectric Ericsson cycle

Authors	Material	T_{low} (°C)	T_{high} (°C)	E_{low} (MVm^{-1})	E_{high} (MVm^{-1})	f (Hz)	Electrical output	η_E	References
Olsen et al.	PLZT	145	178	0.4	3.2	0.13	17 Wl^{-1}	0.42	[67]
Olsen et al.	P(VDF-TrFE) 73/27	23	67	22.5	53	/	30 Wl^{-1} cycle^{-1}	/	[81]
Nquyen et al.	P(VDF-TrFE) 60/40	70.5	85.3	20.2	74	0.12	10.7 Wl^{-1}	0.053	[62]
Cha et al.	P(VDF-TrFE) 56/44	40	100	20	50	1	110 Wl^{-1}	/	[80]

More than 25 years after the work of Olsen and his coworkers, Nquyen et al. [62] built a device similar to that of the Olsen group. In general, the design of the device was almost identical to that of the Olsen group, with only small differences. As the pyroelectric material the P(VDF-TrFE) 60/40 polymer was chosen, since it possesses a low-temperature transition and has a very high dielectric strength. The polymer films were attached to a rectangular supporting structure made of mica plate and 38 of such pyroelectric elements were put on top of each other. Between two individual pyroelectric elements Teflon strips were inserted to create a void for the fluid flow. To simulate the heat source an electrical heater was used. For the heat sink a copper tube bent into a helical shape, through which water was flowing, was applied. A maximum power density of 10.7 W/L of the pyroelectric material was obtained at a frequency of 0.12 Hz and an oscillating temperature between 70.5 and 85.2 °C of the working fluid. The pyroelectric material was subjected to a low electric field of 20.2 MVm^{-1} and to a high electric field of 73.9 MVm^{-1}. An efficiency of 0.053 % was achieved.

In Table 10.11, the main characteristics of the above-presented pyroelectric energy converters working according to the pyroelectric Ericsson cycle are gathered. However, it was only Olsen et al. [67] who provided the information about the absolute power output, which amounted to 30 mW. This means that approximately 1,500 devices presented by Olsen and his coworkers would be needed to power one 50-W light bulb. Nevertheless, with new discoveries in material science and the further development of the systems design, the performance of the devices could be improved. Theoretical estimations of the performance of an ideal pyroelectric energy converter indicate that the device's efficiency could come close to the Carnot efficiency of a heat engine [67].

From the above results, it is obvious that so far research has only tried to prove the concept of pyroelectric harvesting. For efficient pyroelectric energy conversion, some of the main issues that should be considered in future research and development are:

- A large temperature span between the heat source and heat sink is required in order to produce good thermodynamic efficiency from the power cycle. As a result, this may require the layering of pyroelectric materials in the direction of the temperature gradient of the fluid flow. Today, this is a well-known technique in magnetic refrigeration at/near ambient temperature.

- The problem of low frequency in most case relates to an inefficient heat-transfer mechanism. Therefore, like in the case of magnetic refrigeration and magnetic power generation, we should consider the application of a thermal diode mechanism combined with a pyroelectric material. Such a principle has recently been included in a patent application by the authors of this book in collaboration with the Josef Stefan Institute from Ljubljana, Slovenia (see also the chapter on thermal diode mechanisms). This kind of mechanism could provide operating frequencies in the range 20–200 Hz.
- All mechanical movements (moving parts, except for those that can be used on the microscale as thermal diodes) should be strictly avoided.
- Besides the Ericsson (Olsen) cycle, other thermodynamic cycles that employ regeneration should also be considered and evaluated.

10.2 Barocaloric Energy Conversion

In this subsection, we discuss barocaloric energy conversion. The domain of barocaloric energy conversion has not so far received much attention from the scientific community, at least to the same extent as magnetocaloric and electrocaloric energy conversion. For example, up to 2014, a total of 38 papers were found when searching for the word "barocaloric" in the online database Web of Science. In contrast, when searching for the words "electrocaloric" and "magnetocaloric", 273 and 2,876 results were found, respectively. Therefore, only a brief description of the barocaloric effect is given in this chapter. In addition, some materials that exhibit the barocaloric effect and their properties are presented.

10.2.1 Introduction to the Barocaloric Effect and Barocaloric Materials

Barocaloric energy conversion is based on the so-called barocaloric effect. The barocaloric effect is a property of some magnetic materials and is expressed as the temperature change of the material upon varying the pressure [82]. Analogously with the magnetocaloric effect, the material exhibiting the barocaloric effect heats up when subjected to an increase in the external pressure and cools down as the external pressure is removed.

The barocaloric effect can be expressed as the adiabatic temperature or isothermal entropy change of the material [82]. The isothermal entropy change can be written as [82]:

$$\Delta s_{\text{ist}} = \int_{p_1}^{p_2} \left(\frac{\partial s}{\partial p}\right)_T \text{d}p \tag{10.17}$$

where p is the pressure. Furthermore, the partial derivate of the entropy with respect to the pressure can be expressed using the Maxwell relation $(\partial s/\partial p)_T = -(\partial V/\partial T)_p$ and the isothermal entropy change can be expressed as [83]:

$$\Delta s_{\text{ist}} = -\int_{p_1}^{p_2} \left(\frac{\partial V}{\partial T}\right)_p \text{d}p \tag{10.18}$$

Moreover, the adiabatic temperature change can be expresses as:

$$\Delta T_{\text{ad}} = -\int_{p_1}^{p_2} \frac{T}{c_p}\left(\frac{\partial V}{\partial T}\right)_p \text{d}p \tag{10.19}$$

where c_p is the specific heat at constant pressure. However, the volume changes of barocaloric materials are very small and it is difficult to measure them accurately [82]. Therefore, information about the adiabatic temperature change and the isothermal entropy change of the material calculated on the basis of Eqs. (10.18) and (10.19) would not be reliable [82]. Instead, direct measurement techniques, for example, the calorimetric technique, are used to measure barocaloric effects [82]. In Table 10.12, some barocaloric materials with their properties are presented. The maximum specific cooling capacity ($q_{r,\max}$) is defined with Eq. (10.10). It is clear from Table 10.12 that some barocaloric materials, for example, NiMnIn [89], have good cooling properties at room temperature. NiMnIn possesses its maximum specific cooling capacity in the range of bulk and thick-film ceramics presented in Sect. 10.1.2. However, to the best of our knowledge, no energy conversion device based on the barocaloric effect has been presented so far. It was only Oliveira et al. [90] who in their paper discussed a possible concept of a barocaloric cooling

Table 10.12 Barocaloric properties of some materials

Material	ΔT_{ad} (K)	Δs_{is} ($\text{Jkg}^{-1}\text{K}^{-1}$)	T_{ref}^{a} (K)	p (GPa)	$q_{r,\max}$ (Jkg^{-1})	Measurement procedure	References
$Pr_{0.66}La_{0.34}NiO_3$	0.1	/	350	0.5	/	Direct	[84]
HoAs	0.35	/	7	0.3	/	Direct	[85]
CeSb	2	/	21	0.52	/	Direct	[86]
$Ce_3Pd_{20}Ge_6$	0.75	/	4.4	0.3	/	Direct	[87]
$EuNi_2(Si_{0.15}Ge_{0.85})_2$	0.4	37	60	0.34	2,213	Direct	[88]
NiMnIn	4.5	24.4	273	0.26	6,606	Direct	[89]

[a] T_{ref}—temperature at which ΔT_{ad} and Δs_{is} were measured

device. However, as was emphasized in the paper, the concept was only used to illustrate the barocaloric refrigeration process and many technical problems should be solved before building a real prototype [90].

10.3 Elastocaloric Energy Conversion

In this subsection, the elastocaloric effect, which is associated with shape-memory and superelastic effect, is discussed. The fundamental elastocaloric (superelastic) behaviour is shown and the basic equations for the estimation of the elastocaloric effect are introduced. Finally, so far evaluated elastocaloric materials and its properties are presented and the some design concepts of the elastocaloric cooling device are reviewed.

10.3.1 Introduction to the Elastocaloric Effect

The elastocaloric effect (EsCE), sometimes referred to as the thermoelastic effect, is a physical phenomenon associated with the entropy and/or temperature changes of certain materials subjected to an external mechanical stress. The EsCE is closely related to the well-known shape-memory or, better, superelastic effect [91]. In general, we distinguish between shape-memory alloys and polymers.

Shape-memory alloys undergo a reverse martensitic transformation between the austenitic (cubic) and the martensitic (monoclinic) phases (see Fig. 10.24). The transformation can be induced by a temperature change or by the application of an external stress. The temperature-induced transformation (shape-memory effect) is observed by cooling the material from the austenitic to the (twinned) martensitic phase, where it becomes softer and easier to deform. When reheated back it undergoes the reverse martensite-austenite temperature-induced transformation and is restored to its original shape, acting as though it has memory. Similarly, a stress-induced transformation (superelastic effect) is observed when the material is loaded in the austenitic phase. During loading, at a certain critical stress, the transformation to the martensitic phase begins, which causes large strains in the material at almost constant stress, until it fully transforms to a de-twinned martensite. Upon unloading, reverse martensite-austenite transformations occur and the material returns to its original state. Both transformations are associated with a hysteresis in the transformation region. A schematic representation of the shape-memory and superelastic effects is shown in Fig. 10.24, together with the typical shape-memory (down-left) and superelastic (down-right) behaviours.

In order to generate the EsCE a stress-induced transformation is required. The stress-induced forward austenitic-martensitic transformation is exothermic, while the reverse transformation is endothermic. In some alloys, the latent heat released (absorbed) during the transformation can be as high as 20 Jg^{-1} [92]. If the

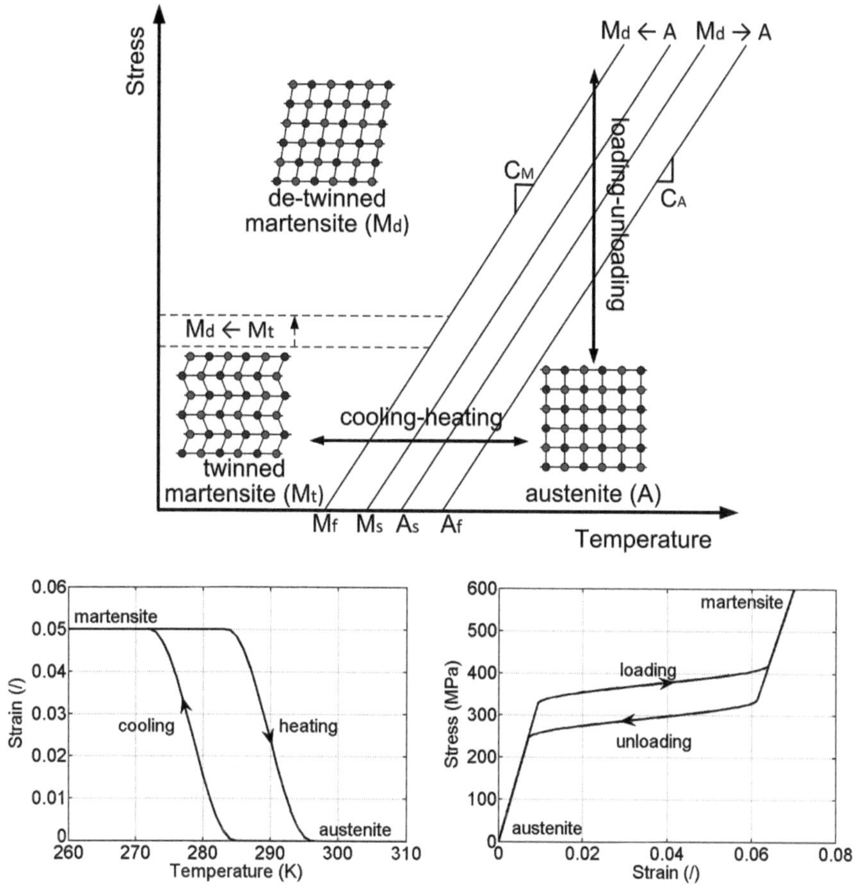

Fig. 10.24 Schematic representation of the shape-memory and superelastic effects in the stress–temperature phase diagram (*up*), together with typical shape-memory (*down-left*) and superelastic (*down-right*) behaviours

transformation is performed quickly enough by means of a high stress or strain rate (adiabatically or close to it), the material heats up when loaded (transformed from austenite to martensite) and cools down when unloaded (transformed from martensite to austenite).

Even though the first shape-memory effect was observed already in the 1930s in Au–Cd and Cu–Zn alloys, the breakthrough occurred in 1963, when what is still the most widely used Ni–Ti shape-memory alloy was discovered. In subsequent years, a number of shape-memory and superelastic alloys were developed and characterized, which can, in general, be divided into three groups [91]:

- Ni–Ti-based alloys doped with Cu, Co, Nb, Pd, etc.
- Cu-based alloys doped with Al, Ni, Zn, Mn, etc.
- Fe-based alloys doped with Mn, Si, Ni, Pd, etc.

These alloys are already applied in different applications in various sectors where the recovery (shape-memory effect) and superelasticity can be used. These are, for example, different types of linear and rotary actuators applied in the automotive, aerospace and robotics industries; in medical applications, where due to its biocompatibility Ni–Ti is commonly used as orthopaedic implants, stents and dental braces; as a heat engine; and in various different engineering (fasteners, seals, connectors, clamps, valves, temperature controls, etc.) and practical applications (glasses frames, underwire bras, cellphone retractable antenna, etc.) [93]. The heat engine was one of the first applications of Ni–Ti alloys as early as the 1970s. It converts heat energy into mechanical and electrical energy by exploiting the recovery force of the shape-memory material. It consists of a shape-memory material (usually in the form of a wire or coil or better many coils), which alternatively moves (rotates) between high- and low-temperature media. The wires entering the low-temperature medium become soft and are forced to bend. In the next step, they enter the high-temperature medium, where they transform back to the austenite phase, tend to straighten back to the original shape, and due to the recovery force they generate the mechanical energy (movement).

The first continuously operated shape-memory-alloy heat engine was developed by R.M. Banks from the Lawrence Berkeley Laboratory of the University of California in 1973. In later years, various different design concepts were developed and tested. However, in order to be practical, these engines would typically be required to produce hundreds of kilowatts. This was proven to be much more difficult than expected, as for all of these devices the scale-up stopped at about one kilowatt. In other words, the inventions at that time failed for both engineering and economic reasons. However, the shape-memory-based heat engine has various advantages, from its simple design, potential passive operation (only requires the appropriate high- and low-temperatures media) and applicability to different operating conditions, so there might still be a promising and intriguing future for this technology. More details on shape-memory heat engines can be found in, e.g., [94–96].

On the other hand, as explained in [97], polymeric materials are also capable of shape-memory and superelastic effects, although the mechanisms responsible differ dramatically from those of metal alloys. In contrast to shape-memory alloys, polymers do not achieve any shape-memory and superelastic behaviour through the martensitic transformation, but rather through a variety of physical means; the underlying, very large extensibility being derived from the intrinsic elasticity of polymeric networks. This is related to the entropy change caused by the second-order phase transition between the soft and hard phases due to the polymeric chain orientation change. A typical shape-memory polymer is natural rubber. In general, they exhibit much larger transformation strains (up to 200 % or more), lower densities and thermal conductivities as well as lower costs compared to the shape-memory alloys and are therefore more suitable for various applications. However, their potential for application in an elastocaloric cooling (or heat pumping) device is somewhat lower (due to the above-mentioned properties and also the lower EsCE), and thus further focus will be put on the shape-memory alloys.

Even though the shape-memory and superelastic effects are known for more than 50 years and already applied in various areas, the EsCE was not seriously proposed for application in a cooling (or heat pumping) device until recently.

In general, the EsCE is analogous to the MCE and the ECE in detail presented in the previous chapters. Here, instead of a magnetic field or an electrical field change, a change of the mechanical stress is required, which causes strain changes in the materials (instead of magnetization and polarization changes). Therefore, the Maxwell relation can be written in the following form:

$$\left(\frac{\partial s}{\partial \sigma}\right)_T = \frac{1}{\rho}\left(\frac{\partial \varepsilon}{\partial T}\right)_\sigma \tag{10.20}$$

where T, s, σ, ε and ρ are the temperature, entropy, stress, strain and density, respectively. Furthermore, the isothermal entropy change and the adiabatic temperature change of the EsCE loaded from the initial (σ_1) to the final stress (σ_2) can be written as follows:

$$\mathrm{d}s = \frac{1}{\rho}\left(\frac{\partial \varepsilon}{\partial T}\right)_\sigma \mathrm{d}\sigma \rightarrow \Delta s_{\mathrm{ist}} = \frac{1}{\rho}\int_{\sigma_1}^{\sigma_2}\left(\frac{\partial \varepsilon}{\partial T}\right)_\sigma \mathrm{d}\sigma \tag{10.21}$$

$$\begin{aligned}\mathrm{d}T &= \frac{T}{c_\sigma}\mathrm{d}s \rightarrow \mathrm{d}T = -\frac{1}{\rho}\frac{T}{c_\sigma}\left(\frac{\partial \varepsilon}{\partial T}\right)_\sigma \mathrm{d}\sigma \rightarrow \Delta T_{\mathrm{ad}} \\ &= -\frac{1}{\rho}\int_{\sigma_1}^{\sigma_2}\frac{T}{c_\sigma}\left(\frac{\partial \varepsilon}{\partial T}\right)_\sigma \mathrm{d}\sigma\end{aligned} \tag{10.22}$$

Figure 10.25 shows a schematic example of the first-order elastocaloric phase transition at two different cases, both isothermally. Figure 10.25a shows a stress–strain dependence diagram for different applied stresses during heating and cooling, while Fig. 10.25b shows a strain–temperature dependence diagram at different temperatures during loading and unloading. The great majority of so far known elastocaloric materials undergo a first-order phase transition, which is associated with hysteresis behaviour by different mechanical behaviours of the material during loading and unloading or heating and cooling, as shown in Fig. 10.25. This further results in different EsCE properties (Eqs. 10.20 and 10.21) for forward and reverse transformations. The differences are the irreversibility losses associated with the hysteresis behaviour of the transformation, which represented an additional heat source of the transformation [98].

In addition to the Maxwell relation, the EsCE is often described and characterized with the Clausius–Clapeyron relation, written as follows:

$$\left(\frac{\partial \sigma_{\mathrm{crit}}}{\partial T}\right) = -\rho\left(\frac{\partial s}{\partial \varepsilon}\right) \tag{10.23}$$

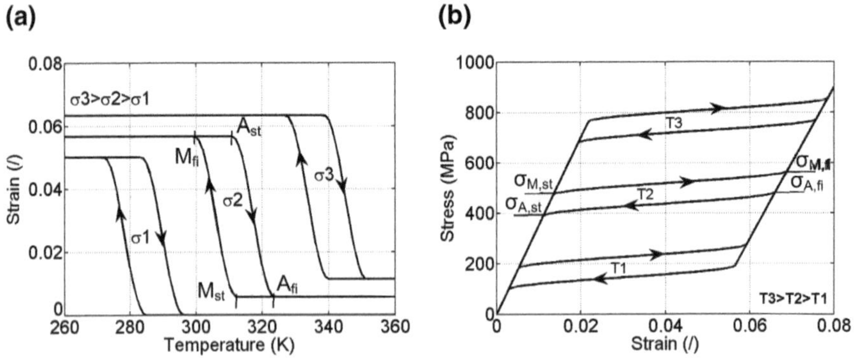

Fig. 10.25 **a** Strain–temperature dependence diagram for different applied stresses during heating and cooling, **b** Stress–strain dependence diagram at different temperatures during loading and unloading

where $(\partial\sigma_{\text{crit}}/\partial T = C_M \text{ and } C_A)$ is the slope of the transformation line in the stress–temperature phase diagram, as schematically noted in Fig. 10.24. These transformation lines are constructed based on the critical stresses (or temperatures) required to start (and finish) the transformation at a particular temperature (stress)—see Fig. 10.25. Equation (10.23) can be further used to estimate the isothermal entropy change and the adiabatic temperature change, as shown in Eqs. (10.21) and (10.22) for the Maxwell relation.

As explained in, e.g., Moya et al. [99], the Clausius–Clapeyron relation is only valid for the first-order transitions and does not account for the other caloric effects that arise in each of the interconvertible phases. However, these effects are normally much smaller than caloric effects during transitions and are often neglected and under this assumption this relation is equivalent to the Maxwell relation.

10.3.2 Elastocaloric Materials

In general, all superelastic materials can also be considered as elastocaloric materials. Of course, the transformation temperatures (in particular the austenitic finish temperature—A_{fi}) should be above the refrigeration (heat sink) temperature, but at the same time not too high in order to decrease the required stress to start the transformation. However, as explained in Moya et al. [99], the first EsCE was detected in Indian rubber, which is a shape-memory polymer, already in early nineteenth Century. That makes the EsCE the oldest-known ferroic caloric effect. Some 50 years later, Joule reported on the EsCE in some metals and dry woods, where he observed small temperature changes under the applied stress caused by the reversible elastic heat. However, the first analysis of the EsCE and the related adiabatic temperature changes on the nowadays known superelastic (shape-memory) materials were performed in 1980s, mostly on Cu-based and Ni–Ti alloys [100–103]. They measured up to 12 K of adiabatic temperature changes for the Ni–Ti alloy [102] and Cu–Zn–Sn [100] and up to 15 K for the Cu–Al–Ni alloy [101]. However, their research was not focused on the application of these materials in a cooling or heat pumping device, but rather on an analysis of different loading conditions and the origin of these thermal effects. The first EsCE proposed for cooling applications was published in 1992 by Nikitin et al. [104] in a polycrystalline $Fe_{49}Rh_{51}$ alloy. They measured a negative adiabatic temperature change of 5.2 K under the stress removal and indirectly estimated 8.7 K using the Clausius–Clapeyron relation.

Between 2008 and 2014, the group at the University of Barcelona published three papers on the EsCE of single-crystalline $Cu_{68}Zn_{16}Al_{16}$, where they measured a negative adiabatic temperature change of about 6 K in the temperature range between 200 and 350 K [105], estimated it to 15 K using the Clausius–Clapeyron relation [106], as well as analysing the homogeneity of the EsCE distribution along the sample [107]. In 2012, Cui et al. [108] presented the EsCE on polycrystalline Ni–Ti wires. They measured a positive adiabatic temperature change of 25.5 K during the mechanical loading and a negative one of 17 K during the unloading. Furthermore, Ossmar et al. [109] measured 16 K of negative adiabatic temperature change under stress removal (unloading) on $Ni_{50.4}Ti_{49.6}$ thin films. Bechtold et al. [110] analysed the EsCE and the functional stability of the $Ti_{54.9}Ni_{32.5}Cu_{12.6}$ thin film and compared it with the $Ni_{50.4}Ti_{49.6}$ alloy. It was shown that adding Cu to the Ni–Ti alloy strongly increases the stability of the superelastic behaviour with no training effect during the initial cycles, but it also reduces the EsCE (negative adiabatic change of 6 K).

It should be noted that all the above-presented elastocaloric materials exhibit the first-order phase transition, which is related to the hysteresis behaviour and irreversibilities (the difference between the positive adiabatic temperature change and the negative one). However, Xiao et al. [111] showed the EsCE of the single-crystalline $Fe_{68.8}Pd_{31.2}$ with the second-order, continuous structural phase transition and near-zero hysteresis. They measured an adiabatic temperature change of about 2.5 K, but a near-zero hysteresis can be its great advantage. Furthermore, Guyomar et al. [112] analysed the EsCE of the shape-memory polymer natural rubber and measured an adiabatic temperature change of 10 K. The details about some of the most interesting elastocaloric materials and the related EsCE are collected in Table 10.13. It should be noted that in recent years a lot of work was performed on the thermal effects associated with the superelastic (elastocaloric) behaviour, where they mostly studied the impact of the strain rate on the temperature changes of the material; however, in most cases on the virgin samples (with not fully repeatable behaviour) as well as not adiabatically (with lower strain-rates), e.g., [113, 114]. Therefore, these studies as well as the earlier similar studies [100–103] are not presented in Table 10.13.

Table 10.13 The details on the EsCE of some elastocaloric materials

Material	T (K)	ΔT_{ad} (K) measured	ΔT_{ad} (K) estimated	Δs_{is} (Jkg^{-1}K^{-1}) estimated	Applied field ($\Delta\sigma$ (Mpa) or $\Delta\varepsilon$ (%))	References
$Fe_{49}Rh_{51}$	305 K	5.2 K (unloading)	8.7 K (unloading)	13 Jkg^{-1}K^{-1} (unloading)	529 MPa	[104]
$Cu_{68}Zn_{16}Al_{16}$	200–350 K	5 K (at 225 K) and 7 K (at 325 K) both for unloading	15 K (at 300 K) for loading	16 Jkg^{-1}K^{-1} (between 225 and 325 K) for loading	275 MPa	[105], [106]
Ni-Ti	295 K	25.5 K (loading), 17 K (unloading)	/	/	650 MPa	[108]
$Ni_{50.4}Ti_{49.6}$	Near room temp.	16 K (unloading)	/	/	5 %	[109]
$Ti_{54.9}Ni_{32.5}Cu_{12.6}$	346 K	5 K (loading) 6 K (unloading)	/	/	2 %	[110]
$Fe_{68.8}Pd_{31.2}$	260 K	2.5 K (loading and unloading)	3.5 K (loading and unloading)	/	100 MPa	[111]
Natural rubber (ASTM D200 AA)	297 K	10 K (loading)	/	/	70 %	[112]

10.3.3 Review of Design Concepts

Even though the EsCE is the oldest-known ferroic (caloric) effect, its usage in prototypes for cooling (or heat pumping) devices is in the earliest stage of development. However, the first and still the only up-to-date presented prototype exploiting the EsCE was developed by DeGregoria [115] in 1994. It is based on a shape-memory polymer (natural rubber) in the form of thin layers (foils) constructed into a regenerator with a thin spacing for the counter-flow heat-transfer fluid (analogous to the AMR in magnetic refrigeration). In order to balance and recover the force, four such regenerators were applied into the acentric rotary system, where two of them were loaded (stretched) and two unloaded (unstretched) at the particular time. Air was used as the heat-transfer fluid. The authors published 19 K of temperature span (with the hot-side temperature of 298 K). Unfortunately, other more detailed results are not available, most probably due to the technical problem associated with the fatigue life of the rubber, especially at the parts connected to the supporting structures, where the stresses are the largest [116].

In 2012, a group from the University of Maryland [117] patented various different design ideas exploiting the EsCE, from a single-stage device with a single elastocaloric material to more sophisticated concepts with elastocaloric porous structures using a heat-transfer fluid. Some of them are already in the prototyping phase, but more detailed information is not yet available.

More detailed experimental or theoretical (numerical) analyses of the performance characteristics of the elastocaloric device in order to estimate the available cooling power, temperature span and efficiency, were not yet performed. Since the EsCE is significantly larger compared to the MCE and in the most cases also the ECE, its potential for application in a cooling (or heat pumping) device is therefore very large. According to a report by the US Department of Energy on alternative cooling technologies from 2014, it has the largest potential among all the alternative non-vapour-compression HVAC technologies [118]. Currently, the major limitation of this technology is in the limited fatigue life of the elastocaloric (superelastic) materials. It is currently not possible to apply up to 108 loading cycles, which would be required for 10 years of operating lifetime for such a device, without causing cracks and failure of the material. However, some recent results on the elastocaloric Ni–Ti–Cu–Co alloy are very promising and show no fatigue, even after 106 cycles, which is an important step towards the application [119]. Furthermore, in order to increase the efficiency, the mechanical hysteresis should be reduced as much as possible. This would further reduce the required stress for the transformation and therefore allow the simpler application of this technology on the micro as well as larger scales.

References

1. Kao KC (2004) Dielectric phenomena in solids. Academic Press, London
2. Young JS (2012) Indirect measurement of the electrocaloric effect. Dissertation, University of Cambridge
3. Pirc R, Kutnjak Z, Blinc R et al (2011) Electrocaloric effect in relaxor ferroelectrics. J Appl Phys 110(7):074113
4. Valant M (2012) Electrocaloric materials for future solid-state refrigeration technologies. Prog Mater Sci 57(6):980–1009
5. Lu S, Rožič B, Zhang Q et al (2010) Comparison of directly and indirectly measured electrocaloric effect in relaxor ferroelectric polymers. Appl Phys Lett 97(20):202901
6. Scott JF (2011) Electrocaloric materials. Annu Rev Mater Res 41(1):229–240
7. Wiseman G, Kuebler JK (1963) Electrocaloric effect in ferroelectric rochelle salt. Phys Rev 131(5):2023–2027
8. Mischenko A, Zhang Q, Scott J et al (2006) Giant electrocaloric effect in thin-film $PbZr_{0.95}Ti_{0.05}O_3$. Science 311:1270–1271
9. Ožbolt M, Kitanovski A, Tušek J et al (2014) Electrocaloric refrigeration: thermodynamics, state of the art and future perspectives. Int J Refrig 40:174–188
10. Niemann R, Heczko O, Schultz L et al (2014) Inapplicability of the Maxwell relation for the quantification of caloric effects in anisotropic ferroic materials. Int J Refrig 37:281–288
11. Goupil FL, Berenov A, Axelsson A-K et al (2012) Direct and indirect electrocaloric measurements on <001> $PbMg_{1/3}Nb_{2/3}O_3$-$30PbTiO_3$ single crystals. J Appl Phys 111 (12):124109
12. Rožič B, Malič B, Uršič H et al (2010) Direct measurements of the giant electrocaloric effect in soft and solid ferroelectric materials. Ferroelectrics 405(1):26–31
13. Kutnjak Z, Rožič B (2014) Indirect and direct measurements of the electrocaloric effect. In: Correia T, Zhang Q (eds) Electrocaloric materials. Springer, Berlin Heidelberg, pp 147–182
14. Moulson AJ, Herbert JM (2003) Electroceramics: materials, properties, applications. Wiley, New York
15. Qian X-S, Ye H-J, Zhang Y-T et al (2014) Giant electrocaloric response over a broad temperature range in modified $BaTiO_3$ ceramics. Adv Funct Mater 24(9):1300–1305
16. Zheng X-C, Zheng G-P, Lin Z et al (2012) Electro-caloric behaviors of lead-free $Bi_{0.5}Na_{0.5}TiO_3$-$BaTiO_3$ ceramics. J Electroceram 28(1):20–26
17. Jiang X, Luo L, Wang B et al (2014) Electrocaloric effect based on the depolarization transition in $_{(1-x)}Bi_{0.5}Na_{0.5}TiO_{3-x}KNbO_3$ lead-free ceramics. Ceram Int 40(2):2627–2634
18. Bai Y, Zheng G-P, Shi S-Q (2011) Abnormal electrocaloric effect of $Na_{0.5}Bi_{0.5}TiO_3$–$BaTiO_3$ lead-free ferroelectric ceramics above room temperature. Mater Res Bull 46(11):1866–1869
19. Rožič B, Kosec M, Uršič H et al (2011) Influence of the critical point on the electrocaloric response of relaxor ferroelectrics. J Appl Phys 110(6):064118
20. Rožič B, Kutnjak Z (2009) Giant electrocaloric effect in ferroelectric relaxor materials. In: Proceedings of the 2009 SEM annual conference and exposition on experimental and applied mechanics, Albuquerque New Mexico
21. Shaobo L, Yanqiu L (2004) Research on the electrocaloric effect of PMN/PT solid solution for ferroelectrics MEMS microcooler. Mater Sci Eng, B 113(1):46–49
22. Xiao D, Wang Y, Zhang R et al (1998) Electrocaloric properties of $(1-x)Pb(Mg_{1/3}Nb_{2/3})O_3$-$xPbTiO_3$ ferroelectric ceramics near room temperature. Mater Chem Phys 57(2):182–185
23. Valant M, Dunne LJ, Axelsson A-K et al (2010) Electrocaloric effect in a ferroelectric $Pb(Zn_{1/3}Nb_{2/3})O_3$-$PbTiO_3$ single crystal. Phys Rev B 81(21):214110
24. Bai Y, Zheng G-P, Ding K et al (2011) The giant electrocaloric effect and high effective cooling power near room temperature for BaTiO3 thick film. J Appl Phys 110(9):094103
25. Kar-Narayan S, Mathur N (2010) Direct and indirect electrocaloric measurements using multilayer capacitors. J Phys D Appl Phys 43(3):032002

26. Kar-Narayan S, Crossley S, Moya X et al (2013) Direct electrocaloric measurements of a multilayer capacitor using scanning thermal microscopy and infra-red imaging. Appl Phys Lett 102(3):032903
27. Kar-Narayan S, Mathur N (2009) Predicted cooling powers for multilayer capacitors based on various electrocaloric and electrode materials. Appl Phys Lett 95(24):242903
28. Shebanovs L, Borman K, Lawless W et al (2002) Electrocaloric effect in some perovskite ferroelectric ceramics and multilayer capacitors. Ferroelectrics 273(1):137–142
29. Rožič B, Malič B, Uršič H et al (2012) The giant electrocaloric effect in inorganic and organic ferroelectric relaxor systems. Ferroelectrics 430(1):98–102
30. Peng B, Fan H, Zhang Q (2013) A giant electrocaloric effect in nanoscale antiferroelectric and ferroelectric phases coexisting in a relaxor $Pb_{0.8}Ba_{0.2}ZrO_3$ thin film at room temperature. Adv Funct Mater 23(23):2987–2992
31. Lu S, Rožič B, Zhang Q et al (2010) Organic and inorganic relaxor ferroelectrics with giant electrocaloric effect. Appl Phys Lett 97:162904
32. Saranya D, Chaudhuri A, Parui J et al (2009) Electrocaloric effect of PMN-PT thin films near morphotropic phase boundary. Bull Mater Sci 32(3):259–262
33. Mischenko A, Zhang Q, Whatmore R et al (2006) Giant electrocaloric effect in the thin film relaxor ferroelectric $0.9PbMg_{1/3}Nb_{2/3}O_3$-$0.1PbTiO_3$ near room temperature. Appl Phys Lett 89(24):242912
34. Correia TM, Young JS, Whatmore RW et al (2009) Investigation of the electrocaloric effect in a $PbMg_{2/3}Nb_{1/3}O_3$-$PbTi0_3$ relaxor thin film. Appl Phys Lett 95(18):182904
35. Correia T, Kar-Narayan S, Young J et al (2011) PST thin films for electrocaloric coolers. J Phys D Appl Phys 44(16):165407
36. Parui J, Krupanidhi SB (2008) Electrocaloric effect in antiferroelectric $PbZrO_3$ thin films. Phys Status Solidi (RRL)–Rapid Res Lett 2(5):230–232
37. Tušek J, Kitanovski A, Zupan S et al (2013) A comprehensive experimental analysis of gadolinium active magnetic regenerators. Appl Therm Eng 53(1):57–66
38. Li X, Qian X-S, Haiming G et al (2012) Giant electrocaloric effect in ferroelectric poly (vinylidenefluoride-trifluoroethylene) copolymers near a first-order ferroelectric transition. Appl Phys Lett 101(13):132903
39. Lu SG, Rožič B, Zhang QM et al (2011) Enhanced electrocaloric effect in ferroelectric poly (vinylidene-fluoride/trifluoroethylene) 55/45 mol % copolymer at ferroelectric-paraelectric transition. Appl Phys Lett 98(12):122906
40. Liu P, Wang J, Meng X et al (2010) Huge electrocaloric effect in Langmuir-Blodgett ferroelectric polymer thin films. New J Phys 12(2):023035
41. Neese B, Lu SG, Chu B et al (2009) Electrocaloric effect of the relaxor ferroelectric poly (vinylidene fluoride-trifluoroethylene-chlorofluoroethylene) terpolymer. Appl Phys Lett 94 (4):042910
42. Chen X-Z, Li X, Qian X-S et al (2013) A polymer blend approach to tailor the ferroelectric responses in P(VDF–TrFE) based copolymers. Polymer 54(9):2373–2381
43. Barclay JA, Steyert WA (1982) Active magnetic regenerator. US Patent No. 4.332.135
44. Yu B, Liu M, Egolf PW et al (2010) A review of magnetic refrigerator and heat pump prototypes built before the year 2010. Int J Refrig 33(6):1029–1060
45. Sinyavsky Y, Brodyansky V (1992) Experimental testing of electrocaloric cooling with transparent ferroelectric ceramic as a working body. Ferroelectrics 131(1):321–325
46. Tomc U, Kitanovski A, Ožbolt M et al (2013) Method for electrocaloric energy conversion. Patent Application P003397EP
47. Ožbolt M, Kitanovski A, Tušek J et al (2014) Electrocaloric vs. magnetocaloric energy conversion. Int J Refrig 37:16–27
48. Guo D, Gao J, Yu Y-J et al (2014) Design and modeling of a fluid-based micro-scale electrocaloric refrigeration system. Int J Heat Mass Transfer 72:559–564
49. Gu H, Craven B, Qian X et al (2013) Simulation of chip-size electrocaloric refrigerator with high cooling-power density. Appl Phys Lett 102(11):112901

50. Epstein RI, Malloy KJ (2009) Electrocaloric devices based on thin-film heat switches. J Appl Phys 106(6):064509
51. Karmanenko SF, Pakhomov OV, Prudan AM et al (2007) Layered ceramic structure based on the electrocaloric elements working as a solid state cooling line. J Eur Ceram Soc 27 (8):3109–3112
52. Es'kov A, Karmanenko S, Pakhomov O et al (2009) Simulation of a solid-state cooler with electrocaloric elements. Phys Solid State 51(8):1574–1577
53. Jia Y, Ju YS (2012) A solid-state refrigerator based on the electrocaloric effect. Appl Phys Lett 100(24):242901
54. Gu H, Qian X, Li X et al (2013) A chip scale electrocaloric effect based cooling device. Appl Phys Lett 102(12):122904
55. Chukka R, Shannigrahi S, Chen L (2012) Investigations of cooling efficiencies in solid-state electrocaloric device. Integr Ferroelectr 133(1):3–8
56. Plaznik U, Kitanovski A, Malič B et al (2014) Small scale electrocaloric cooling device with an active heat regenerator. In: Abstracts of the 6th IIF-IIR international conference on magnetic refrigeration, Victoria, 7–10 Sept 2014
57. Lang SB (2005) Pyroelectricity: from ancient curiosity to modern imaging tool. Phys Today 58(8):31–36
58. Lang SB (2004) A 2,400 year history of pyroelectricity: from ancient Greece to exploration of the solar system. Br Ceram Trans 103(2):65–70
59. Olsen RB (1982) Ferroelectric conversion of heat to electrical energy a demonstration. J Energy 6(2):91–95
60. Kouchachvili L, Ikura M (2007) Pyroelectric conversion-effects of P (VDF–TrFE) preconditioning on power conversion. J Electrostat 65(3):182–188
61. Lee FY, Jo HR, Lynch CS et al (2013) Pyroelectric energy conversion using PLZT ceramics and the ferroelectric–ergodic relaxor phase transition. Smart Mater Struct 22(2):025038
62. Nguyen H, Navid A, Pilon L (2010) Pyroelectric energy converter using co-polymer P (VDF-TrFE) and Olsen cycle for waste heat energy harvesting. Appl Therm Eng 30 (14):2127–2137
63. Mane P, Xie J, Leang KK et al (2011) Cyclic energy harvesting from pyroelectric materials. IEEE Trans Ultrason Ferroelectr Freq Control 58(1):10–17
64. Kumar P, Sharma S, Thakur OP et al (2004) Dielectric, piezoelectric and pyroelectric properties of PMN–PT (68:32) system. Ceram Int 30(4):585–589
65. Cuadras A, Gasulla M, Ferrari V (2010) Thermal energy harvesting through pyroelectricity. Sens Actuators, A 158(1):132–139
66. Navid A, Lynch CS, Pilon L (2010) Purified and porous poly (vinylidene fluoride-trifluoroethylene) thin films for pyroelectric infrared sensing and energy harvesting. Smart Mater Struct 19(5):055006
67. Olsen RB, Briscoe JM, Bruno DA et al (1981) A pyroelectric energy converter which employs regeneration. Ferroelectrics 38(1):975–978
68. Zhang Q, Agbossou A, Feng Z et al (2011) Solar micro-energy harvesting with pyroelectric effect and wind flow. Sens Actuators, A 168(2):335–342
69. Ravindran SKT, Huesgen T, Kroener M et al (2011) A self-sustaining pyroelectric energy harvester utilizing spatial thermal gradients. In: 2011 16th international solid-state sensors, actuators and microsystems conference (TRANSDUCERS)
70. Ravindran SKT, Kroener M, Woias P (2012) A bimetallic micro heat engine for pyroelectric energy conversion. Procedia Eng 47:33–36
71. Sebald G, Lefeuvre E, Guyomar D (2008) Pyroelectric energy conversion: optimization principles. IEEE Trans Ultrason Ferroelectr Freq Control 55(3):538–551
72. McKinley IM, Kandilian R, Pilon L (2012) Waste heat energy harvesting using the Olsen cycle on $0.945Pb(Zn_{1/3}Nb_{2/3})O_3$–$0.055PbTiO_3$ single crystals. Smart Mater Struct 21 (3):035015
73. Lee FY, Goljahi S, McKinley IM et al (2012) Pyroelectric waste heat energy harvesting using relaxor ferroelectric 8/65/35 PLZT and the Olsen cycle. Smart Mater Struct 21(2):025021

74. Navid A, Pilon L (2011) Pyroelectric energy harvesting using Olsen cycles in purified and porous poly (vinylidene fluoride-trifluoroethylene)[P (VDF-TrFE)] thin films. Smart Mater Struct 20(2):025012
75. Olsen RB, Brown DD (1982) High efficieincy direct conversion of heat to electrical energy-related pyroelectric measurements. Ferroelectrics 40(1):17–27
76. Lee F (2012) Experimental and analytical studies on pyroelectric waste heat energy conversion. Master Thesis, UCLA
77. Sebald G, Pruvost S, Guyomar D (2008) Energy harvesting based on Ericsson pyroelectric cycles in a relaxor ferroelectric ceramic. Smart Mater Struct 17(1):015012
78. Khodayari A, Pruvost S, Sebald G et al (2009) Nonlinear pyroelectric energy harvesting from relaxor single crystals. IEEE Trans Ultrason Ferroelectr Freq Control. 56(4):693–699
79. Kandilian R, Navid A, Pilon L (2011) The pyroelectric energy harvesting capabilities of PMN–PT near the morphotropic phase boundary. Smart Mater Struct 20(5):055020
80. Cha G, Ju YS (2013) Pyroelectric energy harvesting using liquid-based switchable thermal interfaces. Sens Actuators, A 189:100–107
81. Olsen RB, Bruno DA, Briscoe JM et al (1985) Pyroelectric conversion cycle of vinylidene fluoride-trifluoroethylene copolymer. J Appl Phys 57(11):5036–5042
82. de Oliveira NA, von Ranke PJ, Troper A (2014) Magnetocaloric and barocaloric effects: Theoretical description and trends. Int J Refrig 37:237–248
83. Zou JD (2012) Magnetocaloric and barocaloric effects in a $Gd_5Si_2Ge_2$ compound. Chin Phys B 21(3):037503
84. Strässle T, Furrer A, Lacorre P et al (2000) A novel principle for cooling by adiabatic pressure application in rare-earth compounds. J Alloys Compd 303–304:228–231
85. Strässle T, Furrer A, Altorfer F et al (2001) HoAs: a model compound for the cooling by the barocaloric effect. J Alloys Compd 323–324:392–395
86. Furrer A, Strässle T, Temprano DR (2001) New excitement with crystal-field excitations. J Alloys Compd 323–324:649–653
87. Strässle T, Furrer A, Dönni A et al (2002) Barocaloric effect: the use of pressure for magnetic cooling in $Ce_3Pd_{20}Ge_6$. J Appl Phys 91(10):8543
88. Hossain Z, Strässle T, Geibel C et al (2004) First-order valence transition and barocaloric effect in $EuNi_2(Si_{1-x}Ge_x)_2$. J Magn Magn Mater 272–276(3):2352–2354
89. Mañosa L, Gonzalez-Alonso D, Planes A et al (2010) Giant solid-state barocaloric effect in the Ni-Mn-In magnetic shape-memory alloy. Nat Mater 9(6):478–481
90. De Oliveira NA (2011) Barocaloric effect and the pressure induced solid state refrigerator. J Appl Phys 109(5):053515
91. Otsuka K (1998) Shape memory materials, C.M. Wayman (ed). Cambridge University Press, Cambridge
92. Shaw JA, Churchill CB, Iadicola MA (2008) Tips and tricks for characterizing shape memory alloy wire: Part 1–differential scanning calorimetry and basic phenomena. Exp Tech 32(5):55–62
93. Jani JM, Leary M, Subic A et al (2014) A review of shape memory alloy research, applications and opportunities. Mater Design 56:1078–1113
94. Wakjira JF (2011) The VT1 shape memory alloy heat engine design. Master Thesis, Virginia Polytechnic Institute and State University
95. Salzbrenner R (1984) Shape memory heat engines. J Mater Sci 19:1827–1835
96. McCormick PG (1986) Shape memory effect heat engine performance. Appl Energy 24:221–243
97. Liu C, Qin H, Mather PT (2007) Review of progress in shape-memory polymers. J Mater Chem 17:1543–1558
98. Ortin J, Planes A (1989) Thermodynamics of thermoelastic martensitic transformation. Acta Metall 37(5):1433–1441
99. Moya X, Kar-Narayan S, Mathur ND (2014) Caloric materials near ferroic phase transitions. Nat Mater 13:439–450

100. Rodriguez C, Brown LC (1980) The thermal effect due to stress-induced martensite formation in β-CuAlNi single crystals. Metall Trans A 11A:147–150
101. Brown LC (1981) The thermal effect in pseudoelastic single crystals of β–CuZnSn. Metall Trans A 12A:1491–1494
102. Mukherjee K, Sircar S, Dahotre NB (1985) Thermal effects associated with stress-induced martensitic transformation in a Ti-Ni alloy. Mater Sci Eng 74:75–84
103. McCormick PG, Liu Y, Miyazaki S (1993) Intrinsic thermal-mechanical behaviour associated with the stress induced martensitic transformation in NiTi. Mater Sci Eng, A 167:51–56
104. Nikitin SA, Myalikgulyev G, Annaorazov MP et al (1992) Giant elastocaloric effect in FeRh alloy. Phys Lett A 171:234–236
105. Manosa L, Jarque-Farnos S, Vives E et al (2013) Large temperature span and giant refrigerant capacity in elastocaloric Cu-Zn-Al shape memory alloys. Appl Phys Lett 103:211904
106. Bonnot E, Romero R, Manosa L et al (2008) Elastocaloric effect associated with the martensitic transition in shape-memory alloys. Phys Rev Lett 100:125901
107. Vives E, Burrows S, Edwards RS et al (2011) Temperature contour maps at the strain-induced martensitic transition of a Cu–Zn–Al shape-memory single crystal. Appl Phys Lett 98:011902
108. Cui J, Wu Y, Muehlbauer J et al (2012) Demonstration of high efficiency elastocaloric cooling with large ΔT using NiTi wires. Appl Phys Lett 101:073904
109. Ossmer H, Chluba C, Krevet B et al (2013) Elastocaloric cooling using shape memory alloy films. J Phys: Conf Ser 476:012138
110. Bechtold C, Chluba C, Lima de Miranda R (2012) High cyclic stability of the elastocaloric effect in sputtered TiNiCu shape memory films. Appl Phys Lett 101:091903
111. Xiao F, Fukuda T, Kakeshita T (2013) Significant elastocaloric effect in a Fe-31.2Pd (at. %) single crystal. Appl Phys Lett 102:161914
112. Guyomar D, Li Y, Sebald G et al (2013) Elastocaloric modeling of natural rubber. Appl Therm Eng 57:33–38
113. Churchill CB, Shaw JA, Iadicola MA (2010) Tips and tricks for characterizing shape memory alloy: Part 4–thermo-mechanical coupling. Exp Tech 34(2):63–80
114. Pieczyska E (2010) Activity of stress-induced martensite transformation in TiNi shape memory alloy studied by infrared technique. J Mod Opt 57(18):1700–1707
115. DeGregoria AJ (1994) Elastomer bed. International Patent WO 94/10517
116. Fischer SK, Tomlinson JJ, Hughes PJ (1994) Energy and global warming impacts of not-in-kind and next generation CFC and HCFC alternatives. Oak Ridge National Laboratory, AFEAS and US Department for Energy
117. Cui J, Takeuchi I, Wuttig M et al (2012) Thermoelastic cooling. US Patent Application Publication US 2012/0273158 A1
118. Goetzler W, Zogg R, Young J et al (2014) Energy savings potential and RD&D opportunities for non-vapor-compression HVAC technologies. Navigant Consulting Inc., prepared for U.S. Department of Energy
119. Chluba C, Lima-de-Miranda R, Kienle L et al (2014) On the role of precipitates for the functional fatigue in TiNiCu films. In: International conference on martensitic transformation (ICOMAT), 2014, Oral presentation

Appendix

See Tables A.1, A.2 and A.3.

Table A.1 Nomenclature

Symbol	Unit	Description
A	m^2	Area (surface)
a	m, /	Thickness, fraction of refrigerant seen by a single high-field region
a_0	/	Geometry factor for effective heat-transfer coefficient
B	$T = Vs\ m^{-2}$	Magnetic flux density, magnetic induction
b	/	Number of high-field regions
Bi	/	Biot number
C	/	Clausius–Clapeyron factor
c	$J\ kg^{-1}\ K^{-1}$ $(J\ m^{-1}\ K^{-1})$	Specific heat capacity
D^d	/	Fluid dispersion factor
d	m	Diameter, thickness
E	$V\ m^{-1}$, J	Electric field (electric field intensity), exergy
$\dot{E}$	W	Exergy flux
e	$J\ m^{-3}\ (J\ kg^{-1})$	Specific exergy
$\dot{e}$	$W\ m^{-2}$	Specific exergy flux
F	N	Force
f	$Hz = s^{-1}$, /	Frequency
f_F	/	Fanning friction factor
g	$m\ s^{-2}$	Gravitational acceleration
Gz	/	Graetz number
H	$A\ m^{-1}$	Magnetic field intensity
h	$J\ m^{-3}\ (J\ kg^{-1})$	Specific enthalpy
I	A	Electric current
I_p	A	Pyroelectric current
J	$A\ m^{-2}$	Current density

(continued)

A. Kitanovski et al., *Magnetocaloric Energy Conversion*,
Green Energy and Technology, DOI 10.1007/978-3-319-08741-2

Table A.1 (continued)

Symbol	Unit	Description
j	/	Colburn heat-transfer factor
k	$J\ K^{-1}$	Boltzmann constant
K_1	/	Leakage coefficient
K_2	/	Loss factor
L	m	Length
M	$A\ m^{-1}$	Magnetization
M_0	$A\ m^{-1}\ kg^{-1}$	Specific magnetization
M*	/	Figure of merit for the permanent magnet assembly
m	kg, $A\ m^2$	Mass, magnetic moment
$\dot{m}_f$	$kg\ s^{-1}$	Mass flow rate
N	/, /	Demagnetization factor, number of wire turns in the coil
Nu	/	Nusselt number
P	W, $As\ m^{-2}$, /	Power, polarization, fraction of AMR cycle where magnet is in use
p	P, V s, $As\ m^{-2}$	Pressure, magnetic pole, pyroelectric coefficient
Pr	/	Prandtl number
Δp	Pa (bar)	Pressure drop
q	$J\ m^{-3}$ ($J\ kg^{-1}$)	Specific heat
$\dot{q}$	$W\ m^{-2}$, $W\ kg^{-1}$	Specific heat flux, heat flux density
Q	J	Heat
$\dot{Q}$	W	Heat power, heat flux
R	$m^2\ K\ W^{-1}$	Thermal resistance
Re	/	Reynolds number
r	m	Radius, distance
s	$J\ m^{-3}\ K^{-1}$ ($J\ kg^{-1}\ K^{-1}$), m, m	Specific entropy, thickness of surfactant layer, thickness
Δs	$J\ kg^{-1}\ K^{-1}$	Specific entropy change
St	/	Stanton number
T	K, Nm	Temperature, magnetic torque
$\bar{T}$	K	Average temperature
t	s, $VsA\ m^{-3}$	Time, torque density
U	/	Utilization factor
u	$J\ m^{-3}$ ($J\ kg^{-1}$)	Specific internal energy
V	m^3	Volume
v	$m\ s^{-1}$, $m^3\ kg^{-1}$	Velocity, specific volume
W	J	Work (energy)
w	$J\ m^{-3}$($J\ kg^{-1}$)	Specific work (specific energy)
$\dot{w}$	$W\ m^{-2}$	Rate of specific work

(continued)

Table A.1 (continued)

Symbol	Unit	Description
Δx	/	Fluid strokes length (the fraction of fluid that is displaced with respect to all the volume of fluid in the AMR)
α	W m^{-2} K^{-1}, /, /	Heat-transfer coefficient, ratio between magnetic and thermal energy of particles, ratio of thermal conductivity between solid and liquid
β	degree	Angle between vorticity and field direction
$\dot{\gamma}$	s^{-1}	Shear rate
δ	m	Diameter, thickness
ε	/	Porosity, strain
ε_0	As V^{-1} m^{-1}	Electric permittivity of vacuum
η	/, Pa s	Efficiency, dynamic viscosity
ϑ	°C	Temperature
κ	W m^{-1} K^{-1}	Thermal conductivity tensor
Λ	$T^{2/3}$	Magnetic characterization parameter
λ	W m^{-1} K^{-1}	Thermal conductivity
λ_F	m	Spin relaxation length
μ	Vs A^{-1} m^{-1}	Magnetic permeability
μ_0	Vs A^{-1} m^{-1}	Magnetic permeability of vacuum
ν	m^2 s^{-1}	Kinematic viscosity
ξ	/, /	Exergy efficiency, Riemann zeta function
π	V K^{-1}	Peltier coefficient
ρ	kg m^{-3}	Density
ρ_A	Vs m^{-2}	Surface density of magnetic poles
σ	As m^{-2}, MPa	Surface charge density, stress
τ	s, /, Pa	Time period, relative time, shear stress
τ_0	Pa	Yield stress
ϕ	degrees, /	Angle, fraction
χ	/	Susceptibility

Table A.2 Subscripts

Symbol	Description
A	Austenite
AC	Cold side element A
AB	Path A-B
ad	Adiabatic
AH	Hot side element A
amb	Ambient
AMR	Active magnetic regenerator
app	Apparent

(continued)

Table A.2 (continued)

Symbol	Description
appl	Applied
ATD	Thermal diode A
B	Bingham
BTD	Thermal diode B
c	Conduction, coercivity, cold, cooling
C	Curie, Carnot, cold, cooling
CA	Cold part of thermal diode A
CB	Cold part of thermal diode B
CB	Cold part of thermal diode C
cool	Index of magnetic characterization parameter
ci	Intrinsic coercivity
const	Constant
cool	Index of magnetic characterization parameter
corr	Corresponding
crit	Critical
CTD	Thermal diode C
D	Device
dem	Demagnetization
dw	De-twinned
E	Constant electric field, electrical
eff	Effective
el	Electric
f	Fluid
fi	Final, finish
f-m	Fluid-material
gap	Air gap
H	Constant magnetic field, hot, heating
h	Hot, heat sink, hydraulic
HA	Hot part of thermal diode A
HB	Hot part of thermal diode B
HC	Hot part of thermal diode C
high	High-field region
HT	Heat transfer
i	Initial
in	Inlet, internal
is	Isothermal
L	Liquid phase
low	Low-field region
m	Mass, magnetic, magnetization
mag	Magnetization, (permanent) magnet

(continued)

Table A.2 (continued)

Symbol	Description
max	Maximum
mc	Magnetocaloric
mcm	Magnetocaloric material
MCE	Magnetocaloric effect
min	Minimum
out	Outlet, external
p	Constant pressure (isobaric), particle
Peltier	Peltier module
q	Specific heat
R	Refrigeration
r	Relative, remanent, rectification
ref	Reference
reg	Regeneration
rel	Relative
rem	Remanence
rot	Rotational
S	Spin, saturation
s	Solid, isentropic
st	Start
sat	Saturation
stat	Static
surr	Surrounding
T	Isothermal, constant temperature
t	Technical
tw	Twinned
TD	Thermal diode
V	Volume
VFF	Volume fraction of the particles including their surfactant layer
vis	Viscous
w	Wall
0	Gap, external
25	The maximum value of figure of merit for the permanent magnet assembly
π	Heating power of Peltier effect
σ	Constant stress

Table A.3 Abbreviations

Abbreviation	Description
A	Austenite
AER	Active electrocaloric regenerator
AMR	Active magnetic regenerator, active magnetic regeneration
CHEX	Cold heat exchanger, heat source heat exchanger
COP	Coefficient of performance
ECE	Electrocaloric effect
EsCE	Elastocaloric effect
EDL	Double electric layer
EICE	Electrowetting on insulator-coated electrodes
ELE	Electrowetting on line electrodes
ER	Electrorheologic
EWOD	Electrowetting on dielectric
FF	Ferrofluid
FHD	Ferrohydrodynamics
GM	Gifford–McMahon
GMR	"Giant" magnetoresistance effect
HHEX	Hot heat exchanger, heat sink heat exchanger
HIT	Hetero-structured integrated thermionics
HTS	High-temperature superconducting
JT	Joule–Thomson
LCA	Life cost assessment
LTS	Low-temperature superconducting
M	Martensite
MC	Multilayer capacitor
MCE	Magnetocaloric effect
MCM	Magnetocaloric material
MHD	Magnetohydrodynamic
MR	Magnetorheological
MRF	Magnetorheological fluid
MVE	Magnetoviscous effect
MWCNT	Multiwall carbon nanotubes
PE	Pyroelectric element
SMES	Superconducting magnetic energy-storage systems
TD	Thermal diode

MIX
Papier aus verantwortungsvollen Quellen
Paper from responsible sources
FSC® C105338

If you have any concerns about our products,
you can contact us on
ProductSafety@springernature.com

In case Publisher is established outside the EU,
the EU authorized representative is:
Springer Nature Customer Service Center GmbH
Europaplatz 3, 69115 Heidelberg, Germany

Printed by Libri Plureos GmbH
in Hamburg, Germany